Science Horizons Year Book

1996

NEW YORK   TORONTO   SYDNEY

SCIENCE HORIZONS YEAR BOOK
1996

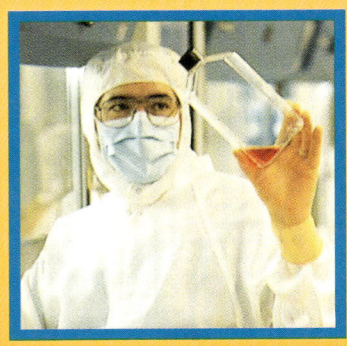

Published 1996 by P. F. Collier

This book is also published under the title Science Annual 1997

Copyright © 1996 by Grolier Incorporated

Library of Congress Catalog Card Number 64-7603

ISBN 1-57161-092-8

All rights reserved. No part of this book may be reproduced or transmitted in any form or by any means, electronic or mechanical, including photocopying, recording, or by any information storage and retrieval system, without permission in writing from the Publisher.

Printed in the United States of America

STAFF

PUBLISHER   PHILIP FRIEDMAN
EDITOR IN CHIEF   JOSEPH M. CASTAGNO
DIRECTOR, ANNUALS   DORIS E. LECHNER
SENIOR EDITOR   PETER A. FLAX
ART DIRECTOR   JEAN ZAMBELLI

EDITORIAL

MANAGING EDITOR   KENNETH W. LEISH
PRODUCTION EDITOR   JEANNE A. SCHIPPER
EDITORIAL ASSISTANTS   KAREN A. FAIRCHILD
   JONE E. BALTZELL
COPY EDITORS   DAVID M. BUSKUS
   MEGHAN O'REILLY
CONTRIBUTING EDITORS   JESSICA SNYDER SACHS
   ELAINE PASCOE
DESIGN DIRECTOR   NANCY A. HAMLEN
DIRECTOR, PHOTO RESEARCH   LISA J. GRIZE
PHOTO RESEARCHERS   JOAN R. MEISEL
   ELISSA G. SPIRO
ASSISTANT PHOTO RESEARCHER   ELIZABETH M. SEE
MANAGER, PICTURE LIBRARY   JANE H. CARRUTH
CHIEF INDEXER   PAULINE M. SHOLTYS
INDEXER   LINDA KING
COMPOSITOR   KAREN CALISE
FINANCIAL MANAGER   MARLANE L. MCLEAN
EDITORIAL LIBRARIAN   CHARLES CHANG

MANUFACTURING

DIRECTOR OF MANUFACTURING   JOSEPH J. CORLETT
DIRECTOR, REFERENCE PUBLICATIONS   CHRISTINE L. MATTA
PRODUCTION MANAGER   BARBARA L. PERSAN
PRODUCTION ASSISTANT   JENNIFER K. FISH

CONTRIBUTORS

TERESA AUSTIN, Freelance engineering writer based in Teaneck, NJ
    CIVIL ENGINEERING

MARCIA BARTUSIAK, Freelance writer specializing in physics and astronomy; author, *Through a Universe Darkly* (HarperCollins, 1993)
    CHANGING TIMES

JAMES BISHOP, JR., Freelance writer based in Arizona; author, *Epitaph for a Desert Anarchist: The Life and Legacy of Edward Abbey* (Athenaeum, 1994)
    MOUNTAIN ISLES, DESERT SEAS

JAMES A. BLACKMAN M.D., Professor of pediatrics, University of Virginia, Charlottesville, VA
    ASK THE SCIENTIST: HUMAN SCIENCES

SUSAN BRINK, Senior editor, *U.S News & World Report*
    BEATING THE ODDS

PETER BRITTON, Freelance science writer
    UNDERSEA EXPLORERS

WILLIAM J. BROAD, Science writer, *The New York Times*; author, *Teller's War: The Top-Secret Story Behind the Star Wars Deception* (Simon & Schuster, 1993)
    WATERY GRAVE OF THE AZORES

JANE E. BRODY, Science writer and personal-health columnist, *The New York Times*
    THE WEATHER ON TRIAL

LINDA J. BROWN, Freelance writer based in Bloomsbury, NJ
    DIVERSITY DOWN ON THE FARM

MALCOLM W. BROWNE, Science writer, *The New York Times*
    MERCURY MIRRORS: A NEW SPIN ON TELESCOPES

JEROME GREER CHANDLER, Freelance science writer
    WHY AIRPLANES CRASH

MARY H. COOPER, Staff writer, *CQ Researcher*
    ENERGY

PETER COY, Technology editor, *Business Week* magazine
    coauthor, 3-D COMPUTING: A WHOLE NEW DIMENSION

DONALD W. CUNNINGHAM, Freelance writer based in Washington, D.C.
    ADDING UP TO THE PRESIDENCY
    AUTOMOTIVE TECHNOLOGY
    TRANSPORTATION

GODE DAVIS, Freelance writer based in Warwick, RI
    ASK THE SCIENTIST: EARTH AND THE ENVIRONMENT
    ASK THE SCIENTIST: TECHNOLOGY
    OCEANOGRAPHY

JAMES A. DAVIS, Department of mathematics, University of Richmond, Richmond, VA
    MATHEMATICS

JERRY DENNIS, Freelance writer; author, *The Bird in the Waterfall: A Natural History of Oceans, Rivers, and Lakes* (HarperCollins, 1996)
    ENDANGERED SPECIES
    ZOOLOGY

EDWARD EDELSON, Freelance writer specializing in health and science issues
    TISSUE IN A TEST TUBE

DAVID S. EPSTEIN, Meteorologist and freelance writer
    METEOROLOGY

DIXIE FARLEY, Staff writer, *FDA Consumer* magazine
    VARYING VIEWS ON VEGETARIAN DIETS

ROBERT C. FIERO, JR., Technical services specialist, Grolier Interactive, Danbury, CT
    COMMUNICATION TECHNOLOGY

GREGORY FREIHERR, Contributing editor, *Air & Space* magazine
    5..4..3..2..ABORT!

MARIA GUGLIELMINO, Registered dietitian and exercise physiologist
    NUTRITION

DAVID K. HILL, Freelance science writer based in San Francisco, CA
    STRANGE LIGHTS ABOVE THUNDERSTORMS

ROBERT D. HOF, Correspondent, *Business Week* magazine
    coauthor, 3-D COMPUTING: A WHOLE NEW DIMENSION

ERIN HYNES, Freelance writer based in Austin, TX
    BOTANY

VINOD K. JAIN, Freelance writer based in Bowie, MD
    CHEMISTRY

JAMES F. JEKEL, M.D., Professor of epidemiology and public health, Yale University School of Medicine, New Haven, CT
    PUBLIC HEALTH

WILLIAM B. KARESH, Director, Department of Field Veterinary Studies, Wildlife Conservation Society
    RHINO RELATIONS

CHRISTOPHER KING, Managing editor, *Science Watch*, Institute for Scientific Information, Philadelphia, PA
    THE INTERNET: AN ON-LINE REVOLUTION
    NOBEL PRIZES

LOUIS LEVINE, Department of biology, City College of New York, New York, NY
    BIOLOGY
    BIOTECHNOLOGY

RICHARD LIPKIN, Freelance science writer based in Riverdale, NY
    ELECTRONICS

MICHAEL LIPSKE, Freelance writer; former senior editor, *International Wildlife* magazine
    IN PURSUIT OF PIGEONS

THERESE A. LLOYD, Freelance science writer based in Bedford, MA
    PHYSICS

VINCENT LYTLE, Freelance science writer
    AFTER THE ACCIDENT

MARTIN M. MCLAUGHLIN, Freelance consultant; former vice president for education, Overseas Development Council
    FOOD AND POPULATION

Dennis L. Mammana, Resident astronomer, Reuben H. Fleet Space Theater and Science Center, San Diego, CA
    STARBIRTH!
    DAY STAR
    ASK THE SCIENTIST: ASTRONOMY AND SPACE SCIENCE
    ASTRONOMY
    SPACE SCIENCE

Thomas H. Maugh II, Science writer, *Los Angeles Times*
    GENETICS

Richard Milner, editor, *National History* magazine; author, *The Encyclopedia of Evolution: Humanity's Search for its Origins* (Facts on File, 1990)
    BRINGING BACK THE DINOSAURS

Richard Monastersky, Earth sciences editor, *Science News*
    GEOLOGY
    PALEONTOLOGY
    SEISMOLOGY

Greg Morgan, Freelance writer specializing in the history and environment of Tasmania
    YOU TASMANIAN DEVIL, YOU!

David Morrison, Director of Space Science Division, NASA Ames Research Center, Mountain View, CA
    TARGET: EARTH!

Zahid B.M. Niazi, M.D., Senior Clinical Fellow, New York Group for Plastic Surgery and Rehabilitation, Westchester County Medical Center, Valhalla, NY
    coauthor, HYPERBARIC MEDICINE: HIGH-PRESSURE THERAPY

Larry O'Hanlon, Contributing editor, *Earth* magazine
    THE MEASURE OF A MOUNTAIN

Suzanne Oliver, Associate editor, *Forbes* magazine
    WHAT ARE MY CHANCES, DOC?

Lori Oliwenstein, Science writer based in Los Angeles, CA
    A DARWINIAN VIEW OF MEDICINE

Lynn O'Shaughnessy, Freelance writer based in La Mesa, CA
    ASK THE SCIENTIST: PAST, PRESENT, AND FUTURE

Daniel Pendick, Freelance writer specializing in earth science
    A BRAVE (AND WET) NEW WORLD FOR LAUNDRY
    VOLCANOLOGY

David A. Pendlebury, Research analyst, Institute for Scientific Information, Philadelphia, PA
    THE TOMB OF THE BROTHERS
    THE NOBLEST METALS
    ASK THE SCIENTIST: PHYSICAL SCIENCES

Devera Pine, Freelance science writer and editor based in Bethlehem, PA
    AVIATION
    BEHAVIORAL SCIENCES

Abigail W. Polek, Freelance writer and editor based in Pennsylvania
    SCIENCE EDUCATION

David Parry Rubincam, Physicist, NASA-Goddard Space Flight Center, Greenbelt, MD
    coauthor, AMERICA'S FOREMOST EARLY ASTRONOMER

Milton Rubincam II, Former president, National Genealogical Society
    coauthor, AMERICA'S FOREMOST EARLY ASTRONOMER

C. Andrew Salzberg, M.D., Assistant Professor of Plastic Surgery, New York Medical College, Valhalla, NY
    coauthor, HYPERBARIC MEDICINE: HIGH-PRESSURE THERAPY

David Scinto, Freelance writer based in Scarsdale, NY
    CONSUMER TECHNOLOGY

Doug Stewart, Freelance journalist based in Massachusetts
    THE IMPORTANCE OF BEING FLASHY
    THE QUEST FOR WATER

Jay Stuller, Freelance writer based in San Francisco, CA
    STEEL MAGNOLIAS

Jenny Tesar, Freelance science and medical writer; author, *Scientific Crime Investigation* (Watts, 1991) and *Global Warming* (Facts on File, 1991)
    ALIEN INVADERS
    ASK THE SCIENTIST: ANIMALS AND PLANTS
    ENVIRONMENT
    HEALTH AND DISEASE

Gary Turbak, Freelance writer; author, *Pronghorn: Portrait of the American Antelope* (Northland Publishing, 1995)
    WHERE THE BUFFALO ROAM

Scott Weidensaul, Contributing editor, *Country Journal*; columnist, *The Philadelphia Inquirer*; author, *Mountains of the Heart: A Natural History of the Appalachians* (Fulcrum Publishing, 1994)
    SKY READING

Peter S. Wells, Professor of anthropology, University of Minnesota, Minneapolis, MN
    ANTHROPOLOGY
    ARCHAEOLOGY

Jo Ann White, Freelance writer and editor based in Bradenton, FL
    BOOK REVIEWS

Simon Winchester, Freelance writer; author, *River at the Center of the World: A Journey up the Yangtze and Back in Chinese Time* (Henry Holt & Company, 1996)
    AN ELUSIVE MAP OF THE WORLD

Robert Wisner, Professor, Iowa State University; coeditor, *Marketing for Farmers*; author, *World Food Trade and U.S. Agriculture*
    AGRICULTURE

George Wuerthner, Freelance writer and photographer; author, *Yosemite: A Visitor's Companion* (Stackpole Books, 1994)
    THE GOD OF THE WOODS

Carl Zimmer, Senior editor, *Discover* magazine
    CIRCUS SCIENCE

Contents

Features

Animals and Plants — 14

- Where the Buffalo Roam 16
- The God of the Woods 22
- In Pursuit of Pigeons 30
- You Tasmanian Devil, You! 35
- Rhino Relations 38
- The Importance of
  Being Flashy 44
- Diversity Down on the Farm 49
- The Quest for Water 54
- Ask the Scientist 60

Astronomy and Space Science — 62

- Starbirth! 64
- Mercury Mirrors: A New
  Spin on Telescopes 71
- Day Star 75
- 5..4..3..2..Abort! 82
- America's Foremost
  Early Astronomer 89
- Target: Earth! 94
- Ask the Scientist 100

Earth and the Environment — 102

- Strange Lights Above
  Thunderstorms 104
- Mountain Isles,
  Desert Seas 110
- Sky Reading 114
- Alien Invaders 120
- The Measure of a Mountain ... 127
- Ask the Scientist 134

HUMAN SCIENCES 136

BEATING THE ODDS 138
HYPERBARIC MEDICINE:
    THERAPY UNDER PRESSURE ... 145
TISSUE IN A TEST TUBE 151
VARYING VIEWS ON
    VEGETARIAN DIETS 156
WHAT ARE MY CHANCES, DOC? ... 162
DARWINIAN MEDICINE 165
ASK THE SCIENTIST 172

PAST, PRESENT, AND FUTURE 174

THE TOMB OF THE BROTHERS 176
BRINGING BACK THE
    DINOSAURS 184
AN ELUSIVE MAP OF
    THE WORLD 191
WATERY GRAVE OF THE AZORES ... 197
THE WEATHER ON TRIAL 202
ASK THE SCIENTIST 208

Features

Physical Sciences — 208

- Circus Science 210
- The Noblest Metals 216
- A Brave (and Wet) New World for Laundry 222
- Changing Times 226
- Adding Up to the Presidency 233
- Ask the Scientist 236

Technology — 238

- 3-D Computing: A Whole New Dimension 240
- Why Airplanes Crash 247
- The Internet: An On-Line Revolution 254
- Steel Magnolias 262
- After the Accident 270
- Undersea Explorers 275
- Ask the Scientist 280

REVIEWS

Agriculture 284
Anthropology 286
Archaeology 288
Astronomy 291
Automotive Technology 293
Aviation 295
Behavioral Sciences 297
Biology 299
Biotechnology 302
Book Reviews 303
Botany 306
Chemistry 309
Civil Engineering 311
Communication Technology ... 313
Computers 314
Consumer Technology 317
Electronics 321
Endangered Species 323
Energy 325
Environment 327
Food and Population 331
Genetics 333
Geology 336
Health and Disease 338
Mathematics 342
Meteorology 343
Nobel Prize: Chemistry 348
Nobel Prize: Physics 350
Nobel Prize: Physiology
    or Medicine 352
Nutrition 354
Oceanography 357
Paleontology 359
Physics 361
Public Health 364
Science Education 367
Seismology 369
Space Science 371
Transportation 376
Volcanology 378
Zoology 380

In Memoriam 384
Index 387
Acknowledgments 397
Illustration Credits 399

Features
1996

Animals and Plants

■ CHEETAH YOUNG HAVE A DISTINCTIVE MANE ALONG THE NECK AND BACK. THIS EXTRA BIT OF FUR MAKES THE CUBS LOOK LARGER AND MORE FEROCIOUS THAN THEY REALLY ARE, AN ILLUSION THAT HELPS DETER POTENTIAL PREDATORS. BY THE TIME THEY SHED THEIR BABY FUR, CHEETAHS CAN OUTRUN MOST ENEMIES.

CONTENTS

Where the Buffalo Roam 16
The God of the Woods 22
In Pursuit of Pigeons 30
You Tasmanian Devil, You! 35
Rhino Relations 38
The Importance of Being Flashy 44
Diversity Down on the Farm 49
The Quest for Water 54
Ask the Scientist 60

WHERE THE BUFFALO ROAM

by Gary Turbak

The bison's steamy breath curls into the frigid air of Montana's Paradise Valley. Its hooves crunch ploddingly across the frozen prairie. Then, without alarm, the bull halts and slowly turns its massive head to gaze at the huddle of humans 100 yards (90 meters) away.

The rifle's roar rips the icy air. The bison's body shudders briefly, then topples in slow motion to the ground. For a few interminable seconds, its legs flail, then lie still. Peace again reigns in Paradise Valley.

BISON-BORNE DISEASE

But for Yellowstone National Park's northern bison herd, there is no peace. Each winter for nearly 20 years, park bison have wandered into Montana—and into the path of bullets and controversy. Straying onto private land, the errant bison have destroyed fences and eaten grass destined for cattle bellies, but what gets them shot is a disease called brucellosis. Some of Yellowstone's bison carry brucellosis, which causes female bison to abort their firstborn, and which can infect humans with sometimes-fatal undulant fever. Montana ranchers fear that wandering park bison will infect their cattle with the disease.

Brucellosis generally has no long-term effect on wildlife. Over time, an infected population will become somewhat immune to the disease, and among healthy wild populations the loss of some firstborn represents no great tragedy. In species with largely solitary lifestyles, such as elk, the contagion is slow to spread. Very few northern Yellowstone elk have the disease. Bison, however,

16 ANIMALS AND PLANTS

Each winter, hungry bison lumber out of Yellowstone National Park—and into the path of controversy. Such wanderers are often shot over concerns that they could infect local cattle with a disease called brucellosis.

are much more social, which can cause the incidence of brucellosis to be higher.

Like bison, cattle can be excellent incubators for brucellosis, and the disease is capable of killing plenty of calves—animals generally considered more valuable than the young of wild species. Long ago, the agricultural community and the federal government decided that brucellosis should not be tolerated in cattle, and many states embarked on ambitious plans to eliminate the disease. Montana ranchers, for example, recently spent $30 million getting their state certified as brucellosis-free.

Questions arise, however, about the severity of the bison brucellosis threat. For starters, brucellosis-infected elk and other wildlife often pass freely into and out of Yellowstone with hardly a notice. (Lately elk-borne brucellosis has become a concern south of Yellowstone around Jackson Hole, Wyoming, largely due to

Yellowstone National Park, which straddles Wyoming, Idaho, and Montana (map at left), is home to some 4,000 bison. These creatures—along with bears, elk, recently reintroduced wolves, and hundreds of other species—have free rein in the park (below), the nation's largest ecosystem that functions as it did before white explorers arrived.

the congregation of elk at artificial feeding sites.) Also, the exact extent of brucellosis infection among the Yellowstone bison is unknown. One study suggests the incidence may be as low as 12 percent, although Montana officials say it is closer to 50 percent.

Finally, no one knows for sure if free-ranging bison can transmit the disease to cattle. "There's never been a proven case of a bison transmitting brucellosis to cattle in the wild," says Yellowstone Park spokesperson Marsha Karle. It has happened in corrals, but never documented on the open range. Some experts say the brucellosis threat is virtually nonexistent. Others disagree, and the debate continues.

Meanwhile, Montana cattle ranchers say they don't want to be a test case. According to Jim Peterson, executive vice president of the Montana Stockgrowers Association, loss of the brucellosis-free designation would require all Montana cattle to be tested for the disease prior to shipment out of state—at an annual cost of about $10 million. "It would be devastating," he says. With so much money at stake, ranchers remain adamant about keeping Yellowstone bison out of Montana—and so far, the only effective solution has come from the barrel of a rifle. To date, more than 1,500 Yellowstone bison have been shot for trespassing into Montana.

THE BRINK OF EXTINCTION

Roots of the Yellowstone dilemma go back 100 years. When Europeans settled the American West, they encountered bison masses so large as to defy belief—an estimated 60 million of the beasts, the greatest collection of large mammals ever to tread the globe. What followed was one of the blackest episodes in the history of human-wildlife interaction. By 1887, settlers, professional and recreational shooters, the railroads, and the U.S. Army had pushed the bison to the edge of extinction. That year, the American Museum of Natural History sent hunters westward to collect bison specimens. In three months of searching, they found none.

Fortunately, 20 or so bison still survived in the remote reaches of Yellowstone National Park, and others existed in private herds. In 1902, the government bolstered the Yellowstone herd with additional bison purchased from ranchers. (It may have been these imported animals that brought brucellosis to Yellowstone, or park bison could have acquired the disease from nearby cattle.) Under Yellowstone's protective umbrella, the park bison flourished.

During the first half of this century, the National Park Service (NPS)—via sale and slaughter—held the Yellowstone herd to a manageable 500 head. In the 1960s, park policy changed, and Yellowstone became a "hands-off" zone, where authorities allowed nature to run its course. This was great for the bison, and their numbers soon mushroomed to nearly 3,000—the nation's largest free-roaming herd (divided by park geography into three distinct groups).

Few animals have endured a more devastating encounter with humans than have bison. During the 19th century, the arrival of settlers, soldiers, railroads, and fur traders precipitated a mass slaughter that left the species—which had once numbered some 60 million—at the brink of extinction.

MONTANA BECKONS

In the mid-1970s, wanderlust and the search for winter food sent a few members of Yellowstone's expanding northern herd into Montana. They found the grazing good and the snow light. At first, authorities successfully hazed most of the errant bison back into the park, choosing to kill only 13 hard-core wanderers through 1983. Each succeeding winter, however, the bison returned, with more of their friends. As the annual migration grew, more bison met bullets—a total of 190 between 1984 and 1988.

For a few years, Montana ill-advisedly allowed closely controlled hunters to kill the wandering bison, and, in the winter of 1988-89, a firestorm of publicity erupted when 569 animals were shot. Today Montana game wardens, sometimes assisted by NPS employees, do the job. The annual kill has ranged from the 1988-89 high to just four dead bison the following year. During the winter of 1994-95, authorities shot 412 animals. The meat is not wasted. With simple precautions, butchers can process meat from brucellosis-infected bison without contracting undulant fever,

Brucellosis strikes bison and other cattle species, causing females to abort their firstborn. Once a calf (above) is born, it is unlikely to be harmed by the disease. Other Yellowstone denizens such as elk (below) also carry brucellosis; these animals, however, pass into and out of the park with little fanfare.

Critics charge that hands-off management policies in Yellowstone should be reversed so that bison can be inoculated with a brucellosis vaccine already in use elsewhere (right).

and cooking kills the bacteria that cause the disease. Native American peoples often receive much of the meat.

NO EASY ANSWERS

Although winters with heavy kills temporarily reduce the northern herd's size, the park bison are not in danger. "The shooting represents no threat to Yellowstone bison populations—not even to the northern herd," says Karle. The northern herd numbers about 700, with approximately 3,300 more bison in the park's two other herds. (The total U.S. bison population, public and private, approaches 100,000.)

Even so, the public has rallied against the killing of America's shaggy wild icon, and

More than 2,000 U.S. ranches are now raising bison as livestock. The animals are noted for their hardy constitutions and their low-fat meat.

fingers of blame point in all directions: at the NPS, for allowing its wildlife to spill over park boundaries; at Montana, for treating the problem with bullets; at animal-rights groups, for pandering to public emotion; at ranchers, for failing to accommodate a few bison; and at snowmobilers, for packing down the trails that permit the annual exodus.

Finding fault is easy. Solutions are more difficult to come by, although plenty have been tried: chase the bison to wherever they need to be; fence the park boundary; truck the wanderers away; feed the bison inside the park; tolerate the nomads in Montana; vaccinate the entire herd; eliminate all park bison and start over with clean stock; kill bison within the park; and manage numbers with birth control.

The majority of wildlife officials write off fencing, trucking, and artificial feeding as too expensive or impractical. Hazing may be a good idea, but has proven difficult. "We've tried helicopters, horses, trucks, dogs, and rubber darts, but nothing has kept these bison in the park," laments Ron Aasheim, spokesperson for the Montana Department of Fish, Wildlife, and Parks, the state agency charged with bison control.

Many other solutions—vaccination, birth control, and in-park culling—are contrary to the park's hands-off management plan, a

Although bison usually find peace and ample feeding grounds within Yellowstone's boundaries, the animals will likely continue to stir controversy whenever they seek greener pastures outside the park.

canon the NPS is being pressured to change. Montanans, with plenty of justification, see the situation as a Yellowstone problem that has been unfairly dumped on them. "The park's bison should be managed so they remain compatible with the habitat inside Yellowstone," says Aasheim. The Montana cattle industry agrees. "The solution lies within the park," says Peterson. "Yellowstone's so-called natural regulation process is not working."

Ridiculous Choices?

Another vocal player in this drama is the animal-rights community, led by the Fund for Animals. Opposed to any solution that would reduce bison numbers, the fund's proposal would force Montana ranchers to accept bison on the open range. "Landowners assumed certain risks when they moved there, and if they can't live in harmony with the bison, perhaps they should live elsewhere," says D.J. Schubert, the fund's director of investigation.

Montana feels pinched between the NPS' laissez-faire bison management and the brucellosis-eradication directives of the U.S. Department of Agriculture. "Montana is restricted to ridiculous choices because two federal agencies have conflicting regulations and policies," said Montana Governor Marc Racicot last winter. "We don't want to kill the nation's bison, but we cannot allow the livestock industry in Montana to be irreparably harmed."

In January 1995, Montana asked the federal courts to intervene; a year later, the parties appeared close to an agreement. Montana is expected to allow some park bison to enter public land in the state as long as cattle are not threatened with brucellosis. In addition, capture and removal will replace slaughter as the first line of attack against offending bison. For its part, the NPS will cooperate with the in-park capture—and perhaps the slaughter—of some wandering bison. An environmental-impact assessment planned for completion in 1997 may build on these temporary solutions. Now the next move belongs to the bison.

ANIMALS AND PLANTS 21

The God of the Woods

by George Wuerthner

Walter Fry came to California to log its giant trees. After cutting down a huge sequoia whose tree rings revealed an age of over 3,000 years, Fry lay down his saw and quit logging forever. He spent the rest of his life trying to save the trees he had once treated so contemptuously. Fry was among the first to sign a petition to set aside the finest groves as a national park, and he eventually became the first civilian superintendent of Sequoia National Park.

THE MAGNIFICENT BIG TREES

It is not difficult to understand what changed Fry from a logger to a preservationist. Giant sequoias are so imposing, so strikingly beautiful, that they command reverence. When walking through a sequoia grove, one talks in hushed tones, as if entering a church, as if chatter would disturb the serenity.

the Grizzly Giant in Yosemite National Park is 30.7 feet (9.4 meters) across at ground level; the General Sherman Tree in Sequoia National Park is more than 12 feet (3.7 meters) across, 200 feet (61 meters) above its base. Indeed, some of the branches on these outsized specimens are 6 feet (2 meters) in diameter and more than 150 feet (46 meters) long, making them larger than most trees.

Although the tallest coast redwoods exceed the sequoia in height, some Big Trees still tower over 300 feet (90 meters) tall. Laid down upon a football field, such a tree would span the end zones. It is easy to see why the sequoia's species name is *giganteum*.

A LONG-HIDDEN SPECIES

Part of the giant sequoia's allure has to do with its relative scarcity. Although members of Joseph Walker's 1833 cross-Sierra expedi-

Mariposa Grove (left) in Yosemite National Park contains about 500 mature giant sequoias, some that tower hundreds of feet above the ground. At right, naturalist John Muir (left), who led the fight against logging sequoias, and landscape painter William Keith are dwarfed by the mammoth trees.

John Muir, high priest of the Sierra, described the sequoia as "the god of the woods." In the Kaweah Basin within what is now Sequoia National Park, Muir met magnificent clusters of the Big Trees he described as "pure temples," where every tree "seemed religious and conscious of the presence of God." Muir can be forgiven his anthropomorphizing. These are no ordinary trees.

So exceptional and monumental are the largest sequoias that many individual specimens have names: Mariposa Tree, General Sherman Tree, Fallen Monarch, General Grant Tree, Mather Tree, Bachelor, and the Three Graces. The Boole Tree in the Converse Basin measures 35.7 feet (10.9 meters) in diameter;

tion did record in their journals that they had seen sequoias, their comments remained unnoticed until the 1900s. It was not until 1852, when A.T. Dowd stumbled upon the Calaveras Grove while pursuing a wounded bear, that the Euro-American community learned of the tree's existence.

How could such a large tree remain hidden so long? The sequoia's native range is a narrow 250-mile (400-kilometer)-long by 15-mile (24-kilometer)-wide belt between 3,000 and 8,900 feet (900 and 2,700 kilometers) in elevation along the western slope of the Sierra Nevada. Even within this belt, the sequoia's occurrence is spotty at best. The tree is found in 75 groves that occupy only

ANIMALS AND PLANTS 23

36,000 of the 19 million acres encompassed by the Sierra Nevada. The groves range in size from 1 acre to 4,000 acres (0.4 to 1,620 hectares). Ninety percent of the trees are found in the southern third of its range, culminating with Deer Creek Grove in Tulare County. The northernmost grove consists of six trees in Placer County.

Sequoias were once far more widely distributed. They and their close relatives—the dawn redwoods of Asia, the coast redwoods of California, and the pond and bald cypresses now restricted to the swamps of the southeastern United States—were once found across North America.

Fossil remains of *Sequoiadendron*, along with oaks and conifers, show that these trees were once abundant in Nevada under a more benign climate, with 25 to 35 inches (65 to 90 centimeters) of annual precipitation. The rise of the Sierra Nevada during the past 10 million years led to a drying of Nevada and a subsequent shift in sequoia distribution westward. Big Trees arrived in the Golden State approximately 7 million years ago.

CARE AND FEEDING

Despite the shrinkage in its natural range, giant sequoias apparently can survive—with horticultural care—under a wide range of conditions. Sequoias can thrive in Boise, Idaho, and are popular in Europe as timber producers and ornamental trees. Big Trees can be found growing in 25 European countries from Norway south to the Mediterranean Sea and east to the Black Sea.

Nevertheless, the sequoia's patchy distribution pattern in the Sierra remains a puzzle. John Muir, along with Brownie, a recalcitrant burro, undertook the first systematic survey of their distribution in 1875. After traveling from Yosemite south to the Kaweah River, Muir documented the narrow elevational restrictions and irregular occurrence of sequoias and attributed them to past glaciation. He wrote: "Just where, at a certain period in the history of the Sierra, the glaciers were not, there the sequoia is, and just where the glaciers were, there the sequoia is not."

Recent research, particularly by Paul Zinke and Alan Stangenberger at the University of California at Berkeley, has confirmed that Muir's initial observations were at least par-

Only 1 in 1 million sequoia seeds germinates; of these, only 1 in 10,000 survives long enough to reach the age achieved by the fallen giant below. The average sequoia sapling grows at a rate of about 2 feet per year.

The Name Game

For a long time, the species name—*giganteum*—was the only thing botanists could agree upon. For years, the Big Tree was assigned to the genus *Sequoia*. More recently, taxonomists have assigned the tree to its own genus, *Sequoiadendron*. The name *Sequoiadendron giganteum* is only the latest rendition in an almost comical cross-Atlantic rivalry.

In 1852, American botanist Albert Kellogg, whose name is commemorated by the black oak (*Quercus kelloggii*), collected some sequoia specimens. Kellogg intended to assign the species the generic name *Washingtonia*, to honor the first American president. However, before he got around to publishing his findings, English botanist John Lindley named the trees after the Duke of Wellington, giving the official name of *Wellingtonia gigantea*. Chauvinistic American botanists were outraged. Unfortunately for the British, and fortunately for American nationalists, the name *Wellingtonia* had previously been assigned to another group of tropical hardwoods. Although some English still refer to the giant sequoia as "Wellingtonia," the scientific community, recognizing that there could be only one name, transferred the Big Trees to the genus *Sequoia*, which includes the coast redwood as well. The name honors a Cherokee Indian, Sequoya, who in 1821 invented an 86-character alphabet for his native language. Later research revealed significant differences between the redwood and the sequoia, and today most botanists give the tree its own generic name of *Sequoiadendron*.

The genus name Sequoia honors Sequoya, a Cherokee Indian who invented an 86-character alphabet for his native language.

---

tially correct. Sequoia groves are restricted to ridge tops or gentler mid-elevation slopes. They avoid both the heavy clay soils of the lower elevations and the sandy and stony soils found on glaciated slopes and canyon bottoms. Giant sequoias tend to grow in gentle swales with deep, rich, moist soils, although a few trees seem to grow nicely on rocky, apparently dry ridges.

Along with the soil's moisture content, long-term climatic trends may also play a role. Paleobotanist Daniel Axelrod believes the current distribution can be partially attributed to a dry, warm climatic regime that dominated western North America between 8,000 and 4,000 years ago and shrank the sequoia's range in the Sierra Nevada to only a few sites.

Scott Anderson, a professor at Northern Arizona University in Flagstaff who has studied the pollen record of Log Meadow in Sequoia National Park, confirms that, in contrast to the longevity of individual trees, sequoias may have been in their present range for only a few thousand years. He says the aged individuals found today may be "direct descendants—third or fourth generation—of initial pioneers."

The Perfect Forest Tree

Retired associate Forest Service chief Douglas Leisz finds it hard to reconcile the sequoia's limited distribution with his belief that it's probably close to the perfect forest tree. He has written: "If we were able to design the ideal conifer for our Sierra Nevada forests, it might look a lot like the giant sequoia."

Take growth, for example. The sequoia is among the fastest-growing trees on Earth, especially when it is not overshadowed by other trees. Giant-sequoia seedlings and saplings consistently outperform native competitors in height, diameter, and dominance. Recent studies near Georgetown in the Sierra demonstrate that young sequoias rapidly outpace other forest trees in height. The Big Tree's upward spurt is 20 percent faster than that of its nearest competitor, the ponderosa pine, and is nearly double that of any other native species.

With a height of 273 feet, a thickness of 36.5 feet at its base, and an estimated weight of 1,385 tons, the General Sherman Tree in Sequoia National Park is the biggest of the biggest.

Height is not the only remarkable attribute. The growth in the girth of a young sequoia typically averages three times the rate of Douglas fir, sugar pine, white fir, and other species that live beside it in the mixed-conifer forests. This rapid growth gives the sequoia a tremendous advantage in overtopping other species.

While in most tree species, rapid growth typically results in a short life, the giant sequoia is among the longest-lived species on Earth. Muir reported finding a tree with 4,000 annual rings. Although no trees that old have been confirmed by today's more accurate aging methods, the dead but well-preserved "Muir snag" in the Converse Basin was found to be 3,500 years old. And the ring count of a stump left by loggers in the Converse Basin Grove revealed the tree was 3,232 years old before its untimely end.

There are several reasons for this longevity. The rose-colored bark and wood are nearly rot- and insect-proof. Impregnated with natural bug- and fungus-resistant chemicals, sequoias are able to repel many forest pathogens. Muir once observed with only a little hyperbole, "Build a house of Big Tree logs on granite and that house will last about as long as its foundation." Near the Nelder Grove in the Sierra National Forest, Muir found a Big Tree whose wood was still sound, although a fir some 380 years old was growing on top of it. Muir estimated that the fallen giant had lain on the ground for nearly 1,000 years. National Biological Service ecologist Tony Caprio remarks, "I've been surprised at how few things chew on sequoia trees."

Nevertheless, the sequoia does occasionally succumb to fungus and insect attack, typically after fire scarring provides an entry. Indeed, it is the gradual weakening of its base and relatively shallow roots, combined with heavy wind or

Giant Sequoia vs. Coast Redwood

The coast redwood and the giant sequoia share many characteristics—including huge size, adaptation to fire, and a restricted distribution—but they are distinct species, each with its own attributes.

	GIANT SEQUOIA	**COAST REDWOOD**
Distribution	Western slope of the Sierra Nevada	Near California coast from Oregon border to Big Sur
Size at Base	30 feet (9 meters)	18 feet (5 meters)
Maximum Diameter	35 feet (11 meters)	23 feet (7 meters)
Maximum Height	310 feet (95 meters)	370 feet (113 meters)
Bark	Rich red-brown	Weathered gray-red
Cones	2 to 3 inches (5 to 8 centimeters) long	Nutlike, 1 inch (3 centimeters) long
Reproduction	Seeds	Seeds and sprouting from root crowns
Shade Tolerance	Young trees not tolerant	Moderately tolerant

slope failure, that leads to the eventual toppling of most mature, yet still-living, sequoias.

FRIENDLY FIRE

Although fires may provide the means for the tree's ultimate death, sequoias seemingly cannot survive without them, either. Fires are a natural and essential ecological process that maintains functioning Sierran forests. As a rule, most fires pose little threat to full-grown sequoias, whose bark is up to 2 feet (.6 meter) thick and remarkably fire-resistant. Low-intensity fires creep through the forest like an oozing, crackling lava flow. Only occasional-

Controlled burns play an essential role in maintaining a healthy stand of sequoias. Residual fire scars on the trunk may persist for years.

ly, when the fire manages to leap into the upper branches of small firs or to burn into a pile of accumulated branches, will the flames flare up and blaze like a torch. But the sequoia's protective bark and self-pruning bole with its few low branches ensure that mature trees are rarely killed by fire.

Reconstruction of fire history within the Sequoia National Forest shows that tiny blazes occurred in one or more sequoia groves every year. This helps to kill competing trees, as well as create the bare-mineral soil that sequoia seeds need for successful germination.

However, Jeff Manley, natural-resource specialist at Sequoia–Kings Canyon National Park, says that light fires may not be enough. "Friendly flame is great to keep fuels down, if what we are mainly concerned about is hazard reduction. If we are serious about maintaining the groves, we are going to have to accept that there are going to be high-intensity burns in the stand. That means burning down some giant sequoias once in a while."

Ecologist Nathan Stephenson of the National Biological Service agrees. "Not only do sequoias need sunlight for successful seedling establishment and growth, but they need the reduced competition for water that results when most other trees are killed," he says. In addition, fires dry out sequoia cones, releasing a shower of small seeds upon the recently burned ground.

In research conducted on the sequoia's response to intense fire, Linda Mutch, an ecologist also with the National Biological Service, found that seedling establishment was directly proportional to fire severity. Where fires were hot enough to kill most of the mature trees, she found far more young trees. In addition, Mutch found that adult sequoia trees tended to have larger growth spurts after a fire.

Although the ecological role of fire was recognized in the 1960s, agency fire-suppression policies have been slow to change. Even today, on most non–national park forests in the Sierra, full suppression of most fires is still the norm, not the exception. Political as well as social barriers, including the effect of smoke upon smog-filled valleys and the potential loss of homes and lives, make the reintroduction of fire into the ecosystem difficult.

Some critics of agency fire policy note that an abundance of small sequoias can be found even in groves where few or no fires have burned. But Stephenson cautions that appearances can be deceptive. Even though there appear to be many small trees in the understory of major groves that might indicate recent regeneration, in reality such trees may be hundreds of years old. Aging of many of these understory trees reveals that

The brittle wood of the sequoia, useless for lumber, has in the past been used for shingles or, in the case of the tree above, for fence posts.

"most of them came in after the last major fires in the 1800s," Stephenson says. "There is a tremendous gap in regeneration in this century that is a magnitude 10 times over what prevailed in the past."

There are other reasons as well for maintaining fires within the ecosystem. Stephenson notes that fires aid nutrient cycling; mobilize nutrients; change soil properties, like its ability to hold water; and kill soil pathogens that might otherwise destroy sequoia seedlings. Stephenson says, "We don't know the long-term consequences of fire exclusion. Locally intense fires and the occasional large fire were both part of sequoia ecosystems. The conservative thing to do is to maintain the processes that maintained the sequoia groves in the past."

Sequoia-Preservation Efforts

While fire exclusion poses a threat to sequoia regeneration today, in the past the greater concern was logging. The first efforts at sequoia preservation began quite early with the 1864 deeding of the Mariposa Grove to the state of California for protection (this grove was eventually incorporated into Yosemite National Park). Yosemite, Sequoia, and General Grant national parks were all established in 1890 in large part to protect clusters of sequoias. But, helped by multiple land-acquisition laws such as the Homestead Act and the Timber and Stone Act, giant-sequoia groves were often targeted for private acquisition and timber harvest.

The brittleness of sequoia wood meant that logging was particularly wasteful; trees had a tendency to shatter upon impact. The typical method of felling the huge trees was to use dynamite and blow them up. Many giant sequoias ended up as garden lathing.

Appalled at the wanton waste, Muir and others lobbied for protection of the entire southern Sierra forestlands. In 1893, President Benjamin Harrison established, by executive order, the Sierra Forest Reserve. (The Sierra Forest Reserve is now divided up among the Sierra National Forest, Sequoia National Forest, Inyo National Forest, and Sequoia–Kings Canyon National Park.) The reserve designation precluded sale or private acquisition of the remaining public domain in the southern Sierra. Other forest reserves, precursors of our modern national-forest system, soon followed.

After the turn of the century, public reacquisition of privatized groves began. In 1935, for example, some 20,000 acres (8,100 hectares), including the Converse Basin—largest of all Sierra Big Tree groves prior to being logged—and 10 other sequoia groves were acquired by the Sequoia National Park. Today 90 percent of the giant-sequoia groves are in public ownership. Fourteen percent are managed by the state of California; the National Park Service (NPS) controls 34 percent, while the remaining 42 percent—including the Sequoia, Stanislaus, Tahoe, and Sierra national forests—are managed by the Forest Service.

New Threats to the Big Trees

The most recent controversy centers upon sequoia management on national-forest lands, particularly the Sequoia National Forest. In the 1980s, the Forest Service embarked on a "whitewood" logging program (whitewoods are nonsequoia species like pine and fir) to cut old-growth timber within sequoia groves. The harvest was justified for "grove enhancement" as a means of stimulating sequoia regeneration. Although no mature sequoias were logged, many people were appalled that the sanctity of the groves was violated by logging roads and the removal or reduction of the large woody debris and snags, and a pruning of the multilayered canopy. There was also concern that the new logging roads would interrupt subsurface water flow.

Logging is no ecological substitute for fire. Even though it is possible to obtain sequoia regeneration through logging, regrowing trees on a logged site is not the same as maintaining a forest ecosystem. The lack of fires affects nutrient cycling, soil pathogens, and the distribution, age, and structure of the forest. In essence, the Forest Service was preserving large specimens like museum pieces, while removing the ecological context and processes that created and continue to maintain them. The Sierra Club filed suit and sought an injunction on nine timber sales. After 16 months, the Ninth Circuit Court of Appeals granted an injunction—not because it was illegal to log in sequoia groves, but because the Forest Service failed to produce an environmental-impact statement. By the time the court granted the injunction, most of the affected timber had been logged.

Despite the injunction and public outcry, the Sequoia National Forest released a new land-management plan in 1988 that called for similar treatment of another 9,300 acres (3,800 hectares). The Sierra Club, National Audubon Society, Save the Redwoods League, and others appealed the plan. John Dewitt, executive director of the Save the Redwoods League, says that in 1990 a settlement was reached that directed the Forest Service to manage sequoia groves as natural areas. No logging would be permitted within a 500-foot (150-meter) buffer of a grove.

Because of the perceived failure of the Forest Service to adequately protect sequoia ecosystems, some environmentalists have called for the transfer of all sequoia groves to national-park protection. U.S. Representative George Brown of San Bernardino, California, introduced legislation to create a Sequoia National Preserve that would prohibit logging in giant-sequoia groves and on surrounding lands. Resistance to the idea by U.S. Representative Calvin Dooley, who represents the local area, has stalled consideration of the bill.

Dan Taylor of the National Audubon Society summed up what many environmentalists felt when he wrote: "The sequoia cannot truly be protected when it alone is the focus of society's consideration. What's at stake is more important: the health of an entire regional ecosystem. It is accurate to call the giant sequoia one of the crown jewels of the Sierra Nevada. But what purpose are crown jewels without a crown in which to place them?"

The beautiful Nelder Grove of sequoias became part of the Sierra National Forest in 1928. Some environmentalists are calling for the transfer of all sequoia groves to national-park protection.

IN PURSUIT OF Pigeons

by Michael Lipske

We've all met those intrepid scientists (if only in magazine stories and television documentaries) who study rare animals in the world's far corners—those muddy-booted field researchers who brave monsoons or quicksand to seize the truths that may prolong the lives of gorillas, tigers, and other noble beasts.

Then there is Louis Lefebvre. Mud-free, middle-aged, something of a Renaissance man, this French-Canadian biology professor at McGill University launches research expeditions within a few minutes' walk of his office. On the McGill campus in deepest downtown Montreal, Quebec, he explores the ways of animals that Woody Allen and Winston Churchill called "rats with wings."

URBAN DENIZENS

Louis Lefebvre studies pigeons. Common, everyday, fecund, soot-shaded, park-bench-spattering street pigeons. "I think they're neat little animals," he says without embarrassment. "I also like the kind of people that like pigeons—this sort of marginal type of person." By which he means those urbanites, often the elderly or loners or both, who go out of their way to feed pigeons. Perhaps the researcher is just keeping his eye on his goal: "I'm interested in why pigeons do so well. So the more there are, the happier I am."

A 44-year-old Montreal native, Lefebvre has published two novels (neither of them about pigeons) as well as poetry. As a scientist, he is curious about the advantages animals derive from group living. To investigate the question, he has studied pigeon-flock dynamics and has looked at how new information or behavior (how to capitalize on new food sources, for example) is transmitted between the birds. In experiments bridging the world of the laboratory and of the street, he has examined how the birds compete through imitation and speed instead of aggression.

Seeing live pigeons in the park will never quicken the bird-watcher's pulse quite like a rare warbler flashing through on migration. But give pigeons credit. Every day, these birds perform feats demanding adaptability, intelligence, and pluck. Namely, they survive on today's urban mean streets.

The ubiquitous street pigeon is so common as to go practically unnoticed. Nevertheless, scientists have discovered that these very ordinary birds exhibit some very extraordinary types of behavior.

"They have specifically adapted to that niche of living in cities, treating buildings as if they were cliffs [the birds' wild ancestors roosted in crevices on rocky outcrops], treating people as if they were any normal source of food," says Lefebvre. "Whatever type of environment you decide to colonize, whether it's the romantic wilds of Costa Rica

30 ANIMALS AND PLANTS

or just a plain city square, the fact that you're doing well in it is something that I find admirable." The practical biologist has another reason to study the pigeon. "It's available—you just step out of your lab, and you've got your research animal."

And how. Lefebvre guesses that 50,000 to 100,000 pigeons dwell in Montreal. But some cities have many, many more pigeons. American physics professor and author James Trefil, in his book *A Scientist in the City*, writes that in pigeon-carpeted Trafalgar Square in London, England, and Piazza San Marco in Venice, Italy, "The weight of the living pigeons is probably greater than the total weight of living animals on any comparably sized wild habitat."

FLOCK DYNAMICS
For thousands of years, chubby little *Columba livia*, a.k.a. the rock dove, a.k.a. the feral city pigeon, has strutted along the urban human's boulevards and backstreets. Pigeons in North American cities are descended from birds brought across the Atlantic as food by colonists as early as the 1600s. Their ancestors also include escapees from homing and racing flocks and even the occasional show bird.

As far back as the 14th century, relations between urban people and these urban birds could turn tense, with the bishop of St. Paul's Cathedral in London lamenting that folks hurling rocks at the rock doves were also smashing cathedral windows. Acid in pigeon droppings can corrode paint and stone, and the guano plugs roof drains, causing buildings to flood with rainwater.

Regardless of the season, London's Trafalgar Square (above) teems with huge flocks of pigeons. Although many people love street pigeons, others detest the birds—and their apparent lack of cleanliness.

With the birds able to breed four to five times a year, urban pigeon populations can get out of hand. City officials around the world have fought to halt the bird boom, enticing pigeons to eat food treated with contraceptives, jolting them off buildings with electrified wire, scaring them with plastic owls, and shooing them to that big city up yonder by deploying marksmen with air rifles.

ANIMALS AND PLANTS   31

Urban property owners often "pigeon proof" ledges and windowsills to deter the birds from perching on—and ultimately soiling—their buildings.

Some of Lefebvre's findings could aid trappers arranging an ambush, such as the discovery that city pigeons stick to a bimodal schedule—early morning and late afternoon—for peak feedings. The pigeons do so despite having to rely on unpredictable food sources and having to visit several feeding areas.

Lefebvre also has looked into how tightly knit Montreal pigeon flocks are. "There's very little known about flock dynamics," he says. "When you see 50 pigeons in a place day in and day out, is this the same group of pigeons?" He says the question is important "because information is not going to be transmitted at the same rate if your groups are closed or if your groups are just temporary aggregates of animals that move around a lot."

To answer the question, Lefebvre captured a wild flock in the city and tagged the birds. His follow-up observations revealed that about two-thirds of the pigeons in a given flock are regular residents at the place where they are seen, and they are sure to be on hand about 75 percent of the time. The remaining birds in the flock are just visiting and have their own primary residence elsewhere in the city.

Lefebvre explains that even when pigeons have enough food at their primary area, they frequently are out sampling and exploring areas close by. "Because they're always moving around, they're able to assess whether food quality or food amounts are greater at another place," he says. Also, the changing mix of resident and visitor birds—the endless fluxion of the flock—enhances social learning, or the exchange of information between pigeons. When one bird masters an entirely new food source, visitor birds can observe the new behavior and bring it home to their primary flock.

Scores of laboratory experiments have shown that pigeons are skilled at a type of learning that involves watching other pigeons perform a task that leads to food. Lefebvre has carried such studies an extra step by capturing feral pigeons in Montreal, training the birds in some classic laboratory task, and then releasing the pigeons on the street.

In one such study, four wild city pigeons were trained to pierce the paper cover on a box containing seed. Lefebvre then released the educated birds back to their old flock, put out a covered seed box, and plotted the spread of the piercing behavior among "naive" city birds. Within a month, 24 street pigeons had learned the new skill.

Homing pigeons are trained to carry messages. Their uncanny ability to return to their home lofts from great distances made these birds especially valuable during times of war.

Many people derive great pleasure from pigeon breeding. Exhibits aside, most breeders allow the birds to fly away and return at will or, at the very least, house the pigeons in a cage large enough for them to fly about.

RESOURCE PARTITIONING

Pigeons seem to do better than humans at some tasks. For example, Lefebvre confesses that when he visits the supermarket, he has a hard time deciding whether to stand in a checkout line with relatively few people or take a chance on a line with more people but a faster cashier. Like many of us, he often chooses poorly and lands in a slower line.

Presenting Montreal's pigeons with a similar challenge, he and a colleague threw bread to a flock of the birds at different rates. Using a technique developed by a researcher who studied duck behavior in London, one experimenter tossed bread every five seconds, while the other tossed every 10 seconds. Lefebvre notes that at first the flock divided equally among the bread tossers, yet within moments the pigeons had rearranged themselves so that two-thirds of the birds were standing around the faster tosser.

This behavior suggests that on some level, pigeons understand it is best not to waste energy squabbling over resources. This was demonstrated to Lefebvre in a pigeon tragedy.

To catch wild flocks, the biologist and graduate student Luc-Alain Giraldeau put out seed laced with sedatives. Once, about a decade ago, Lefebvre and his team miscalculated the drug dose. Fifty-seven pigeons died on the spot. Instantly rescripting the day's research plan, the pigeon experimenters cut open each dead bird's crop (the saclike

Although initially sightless and essentially featherless, baby pigeons, called squabs, develop quite rapidly. Most leave the nest in two to three weeks.

extension of the esophagus in which pigeons store food) in order to analyze pigeon seed preferences.

The impromptu surgery led to a striking discovery. Although the food put out for the pigeons had contained seven types of seed, and the birds apparently had been pecking away at the pile with mad abandon, accord-

ANIMALS AND PLANTS 33

ing to Lefebvre, "Every single one of those pigeons had eaten a different mix. There was only corn in some of their crops. In others, there was only wheat. In some of them, there was only vetch."

Lefebvre says that instead of a "chaotic group of individuals that's randomly picking whatever seed it can have access to," the pigeons were actually behaving like specialists and practicing a form of resource partitioning in which each bird focused on a single kind of seed. "By specializing, they can be a lot quicker at what they do," he says. "They're always going for the same color, always using the same beak-opening size, and they're avoiding the competition of the other guy." For each member of the flock, it is the ultimate win-win situation.

City pigeons have adapted remarkably well to their urban lifestyle. Pigeons typically greet faithful feeders with obvious enthusiasm; once food distribution begins, the birds scramble for every last morsel.

Scramble Competition

Even in a flock feeding on the same kind of food—say, bread tossed by a Montreal pigeon lover—the birds do not attack one another in order to increase their share. They compete through speed, or what Lefebvre calls scramble competition. "You just try to get the food into your stomach before it gets into the stomach of the other guy," he says.

Pigeons are not the only adaptable animals in cities. "I think humans adapt and conform to pigeons to a certain extent," Lefebvre says. "People learn very quickly that pigeons will like certain seed types and not others, and people will learn the times at which pigeons are more likely to come for food." Faithful feeders are rewarded, however, for the birds usually recognize the people who provide for them and approach their benefactors enthusiastically.

Watching a pigeon flock one recent fall afternoon on the McGill campus, Lefebvre says that although he looks favorably on Montrealers who feed the birds, he also steers clear of them. "They often have ownership claims to a particular flock, and they feel that you're a threat to that. We just sort of stay away from them and make sure that we don't interfere."

Down at Lefebvre's feet, a pigeon has begun pecking at a mushroom, and several more birds wander over to investigate the new food source. A metal sign on the grass says, "Please refrain from feeding the pigeons in this area." Traffic sounds waft in from Sherbooke Avenue just beyond McGill's Roddick Gate.

On the other side of the lawn, a lone man stands at a park bench. Dozens of pigeons gather in a close circle by his feet. The man spreads bits of bread on the bench and pavement, then stuffs a paper sack into his coat pocket. Trailed by fluttering birds, he strolls along the path. Smiling a little guiltily as he passes, the pigeon feeder fades into the crowds of people on the avenue.

You Tasmanian Devil, You!

by Greg Morgan

It spits, hisses, and snarls, fights with unusual aggression, and eats most anything in its path. The real Tasmanian devil *(Sarcophilus harrisii)* is as feisty and bizarre as its cartoon incarnation. When excited, the devil's blood-engorged ears turn purple. The muscular head and neck make up one-fifth of the body weight, adding to the animal's stunted, brutish appearance.

At about 20 pounds, the Tasmanian devil holds the distinction of being the world's largest carnivorous marsupial. The creature's powerful jaws rival those of the hyena and even of the great white shark.

Devil Data

The Tasmanian devil is the world's largest carnivorous marsupial. Its pouch is smaller than a kangaroo's, and the opening faces the rear of its body. About the size of a small dog, with thick black fur splashed with white across the chest or rump, adult male devils weigh about 20 pounds (9 kilograms). Fat is stored in the animal's plump tail. The Tasmanian devil walks on its toes, and its feet have pads between the toes and heels to aid in climbing.

The devil breeds once a year, with gestation lasting only 18 to 21 days. At birth, up to 20 minuscule devils, no bigger than grains of rice, race from the birth canal to the pouch. Only the first four to reach and cling to a nipple survive. They nurse for 90 to 100 days, and then the mother weans them for up to five months, leaving the young in a grass-lined den while she finds food.

What's for Dinner?

Much of the devil's notoriety stems from how and what it eats. Nick Mooney, a biologist with the Tasmanian Department of Parks, Wildlife and Heritage, believes that the devil's huge head and shoulder proportions are adaptations for feeding on large animals. Despite their small size, devils have powerful jaws that rival even those of great white sharks and hyenas. "There is no doubt devils are immensely strong in terms of chewing," Mooney says. "The skull is extremely robust. The short, wide jaw gives tremendous purchase and biting power." The skulls of these tough creatures have been known to survive shattered bones and .22-caliber-bullet wounds.

Devils may travel up to 10 miles (16 kilometers) in one night, scavenging any carrion they encounter and hunting for wallabies, rabbits, and other animals. Their Latin genus name, *Sarcophilus*, means "flesh lover," but Mooney has removed from devil droppings the remnants of prodigious gastronomic feats. The list of items reads

Warner Brothers exaggerated various traits of the Tasmanian devil to create an animated character noted for its whirlwind arrivals and copious drooling.

like the ingredients of some bizarre witch's brew: a woolen sock; part of a pet collar; 27 echidna quills; cattle ear tags and rubber lamb "docking" rings; the head of a tiger snake; bits of foil, plastic, and Styrofoam; half a pencil; the dorsal spine from a leather-jacket fish; a boobook-owl foot; a cigarette butt; a steel-wool pot scrubber.

At a carcass, devils devour everything they can—fur, skin, meat, organs, teeth, and bones, except for the central part of the skull, the pelvis, and the ends of large leg bones. There are accounts of dead cows all but disappearing overnight. The size of a carcass seems to determine how many devils can feed at once. Usually only two or three will feed on a dead wallaby, but 22 have been observed eating a dead cow. Mooney muses that, with so many crowding around a carcass, "it is probable that the devil's long whiskers assist in spacing, rather like 'touch parking.'"

A casual observer sees only a continuous feeding frenzy marked by constant growling, but, according to Mooney, "Complex rituals are in action, minimizing physical contact by creating order from the stress of normally solitary animals feeding together." Their fierce reputation belies the fact that devils are actually shy.

Biologist David Pemberton of the Tasmanian Department of Parks, Wildlife and Heritage recently recorded 11 distinct vocalizations—including three types of growls, a bark, a whine, a snort, a yip, a crescendo, a click, a humph-growl, and a shriek. All of these may be heard when devils congregate around a carcass. Pemberton says that "if another devil arrives and approaches the carcass, only one of the feeders responds to it."

Tasmanian devils are both scavengers and hunters. Although usually solitary, they often come together to devour a carcass (below). What appears to be a feeding frenzy may actually be quite ritualized.

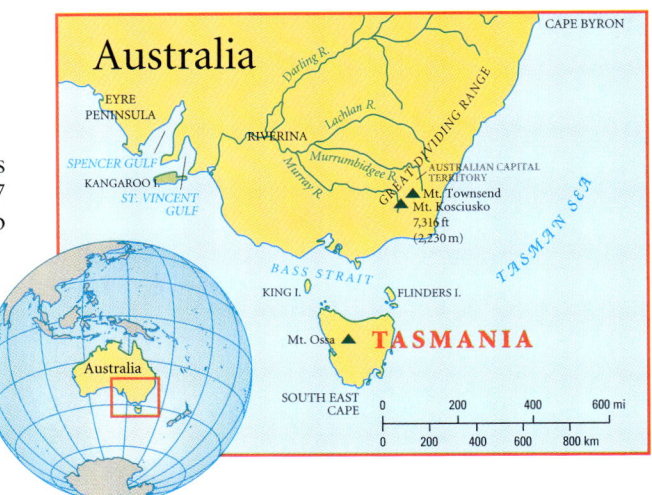

No more than four young devils (above) survive from each litter. The creatures now live only on the island of Tasmania (see map), although they once ranged throughout Australia.

The newcomer will approach a carcass tentatively, displaying a variety of postures and vocalizations. If it likes its chances, it may rush in and ram a feeder to displace it. If this fails, it turns and flees with its tail up. The feeder usually gives chase, biting the rump of the fleeing intruder. On one occasion, Pemberton observes, "The pursuit continued for about 100 meters in a zigzag path, with both animals constantly vocalizing."

Pemberton has found that devils eat up to 40 percent of their own body weight (compared to 17 to 23 percent for lions, and 18 to 27 percent for hyenas) at each meal, which occurs every three days or so. Mooney believes that "they may regurgitate food for their den young. If they had to run with a lump of meat, they'd probably lose it in competition."

Expanding Its Niche

The devil today appears as numerous as ever. Mooney attributes this partly to the tragic demise of a larger marsupial carnivore, the thylacine, whose disappearance enabled the devil to expand its ecological niche on the island. The thylacine, or Tasmanian wolf *(Thylacinus cynocephalus)*, has not been seen on the island since the early 1930s. With its natural prey on the decline in the 19th century, thylacines turned to killing sheep and poultry and became the target of bounty hunters.

Mooney says, "While there is no hard evidence for this, it seems that the devil has expanded its niche in the absence of competition from the thylacine. . . . Its numbers have steadily increased over the same period, and [it] is now common throughout Tasmania."

Farmers are ambivalent about Tasmanian devils. They lose lambs and other old or sick stock to the animals. Says Mooney: "The bleating of a newborn lamb could be as good as a dinner gong for a devil." On the other hand, farmers know that devils regulate the population of pasture-razing rabbits and, apparently oblivious to maggots and bacteria, dispose of flyblown sheep.

At different times since the European colonization of Tasmania nearly 200 years ago, the fate of devils has appeared to hang in the balance. For a short period this century, a bounty was paid by a large livestock company. At other times, their numbers have been decimated by viral infections. In recent years, there have been fears about the cumulative

The Tasmanian devil is a shy creature—especially around humans. Although its numbers are increasing, sightings are rare, primarily because the animal rarely ventures out during the day.

effects of herbicides on devils eating herbivorous prey. Conservationists and other Tasmanians now realize that the future of this international icon of their island cannot be left to chance. "Unfortunately, though," says Mooney, "it's embarrassing how little is being done. . . . I'd be delighted to hear from any reader interested in working on devils." This is one devil to whom, in order to ensure the creature's survival, Tasmanians should be prepared to sell their souls.

Rhino Relations

by William B. Karesh

The long, dry grass stems swished, crunched, and snapped as Kes Smith and I crept forward. Stopping, barely breathing, we watched the rhinos' ears turn to catch our sounds. But the animals did not move from their shady spot under a tree in the midday equatorial sun. A slight breeze was keeping our scent away from them. Slowly, we inched ahead to within 35 yards (32 meters). One animal stood up, acting nervous, almost suspicious. We raised rifle and crossbow, nodded to one another, and fired. Startled, the rhinos spun around and, lucky for us, ran off in the opposite direction. We sprinted through the grass, searching for the biopsy darts that had fallen to the ground. If all had gone according to plan, the darts would carry valuable skin-tissue samples from the animals, which could be used to analyze the genetic diversity among the white rhinos in Zaïre's Garamba National Park.

Kes and her husband Fraser have worked in Garamba for more than 10 years, redeveloping the park to protect the last population of northern white rhinos in the wild. The Smiths had asked me to help them obtain tissue samples from the rhinos in the safest manner possible—the biopsy dart, which I helped develop in 1987. Besides involving an extremely rare species, this cooperative project would forge yet another link in a chain that has connected the Wildlife Conservation Society (WCS) to Garamba National Park for more than 85 years.

Past Progress

It all began in 1909, when Herbert Lang of the American Museum of Natural History; Henry Fairfield Osborn, soon to be president of WCS (then called the New York Zoological Society); Theodore Roosevelt; and others were exploring the northeastern part of the Belgian Congo, now Zaïre. Perhaps the most significant difference between their expedition and ours was that they used bullets and spears to obtain their specimens, and we used darts. We both shared an interest in the northern white rhinoceros, *Ceratotherium simum cottoni*, a rare race of white rhino. In those days, more than 1,000 roamed the savannas where the Belgian Congo met Sudan and Uganda. Today fewer

Garamba National Park in Zaïre is the only place where the northern white rhino survives in the wild. An international conservation team is urgently working to save the creature from extinction.

than three dozen of these magnificent creatures graze there, separated by 1,500 miles (2,400 kilometers) from their more numerous cousins, the southern white rhinos.

The Garamba region and its wildlife have had a convoluted history. More than 100 years ago, King Leopold II of Belgium decided to try to duplicate the success of Carthaginian general Hannibal, who, in 218 B.C., led an army across the Alps. His force included trained African elephants. The king had four trained Asian elephants shipped to the coast of East Africa, and, along with 600 porters, they set out to walk to the Congo. They never made it. But the king's efforts to work with African elephants did not end there, and by 1927 more than 50 elephants were trained and stationed at Garamba. Over the next decades, these animals provided the power to maintain roads, haul goods and supplies, and cultivate the land—a virtually unheard-of concept on the continent, and a perfectly appropriate technology for the remote reaches of Africa.

Trained elephants not only made the region notable, they made it manageable. In 1938, nearly 2,000 square miles (5,000 square kilometers) of this beautiful, wild, undulating long-grass savanna were declared a national park by the colonial Belgian government. Another 3,000 square miles (8,000 square kilometers) bordering three sides of the park were protected as wildlife reserves with sustainable use of natural resources. Of primary importance was the northern white rhino and the northern savanna giraffe, *Giraffa camelopardalis congoensis*. Garamba is still the only place where these species occur in Zaïre.

Growing Threats

But times change. Following independence from Belgium in 1960, the new government was left to support its parks during a period of intense turmoil. Simba rebels were active in the Garamba region; to survive, the elephant trainers and their charges took refuge in the bush. When they finally returned to Garamba, the park retained only a shadow of its past glory, and for 20 years the staff was pretty much left to their own devices. By the second half of the 1970s, the area was thought to have 400 to 800 rhinos and between 15,000 and 30,000 elephants.

The demand for elephant ivory and rhino horn, however, was beginning to reach even this remote site. Recognizing the threat to rhinos, the Wildlife Conservation Society sent Kes Hillman Smith, with Ian Grimwood of the World Wildlife Fund (WWF), to survey rhinos throughout Africa. By 1982, the number of Garamba rhinos had fallen precipitously, and the conservation of the northern white rhino was given highest priority by the World Conservation Union's African Rhino Specialist Group. About the same time, the United Nations Education, Science, and Cultural Organization (UNESCO) recognized the importance of Garamba and named it a World Heritage Site.

International Help

A coalition of international conservation groups, led by the WWF and the Frankfurt Zoological Society, was organized to restore the park and protect the remaining wildlife. At the same time, however, poachers were hard at work. When Kes and Fraser arrived in

The number of northern white rhinos has doubled—to 30—since 1984. Skin-tissue samples (above) are used to analyze the population's genetic diversity.

ANIMALS AND PLANTS 39

Thanks to intensive conservation efforts, Garamba's elephant population has more than doubled—to about 11,000—in little more than 10 years.

Garamba in 1984 with their colleagues Charles and Jan Mackie, there were only 15 rhinos and roughly 4,500 elephants left. Spurred by the desperate situation, the Institut Zaïrois pour la Conservation de la Nature (IZCN) appointed a new, aggressive head of the park, Muhindo Mesi Habuye, and a sharp, enthusiastic rhino-protection officer, Mbayma Atalia. Together, they built a team that began to restore the run-down buildings, train the guards and staff, and build roads, river crossings, and outposts. They started a tough antipoaching program and began working with people in the surrounding communities. They also instituted research and monitoring programs to understand the ecology of the unique grassland and to determine the numbers and types of animals it could support. These ambitious efforts paid off. In 10 years, the rhino population doubled, and the elephant population increased to 11,000. Hundreds of species of birds live in the park or use it as a migration stop.

There is still a tremendous challenge to keep the park running and to protect its inhabitants. Civil disturbance and poverty

Wildlife officials must first sedate a rhinoceros before obtaining a skin-tissue sample or placing a radio-equipped collar around the gigantic creature's neck. Below, a specially designed biopsy dart in the rhino's right foreleg obtains the needed skin-tissue sample. Above, the groggy rhinoceros, radio collar securely in place, struggles to its feet.

Garamba National Park is the only reserve in Africa where tourists ride trained African elephants (below) to see the wildlife. Revenues from tourists help to pay the park guards (left), who protect the endangered animals from poachers.

have resulted in extinction of the northern white rhino in neighboring Sudan and Uganda. After 25 years of civil war in Sudan, tens of thousands of refugees and rebels in need of food live near Garamba's boundary. Park guards routinely risk, and sometimes give, their lives to protect the animals. Zaïre's economy has yet to come out of its slump, and government agencies do not have the funds to protect and manage a reserve the size of Garamba. Funding from international organizations has fallen, partly in response to changes in political alliances and diplomatic issues.

PROTECTION EFFORTS

Most of the wildlife in the park can only wait to see what fate brings, but some of the elephants are participating in protection efforts. Three of the four trained elephants that hid in the forest during the Simba rebellion still live and work around park headquarters. Instead of construction and agriculture work, as they did in the past, they now carry tourists

A century ago, more than 1,000 northern white rhinos roamed the savannas. Today, those that remain are separated by 1,500 miles from their more numerous cousins, the southern white rhinos.

safely past crocodiles and hippos to explore the wonders of Garamba. Tourism revenues help support IZCN and demonstrate the economic benefits of conserving wildlife. The fee for one tourist for one day ($50) covers the monthly salaries for 10 park guards. Recent threats of violence and disease have slowed the flow of foreign tourists to a trickle, but the remoteness of Garamba protects it from the human tragedies of Central Africa. Adventurous tourists who come experience a piece of wild Africa unlike anywhere else.

The Wildlife Conservation Society continues to participate in the unfolding story. Efforts are under way to link Garamba with the work of WCS biologists John and Terese Hart in the vast Ituri Forest to the southeast. Exchanges between the researchers and the Zaïrean staffs of these two areas is pushing conservation activities forward. The society's field veterinary program provides services for both projects. The society has also been helping with care for the trained elephants in Garamba, and with radio-collaring wild elephants to determine the quantity and quality of available habitat. Similar work has been conducted on the rhinos, with the assistance of Peter Morkel, a wildlife veterinarian from Namibia.

Because Garamba abuts Sudan and Uganda, it is a potential entry point for diseases. Examinations of kob antelope, African buffalo, and elephants have provided critical information, and the Garamba staff has been trained to handle a disease crisis should one occur.

International wildlife organizations are working together to develop a long-term strategy to support Garamba and the rhinos. The future of Garamba National Park, like its past, depends on the interest and involvement of people around the world.

ANIMALS AND PLANTS

The Importance of Being FLASHY

by Doug Stewart

Peacocks' tails have long baffled evolutionary biology. Clearly peahens love them—the bigger and brighter, the better. But it is hard to imagine what possible advantage these eye-catching but burdensome appendages offer either cocks or hens in the grim business of survival.

MORE THAN A PRETTY TAIL

A study now shows that there is nothing capricious about a finicky peahen's habit of sizing up a suitor by his tail. In a recent experiment, Marion Petrie, Ph.D., a zoologist at the University of Oxford in England, placed a different peacock in each of eight pens containing a few randomly chosen peahens. She then charted the progress of the 350 chicks that hatched. Bizarre as the result may sound, the chicks that grew the fastest and survived the longest when released into a British park were precisely those whose fathers' trains had the largest eyespots.

Petrie had expected as much. "We think an elaborate train acts as a sort of handicap," she says. "Only the best-quality males can afford to carry it around, so females are using the train as a cue for male quality."

In other words, if a peacock can not only strut his stuff but also find food and evade predators, all the while dragging around a bigger and more conspicuous tail than his rivals, he must indeed be hot stuff. By fanning out his tail at opportune moments, he is bran-

For a lonely peacock, a dazzling display can go a long way toward getting a disinterested female interested. Beyond impressing the peahen with his plumage, the male is also proclaiming himself a healthy specimen.

Making the most of its extravagant mating plumage, the red bird of paradise goes head over heels to attract a female during its extraordinary courtship ritual.

dishing a Day-Glo, billboard-sized self-advertisement: LOOK WHAT I CAN DO! And since any animal's health and fitness depend partly on the genes it inherits, a fitter-than-average peacock is likely to produce fitter-than-average chicks.

Naturalists have long agreed that among animals that use unusual ornaments or behaviors to win mates, the individuals with the most attention-getting displays tend to be the most successful. What scientists are now beginning to resolve is *why* this should be so. In the past 15 years, a flurry of research has suggested that beneath the apparent silliness of so many of the animal kingdom's courtship displays—the greater bird of paradise's imitation of a feather duster, the synchronized swimming of great crested grebes, the dance of the fruit fly—hidden signals are being sent and received. As with the peacock, these messages are about who will make a good parent.

Survival of the Cutest

In *Origin of Species*, Charles Darwin conceded that natural selection alone—survival of the fittest, in simple terms—could not account for flamboyant ornaments, which are found mainly in birds and fish. He suggested a second, parallel mechanism, which he called sexual selection. This includes selection for horns and other weapons that rival males use to batter one

The outlandishly long tail feathers grown by the male widow bird during mating season, while vitally important for attracting females, impede the bird in its more mundane activities.

another during the breeding season in a process that determines dominance and thereby access to females. Darwin also described a form of sexual selection that was "of a more peaceful character": female choice. If, over thousands of generations of birds, females chose "the most melodious or beautiful males, according to their standard of beauty," the males of the species would find themselves slowly evolving to meet that standard. The ugly and tone-deaf would go childless, and their frillier rivals would multiply. As to how and why a hen, let alone an insect, would hew to a "standard of beauty," Darwin did not venture to speculate.

Contemporary researchers have begun to fill in the answers. Evolution works in small steps, of course, and what was merely practical yesterday can become beautiful (to the wooed female) today, the scientists say. The elk first grew antlers to fight and defend itself, while the mane of the lion and the ruffled cape of the hamadryas baboon protected their necks from lethal bites. Gradually, females of each species evolved an innate preference for these now-ornamental armaments. To the females, big antlers and heavy manes now mean "marriage material."

In other cases, perhaps most cases, the researchers conclude, the reverse may have happened. The females somehow acquired an inborn preference for a particular trait, and the male population over time adapted to that preference by developing the trait.

Several ingenious experiments in the past 15 years have supported this idea. When Malte Andersson, Ph.D., of the University of

In a troop of hamadryas baboons, the most majestically maned male is doted upon by the females, who instinctively know that his offspring will be strong and healthy.

Göteborg in Sweden glued extra tail feathers to male African long-tailed widow birds, he found more females nesting in their territories than in those of males with shorter tails. Since long tails otherwise serve no useful purpose, Andersson concluded that the females' preference for them was driving the evolution of the trait.

WHY SO GARISH?

As to why this preference arose in the first place, the mainstream assumption has long been that garish, highly specific ornaments simply say two things: "Here I am!" and "I'm just your species!" But catching a passing female's attention is one thing. Winning her over is another. Many biologists today find the look-at-me scenario too simple to explain many of the intricacies of wildlife courtship.

The theory now in vogue is that ornaments are a cue to fitness, as Marion Petrie surmises with her peacocks' tails. Likewise for species that engage in elaborate, and exhausting, courtship rituals: if a sedge warbler can sing its song repeatedly without missing a note, or a Panamanian mosquito can waggle its proboscis and wave its iridescent leg paddles without missing a step, that performance in itself is a badge of male fitness.

Some researchers suspect that courtship displays do more than advertise

---

When Females Play the Game

In the wild, most show-offs are male, but not all.

"When there's competition among members of one sex for the other sex—like a dance with 20 men and only four women—there's pressure on the men to put out their best," says Ian Jones, Ph.D., a biologist at Simon Fraser University in Burnaby, British Columbia. "In nature, they become ornamented."

In species that naturally match one male with more than one female, the most-successful males often amass harems, while the losers face a life of celibacy. The pressure is on, evolutionarily speaking, not to lose. Having a choice of suitors, the females can afford to be drab, as so many are in the natural world.

But when each sex actively competes for mates, both sides evolve ornaments. This, says Jones, explains why both sexes of a seabird he has studied—a Bering Sea native called the crested auklet—have sprays of feathers protruding oddly from their foreheads. And in the relatively few species where the female chases the male (birds such as phalaropes, for instance), she is brightly colored, while he is dull.

A flourish of feathers adorns the foreheads of both female (left) and male crested auklets.

What Large Antlers Tell a Mate

Among deer, females that choose well-antlered males may not simply be swooning over a macho display. According to Valerius Geist, Ph.D., at the University of Calgary in Alberta, they are instinctively protecting their future offspring.

"If you are an ungulate living in the open plains, you are likely to be chased by your predators, not stalked and ambushed," Geist explains. Early running ability is essential. "To survive, the young must be born large and grow rapidly." The species therefore evolves an ability to divert an unusual amount of nutrients into milk.

In males, the excess nutrients are available for other uses. Geist theorizes that they wind up in antlers. "Antlers are a proxy for success at feeding," he says.

If his theory is true, then large antlers and rich milk should go together, along with large newborns and superb running ability. And they do, says Geist. "Of all deer, caribou young are the most highly developed—they can run a few minutes after birth—their mothers have the richest milk, and their fathers have the largest relative antler size of all living deer."

When it comes to antlers, size means everything. As with other males in the deer family, a Rocky Mountain elk (above) that sports a big rack will likely sire large newborns.

an animal's ability to carry a burden or perform a taxing chore. Marlene Zuk, Ph.D., a behavioral ecologist at the University of California at Riverside, argues that the changeable appearance of ornaments like a red jungle fowl's comb are an up-to-the-minute indicator of whether an animal is infested with parasites. "If there is a trait that you can only produce if you're free of parasites, then a female who chooses you for that trait is more likely to have successful offspring," she says.

When Big Gets Bigger

Whatever the reason for their existence, a species' ornaments tend to grow more exaggerated as they evolve. Displays that once were merely colorful become outlandish. If a female has an innate liking for a certain male trait—big antlers, say, or a long tail—and accepts the overtures of only big-antlered or long-tailed males, then her daughters are likely to inherit the preference while her sons are apt to inherit the trait. Over many generations, the two tendencies become intertwined, and each becomes more pronounced.

Eventually the costs begin to outweigh the advantages—the antlers get stuck in branches; the tail makes flight unwieldy—and the evolutionary pressure abates, the scientists say. Some species get around the problem by keeping these added burdens temporary. For example, the male cuckoo wrasse fish, which ranges from Italy to Norway, becomes a Technicolor extravaganza only during mating season; the rest of the year, it is a dull, reduced-risk blue-green. Widow birds jettison their long tails when they are through courting for the season. Elk and caribou shed their antlers at the end of the annual rut.

The male bowerbird, perhaps aware that his plumage is less than sensational, carefully gathers and arranges an array of tantalizing baubles in hopes of attracting curious females to his territory.

Some species avoid burdening themselves by keeping their finery detached. The male satin bowerbird of New Guinea, itself a fairly nondescript creature, entices females by building elaborate bowers of grass that it paints with berry juice and decorates with miscellaneous blue decorations (including fresh flowers, changed daily). Rival males often raid one another's bowers and steal decorations, which may be how an elaborate bower signifies that its defender is fit.

SOCIETY COURTING

In species that live in socially cohesive groups, primates among them, males are not under pressure to stand out by being flashy, researchers say. Instead, these animals develop other ways to make themselves desirable. (An exception is the male mandrill. No one is sure why the stunning blue markings on its face have evolved.) In judging a mate, female cotton-top tamarins, for instance, are impressed by baby-sitting displays. Cotton-top tamarins are a species of New World macaque that typically bears twins. Twins are more than a mother can handle alone, so males in the group help out with child care.

In the late 1980s, Eluned Price, Ph.D., then a primatologist at Scotland's University of Stirling, studied courtship strategies in a troop of captive cotton-tops. "I happened to notice that males would pick up an infant, then rush to the female and try to mate with her," she recalls. When Price started keeping count, she found that females gave the green light more readily to child-carrying suitors. It was as if the behavior conjured images of the males as good fathers, according to Price. As a sexual ploy, it may have been transparent. But it worked.

Hidden Desires: The Power of Food?

Sometimes females pine for male ornaments they have never seen before.

Among male green swordtail fish, sword-shaped tails are a potent sexual lure—the longer the better. In the past five years, Alexandra Basolo, Ph.D., a biologist now at the University of Nebraska in Omaha, chose a closely related but swordless species, the southern platyfish, and attached fake swords to the rear ends of a group of males. The female platyfish went wild.

The common ancestor of swordtails and platyfish were swordless, so an ancestral sword fetish cannot explain the results. One of Basolo's conjectures is that the females are innately sensitive to certain shapes—maybe those of food. "If a male happens to develop an ornament that mimics that food item," says Basolo, "he may have an advantage in capturing a female's attention."

Male Panamanian grass anoles, lizards akin to pet-shop chameleons, exploit a similar sensory bias. The male lures mates by jerkily bobbing his head and unfolding a decorative throat fan. His instinctive movements, says Leo Fleishman, Ph.D., a biologist at Union College in Schenectady, New York, mimic those of crawling insects, the lizard's favorite food.

Diversity Down on the Farm

by Linda J. Brown

The San Clemente goat, the Dutch Belted cow, the Navajo-Churro sheep, the Suffolk Punch horse, the Gloucester Old Spot pig, the Pilgrim goose. Sound familiar? Probably not. These animals were common on farms for hundreds of years, but today their numbers are so few that they are in danger of disappearing entirely. And they are not alone. Nearly 80 livestock breeds in the United States are in decline or perilously close to extinction.

The diversity of farm-animal breeds has quickly eroded in the past 50 years, in North America and worldwide. More than half of the 15 breeds of swine raised in North America in the 1930s have vanished, and just three breeds now account for 75 percent of all registered pigs. And while there were six major dairy-cattle breeds a few decades ago, today the Holstein, bred to produce massive quantities of milk, dwarfs all other breeds in numbers both in America and abroad.

As livestock breeds disappear, so do the hereditary traits carried in their genes. That is cause for concern because no one can predict what traits may prove useful for science or agriculture in the future. As a result, there are new efforts to protect rare livestock breeds and preserve genetic diversity on the farm.

WHAT'S BEHIND THE DECLINE?

Why is the genetic base for livestock under assault? "The industrialization of agriculture has selected our livestock breeds for uniformity and for single production characteristics," says Don Bixby, D.V.M., executive director of the American Livestock Breeds Conservancy (ALBC). The ALBC, founded in 1977 and based in Pittsboro, North Carolina, is a private, nonprofit organization dedicated to protecting the genetic diversity of American livestock.

Today's agricultural techniques call for animals that grow rapidly, mature quickly, reproduce early, and thrive on a high-grain diet, Bixby says. Over the past half century,

The increasing rarity of Toulouse geese (above), Muscovy ducks (below), and dozens of other diverse livestock breeds is due in large part to the high production demands of modern agriculture.

ANIMALS AND PLANTS 49

Many modern farms and ranches concentrate on livestock breeds that grow rapidly, mature quickly, reproduce early, and thrive on a high-grain diet (left). As a result, the Belted Galloway cow (below) and other cattle breeds considered commercially inefficient are in danger of dying out.

livestock producers have selected and bred animals that have these characteristics. The animals are often intensively managed—fed growth enhancers and antibiotics and kept in climate-controlled housing—making traits such as the ability to forage or to withstand harsh weather meaningless. The result: breeds chosen or developed to meet high production characteristics have come to dominate each livestock species, while others have disappeared.

This contrasts sharply with traditional farming practices, in which farmers relied upon many breeds, most of which served more than one purpose. (Chickens, for example, provided both eggs and Sunday dinner; today layers and broilers are raised separately.) In fact, the historical trend in agriculture was toward increased diversity in farm animals, as farmers aimed to develop animals particularly suited to their region, market, and farming methods. Today agricultural products compete in national and international markets, and that has prompted a drive for uniformity and predictability.

Why Diversity Matters

If a few superproducing livestock breeds provide us with all the food and animal products we need, why worry about preserving rare and endangered breeds? There are three important reasons.

For one, we may be putting our food supply at risk. As breeds disappear, so do significant traits such as maternal instincts, fertility, disease resistance, and tolerance of heat or cold. No one can know the extent of the traits that are lost when a breed slips away. There is also no way to predict which traits may be needed. Evolving pests and diseases, global warming, and changing market demands may call for different characteristics.

Parasites, for example, have long plagued sheep. Today's ranchers rely on modern parasiticides to keep them at bay. Parasites have a way of developing resistance to chemicals, however, and in recent years parasites unaffected by drugs have afflicted large numbers of sheep in Australia. Now drug-resistant parasites are appearing in the United States as well. But two rare breeds—the Gulf Coast native sheep and Caribbean haired sheep—exhibit incredible genetic parasite resistance. Developing breeding programs that make use of this natural resistance may prove a better way to deal with parasites than developing new drugs, which may work only temporarily. Were these breeds to become extinct, this possibly lifesaving trait might be lost.

Second, we need to protect the full genetic complement of farm animals for its potential to further scientific and medical knowledge, as many of the rare breeds are biologically unusual. Already, the Ossabaw Island hog is a research model for non-insulin-dependent diabetes, and the myotonic goat is a model for human myotonia congenita (a muscular disease). It is impossible to predict what role other species might play in future research.

Finally, these rare animals are part of our heritage. They "represent an important historic and aesthetic remnant of our past," says Bixby. "All of these breeds were developed as partners for humans in some endeavor."

Navajo-Churro sheep, for instance, were brought to this country by Spanish explorers around 1520 for their meat. But their fleece was discovered and embraced by Native Americans in the Southwest, who used it in weavings. The sheep thus helped shift the Native American culture from hunting and gathering to herding and weaving. By the 1970s, however, the breed had almost disappeared, after barely surviving two rounds of slaughter by the U.S. government. A professor at Utah State University in Logan took on conservation of the sheep, and since then their numbers have slowly climbed. They are being restored to tribal flocks, and they once again furnish the raw material for unique tapestries.

PRESERVING NOAH'S ARK
How can people pull these endangered farm breeds, many with such rich cultural histories, back onto solid ground? The problem is being attacked on several fronts.

The ALBC has taken a leading role in protecting nearly 100 breeds of livestock from extinction. The group offers technical support

Unusual sheep breeds may ultimately survive because each has a characteristic type of wool. The dreadlocked texture of the Cotswold's wool (above), for instance, is prized by hand weavers. The Jacob sheep (left) is noted for the color of its wool.

to breeders, breed associations, and farmers who raise these animals. It has also gathered census figures for asses, cattle, goats, horses, sheep, and swine, and is completing a census for poultry.

Using these figures, the ALBC classifies breeds in trouble as critical (fewer than 200 North American registrations and fewer than 2,000 estimated globally), rare (fewer than 1,000 in the United States and 5,000 worldwide), watch (fewer than 2,500 in the United States and 10,000 worldwide), or study (breeds that are of genetic interest but lack documentation or definition). The group has worked with the U.S. Department of Agriculture (USDA) to set up a national livestock database that will maintain up-to-date information on breed populations and characteristics.

The USDA's Agricultural Research Service has plans for a gene bank but no funding as yet. Not wanting to depend on government support, the ALBC began its own gene bank years ago, using cryogenic preservation to freeze and store semen, embryos, and genetic material. So far, the bank includes 15 breeds of cattle and two to four breeds each of sheep, pigs, and goats. "The gene bank broadens the genetic pool because it means that animals that are long dead can still participate in current reproduction programs," says Bixby. "And if a natural disaster decimates a particular breed, then we have the genetics available."

The decline of livestock diversity is not a problem just in North America. To address the problem globally, Rare Breeds International (RBI) was formed in 1991. RBI allows members from different countries to share ideas about conservation. The organization also operates international livestock data banks in Rome, Italy, and Hanover, Germany.

Although generally eschewed by most American pig farmers, Ossabaw Island hogs (above) are sought by medical schools as research models for non-insulin-dependent diabetes.

TO MARKET, TO MARKET

A critical step in boosting the populations of rare breeds is to identify a commercial value for each. Most of these rare breeds have some exceptional characteristics, and the trick is to find the market niche for their attributes. For example, sheep such as the Leicester Longwool, Cotswold, and Karakul are valued for the unusual texture of their fleece, which finds favor with hand spinners and weavers. Marsha VanValin and her husband, Tad, raise Cotswolds and five other breeds of rare sheep on their farm in Sullivan, Wisconsin. "People often compare the Cotswold's wool to dreadlocks. It looks like a carpet coming when they run after you with their long fleece blowing in the breeze," says Marsha.

The wool of other sheep breeds—the Jacob, Shetland, and Black Welsh Mountain—is valued for its natural range of colors. The surge of interest in free-range, organic beef and lamb has also been a boon for many of the older breeds, which are good foragers and grazers.

In Corvallis, Oregon, David Holderread raises a variety of rare ducks and geese, from Embden and Toulouse geese to Chocolate Muscovy and Welsh Harlequin ducks. He sells their eggs as food and art—artists buy infertile eggs and create egg art with the shells. And he sells his geese for their meat and for their ability to keep grass trimmed and act as guards. People raising chickens or small animals can use guard geese to warn off predators like raccoons, skunks, and opossums. "As soon as geese see something unusual, they walk toward it and talk about it," says Holderread. "Predators do not like to work under conditions where they have color commentary going on."

Some rare breeds have no obvious commercial-market niche. "Those are the ones we're particularly worried about," says Bixby. Rare animals that are small, such as the Barbados goat, do not fit industry criteria as meat-producing animals. Likewise, pigs that were bred for fatness, like the Saddleback hog, do not fit today's trend toward leaner swine.

Especially when there is no great market value for a critically endangered breed, the breed's survival may rest in the hands of a single person. Holderread, for instance, raises a very rare chicken called the Ancona. "If we quit offering them to the public, they would disappear from North America within five years," he says. Rare breeds can be difficult to find, and Holderread works hard to disperse his birds. "If they're available, that really increases their chance of survival," he says.

Mark Fields, who raises rare White Park cattle, Mulefoot hogs, and several rare poultry breeds in Clark, Missouri, managed to acquire his hogs from one of the few people

Get Close to a Goat

Historic sites, nature centers, zoos, agricultural museums, and demonstration farms that raise and promote endangered breeds of livestock can spark interest in and empathy for these animals' plight. At Colonial Williamsburg in Virginia, for example, various rare breeds of farm animals show how vital livestock was in the 1700s. "We're losing our rural heritage, and places like Williamsburg are becoming more important as an area you can visit and get close to a horse or a sheep or a cow," says Richard Nicoll, manager of livestock at Williamsburg.

Here is a sampling of places where you can view these animals firsthand:

- Plimoth Plantation, Plymouth, Massachusetts: Living history museum that documents life in the Pilgrim settlement around 1627. Features Kerry and Milking Devon cattle, Wiltshire Horn and Dorset Horn sheep, San Clemente goats, Tamworth pigs, and Red Dorking chickens.
- Lake Farmpark, Kirtland, Ohio: Open-air museum on more than 200 acres (80 hectares), designed to help people understand how farmers use land, animals, and plants to produce food, fiber, and other products. Rare breeds include Suffolk and Black Clydesdale horses, Dutch Belted cows, Hereford hogs, and White Wyandotte chickens.
- Mount Vernon, Virginia: The home of George Washington has an ambitious new agriculture program. Rare breeds include Milking Devon cattle, Ossabaw Island hogs, Hog Island sheep, and Dominique chickens.

A San Clemente goat is milked at the Plimoth Plantation (above), one of several "living-history" museums in the United States where livestock breeds prominent in colonial times are exhibited.

- Greenfield Village, Dearborn, Michigan: Outdoor museum that features wrinkly Merino sheep, Poland China hogs, Shorthorn cows, and Percheron draft horses.
- Colonial Williamsburg, Virginia: Historic village that depicts life in the late 1700s. Features Leicester Longwool sheep, Milking Devon cows, American Cream draft horses, and Dominique chickens.

To Learn More:
For further information about rare farm animals, contact the American Livestock Breeds Conservancy, P.O. Box 477, Pittsboro, NC 27312; (919) 542-5704.

---

in the country who had any left. This man swore he would take his hogs to his grave because they had been mismanaged. In the early 1990s, Fields gained his trust and bought some pigs. Now Fields has shipped Mulefoot hogs as far as Georgia and Iowa.

Affordability is another factor in the efforts to save these animals. Marsha VanValin paid up to $2,000 for her first Shetland ewes. She says, "That's crazy. We're selling them for $250 to $300 now. The only way to have these breeds catch on and multiply is to make them affordable."

What is the outlook for endangered livestock? Some, like VanValin, believe that "small holders are the people in this country who are keeping these breeds alive." Bixby sees additional opportunities for rare farm animals in a trend toward "sustainable agriculture," using less-intensive livestock management. This kind of low-input farming harkens back to the past and is tailor-made for the rare-heritage breeds' strong traits of hardiness and self-reliance. Perhaps, he and others believe, we can learn from the past and go forward with a full barnyard ark.

THE QUEST FOR WATER

by Doug Stewart

The Texas horned toad is one of the fiercer-looking inhabitants of the arid American Southwest. When threatened, the reptile (actually not a toad but a lizard) splays its legs wide, lowers its head menacingly, and arches its spike-covered back like a tank ready for war. Oddly, the reptile assumes much the same posture during desert rains.

"It's a rain-harvesting stance," explains biologist Wade Sherbrooke of the American Museum of Natural History's Southwestern Research Station in Portal, Arizona. In the desert, a light rainfall normally disappears into the sand like water into a sponge. And *Phrynosoma cornutum*'s unusually wide body is perfectly shaped to catch the moisture before it is lost. The water is thence drawn—by capillary action—through hair-thin channels between the reptile's scales all the way to its mouth. The creature does not have to move, just swallow.

Sherbrooke has observed horned toads do something stranger still: "When the rain stops, they drop down and rub their bellies on the moist sand." He assumed at first they were scent-marking. But based on his recent studies of the equally bizarre-looking Australian thorny devil, Sherbrooke now believes that the horned toad is using its water-channeling hide to suck moisture right out of the sand and, in defiance of gravity, carry water up to its mouth.

An Unending Effort

The toad's water-gathering technique is only one of the many adaptations animals have evolved to wring water from even the most inhospitable-seeming habitats. From a purely physiological point of view, life for most wild creatures is an unending effort to take in water. They manage to do so in an astounding variety of ways, while also struggling to avoid losing water too rapidly through evaporation, respiration, or elimination.

Every animal is, in essence, a watery solution inside a not-always-watertight bag of skin. And maintaining that solution at just

Life for most animals in the wild is an unending effort to locate and take in a single essential commodity: water. A jaguar in the rain forests of Belize (left) is fortunate to have an ample supply readily available.

the right strength is crucial to life itself. Starting in the 1950s, physiologist Knut Schmidt-Nielsen at Duke University in Durham, North Carolina, carried out a series of groundbreaking studies of how different creatures—from birds to camels—maintain their water balance. For land animals, the key struggle is avoiding dehydration; freshwater fish, on the other hand, have to keep water out, not in.

Whereas Schmidt-Nielsen looked at organisms as a whole, in habitats wet and dry, research in the high-tech 1990s usually proceeds on a finer scale. Says comparative physiologist Vaughan Shoemaker of the University of California at Riverside: "There's been a progression from the naturalist who asks, 'How does this animal solve its problems?' to the level of researchers who ask, 'What control mechanism turns on a salt gland or turns down a kidney?'" Underlying every insight is the constant theme of how particular adaptations, both physical and behavioral, allow an animal to flourish in its particular habitat.

Drinking Problems

For birds and mammals exposed to the full dehydrating effects of sun and wind, the most obvious adaptation is nomadism. Unlike seasonal migrants, nomads are opportunists. Strong-flying birds like sand grouse, perhaps lured by distant clouds, will fly as far as 60 miles (97 kilometers) to drink from a pond of rainwater. In arid regions, large grazing animals such as camels and antelope often depend on widely spaced watering holes, and so must be able to go for days or longer without drinking. Finding water, they will drink their fill more as a precaution than out of thirst.

But water holes are not always the lifesavers they may seem to be. For decades, wildlife managers in dry parts of the U.S. West have maintained human-made watering units in remote areas, figuring that these should help wildlife—especially game animals such as quail, deer, and pronghorn—to survive droughts.

To test this assumption, biologist Bruce Thompson and his colleagues at New Mexico's Cooperative Fish and Wildlife Research Unit recently tallied numbers of species near watering sites in the vicinity of the parched White Sands Missile Range. Their study, published in 1994, showed no evidence of a boost. "People think, 'It sure is hot. It sure is dry. I'll bet those animals need a drink,'" Thompson says. But a huge number of species in the world's arid regions, from sand rats to giraffes, can get all the moisture they need from the plants they eat. Higher up the food chain, carnivorous snakes and lizards take in most of their water in the moist flesh of their prey.

Tiny grooves in its scales help the Texas horned toad (above) draw raindrops to its mouth. The lizard can even suck minute amounts of moisture from the sand using its water-channeling hide.

Efficient Insects

That sounds plausible enough (even a meal of ants is almost two-thirds water), but what about flour beetles, which can spend their lives in a bone-dry flour bin? How do they find moisture? In part, they manufacture it. The hydrogen (H) and the oxygen (O) in water (H_2O) are, along with carbon, chemical constituents of all carbohydrates. Rearranging the molecules to make water takes only energy. Humans actually perform this feat as part of metabolism. Flour beetles, clothes moths, and termites convert carbohydrates to water, too, but far more efficiently than we do. (Moreover, even "bone-dry" flour is normally 5 to 10 percent water.)

However they obtain water, terrestrial insects are stingy in letting it go. Their hard

ANIMALS AND PLANTS 55

A spadefoot toad (above) emerges during a brief muddy spell in the Arizona desert. Most of the year, the creature stays buried deep in the desert soil, where it is able to absorb moisture.

outer covering is more watertight than a mammal's hide, which cuts down on evaporation. Also, like birds and reptiles, the bugs reclaim most of the moisture in their waste before releasing it. Isopods like wood lice can actually reverse the usual direction of water flow: they can absorb moisture through the anus, which means they can quench their thirst just by sitting down.

In a pinch, some insects wait out long dry spells almost in a state of suspended animation, like a plant's seeds. One hardy insect, a scale of the genus *Margarodes*, was discovered alive in an entomologist's case 17 years after it had been removed from its native soil.

Adaptive Amphibians

While an insect's "skin" is relatively watertight, an amphibian's is highly porous. Amphibians actually use their skin, not their mouths, to take in water. Although they require moist surroundings to avoid death by dehydration, a surprising number of amphibians do quite well in the arid environment of deserts. The water-holding frog of the Australian outback, for example, keeps extra water available by letting its bladder swell with dilute urine, available for reabsorption. In this state, the frog's body resembles a "knobbly tennis ball," wrote naturalist P. A. Buxton in amazement in 1923.

The camel's adaptations to desert life include the ability to tolerate water losses equal to 25 percent of its body weight. In humans, a 10-percent water loss causes deafness and delirium.

The urine is so watery (amphibian kidneys are not terribly efficient) that Aborigines have used the fluid as a thirst quencher.

A desert-adapted amphibian closer to home is the spadefoot toad, which spends most of the year buried like a stone, cool and moist. Only heavy summer rain clattering on the hard desert above the toads prompts them to emerge. In the shallow ponds of rainwater that appear, they noisily mingle and mate. Their progeny's youth is fleeting: a tadpole needs only 10 days to mature into a four-footed, air-breathing metamorph capable of hopping away, but many desert ponds dry up well before then. "The tadpoles are left in smaller and smaller ponds," says physiologist Shoemaker, an expert on desert toads. "Eventually, you'll see a pancake of dead tadpoles." Still, enough survive in the bigger pools to keep the population going.

While adult spadefoots and other frogs and toads can absorb water while buried, another amphibian, the siren, has evolved a very different mechanism for staying moist during dry spells. Sirens are salamanders that live in southeastern freshwater wetlands called Carolina bays, which are prone to drying up. These large, eel-like creatures are so slimy that field biologists slip dry socks over their hands before trying to grip the things.

The socks may also help the squeamish among the scientists get the job done, for the sirens' slime is mucus. "When their habitat dries up completely, instead of traveling overland to find water, they burrow down into the mud," explains herpetologist Whitfield Gib-

In the parched Namibian desert, the oryx (above) deals with the heat in an unusual way: instead of losing precious water through perspiration, the creature simply allows itself to heat up—to as much as 110°F! The oryx cools back down at night, when the temperature of the dry desert air often drops precipitously.

bons of the University of Georgia's Savannah River Ecology Lab. "They secrete a mucous coating that hardens underground to form a cocoon that's impermeable to water." The sirens remain moist but inert for months and perhaps years, he says, "until it rains again."

Mammalian Strategies

Mammals, of course, are incapable of waterproofing themselves, so they have developed other strategies. The camel can lose more than a quarter of its body weight in water without suffering. By comparison, humans who lose one-tenth their weight in water become deaf and delirious.

Large grazing animals such as cattle, bison, and eland keep evaporation low by seeking shade and staying quiet during midday. Still, they need water to drink. The oryx, by contrast, does not need a drop. The majestically horned East African antelope does not even seek shade during the hottest part of the day. How the oryx pulls off this neat trick was discovered in the late 1960s by physiologists C. Richard Taylor, now at Harvard University, and David Robertshaw, now at Cornell's College of Veterinary Medicine. As the day heats up, they found, the creature slows its metabolism, stops sweating, and simply gets hot—in some cases as hot as 110° F (43° C). "The oryx doesn't bother to fight the heat," says Robertshaw. "It lets its body temperature rise during the day and then dumps the heat at night." It simply waits for the chilly night air to cool its body down, rather than wasting water on sweat earlier in the day. As for water

ANIMALS AND PLANTS   57

intake, the oryx grazes before dawn when even dead grass turns moist with dew.

If any mammal has completely kicked the water habit, it is the tiny kangaroo rat, which resides in arid regions of the U.S. Southwest. With several kangaroo-rat species facing extinction, the little rodent is now the subject of intense study. Everything about the creature is water-conserving. Ounce for ounce, its urine holds 14 times as much material in solution as its blood does. "It's almost like syrup," says zoologist Jack Cranford of Virginia Polytechnic Institute in Blacksburg. "In several species, the urine almost crystallizes when it hits the ground."

The kangaroo rat spends most of its life in a burrow, which it plugs up so the air within stays cool and humid—so humid that the seeds it caches there can swell with one-third their weight in water before the animal devours them. The rat's nasal passages are narrow and cool, so water vapor condenses on their linings when the animal breathes out. Also, like other species in dry habitats, the rat converts some of its food into metabolic water. Cranford reports keeping captive kangaroo rats for two years on a diet of dried seeds and no water.

The albatross (above) and other species of oceanic birds can drink seawater exclusively and without any ill effects, thanks to special glands in their heads that remove the extra salt via hypersaline teardrops.

Drinking at Sea

If the desert is the most challenging environment for obtaining water, is the ocean the easiest? Hardly. There is indeed water, water everywhere—but only one known vertebrate, the primitive hagfish, can drink seawater without somehow removing the excess salt. Still, oceanic birds like albatross and petrels can spend more than a year at sea, drinking salt water freely. So if the kidneys are removing the excess salt, they must be producing 10 gallons of urine for every gallon of seawater drunk—an impossibility. The puzzle was solved by Schmidt-Nielsen in a series of studies in the 1950s. The birds' secret turned out to be a set of salt glands in their heads, a bit like tear ducts that release hypersalty tears, but through the nostrils. A black-backed gull that drinks one-tenth its weight in seawater can get rid of the extra salt in only three hours. Reptiles like sea turtles have similar glands. The most efficient salt glands, logically enough, belong to the species that eat the saltiest food, such as plankton-eating petrels.

Animals that live in salt water have to avoid becoming pickled in the stuff. The fluids inside an oceanic fish are only about one-third as salty as the ocean itself. Osmotic

In an Alaskan freshwater stream, spawning salmon (below) stop drinking and absorb their water through osmosis. When at sea, the fish must drink and desalinate seawater in order to remain hydrated.

58 ANIMALS AND PLANTS

In Botswana (above) and elsewhere in tropical Africa, species large and small gather at watering holes to quench their thirst. Usually, while some animals drink, others act as lookouts against predators. During a drought, some creatures will travel hundreds of miles in search of a reliable water supply.

pressure acts constantly to suck water out of the fish through its skin and gills. To avoid drying out, the fish opens its mouth and drinks seawater. The new problem becomes ridding itself of unwanted salt. The kidneys of bony fish are not up to the task, so the gills take over much of the job.

For freshwater fish, the osmotic pressure is reversed: the fluids inside the fish are saltier than those outside (even freshwater contains some salt), so water tends to seep in through the fish's skin and gills. After entering a freshwater stream, a salmon stops drinking altogether.

Whales and other aquatic mammals avoid dehydration in part by producing extremely concentrated urine, saltier than seawater itself. Aquatic mammals are even sparing in the amount of water they release when nursing. The ever-curious Knut Schmidt-Nielsen once analyzed milk he had coaxed from a seal in the Antarctic's Weddell Sea, and found it had twice the fat content of whipping cream and less than half the water content of a lean hamburger. Humans cannot hope to compete with that kind of water-conservation effort, no matter how many bricks they place in their toilet tanks.

Ask the Scientist

▶ *Do horses really sleep while standing up? What about zebras? If not to rest, why, then, do horses sometimes lie down?*

Yes, horses can sleep standing up, but it is more complicated than that, according to Katherine A. Houpt, director of the Animal Behavior Clinic at Cornell University's College of Veterinary Medicine in Ithaca, New York. There are two distinct types of sleep: slow-wave, or sleep of the mind, and rapid-eye-movement (REM), which is dream sleep or sleep of the body. Horses can slow-wave sleep while standing, but they must lie down to REM sleep because it involves total muscle relaxation.

Horses sleep only a few hours during a 24-hour period, and spend only an hour or so in REM sleep. Zebras are similar in their behavior patterns, but scientists have not measured their brain waves to determine the type of sleep they experience.

▶ *What is the difference between horticulture, floriculture, and botany? Also, my high-school-aged daughter wishes to become a botanist. Are there many job opportunities in that field?*

Botanists are scientists who study plants and their environment. Horticulturists specialize in the growth of fruits, vegetables, and flowers. Floriculturists are horticulturists who focus on the cultivation and management of flowering and other ornamental plants. According to the Botanical Society of America, "the field is so broad" that it offers many different job opportunities. "New positions in botany are expected to increase at an above-average rate through the turn of the century," according to the society, in part because of environmental concerns, the search for new medicines, and the need for increased food supplies to nourish a growing world population.

▶ *Aside from domestic dogs and cats, what mammal is responsible for the most human deaths? Are any birds known to regularly (or even occasionally) attack humans?*

The hippopotamus of Africa is notoriously dangerous, says James Serpell, Ph.D., associate professor of humane ethics and animal welfare at the University of Pennsylvania School of Veterinary Medicine in Philadelphia. They can become very aggressive, and will even attack people in boats; "their big teeth can give an awful bite," Serpell says. In North America, domestic horses and cattle can be dangerous—for instance, a well-directed kick by a horse can be fatal. Bears are potentially dangerous, although, contrary to popular belief, they do not go out of their way to attack people. The danger from cougars has been greatly exaggerated.

The cassowary of Australia and New Guinea is the most dangerous bird; the large, sharp claws on its hind feet can cause severe injury. Some seabirds, such as seagulls and skuas, will attack people during nesting season if they think their nests or young are threatened. But, stresses Serpell, many more people are killed each year by bee and wasp stings than are killed by birds and mammals.

▶ *Are pheasant and quail related? Are these birds raised on ranches for their commercial value? Are they endangered in the wild?*

Worldwide, there are hundreds of species and subspecies of pheasant and quail. These birds, commonly called game birds, are classified in the pheasant family—Phasianidae. They share many physical characteristics, such as short bills and short wings. However, pheasant are large, with very long, pointed tails, while quail are smaller and have short, rounded tails. Both pheasant and quail are raised on ranches for food, according to Robert Robel, Ph.D., professor of environmental biology at Kansas State University in Manhattan. For example, a ranch near Wichita, Kansas, raises about 25,000 bobwhite quail a year; in Europe, ring-necked pheasant is commonly served in restaurants. At least one subspecies, the masked bobwhite quail of Texas and northern Mexico, is classified as endangered.

▶ *Do all social insects have a queen, or is that a position peculiar to bees? Sizewise, is a queen bee the largest bee in a given hive?*

"A colony of truly social insects—whether honeybees, ants, or termites—is really an extended family group, with the queen being the mother," says John Ambrose, Ph.D., professor of entomology at North Carolina State University in Raleigh. "The queen can mate and is basically an egg-laying machine. She has a profound influence on the lives of all the bees in the hive. The other females, which make up a worker caste, cannot mate; in some cases, however, they lay unfertilized eggs, which hatch to produce males."

Normally, the queen is significantly larger than other members of a colony. In a honeybee colony, for example, the queen is about one-and-a-half times as big as the workers. In a termite colony, she may be as much as 100 times as big; indeed, she may be so large that she is unable to move.

▶ *How is it that tigers and leopards can survive in tropical regions with such thick coats? Do they have trouble coping with the heat and humidity?*

Tigers and leopards have several ways to cope with heat and humidity, says John Seidensticker, curator of mammals at the National Zoological Park in Washington, D.C. The coats of animals that live in hot places such as India and Thailand have far fewer underhairs, and thus are not as thick as the coats of animals that live in cooler climes. But move a tiger or leopard from its native habitat to Canada, for example, and it soon grows a thicker coat. Tigers spend the hottest part of the day in shady spots, often lying in water. Leopards are somewhat more active during the midday hours—perhaps because at that time they are less likely to run into tigers and lions. They don't cool off in water; instead, they will lie on shady branches high in trees.

▶ *Is it really true that Ireland has no snakes? Does Ireland have any reptiles at all?*

"There are no endemic species of snakes found in Ireland," says Keith W. Brown, curator of the department of herpetology at the North of England Zoological Society in Chester. "There are, however, large numbers of amateur herpetologists in Ireland, and I can guarantee that there are a number of escaped 'exotic' pet snakes living quite happily in some of the most unlikely parts of that fair isle. Snakes are expert escape artists, and if someone's pet snake escapes, it is very unlikely to be announced, as people tend to panic over the thought of a snake on the loose." The only reptile found throughout Ireland is the viviparous lizard, *Lacerta vivipara*, which, as the name suggests, gives birth to well-developed young. This lizard does not have a placenta; the young are nourished by the food supply of the yolk, but the eggshell fails to develop.

ANIMALS AND PLANTS 61

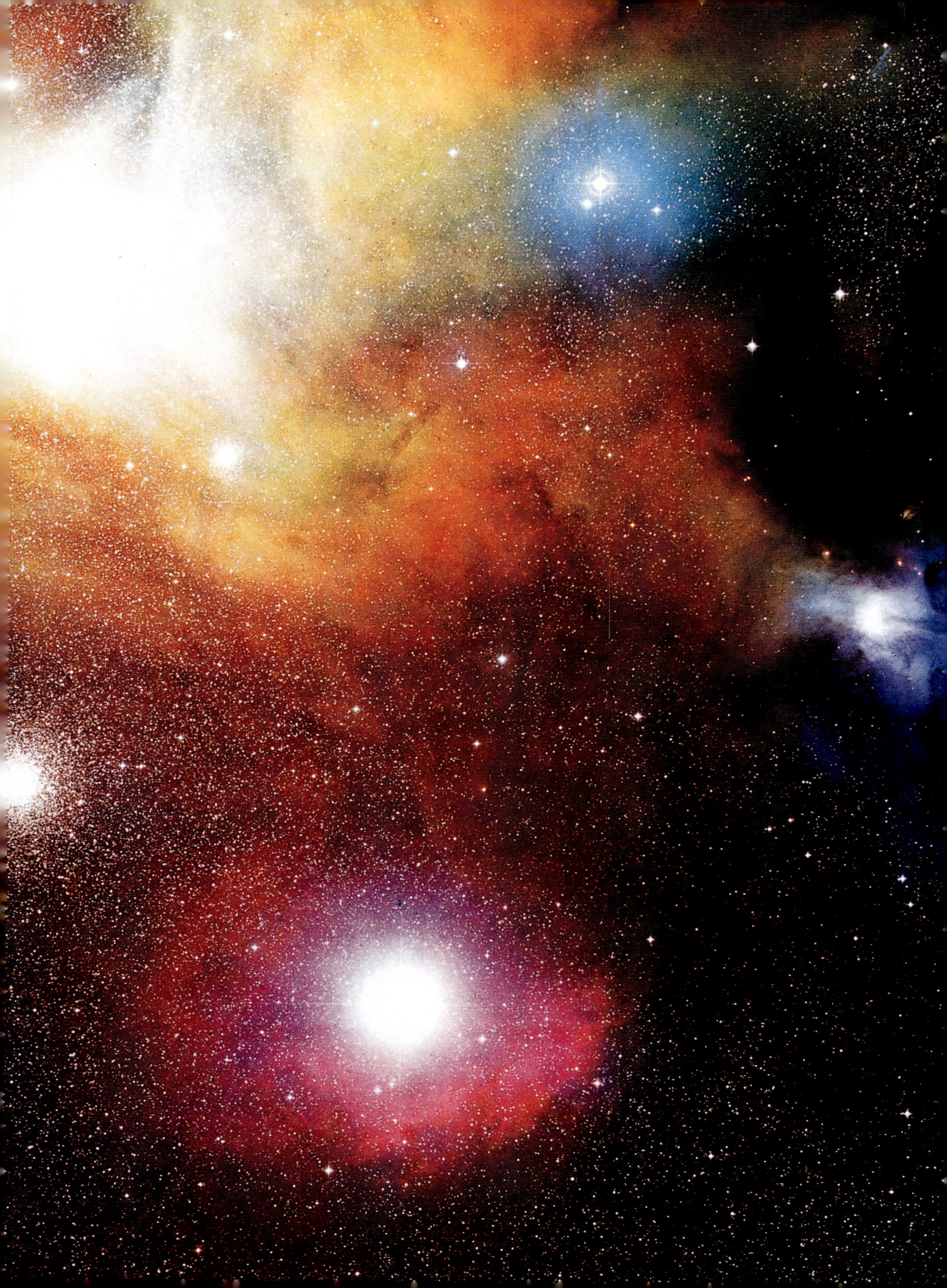

Astronomy and Space Science

▪ The interstellar dust and gas that exist between stars have intrigued astronomers for years. Nearly all information about interstellar matter has been obtained by studying its interaction with radiation. In the constellations of Scorpius and Ophiuchus, these interactions create a stunning array of colors that can be seen by the earth-bound observer.

CONTENTS

Starbirth!	64
Mercury Mirrors	71
Day Star	75
Abort Liftoff!	82
America's Earliest Astronomer	89
Target: Earth!	94
Ask the Scientist	100

STARBIRTH!

by Dennis L. Mammana

In the cold, dark, mysterious clouds of interstellar space, stars are continually being born. At least that's what astronomers have believed for decades. But believing that stars are born within these clouds and actually seeing it happen are two completely different things.

Science cannot work simply on the beliefs of a few. Even if *every* astronomer were to *believe* that stars formed in a particular way, this does not make it so. Science requires that researchers go out and *prove* it. In this way, astronomy is much like a cosmic detective story, where a suspicion exists—in this case, that stars actually come into being inside of clouds of interstellar material—evidence is gathered by a variety of techniques, scientists analyze the data and debate its merit, and a conclusion is drawn based on the strength of the evidence.

The problem with learning the secrets of starbirth is that stars spend much of their formative years deep within thick, dusty clouds so opaque that visible light cannot escape. They begin their lives protected by these stellar "cocoons," out of sight of the probing eyes of earthly astronomers, until they are ready to shine their light across the galaxy. By then, the stars have already formed, and the mystery of their origin remains.

Because the birth of a star has never actually been watched, astronomers have long used their knowledge of physics and mathematics to create theories about how stars might form. With the advent of new technology—infrared and radio telescopes on the ground, and optical telescopes in space—astronomers can now pierce the veil of stellar cocoons, and come another step closer to watching the birth of a star.

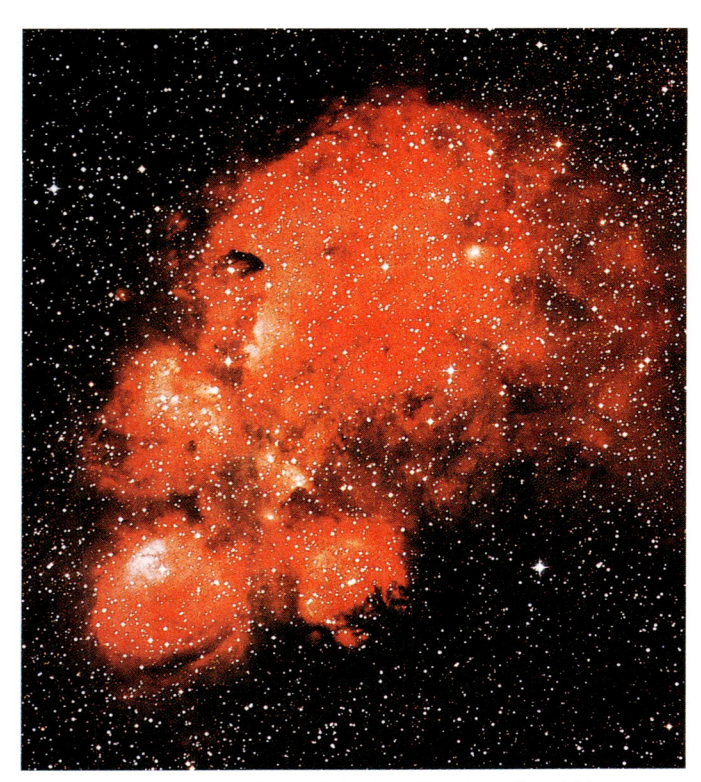

A Star Is Born

Astronomers have long believed that huge clouds of cosmic gas and dust should produce stars in great numbers. Such a cloud is a dark mass trillions of miles across and composed mainly of hydrogen and helium. It also contains molecular grains scattered throughout, and traces of heavy elements blasted into space by exploding stars of ages past.

For eons, a cloud such as this might remain stable—its grains held together by their own gravitation that perfectly balances their natural efforts to escape. But then, within a few million years, things would most likely begin to change.

Inside the cloud, gravity eventually overtakes the particles' individual motions, and the nebula begins to collapse. Where particles happen to be more densely packed, clumps hundreds or thousands of times more massive than the Sun form and begin to spin on their own. Eventually turbulence breaks each vortex into smaller fragments. These one day become cloudy disks out of which new stars, and possibly new planetary systems, can form.

In each of these smaller primordial nebulae, gravity continues to dominate. As a nebula collapses, a fragment spins more rapidly—much as ice-skaters whirl faster as they pull their outstretched arms inward. At the same time, centrifugal force throws the accompanying material outward into a thick, pancakelike disk. Material near the center whirls about in only days, while that farther out takes centuries to revolve once.

Matter continues to fall inward, causing the disk's center to become a bulge more than 19 million miles (30 million kilometers) across. Increasing gravitational pressures slow its collapse and cause the internal temperature to soar to thousands of degrees (Celsius or Fahrenheit—at such temperature, it does not much matter!). A "protostar" is born.

In the spinning disk, planets, asteroids and comets take shape, while the central mass grows larger and hotter. Now, enshrouded by a thick cloud of opaque dust, temperatures within rise from a few thousand to several million degrees, thermonuclear reactions begin, and a star is born.

As it tries desperately to cope with its new power, the newborn star bursts erratically into view. Intense blasts throw into space as much as half the star's mass in only a few million years, and blow away from the inner disk nearly every speck of dust not held down by gravity. Within several million

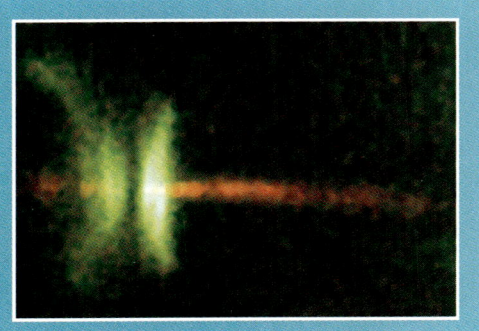

Huge interstellar clouds called nebulae have long been thought to be the birthplace of stars. Using new telescopes, astronomers have begun piercing these cloudy veils to study how stars actually form.

The Hubble Space Telescope has captured images of starbirth in a spinning disk of cosmic debris (above). Scientists believe that the disk's dense center has already collapsed into a star, while the periphery may coagulate into planets.

ASTRONOMY AND SPACE SCIENCE 65

Orion Star Factory

Inside the dense clouds of Orion (below) and other nebulae, a new "protostar" begins to form as hydrogen gas and stellar dust contract under the force of gravity.

Eventually, the dense protostar and its remnant cloud of debris drift away from the larger nebular body (below).

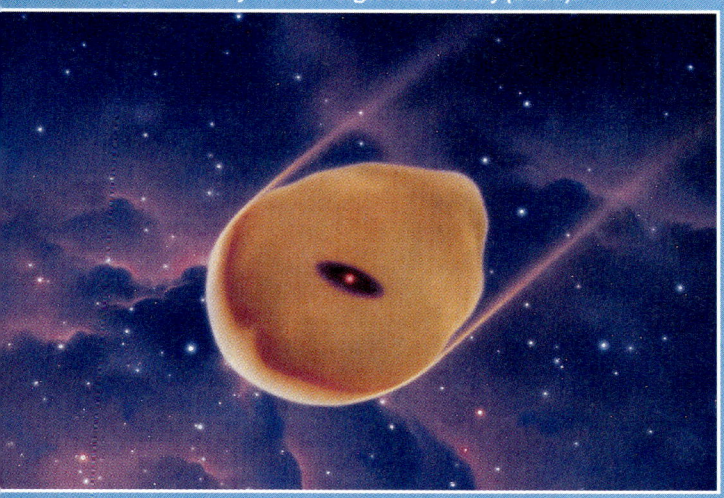

As the star grows more powerful, the accompanying cloud spreads out into a spinning disk (below) that orbits the star.

years, the star's outwardly flowing energy reaches a balance with the inward pull of gravity, and the star settles down to a peaceful and relatively serene life.

CLOSER TO HOME
Until this century, such ideas were only speculation. Astronomers, hoping to understand the process more directly, have turned their attention to a star so close we can feel it: the Sun.

Anyone who has ever basked on a sun-drenched beach knows the power of our star. Deep within its internal cauldron, the Sun generates 400 trillion trillion watts of power each second—enough to illuminate 2,600 Earths filled with 200-watt lightbulbs. At the top of its visible atmosphere—the photosphere—temperatures hover near 10,000° F (5,500° C). Heat-carrying bubbles as large as many states rise and sink in the solar atmosphere, and flares of superheated gas shoot upward for thousands of miles into space.

The Sun's power has been revered by every culture that ever lived on Earth. The ancients entrusted to the Sun the personification of their most powerful gods. The source of this energy, however, was a mystery that had to wait until modern times to be solved.

At the turn of this century, a young German-born physicist named Albert Einstein offered a possible solution. When Einstein calculated his now-famous equation $E=mc^2$ (where E represents energy, m represents mass, and c represents the velocity of light in a vacuum), he showed that matter and energy were merely manifestations of the same phenomenon—that matter was just a very concentrated form of energy and that the two were interchangeable. In other words, matter could be turned into energy, said Einstein.

In fact, a little matter can go a long, long way. For example, 1 gram (0.03 ounce) of matter, if converted completely, would produce 21.5 trillion calories of energy—equivalent to that released by the chemical burning of 660,000 gallons (2.5 million liters) of gasoline. And it is this that makes the Sun shine.

In the core of the Sun, where temperatures soar to more than 10 million degrees, hydrogen atoms race around at incredible speeds. When four hydrogen atoms slam into one another, they fuse to form two atoms of helium. But two helium atoms weigh *less* than the four of hydrogen that created them. This mass does not vanish, but instead is transformed into energy by this "fusion" process and is described by Einstein's equation. The energy released is not very large—barely enough to lift a housefly three-hundredths of a millimeter. But in the Sun, where 10^{38} such reactions take place *every second* (that's 1 followed by 38 zeros!), the total energy production is enormous.

Other stars work the same way, but can produce energy at dramatically different rates. For example, massive bluish-white supergiant stars contain 15 to 20 times more material than the Sun and, in order to hold up their heavy atmospheres, must convert their fuel to energy thousands of times faster. And, even though they have a larger supply of hydrogen to consume, these supergiants do it in only millions of years.

Much more common, however, are the stars known as red dwarfs, which contain only a fraction of a solar mass. With atmospheres that are relatively light, their internal engines burn at a much more leisurely pace, and, as a result, these stars can last for hundreds of billions of years.

WHENCE CAME THE STARS?

If all stars work like the Sun—by the thermonuclear mechanism known as fusion—then perhaps all stars form in similar ways as well, along with planetary families of their own. To find out, astronomers must explore the clouds of gas and dust that permeate interstellar space. Even a small backyard telescope can reveal the cloudy birthplaces of stars.

One of the first astronomers to make note of these nebulae was the 18th-century French comet hunter Charles Messier. In his relentless search for new comets, Messier continually stumbled across these hazy patches and mistook them for comets. To warn other comet hunters of these celestial "nuisances," Messier began listing them as objects for comet hunters to ignore. Ironically, Messier is known today, not for his comet discoveries, but for his catalog of nebulous objects he deemed a waste of time.

Modern telescopes reveal the entries of Messier's catalog, expressed as a number preceded by a capital *M*, to be a variety of objects much more distant than any comet. Some turned out to be clusters of hundreds,

Astronomers have recently identified incredibly dense molecular clouds, scores of protoplanetary disks, and other persuasive evidence of starmaking within the glowing and dusty confines of nebulae.

The Orion nebula is located in the farthest reaches of the spectacular Orion constellation (above). Both can be seen by the naked eye from the Northern Hemisphere during the winter months.

starlight and through dust appear bluish in color. Others glow red—the telltale signs of fluorescing hydrogen gas.

As different as each cloud appears, most share some common features. For example, when studied by radio and infrared telescopes, these hot clouds reveal chemical compositions of 75 percent hydrogen and nearly 25 percent helium, all intertwined with traces of carbon, nitrogen, oxygen, calcium, sodium, and heavier atoms. They appear to be composed of similar materials out of which the Sun and its planetary system coalesced billions of years ago.

Within thicker, darker regions of these clouds, places through which even starlight cannot penetrate, the chemical composition is different. Here, protected from the devastating effects of ultraviolet radiation, more-complex gases can exist as molecules. Many are simple compounds such as methane and ammonia, while others are much more elaborate: ethyl alcohol, formaldehyde, and carbon dioxide—even water. To date, more than five dozen such compounds have been found within the dark clouds of interstellar nebulae—many of them organic in nature.

thousands, or millions of stars within the Milky Way galaxy; others were galaxies far beyond our own.

Still others proved to be gargantuan clouds of gas and dust that appear quite intricately structured in long-exposure photographs. Some appear as thin as smoke twisted and stretched by a breeze. Others are so dense that their great mass blocks from view all behind them, and they reveal themselves only in silhouette. Some scatter

Nor do these clouds contain only gases. A tiny fraction of their mass is actually composed of dust grains about the size of particles in smoke, and made of carbon, iron, and silicates (rocklike materials)—indeed, all the building blocks required to make stars, planets, perhaps even life itself.

Upon closer examination, astronomers have found that, whenever they see a thick interstellar cloud, they also find in its vicinity bluish-white supermassive stars, whose intense heat and high energy radiation disrupt the gas and illuminate it from within. The phenomenon occurs far too frequently to be dismissed as mere coincidence.

THE GREAT ORION STAR FACTORY

One of the largest and nearest such nebulae lies 1,500 light-years away—each light-year stretches about 6 trillion miles (10 trillion kilometers)—in the direction of the constellation Orion, and is known to every amateur astronomer as M42: the Great Nebula in Orion.

Deep within its twisted clouds of gas and dust lie the four bluish-white stars of the Trapezium, the most massive of which weighs 40 times more than the Sun and shines with 300,000 times its brilliance. These stars are known to be extremely young—perhaps only millions of years old—for, if they were much older, their tremendous energy output would have long since exhausted their fuel supplies. Since the stars' motions have not carried them very far in this short time span, they must still lie near the places of their origin.

Could it be that, by peering deep into the heart of nebulae like Orion, astronomers are looking at the birthplaces of stars—just as their theories suggest? Is this the very material out of which stars and planets originate?

Astronomers using ground-based optical telescopes have long gazed into this nebula for answers to such profound questions. Today, with the advent of the Hubble Space Telescope (HST), they are beginning to find clues that their suspicions are correct. Recent photographs of the Great Nebula in Orion have showed clumps of interstellar material, only a fraction of the size of our solar system, embedded within the nebula.

Many are shaped as disks—just what astronomers would expect if their theories are true. Since these clumps of interstellar material are believed to be protoplanetary disks surrounding newborn stars, astronomers have dubbed them "proplyds," a shortened term for the more cumbersome *proto*pl*anetary d*isks. These disks are obviously tilted at different angles to Earth. One striking photograph—resembling an interstellar Frisbee—shows a disk tilted edge-on, hiding the young star at its center.

Astronomers' extensive study of the region has yielded 153 such disks that are glowing due to a torrent of ultraviolet radiation from the nebula's central stars. The existence of so many young stars with protoplanetary disks mathematically increases the likelihood that planetary systems form around newborn stars.

More strong evidence of active starbirth comes not from within the clouds but from *behind* them—out of view of traditional optical telescopes. For studying these regions,

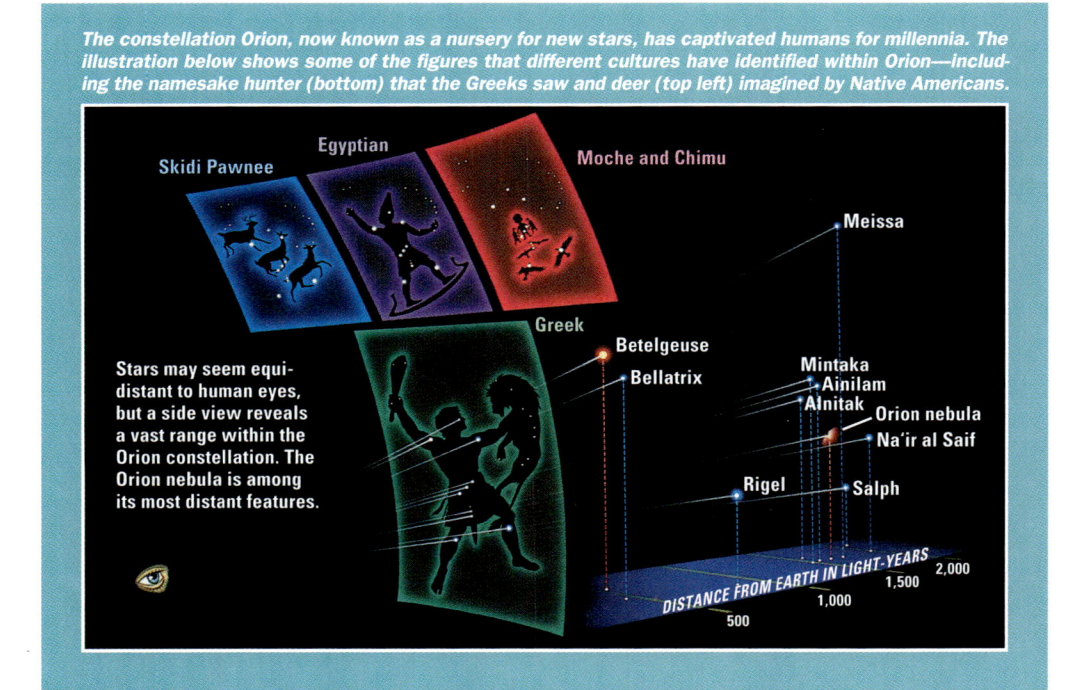

The constellation Orion, now known as a nursery for new stars, has captivated humans for millennia. The illustration below shows some of the figures that different cultures have identified within Orion—including the namesake hunter (bottom) that the Greeks saw and deer (top left) imagined by Native Americans.

astronomers have used infrared and radio telescopes, combined with powerful computers, to peer through the nebulosity and construct a detailed map of the locations, velocities, and densities of more than 20 atomic hydrogen clouds and filaments in the Orion region.

Here they found two great, dark clouds so cold and dense that molecules like carbon monoxide have formed in the shadows of dust particles. Astronomers have dubbed them the Orion Molecular Clouds 1 and 2 (OMC-1 and OMC-2).

These giant molecular clouds contain at least a million atoms of hydrogen gas per cubic centimeter, along with tremendous amounts of dust. And, while this is far less than the 10^{19} atoms per cubic centimeter at sea level in the Earth's atmosphere, it is more than 1,600 times denser than the nebulosity in which the clouds are embedded. Deep within, warm, compact centers seem to suggest that the clouds may now be collapsing. At the core of OMC-1 lie two intriguing objects. One, astronomers believe, is a young star still enshrouded by its dusty cocoon. The other, just to its south, is a cluster engulfed in an expanding shell of gas, and blasting in opposite directions two jets of hot gas at such a rate that it could continue only for a thousand years. This must, therefore, represent a group of even younger stars.

DEEP IN THE HEART OF TAURUS

Another type of cloud that astronomers have found appears far less energetic and violent—this lying in front of the constellations Taurus and Auriga. Photographs of this nebulosity, known as the Taurus-Auriga Complex, have shown tiny, dark, spherical "globules" embedded deep within the luminous cloud. These globules contain masses ranging anywhere from 10 to 10,000 times that of the Sun within a volume of space barely 1 light-year across.

While they appear dark to the eye, globules glow brilliantly to an infrared telescope, indicating that internal temperatures are quite high. Something is obviously happening on their insides, perhaps the rapid collapse of a protostar. In fact, studies of this complex suggest that, over the past 40 million years or so, stars like the Sun have constantly been forming there.

Astronomers now believe that, when a star ignites from within such a stellar cocoon and blows away its protective shroud, the star shines with tremendous light. In recent years, many of these objects have been discovered emerging from such clouds. They are named T Tauri stars, after the first one ever found.

As their newly found energy source tries to reach equilibrium with the weight of the star's atmosphere, T Tauri stars flicker wildly in brightness with periods of several days. In addition, they blast streams of hot, luminous gas outward in two opposite directions—possibly the result of a spinning disk of material falling inward, dragging with it the magnetic field from the interstellar medium and forcing gases out of the stars' poles of rotation.

Perhaps more common, however, are the so-called "naked" T Tauri stars, which outnumber their classical brethren by 10 to 1. These stars show no signs of being surrounded by dusty material, and they can be examined closely without interference from circumstellar "pollution."

Their discovery poses a number of important questions for astronomers studying starbirth. Have these stars moved so rapidly from their places of origin that they left behind their accompanying disks? Did a second star orbiting nearby disrupt their disks entirely? Were these stars spinning so rapidly that they ejected material completely, or so slowly that disks could never form in the first place? Whatever stripped this material from these stars must have done so remarkably quickly, for some naked T Tauri stars are believed to be younger than 1 million years old.

After many centuries of pondering, philosophizing and probing, it appears to astronomers that star formation is a natural—even common—process. There seems to be enough visible matter in our neighborhood of space to produce one new star for every 10 existing ones. And, with the amazing new telescopes, computers, and cameras now being developed, astronomers may one day be able to watch a star "turn on"!

Mercury Mirrors
A New Spin on Telescopes

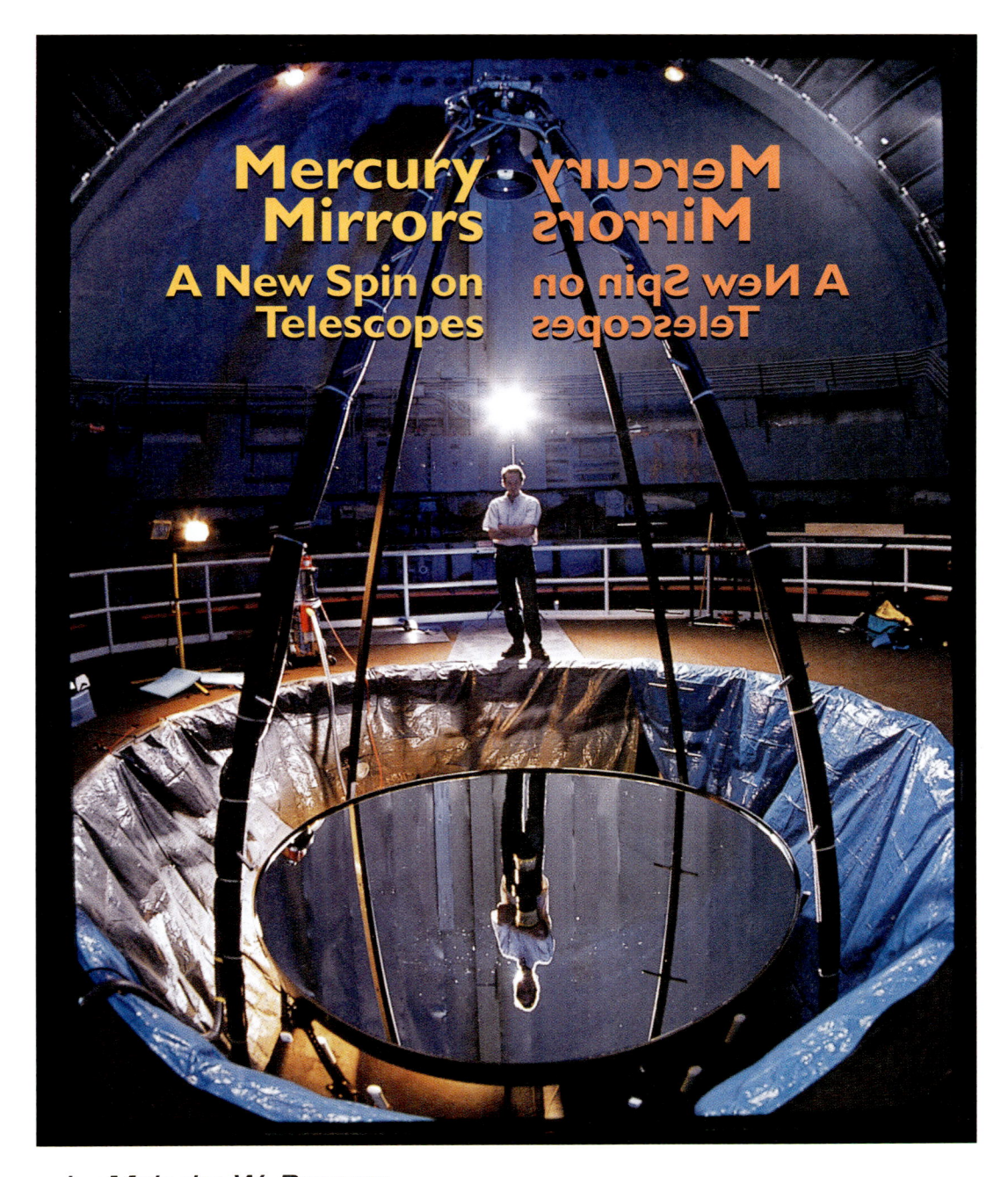

by Malcolm W. Browne

Since Sir Isaac Newton invented the reflecting telescope three centuries ago, astronomical observatories have been largely dependent on huge and increasingly costly light-collecting mirrors made of aluminum-coated glass. But now a new generation of mirrors based on silvery liquid mercury promises to drastically reduce the cost of cutting-edge astronomy.

The new mirrors are little more than rotating dishes containing small quantities of liquid-mercury metal. As the mirror dish at the base of a telescope rotates, centrifugal force counters the force of gravity on the

Liquid-mercury telescopes—which generate images equal in quality to conventional (and far more costly) glass mirrors—have put a new spin on stargazing.

The mercury mirror housed in the Orbital Debris Observatory in southern New Mexico (above) can detect tiny spaceborne objects that threaten satellites and manned spacecraft.

heavy fluid, pushing some of the mercury outward and upward toward the edge, curving its shiny surface into a perfect paraboloid. By adjusting the speed of rotation, and hence the curvature of the mercury's surface, the focal length of the mirror can be adjusted.

New telescopes based on liquid-mercury mirrors must look almost straight up, a limitation for an astronomical observatory. But for many purposes, including astronomical scanning, searching for space debris, and probing the chemistry of Earth's atmosphere, a vertical view is adequate.

Users of the new mercury telescopes describe the quality of their reflected images as better than those of most solid-glass mirrors.

"We are getting image quality from mercury mirrors comparable to that of the glass mirror in the Hubble Space Telescope," says Ermanno F. Borra, Ph.D., of the Université Laval in Quebec, a leading innovator in the development of spinning liquid mirrors. Best of all, a telescope with a liquid-mercury mirror costs less than one-hundredth as much as a telescope with a conventional glass mirror, he says.

Mercury Mirrors In Action

In September 1995, the University of California at Los Angeles (UCLA) began operating a 104-inch (2.6-meter) mercury mirror 100 miles (160 kilometers) below the Arctic Circle near Fairbanks, Alaska. The telescope, the 20th largest in the world, is dedicated to studying the aurora borealis (northern lights) and the ionosphere, the region of the upper atmosphere where the aurora occurs. The telescope, say its builders, cost only $50,000—a small fraction of the cost of a telescope with a glass mirror of the same size.

The UCLA telescope is partly financed by the U.S. military because it may lead to new techniques for signaling submarines and probing battlefields for secret tunnels and bunkers. Users of the telescope will also study the interaction of pulses of radio beams with the ionosphere.

Ralph Wuerker, Ph.D., a UCLA development engineer, said that the instrument would spend the winter of 1995–96 studying ionized nitrogen in the atmosphere about 60 miles (100 kilometers) high, where solar radiation rips away some electrons from the atoms making up gases, leaving them in an electrically charged state. A laser mounted near the telescope fires brief pulses of light upward, tuned to excite the ionized gas to emit a pulse of light of its own. This return pulse is then collected by the mercury-mirror telescope, and its light is analyzed in terms of spectral composition.

Several other mercury-mirror telescopes are in operation.

At Sunspot, New Mexico, the National Aeronautics and Space Administration (NASA) has completed a 118-inch (3-meter)-diameter mercury-mirror telescope designed to collect images of space debris that might hit manned spacecraft with catastrophic results. The telescope can detect orbiting objects as small as 0.5 inch (about 1 centimeter) across as they pass overhead. NASA plans to build a similar telescope at the equator.

Since 1992, the University of Western Ontario at London, Ontario, has operated a 106-inch (2.7-meter) mercury mirror in conjunction with an upward-pointing laser for atmospheric studies, which include the detection of changes in the protective ozone layer.

At Borra's laboratory, a 142-inch (3.6-meter) liquid mirror is under development, and Borra is collaborating with Paul Hickson, Ph.D., of the University of British Columbia at Vancouver in the building of a gigantic 200-inch (5-meter) mercury-mirror instrument. This telescope will have the same light-collecting power as the great 200-inch telescope at Mount Palomar, California, which is the third largest in the world. (Russia has a 236-inch [6-meter] telescope; the Keck Telescope in Hawaii, 393 inches [10 meters] in diameter, is the world's largest.)

Mercury mirrors are so accurate, Borra says, that one of them, a 59-inch (1.5-meter) reflector built in Belgium, is used as a reference standard for testing conventional glass telescope mirrors being sent into space by the European Space Agency (ESA).

AN OLD IDEA

The idea of using a liquid mirror in an astronomical telescope is not new. Isaac Newton himself is believed to have thought of it.

In 1908, Robert W. Wood, Ph.D., of Johns Hopkins University in Baltimore, Maryland, built a primitive observatory with a spinning liquid-mercury mirror. The mirror, a dish 20 inches (51 centimeters) in diameter, was mounted on a bearing attached to a sturdy concrete foundation at the bottom of a deep pit. The mirror was turned by a pulley connected to an electric motor. By adjusting the motor's speed, Wood could change the curvature of the mercury and bring an image of stars directly overhead to focus on a photographic plate mounted at the top of the pit. The system produced blurred images, but worked well enough to photograph the barely separated components of a double star.

Two main obstacles delayed the further development of liquid-mirror telescopes. One was that such telescopes cannot track their targets. They look straight up and can photograph stars and other celestial objects only as streaks, as Earth's rotation carries the objects through their vertical fields of view.

A second formidable problem was that the early bearings and motors used by Wood

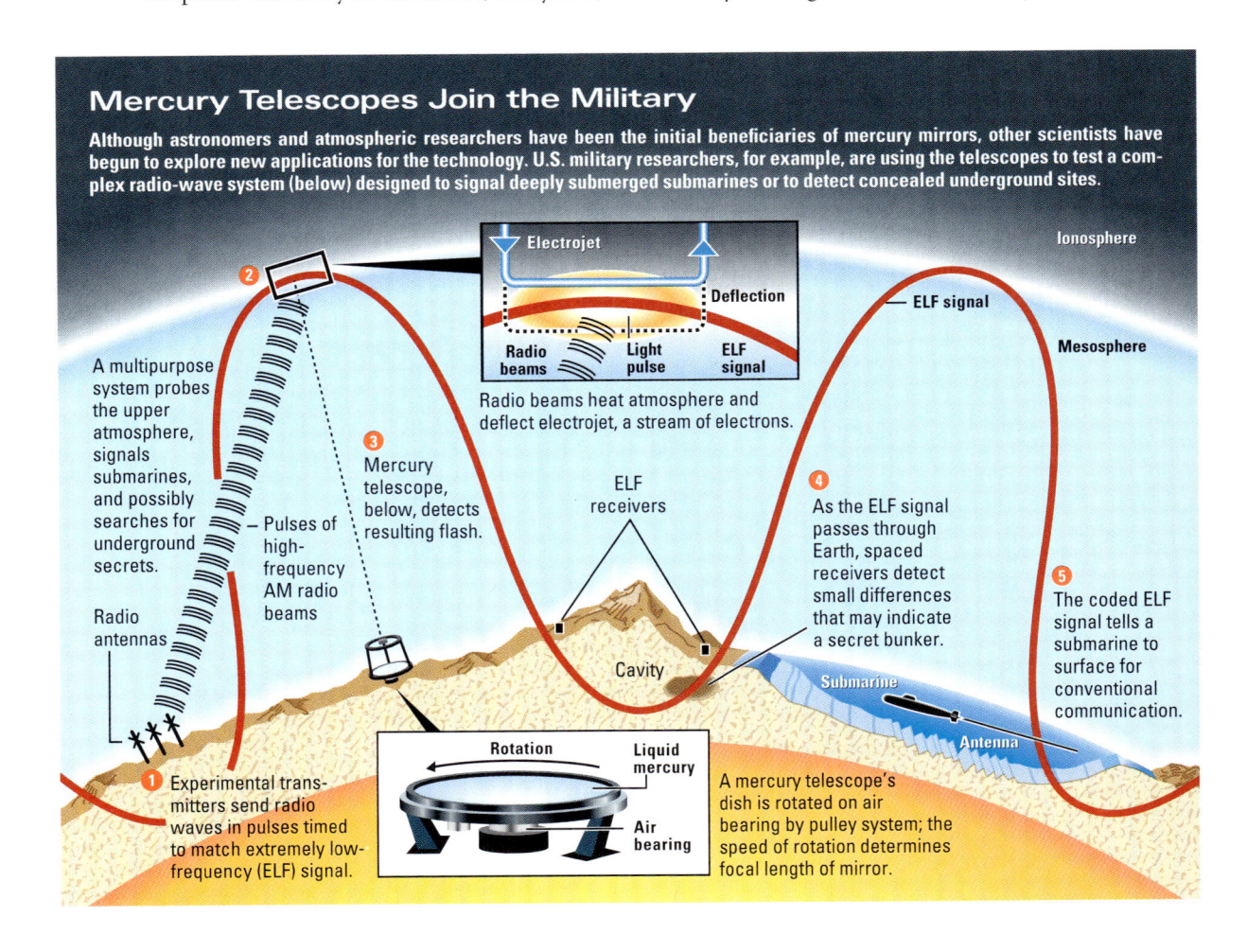

Mercury Telescopes Join the Military

Although astronomers and atmospheric researchers have been the initial beneficiaries of mercury mirrors, other scientists have begun to explore new applications for the technology. U.S. military researchers, for example, are using the telescopes to test a complex radio-wave system (below) designed to signal deeply submerged submarines or to detect concealed underground sites.

Electrojet — Radio beams heat atmosphere and deflect electrojet, a stream of electrons. (Radio beams, Light pulse, ELF signal, Deflection)

Ionosphere — ELF signal — **Mesosphere**

1 Experimental transmitters send radio waves in pulses timed to match extremely low-frequency (ELF) signal.

2 A multipurpose system probes the upper atmosphere, signals submarines, and possibly searches for underground secrets. — Pulses of high-frequency AM radio beams — Radio antennas

3 Mercury telescope, below, detects resulting flash. — A mercury telescope's dish is rotated on air bearing by pulley system; the speed of rotation determines focal length of mirror. (Rotation, Liquid mercury, Air bearing)

ELF receivers — Cavity

4 As the ELF signal passes through Earth, spaced receivers detect small differences that may indicate a secret bunker.

5 The coded ELF signal tells a submarine to surface for conventional communication. — Submarine, Antenna and a few other experimenters were vulnerable to even the slightest vibrations, which tended to produce waves on the mercury surface and distort images.

But in recent years, astronomers have realized that even fixed telescopes that can point only upward are useful. One of the most productive astronomical instruments in the world is the great radio telescope at Arecibo, Puerto Rico, some 1,000 feet (305 meters) across. The telescope cannot be aimed, but it scans the sky overhead as Earth's rotation moves a strip of sky across the instrument's line of sight.

Scientists using mercury telescopes must closely monitor the concentration of highly toxic vapors. Once a mirror is revolving, however, such emissions are negligible.

Technology has enhanced the abilities of fixed optical telescopes as well as radio telescopes. The invention of the charge-coupled device (CCD), a chip somewhat similar to the image chip in a television camera, endows even a fixed telescope with limited tracking ability. As an image sweeps across the chip, a computer follows the image electronically and records it as a single point of light rather than a smeared streak.

The stability of the latest mercury-mirror telescopes has been greatly improved by replacing conventional mechanical bearings with compressed-air bearings, which prevent the turning dish from coming into contact with any solid object that might cause vibrations. Another improvement has been in the quality of drive motors and power supplies, which can now maintain almost-perfect constancy of rotation speed.

Making a Mercury Mirror

The latest mercury mirrors start out as shells cast in approximately paraboloidal shapes from strong, rigid epoxies and Kevlar (a synthetic fiber commonly used in bulletproof vests). To create an accurate starting paraboloid, the finished shell is spun on a turntable at the required speed, and liquid polyurethane is poured in. The liquid polyurethane quickly spreads out in the shape of a paraboloid, and then hardens.

To prepare the mirror for use, its rotation rate is adjusted and mercury is added. The new 102-inch (2.6-meter) mirror built by UCLA in Alaska contains 300 pounds (136 kilograms) of the heavy metal, which is spread out in the revolving dish to a thickness of about one-eighth of an inch (3 millimeters).

Builders of mercury-mirror telescopes acknowledge that mercury vapor is toxic, but once a mirror is spinning, they say, the surface of the mercury is quickly oxidized by contact with air. Although the thin oxidized layer is just as reflective as the bare metal, it forms a barrier preventing the escape of poisonous vapor, and gas masks must be used only while a mirror is being started.

To start a mirror, operators must turn it by hand until it is revolving at about the rate at which the motor operates. They then sweep the pool of heavy mercury at the bottom of the dish outward, until its surface stabilizes into a parabolic curve.

The great weight of even a small quantity of mercury forces telescope builders to design extremely strong mirror shells, but Borra and his associates are experimenting with lighter liquid metals requiring less dish strength. One candidate is liquid gallium.

In an article in *Scientific American*, Borra said that, one day, astronomers may build gigantic and relatively inexpensive liquid-metal telescopes on the Moon, using special liquid-metal alloys with low freezing points.

"The point of all this work is to bring down the cost of astronomy," he says, "and we're making rapid progress."

> Beautiful is your shining forth at heaven's edge,
>
> O living Aton, beginning of life!
>
> When you arise on the eastern horizon,
>
> You fill every land with your beauty.
>
> You are lovely, great and glittering
>
> And go high above every land.
>
> Though you are far away, your rays are on Earth.
>
> Though you fill men's eyes, your footprints are unseen.
>
> —Akhenaton (Amenhotep IV), king of Egypt (ca. 1370 B.C.)

DAY STAR

by Dennis L. Mammana

It has been known by many names. The ancients called it Ra, Shamash, Helios, and Surya. It has been considered the eye of justice, the source of wisdom, the giver of life. All living things respond to its light, bask in its warmth, move to its annual rhythms. It is our nearest star—that to which we owe our very existence: the Sun.

The Sun is, without a doubt, the largest object in our entire solar system, with a diameter of about 864,000 miles (1.4 million kilometers). An incredible 109 Earths would be able to fit across its luminous disk; even more amazing, if the Sun were hollow, it could hold about 1.3 *million* Earths! More massive than all the planets put together, the Sun contains 98 percent of the solar system's material.

THE SUN'S FACE

The Sun's outer visible layer is called the photosphere (meaning "sphere of light"), and has a temperature of 10,000° F (5,500° C). A view with a solar telescope shows the layer to have a mottled appearance. This "rice grain" texture is caused by the turbulent eruptions of hot convection cells (bubbles) as large as the state of Pennsylvania rising from the hotter interior of the Sun.

Scattered about this region of the Sun are dark spots that can often be seen by solar astronomers. These "sunspots," as they are known, are dark depressions on the photo-

Since ancient times, humans have recognized the power of the Sun. Without the light and heat of this star, life on Earth simply would not exist.

this region is relatively thin—only about 1,550 miles (2,500 kilometers) thick—the temperature increases rapidly from 7,200° F (4,000° C) to 1,800,000° F (1,000,000° C). It is here that bright, luminous hydrogen clouds (faculae) often appear above where sunspots are about to form, and violent eruptions (flares) of hot gas burst outward, carrying solar particles into space.

The outermost region of the Sun's atmosphere is known as the corona, the spectacular "crown" that makes its rare appearance to the human eye only during a total solar eclipse. The corona, which is hot enough to emit energetic X rays, is dominated by intense magnetic forces. Hot gas accumulates around these magnetic areas to produce the fascinating shapes we see during a total solar eclipse. In the inner part of the corona, huge clouds of glowing gas erupt from the Sun's upper chromosphere and often appear against the pearly-white atmosphere as luminous red loops many times larger than our own planet. The outer region of the corona stretches well beyond Earth's orbit, and carries hot ionized gas far into our solar system and beyond in a stream called the "solar wind."

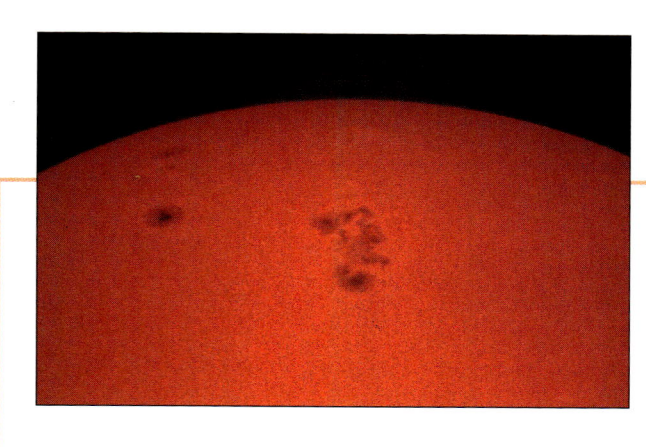

sphere. Although they appear dark, sunspots themselves are actually quite hot, and their temperatures have been measured at around 7,000° F (4,000° C). Sunspot numbers periodically increase and decrease over an 11-year cycle.

Just above the Sun's photosphere lies its chromosphere ("sphere of color"). Solar energy passes through this region on its way out from the center of the Sun. Although

SUNSPOTS

In the past century, the appearance of sunspots—dark features that form on the solar surface (left)—has fluctuated in 11-year cycles. By contrast, sunspots were very rare between 1640 and 1715—an era in which global temperatures were unusually cold. Scientists are now exploring possible links between sunspot activity and the Earth's climate.

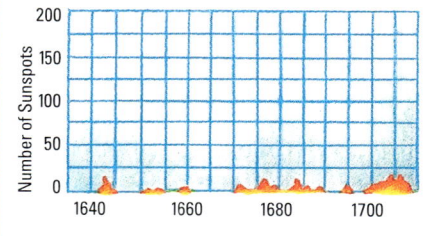

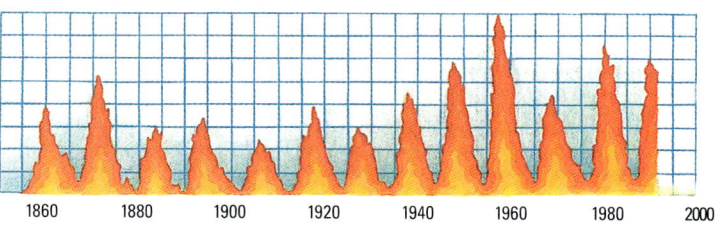

Deep Within . . .

The Sun's tremendous heat and light originate deep within its core, where temperatures soar to 27,000,000° F (15,000,000° C), and the weight of its gases creates a density 12 times that of lead. It might seem that, in such an infernal cauldron, the Sun must be on fire. But even the hottest fire on Earth would be icy compared to the heat at the Sun's core. Instead, the Sun's core is filled with hydrogen gas that undergoes a violent nuclear reaction called thermonuclear fusion.

As hydrogen nuclei race around in the Sun's core, they frequently collide. When four hydrogen nuclei collide, they fuse together to form one alpha particle, or nucleus of the element helium. This new particle contains less mass than the four original particles that gave their lives to form it, and the difference in mass is converted to energy. Some 700 million tons of hydrogen gas are converted into helium each second, releasing the equivalent of 5 million tons of energy. This energy continues to expand outward through the Sun's interior and eventually, after 1 million years or more, reaches the photosphere in the form of light and heat. It is this ancient radiation that we enjoy on a warm, sunny day.

Sun Stats

Mass	1.99×10^{33} grams
Mass (Earth = 1)	332,830
Equatorial radius	432,000 miles
Equatorial radius (Earth = 1)	108.97
Mean density	1.410 grams per cubic centimeter
Mean distance from Earth	93 million miles
Rotational period	25 to 36 days*
Escape velocity	383.79 miles per second
Luminosity	3.827×10^{33} ergs per second
Visual magnitude	-26.8
Mean surface temperature	11,000° F
Approximate age	4.6 billion years

*The Sun's period of rotation at the photosphere varies from about 25 days at its equator to 36 days at its poles. Deep within the Sun, everything seems to rotate with a period of about 27 days.

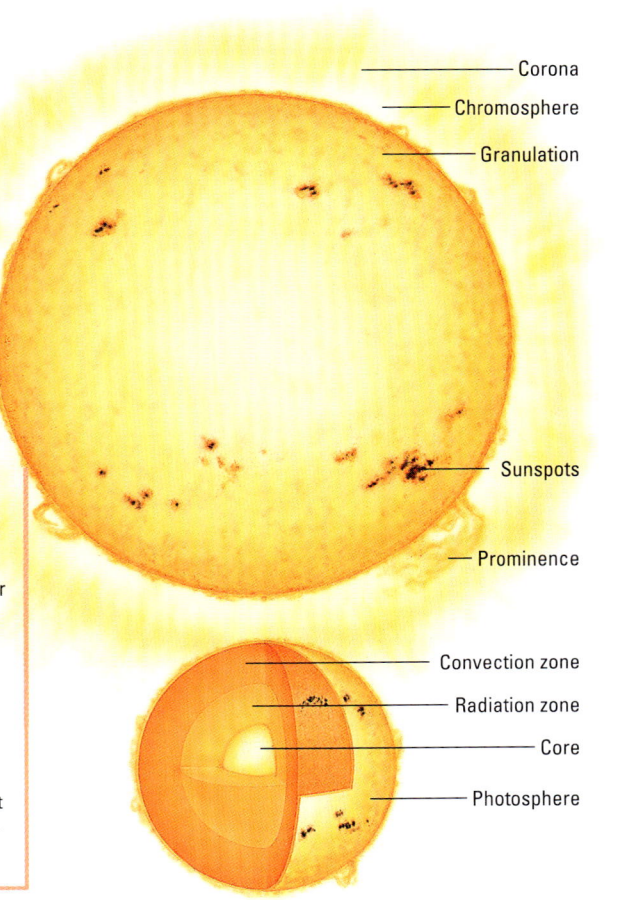

Whence the Sun?

Astronomers who study the Sun, as well as those who watch the birth of other stars around our galaxy, believe that the Sun was born approximately 4.6 billion years ago from a gigantic cloud of gas and dust in interstellar space. As the material in this thick nebula collapsed under its own weight, the temperature and pressure at its core increased until thermonuclear fusion began, and the Sun was born.

It appears that the Sun has enough fuel to shine for another 5 billion years or so. However, at some point, it will run out of fuel, and the Sun will begin to change. At this time, the Sun's core will collapse, and its temperature will rise even higher. The helium "ash" that was created when hydrogen nuclei fused together will then begin to fuse into heavier elements. This will happen very quickly; the resulting energy burst will blow the Sun's photosphere outward into space.

ASTRONOMY AND SPACE SCIENCE

The Sun will become a red giant, and will grow large enough to swallow the orbits of the planets Mercury, Venus, and possibly even Earth. After 1 billion years or so as a red giant, it will collapse into a white dwarf—a star barely the size of Earth itself. Since its fuel will then be gone, the Sun will not be able to create any new energy, but will instead cool and fade slowly over many billions of years.

PROBING THE SUN

As our nearest star, the Sun has always been in the forefront of astrophysics, leading the way to greater understanding of the structure and evolution of stars, galaxies, and the universe. Until recently, measuring the internal structure and dynamics of the Sun and other stars has been impossible, and our understanding of these phenomena has come mainly from theory. However, a discovery made in the 1960s—that millions of distinct, resonating sound waves radiate through the Sun at all times—has begun to change all that. Scientists are beginning to use these waves as probes of the Sun's inside—much as seismologists use the shock waves of an earthquake to study the interior of Earth.

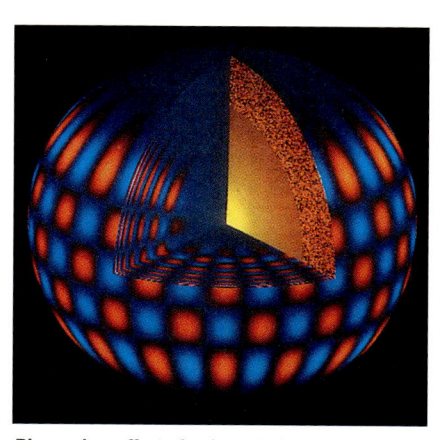

This new field, called helioseismology, promises to revolutionize our understanding of the Sun and distant stars by allowing scientists to measure directly such solar properties as temperature, chemical composition, and motions from just below the Sun's photosphere all the way down to its very core. And a number of efforts have begun to apply these techniques to more-distant stars—a field destined one day to be called astroseismology.

CHEMICAL COMPOSITION

Element	Percent
Hydrogen	92.1 percent
Helium	7.8 percent
Oxygen	0.061 percent
Carbon	0.030 percent
Nitrogen	0.0084 percent
Neon	0.0076 percent
Iron	0.0037 percent
Silicon	0.0031 percent
Magnesium	0.0024 percent
Sulfur	0.0015 percent
All others	0.0015 percent

Pioneering efforts in the relatively new field of helioseismology have focused on studying the internal structure and dynamics of the Sun.

In order to perform such observations, scientists require nearly continuous, uninterrupted observations of the Sun. Since the Sun sets every day from most sites on Earth (except within the Arctic and Antarctic circles), and since cloudy weather often forces astronomers to close their telescopes, scientists have begun efforts to build a global network of solar observatories that can always keep the Sun in their sights.

Several sophisticated programs to do just this are now in operation. One of the most ambitious is known as the Global Oscillation Network Group (GONG). It uses a six-station network of extremely sensitive solar telescopes located in various areas around Earth to obtain nearly continuous observations of the Sun's subtle "five-minute" oscillations. The network also has the ability to analyze and distribute data to all scientific investigators.

There are currently 130 individual members of GONG from 67 different institutions in 20 nations. Observatories that comprise the network are the Big Bear Solar Observatory, the Learmonth Solar Observatory, the Udaipur Solar Observatory, the Observatorio del Teide, the Cerro Tololo Inter-American Observatory, and the Mauna Kea Observatory.

To the delight of scientists, GONG has had several weeklong periods of perfect weather, 24 hours a day—during which the Sun never sets on GONG. The network has provided coverage of the Sun's activities more than 93 percent of the time. Each of the six stations in the network produces more than 200 megabytes of data every day.

JOURNEY TO THE SUN

A most unusual spacecraft journey to study the star to which we owe our very existence began on October 6, 1990. On that day, the Ulysses spacecraft mission, an international project to study the poles of the Sun and the region of outer space above and below the Sun's poles, was deployed from the space shuttle *Discovery*.

All spacecraft that had previously studied the Sun did so in or near the ecliptic plane—the plane defined by Earth's orbit around the Sun. But the Sun's magnetic and electric fields and the solar wind have a strong influence on interplanetary space in that region. Ulysses, an 814-pound (370-kilogram) robot spacecraft, was sent into an orbit at right angles to the ecliptic, as it was built to examine for the first time the regions of the Sun's north and south poles. Ulysses was designed to study three topics of major importance to solar physicists: the Sun; the solar wind; and interstellar space, especially near the solar poles—an area where spacecraft had never visited.

Because no launch vehicle can supply enough energy to lift a spacecraft directly from Earth over the Sun's poles, Ulysses was first sent to Jupiter atop a two-stage inertial-upper-stage (IUS) rocket and a smaller booster engine called the payload assist module (PAM-S). Sixteen months and 575 million miles (925 million kilometers) later, Ulysses swung past Jupiter on February 8, 1992, and used the giant planet's intense gravity to lift itself out of the ecliptic plane and toward the Sun's south pole.

The first high-latitude solar pass began on June 26, 1994, when Ulysses reached 70 degrees south solar latitude. Ulysses spent about four months above that latitude, about 204 million miles (330 million kilometers) from the Sun, studying this previously unseen region of our star. The Sun's gravity then bent the spacecraft's trajectory, and Ulysses crossed the Sun's equator in

The Mauna Kea Observatory (below) in Hawaii is a key participant in the Global Oscillation Network Group (GONG), a network of six international facilities that together can make nearly continuous observations of solar activity.

February 1995 on its way toward the north solar pole in June 1995. Four months after studying this also-never-seen region of the Sun, in October 1995, the spacecraft's primary mission was completed.

After swinging around the Sun, Ulysses will loop by Jupiter once again in April 1998—a distance of 500 million miles (800 million kilometers)—and will use the planet's gravity to swing by the Sun once again, in September 2000. At that time, the Sun will be in its most active sunspot phase, and the solar magnetic field will have reversed polarity. Scientists expect to learn much more about the forces at work within this complex star during the peak of its activity.

ULYSSES MISSION TIMELINE

Event	Date
Launch	October 6, 1990
Upper-stage deployment and firing	October 7, 1990
Ulysses checkout	October 8-15, 1990
First Jupiter gravity assist	February 1992
First solar-polar passage	June 1994
Cross solar equator	March 1995
Second solar-polar passage	June 1995
End of primary mission	October 1995
Second Jupiter gravity assist	April 1998
Third solar-polar passage	September 2000

The Ulysses spacecraft completed its second successful solar flyby in October 1995. The probe, deployed in 1990, has yielded new findings about the Sun's magnetic fields, the solar winds, and the poles of the Sun.

ON BOARD ULYSSES

Ulysses carried nine instruments designed to study the Sun and its environment like never before, and to study interplanetary, planetary, and interstellar phenomena in the Milky Way galaxy and beyond.

A pair of magnetometers measured magnetic fields in space, and measured how the interplanetary magnetic field changes at different latitudes around the Sun. A solar-wind-plasma experiment, called the Solar Wind Observations Over the Poles of the Sun (SWOOPS), studied how electrical particles in the solar wind are distributed around the Sun. A solar-wind ion-composition spectrometer gave scientists insight into how the solar wind is accelerated through the Sun's corona, and how it interacts with plasma in the region.

An energetic-particle-composition experiment measured masses of interplanetary ions and observed helium particles coming from interstellar space. A low-energy charged-particle detector measured elemental abundances of interplanetary ions and electrons. A cosmic-ray and solar-particle instrument helped to search for these out of the ecliptic plane. A unified radio and plasma-wave experiment determined the direction and polarization of distant radio sources and charged particles in the solar wind, and allowed scientists to study waves in clouds of ionized particles.

In a total solar eclipse, the outer region of the Sun's atmosphere—the corona—resembles a radiant halo. The corona is too faint to be glimpsed at other times.

Enhanced ultraviolet photography reveals the fiery eruptions of solar prominences (above). These features are created as hot sheets of glowing gas extend outward from the top of the Sun's chromosphere.

A solar-flare X ray and cosmic gamma-ray-burst experiment measured electrons in solar flares and determined the direction of gamma-ray bursts from the galaxy. And a cosmic-dust experiment provided direct observations of particulate matter and its interaction with solar radiation.

The spacecraft radio, in addition to providing communications with ground control back on Earth, was also used to conduct a pair of experiments. When it was nearly behind the Sun, the radio provided a coronal-sounding experiment to measure the density, turbulence, and velocity of plasma in the Sun's corona. And when it was on the opposite side of Earth from the Sun, the radio enabled scientists to measure unexpected movements of the spacecraft in search of passing gravity waves.

SOLAR SURPRISES

As expected, the Ulysses mission provided some surprises. Scientists found that the gas being continuously carried by the solar wind was flowing very rapidly and very smoothly. In fact, it appeared to move twice as fast from the solar poles as it does from the solar equator.

Surprisingly, the researchers found that the strength of the Sun's magnetic field over the polar region did not increase as much as had been expected. Large-amplitude Alfvén waves are always present in the polar regions, and may help heat and accelerate the solar wind. The scientists also learned that the intensity of cosmic-ray particles arriving from the Milky Way galaxy did not increase as much as they had expected when Ulysses headed toward the Sun's south pole.

With the wealth of data gathered by Ulysses during its recent solar flybys, astronomers and physicists on Earth will have enough to keep them busy until the next flyby occurs after the turn of the century. By then, who knows what new mysteries will have been uncovered!

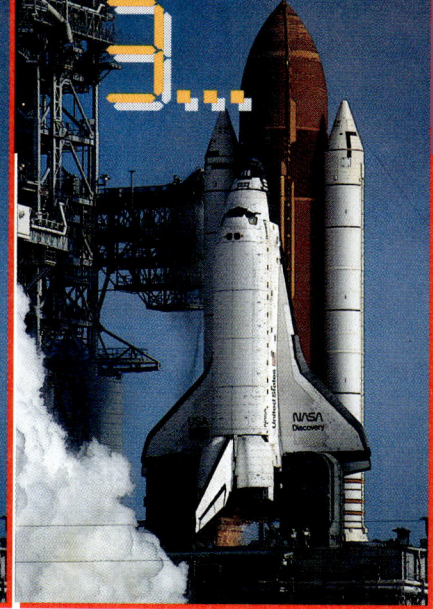

On his first space-shuttle mission, astronaut Daniel W. Bursch was mildly surprised by the violence of the main engine firing. Bursch, a U.S. Navy commander and test pilot, describes a sensation that the shuttle simulator couldn't quite replicate. "It really does feel like these engines are strapped to your back," he says. "On the pad, the engines create a lot of noise, a lot of vibration. You can almost feel the shock waves as they develop out of the engine." Bursch's biggest surprise of the day, however, came seconds later, when the engines shut down.

His first reaction was disbelief. "The first thing that catches your attention is the master alarm," he remembers. "It's very loud, and it's obvious that something's wrong. Five to 10 seconds after it's happened, all the noise has gone away, all the vibration. There's a slight rocking of the vehicle. It's really hard to feel it, but the vehicle continues to sway back and forth."

Shuttle Shutdown

It has happened only five times in shuttle history: the three main engines on the orbiter ignite, computers monitoring them detect a problem, and the space-shuttle onboard computers shut the engines down. June 26, 1984: A main fuel-valve actuator in one of the engines got stuck. July 12, 1985: A chamber coolant valve refused to close. March 22, 1993: An oxidizer purge valve jammed on a chunk of O-ring. August 12, 1993: A faulty sensor indicated abnormal fuel flow. And, almost exactly a year later, less than two seconds before the solid rockets were to ignite, an oxidizer pump overheated. "We are not willing to lift off if we lose redundancy before we get to T-zero," says John B. Plowden, who manages the team that services the shuttle's main engines. "That's the way the system is designed."

The T-zero event is the ignition of the solid-rocket boosters (SRBs), propellant-filled towers that generate 71 percent of the thrust the shuttle needs to leave the ground. "When those SRBs light, there is no recall," says Bruce Bartolini, a launch-team manager with Lockheed Martin Space Operations in Bethesda, Maryland. "You're going flying." The liquid-fuel engines ignite 6.6 seconds earlier than the solids, giving the computers a narrow window in which to call off the launch.

"There's too much stuff going on in too short a time for a human being to make a decision and then take action," Bartolini says. Fifty times a second, a computer on each of the three main engines examines close to 30 critical parameters, including sensor function, fuel pressures, temperature, vibration, fuel-flow rates, and power status. If all three engines reach 90 percent of maximum thrust by T minus three seconds, and

2... ...ABORT

by Gregory Freiherr

On March 22, 1993, when the computer network aboard the space shuttle *Discovery* sensed a problem with the ignition, it shut down the engines (sequence at left). Such a launchpad abort, occurring just seconds before liftoff, creates a whole new set of priorities for the launch team.

all parameters are within limits at T minus zero, the shuttle computers send out commands for pyrotechnics to ignite the SRBs, split the bolts holding the shuttle to the pad, and release the umbilical cord to the external tank. If certain limits are exceeded, the computers command an abort.

Practice Handling an Emergency

A launchpad abort is a safety measure, but it creates a whole new set of problems, since it leaves an enormous amount of potential chemical energy sitting on the pad. "The real key to handling an emergency as serious as an engine abort is practice," says Bartolini. "You have to know your procedures, and you have to be willing to execute them. In other words, you can't sit there and say, 'I hope this never happens. I don't want to ever have to do that.' It's just like flying airplanes. You have normal and emergency procedures—and you had better know your emergency procedures, or you're not going to be doing the normal ones for very long."

To keep the launch teams in practice, the National Aeronautics and Space Administration (NASA) runs a series of simulations at the Kennedy Space Center in Florida, similar to the mission simulations that train astronauts in Houston. Although the space-shuttle countdown is governed by a checklist that fills five volumes and takes three and a half days to execute, the principal training simulation begins at T minus 20 minutes, the point in the countdown when the ground computer network gives the first commands to the computers on the orbiter. (This interaction continues until T minus 31 seconds, when the ground computers hand off the launch sequence to the onboard computers.) Several simulations are run before every launch; the final dress rehearsal, known as the Terminal Countdown Demonstration Test, includes putting the flight crew members on the orbiter and getting them out again. The test always ends with an abort after main-engine ignition.

The Firing Room

One of the first things you notice about the firing room, where the engineers sit during launch, is its impersonality. There are no family photographs, no kids' drawings taped to the consoles, no cartoons stuck on the side of a computer screen, no houseplants, no newspapers, no notepads. It's as naked as a hospital operating room. When this observation is mentioned to Al Sofge, NASA assistant launch director, he shoots back sternly, "This is the firing room. This is where we launch rocket ships." After an instant, he adds, "Dan Marino doesn't have a picture of his kid taped to the side of his helmet."

Sofge's football metaphor is apt. The law of the firing room is concentration; its most frequent activity is drill. Although the room's windows provide a view of the launchpad, the launch-team members rarely see a shuttle liftoff. They read its status in the numbers on their computer screens.

The men and women who orchestrate shuttle launches are perched at workstations. With their backs to the windows, they overlook banks of gray metal consoles with computer screens that fill the basketball-court-size room in front of them. On each bank of consoles, there is a cryptic nameplate: HAZ GAS (hazardous gases), LOX SYS (liquid-oxygen system), MPS/SSME (main propulsion systems/space-shuttle main engines).

ASTRONOMY AND SPACE SCIENCE

calling at one minute, then 45 seconds . . . on down. Other than that, the firing room is extremely quiet. Everybody is looking at their data, hoping that they don't get an anomaly."

VIRTUAL EMERGENCIES

In today's simulation, everyone gets plenty of anomalies. Data-systems engineer Robert Pierce and two math-modeling colleagues have loaded the computers with a variety of

The top priority during any launchpad abort is the safe and rapid evacuation of the astronauts from the disabled shuttle, a procedure that figures prominently in astronaut training. In a typical evacuation drill, astronauts rehearse the trip from the shuttle across the access arm (above left) to the slide-wire basket (above right), which carries them along 1,200-foot-long wires to safety (facing page).

About 200 engineers sit at the consoles, immersed in the illusion of a shuttle countdown. The training goal is to make the monitors look exactly as they would if a real launch were under way. The engineers report to the NASA test director (NTD) and the orbiter test conductor (OTC), who communicate with the flight crew on the shuttle.

"The last command we give the astronaut flight crew is at two minutes and 30 seconds," says Bartolini. At that point, the shuttle begins running solely on internal power, and the OTC tells the astronauts to close and lock their visors and initiate oxygen flow. "He usually gives them a little send-off, and then it gets real quiet. The only talking that's being done is by the ground-launch-sequencer engineer calling out the different milestones as we go on down, and the NTD, who starts

virtual emergencies. "We're really taking a polished team and putting a high gloss on it," Pierce says. "We plan for things that have a likely occurrence of happening. 'Likely' for us space nuts is less than 1 percent. We don't like surprises."

Many of the problems Pierce and his gremlins throw at the launch team occur during a pad abort. After the orbiter's computers command a main-engine cutoff, they grind through the procedures for safing the vehicle: for example, starting a spray of water to disperse unburned hydrogen exhausted from the main engines, sealing the hydrogen and oxygen valves to the engines, and disarming the explosive bolts on the solid rockets. Progress is reflected on computer screens filled with blue, green, yellow, and—to show exceeded limits or other trouble—red or flashing numbers. The red numbers require engineers to respond according to well-documented procedures.

In one simulated emergency, engineers begin to see temperatures in red because the shuttle's ground cooling unit fails. "You don't want to cook your equipment," Pierce explains. The NTD issues an order to activate backup systems, then another to shut down a series of electronic systems on the shuttle that produce heat. An engineer at the environmental-control console manually flips a switch to turn on a chilled-water heat exchanger. Others activate radiators on the inside of the payload-bay doors. At another environmental-control console, a team lowers the temperature of air being pumped into the payload bay by a purge system.

Next the NTD orders staff at the LOX and liquid-hydrogen consoles to prepare to drain the external tank, a precaution in case power to the shuttle must be turned off. The next step is to reestablish power from the ground in order to shut down the onboard fuel cells, which are major heat generators.

"Then we have a decision point," Pierce says. "Are we still hot?" If so, members of the launch team will continue to turn off the shuttle's various systems. Throughout the process, the NTD is getting updates on temperatures from environmental-control engineers. If the temperature doesn't drop to an acceptable range, the NTD will order an emergency power-down and get the crew off the shuttle. Without electrical power on the shuttle, the launch crew no longer sees data from its systems, a situation that would require an emergency egress for the flight crew. "They open the hatch, jump out, run across the arm, and do the slide-wire thing," Pierce says.

WIRES TO SAFETY

"The slide-wire thing" is the astronauts' escape system: seven flat-bottom baskets that slide down 1,200-foot (360-meter) wires to safety. Each basket is made of steel and heat-resistant fiber surrounded by netting, and can carry up to three persons. The baskets slide down wires into catch nets, which drag chains to stop them near a bunker designed to withstand the force of a shuttle explosion.

In a real emergency, the astronauts would take a brisk walk—no more than 50 feet (15 meters)—across the shuttle access arm and fixed service structure to the baskets. Their trip would be complicated by a steady stream of water being sprayed to protect

An M113 armored personnel carrier is used to rescue shuttle astronauts injured during evacuation. All astronauts must learn how to drive the tank (right).

them from flames or heat. To ensure that no one gets lost, crew members are trained to grab a mitt full of each other's space suits. A crew of five, for example, splits into groups of two and three. They would follow a "yellow-brick road"—gold and black chevrons painted on the metal-grate floor—aiming them toward the baskets.

Riding the slide wires has its own risks—ones serious enough that during the abort simulations, NASA fills the baskets with weights and dummies rather than people. But the agency has human-rated the system. George Hoggard, a training officer on the pad rescue team, is one of only three people who have ever ridden in a slide-wire basket at the launchpad. The ride began 195 feet (60 meters) above the ground and ended 21 seconds later. The basket reached a speed of 53 miles (85 kilometers) per hour.

The only part of the ride Hoggard found unnerving came near the end, when the basket slapped the restraining net with a bang. "It was like a shotgun going off," he says. "But nothing hurt, so I figured I was still O.K." The net and drag chain broke free from their poles, as they were designed to do, and the chain dragged through sand to bring the basket to a gradual stop.

The bunker, located about 30 feet (9 meters) from the end of the slide wire, is stocked with water, oxygen, and medical supplies. But if one of the crew is hurt and needs more than first aid, an M113 armored personnel carrier, parked next to the bunker, can be used to get the astronaut to any of several points for evacuation by helicopter.

Several weeks before scheduled liftoff, the crew members take turns driving the M113, an acquired skill. It takes only a minor miscalculation to make a big mistake, as an astronaut discovered last spring when she took a corner too sharply and drove the M113 into a pond behind one of the launchpads.

The exercise isn't designed to turn astronauts into tank drivers; it's part of building a team, says Captain David M. Walker, four-time space veteran and commander of the five-member STS-69 crew, which was launched on September 7, 1995. "It gives us a chance to interact with the fire and rescue people, who are going to be the folks who save our bacon if something goes really wrong," he says.

EMERGENCY-EGRESS PRACTICE

The astronauts practice emergency egress primarily at the Manned Space Flight Center in Houston. The fastest a crew has evacuated the shuttle mock-up there is about two minutes. The exercise begins with a flurry of disconnecting—seat straps, oxygen lines, and communications cords—and culminates with a struggle to get out of a single hatch while wearing a full pressure suit saddled with a parachute and life raft. Engineers in the firing room are taught to be ready to override switches accidentally tripped as the astronauts clamber out of their seats. During simulations at Houston, the astronauts wear old

After a successful liftoff, the flight engineers who seconds earlier were monitoring the shuttle's prelaunch vital signs can now enjoy a splendid view of the craft as it soars spaceward (above).

Following an abort, the shuttle is brought back to the Vehicle Assembly Building (left), where the engines must be removed (above) and serviced. The process can take up to three weeks.

helmets because the visors are commonly scratched and cracked from banging into the mock-up's instrument panels and bulkheads.

Emergency egress is a last resort. Experience has shown that engine shutdown does not require an egress. "In fact, until we really understand what kind of situation we have outside, many times the safest place for [the crew] is inside with the hatch closed," Al Sofge says. "We could egress the crew into a worse situation than they're in. For example, if you had a hypergolic tank rupture on you, and you had a large hypergolic cloud, and that's the only problem you have, and your cloud covers the egress route, you may be better off leaving [the crew] in the vehicle."

ASTRONOMY AND SPACE SCIENCE 87

In addition, the flight crew works during a pad abort, at least initially, switching off the auxiliary power units, disarming the reaction-control system and orbital-maneuvering system, and, most important, shutting down the backup flight software. The most recent abort, on mission STS-68, occurred so close to launch that the backup computer began counting up, as though the shuttle had launched. If the computer had not been shut down, the explosive bolts on the solid rockets might have blown at the one-minute 40-second mark, when the SRBs normally separate from the vehicle in flight. The solid rockets hold the shuttle and external tank upright on the pad. Blow those bolts, and the tank and shuttle fall over.

Engineers in the launch-control center are especially on guard for signs of conditions that could lead to fire or explosion, such as bubbles in the umbilical line that feeds oxygen to the main engines. To remain liquid, oxygen must be kept at -298° F (-183° C). During an emergency shutdown of the main engines, some of the oxygen being pumped to the engines could warm and begin to boil, creating a bubble that could back up through the plumbing and into the external tank. In the process, that bubble would create a void in the 100-foot (30-meter) line leading to the shuttle engines. "When it bursts at the top of the tank, the LOX [liquid oxygen] will come rushing back into the line leading to the shuttle," says John Sterritt, a Lockheed Martin engineer who leads a team of propulsion experts in the firing room. The sudden pressure could cause the external tank to fail, "like popping a paper bag," he says. "With any kind of ignition source, you'd have a real potential for fire." So Sterritt and his team carefully watch data streams that would indicate heating in the oxygen umbilical.

Critical Safing Procedures

All the years of practice, as well as the experience of five pad aborts, have made safing the shuttle almost routine. "The procedures have all been refined; the little discrepancies we noticed in the beginning of the program were changed and tested and put in place," says Greg Katnik, lead flight-structures engineer for NASA. Katnik was an engineer in the firing room 11 years ago when a hydrogen leak caused flames to lick up the east side of the shuttle during the first abort. The launch team manually turned on the Firex water system to disperse the hydrogen and put out the fire. Since then, NASA has programmed its computers to trigger the water system at the start of a pad abort. An engineer on the launch team also pushes a backup button to make sure the engine is flooded with water. Steel plates have been installed under the access arm to keep flames from reaching the astronauts in case they have to cross to the slide-wire baskets.

After an abort, the critical safing procedures take about 10 minutes, according to Bruce Bartolini. "You then launch into several other sequences which get everything secured and get the crews out. So we're really done about 45 minutes after the abort."

"We're prepared for the emergencies," Bartolini says. "I myself, after I give my last command [at] about two minutes, have my checklist tabbed, and I turn to the abort procedure and I'm ready to do it."

He admits, however, "that when the call comes, it's still a surprise." In the case of STS-68, the abort came at T minus 1.9 seconds, so close to launch that the official who announces liftoff said: "We have L . . . abort." "It was the French abort," chuckles Bartolini, "*L'abort*."

He continues, "It was kind of shocking, and then it's . . . you're all business."

It was especially shocking to Daniel Bursch, who was on this mission, too. Because he has experienced main-engine cutoff (MECO) four times, yet flown only twice, his fellow astronauts have dubbed him the MECO Kid. "I'm fully ready for another pad abort," says Bursch. "I said it couldn't happen twice, and it did. Well, it could happen three times."

If it does, the launch team will bring the shuttle back to the Vehicle Assembly Building and spend three weeks changing the engines. Once the engines fire, even for a few seconds, they're removed and serviced. Then the team will send the only reusable launch vehicle operating in the world back to the pad for another try.

AMERICA'S FOREMOST EARLY ASTRONOMER

by David Parry Rubincam and Milton Rubincam II

David Rittenhouse was a craftsman, scientist, and patriot of renown. He was born in 1732 into a distinguished Philadelphia family—his great-grandfather had established the first paper mill in America. Rittenhouse grew up on his parents' farm northwest of that city and took up neither the family papermaking business nor farming. Instead, he made clocks, producing his first one at about age 17. He moved to Philadelphia in 1770, and continued there as a clockmaker until the American Revolution.

MASTER HOROLOGIST

Maturing at his trade, Rittenhouse produced what were then perhaps the finest timepieces in America. According to Brooke Hindle, Rittenhouse's modern biographer, about 40 clocks he sold commercially survive to the present, nearly all of them tall, grandfather-type clocks. The colonial craftsman made astronomical clocks for his own use, sometimes equipping them with temperature-compensating devices to better regulate the beat of the pendulum.

Rittenhouse also made scientific instruments, and was among the first to put a vernier on a magnetic compass; these were often called Rittenhouse compasses. He constructed surveying instruments as well, which—along with his great skill in mathematics—led to several surveying jobs. Among these were helping to establish the New York-New Jersey boundary and surveying the course of the Delaware River.

The finest examples of Rittenhouse's craftsmanship, however, were two orreries, or

Few scientists can match the achievements of David Rittenhouse. Undoubtedly the preeminent American astronomer of his time, Rittenhouse was also a noted inventor, mathematician, physicist, and patriot.

mechanical solar systems. Named for Charles Boyle, the fourth Earl of Orrery, these devices were popular items in the 18th century. They ranged from a simple apparatus showing the relative motions of the Sun, Moon, and Earth to more-complicated ones with all the known planets. A crank often operated the mechanism, which caused the bodies to revolve at the proper speeds in relation to one another.

As might be expected from his clockmaking talent, the Rittenhouse orreries are exquisite creations. Both are unusual in that the plane of the solar system is vertical, rather than horizontal as it had been in earlier designs. The main panel on each device features all of the planets then known to scientists, which explains why Uranus, Neptune, and Pluto are missing. One of the orreries, now at the University of Pennsylvania in Philadelphia, is still in its fine original wooden case; it also contains a lunarium in the right panel, showing in detail the Earth-Moon system. The left panel is blank—it may have been intended to depict Jupiter and Saturn with their retinues of moons. The orrery made for the College of New Jersey (now Princeton University) is displayed in a modern case without side panels.

TRANSIT OF VENUS

Rittenhouse did not confine his interest in the solar system to making mechanical models. Indeed, he is chiefly remembered as an astronomer. His best-known contribution was his observation of the transit of Venus in 1769. Transits were important in those days because they gave the most precise measure of the distance between Earth and the Sun. If a transit is timed by observers at different known locations, the distance to the Sun can be determined by a method called triangulation. From this, the distances to all the planets can be found,

An innovatively designed orrery (left)—a mechanical device that depicts the relative positions and motions of the planets—testifies to Rittenhouse's legendary craftsmanship. The master clockmaker was the first to construct an orrery with the solar system displayed in a vertical plane.

For this work, the Pennsylvania Assembly voted Rittenhouse a cash award of £300, "as a Testimony of the high Sense which this House entertain of his Mathematical Genius and Mechanical Abilities in constructing the said Orrery."

since their relative distances from the Sun have long been known. (See illustration on facing page.)

The 1769 transit was eagerly awaited because another involving Venus would not occur for more than 100 years. Pennsylvania

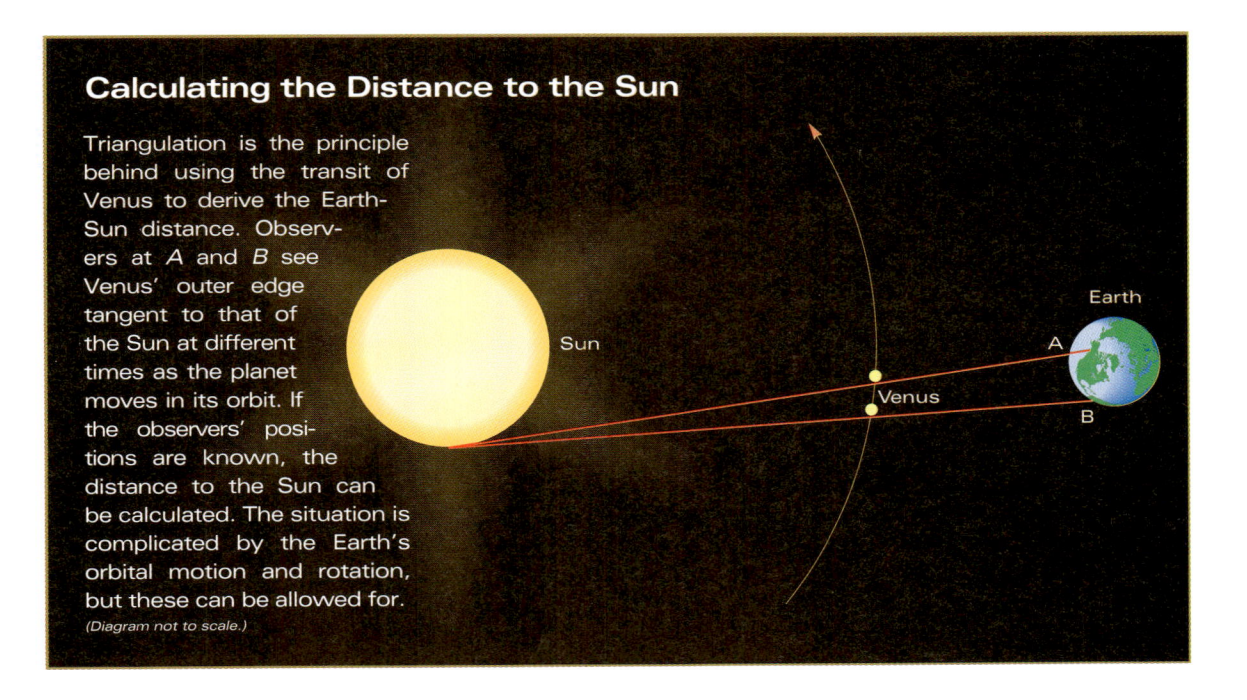

Calculating the Distance to the Sun

Triangulation is the principle behind using the transit of Venus to derive the Earth-Sun distance. Observers at A and B see Venus' outer edge tangent to that of the Sun at different times as the planet moves in its orbit. If the observers' positions are known, the distance to the Sun can be calculated. The situation is complicated by the Earth's orbital motion and rotation, but these can be allowed for.
(Diagram not to scale.)

---

fielded three separate groups of enthusiasts bent on seeing it. Rittenhouse was to observe with William Smith, the provost of the College of Philadelphia, and John Lukens, the surveyor general of Pennsylvania, at the Rittenhouse farm.

Rittenhouse made most of the preparations. He constructed an observatory and assembled telescopes—one from lenses meant for Harvard College that would not reach New England in time for the transit. Rittenhouse also built an astronomical clock to make timings of first and second contact. Thomas Penn, the proprietor of Pennsylvania, donated a reflecting telescope, one of the few instruments not supplied by Rittenhouse.

Transit day, June 3, dawned perfectly clear. That afternoon, all three men took their places—Smith inside the observatory with Penn's reflector, and Rittenhouse and Lukens outside, lying on their backs staring through refractors. Smoked glass on each telescope (a practice not recommended today) diminished the Sun's intensity so it would not blind the observers.

Shortly after first contact, just as Venus' disk began to intrude on the face of the Sun, Rittenhouse apparently fainted. (At least that was what Benjamin Rush reported at the astronomer's funeral years later.) His own account of the transit says nothing about the first few minutes. Yet shortly thereafter, Rittenhouse reported seeing a pyramid of light broaden into a halo on the side of Venus that had not yet entered onto the Sun. This was due to Venus' atmospheric refraction and scattering of sunlight. He also recorded second contact and continued observing until sunset.

Rittenhouse's observations, along with those of his Pennsylvanian colleagues, did not alone determine the distance between Earth and the Sun; that depended upon many widely scattered observations on both sides of the Atlantic. His role in the transit, however, in addition to his orreries and recognized mathematical ability, helped propel him to scientific fame in the colonies.

REVOLUTIONARY ASTRONOMER

The Revolutionary War interrupted Rittenhouse's scientific and clockmaking career. He threw his support to the cause of independence by serving on the Committee of Safety, which was entrusted with the defense of Pennsylvania against the British. Curiously, one of his duties was to procure iron clock weights and exchange them for the lead weights then commonly used, which would be turned into ammunition.

In 1776, Rittenhouse became a member of Pennsylvania's constitutional convention. He chaired the committee on the declaration of rights under the new government. The resulting Revolutionary Constitution served as the basis of local government until Pennsylvania joined the United States in 1790.

Early in 1777, the legislature unanimously elected Rittenhouse treasurer of Pennsylvania. He grappled with the task of helping finance a revolution with inadequate means of raising revenue. When British troops seized Philadelphia, Rittenhouse fled with the rest of the government, becoming treasurer-in-exile in the new capital of Lancaster. When the enemy evacuated Philadelphia the next year, he returned to the city. Although the Revolution continued, Rittenhouse was again able to pursue science. Yet he earned his living as a treasurer, not as a clockmaker.

Rittenhouse is widely credited as being the first American to build a telescope. Among his many other contributions to astronomy were his observations of the orbit of Uranus and the transit and atmosphere of Venus.

An Optical Mystery

In March 1785, Judge Francis Hopkinson wrote to Rittenhouse describing a puzzling optical phenomenon. "Setting at my door one evening last summer, I took a silk handkerchief out of my pocket, and stretching a portion of it tight between my two hands, I held it up before my face and viewed . . . one of the street lamps which was about 100 yards [90 meters] distant; expecting to see the threads much magnified. Agreeably to my expectation I observed the silk threads magnified to the size of very coarse wires; but was much surprised to find that, although I moved the handkerchief to the right and left before my eyes, the dark bars did not seem to move at all, but remained permanent before the eye."

Rittenhouse conducted an investigation. He disposed of Hopkinson's notion that the threads had been magnified; a distant source of light could perform no such feat. Besides, because the "dark bars" did not move when the threads did, they could not be one and the same.

To discover the truth, Rittenhouse did away with the complications of crisscrossing threads by making a 0.5-inch (1.25-centimeter) square of parallel hairs. "And to have them nearly parallel and equidistant, I got a watchmaker to cut a very fine screw on two pieces of brass wire," he wrote in his reply, which appeared in the *Transactions* of the American Philosophical Society along with Hopkinson's letter. "In the threads of these screws, 106 of which made 1 inch [2.5 centimeters], the hairs were laid 50 or 60 in number." He held the square up to a slit in a window shutter that let light into an otherwise-darkened room, with the hairs parallel to the slit. Rittenhouse saw bright lines separated by dark bands, a simplified version of Hopkinson's pattern. This was the first diffraction grating.

"Thinking my apparatus not so perfect as it might be, I took out the hairs and put in others, . . . of these 190 made 1 inch, and therefore the spaces between them were about the 1/250 part of an inch." This time when he held his grating up to the light, he noted a bright central line with about six lines on either side of it. While the central line was colorless, the inner edges of the lines on either side were tinged with blue, and their outer edges with red. Rittenhouse correctly ascribed the side lines to the diffraction of light—or inflection, as it was called then.

Diffraction was not understood in Rittenhouse's day. Early in the 18th century, Isaac Newton speculated in his book *Opticks* that light consisted of corpuscles. He conjectured that those passing near an obstacle (such as a hair) might deviate from their course according to their color. This effect would produce the side lines seen by Rittenhouse. Corpuscles farther from the hairs would be unaffected and pass straight through, producing the colorless central line.

This theory held sway until the early 19th century, when the inadequacies of Newton's theory were revealed and Christian Huygens' rival wave theory triumphed. Diffraction is now known to be caused by the interference of light waves. The bright lines Rittenhouse saw occurred where the waves reinforced each other; the dark bands, where they destructively interfered. "By pursuing these experiments," Rittenhouse wrote, "it is probable that new and interesting discoveries may be made. . . . But want of leisure obliges me to quit the subject for the present."

LATER LIFE

Rittenhouse resigned his position as treasurer of Pennsylvania in 1789, only to become the first director of the United States Mint in 1792. He accepted the job reluctantly because of ill health, but labored diligently until the last year of his life. And when Benjamin Franklin died, Rittenhouse took his place as president of the American Philosophical Society, of which he was a longtime member.

Although busy and in poor health, Rittenhouse remained scientifically active. He made astronomical observations that included charting the position of Uranus after its discovery in 1781, and he conducted scientific experiments. He also wrote some purely mathematical papers. But Rittenhouse never got back to the diffraction grating, and thus made no progress in understanding the nature of light. He certainly had no idea that gratings would one day become the most powerful probes of the compositions of stars.

Rittenhouse died in 1796, when his reputation was at its peak. One of his most ardent admirers, his friend Thomas Jefferson, once boasted: "We have supposed Mr. Rittenhouse second to no astronomer living." A legend even grew up that Rittenhouse had invented calculus independently of Newton and Leibniz! As a result, the astronomer was often asked to oversee matters in which he had no prior experience (such as the minting of coins) in the expectation that he would succeed handily.

A FORGOTTEN PERSONAGE

Rittenhouse's reputation suffered a curious fate. Despite his roles in science and in the American Revolution, his name faded from public memory. His diffraction grating attracted scant notice in Europe. Years later, Joseph von Fraunhofer, having no knowledge of Rittenhouse, constructed his own diffraction grating; its invention was then attributed to the German physicist. A century after Rittenhouse's death, Simon Newcomb, an astronomer at the U.S. Naval Observatory, reworked the old transit observations and threw out Rittenhouse's as being hopelessly off. Even the orrery that Rittenhouse built for Princeton vanished for a time.

For all his true accomplishments, David Rittenhouse was once credited with a feat he did not achieve: piloted flight! A hoaxer (probably Franklin) claimed Rittenhouse and his friends had flown a balloon at about the same time as the Montgolfier brothers. The hoax survived into the 20th century, being cited as an example of independent but simultaneous invention.

Today Rittenhouse's reputation is making something of a comeback. In 1932, Thomas D. Cope of the University of Pennsylvania pointed out that Rittenhouse had indeed invented the first diffraction grating; more and more references now give him priority. Present-day astronomers have a better appreciation of the uncertain timings associated with transits of Venus. The Princeton orrery, lost at the end of the 19th century, surfaced 50 years later. Two biographies of Rittenhouse have appeared since the end of World War II. And one can hope that the hoax about ballooning is finally dead!

The crash of Comet Shoemaker-Levy 9 (S-L 9) into Jupiter in July 1994 was the most widely witnessed event in astronomical history. For a few hectic days, professional astronomers turned every telescope in the world (and above it) toward Jupiter. They were joined by legions of backyard observers who could see Jupiter's Earth-sized impact scars in scopes with apertures as small as 2.5 inches (6.35 centimeters). For years to come, such sights will remain etched in the minds of those who saw the crash firsthand.

This indelible legacy of the Great Comet Crash of 1994 raised humanity's awareness of catastrophic impacts in the solar system. Watching Jupiter suffer hit after hit left scientists and laypeople wondering aloud if the same thing could happen to Earth.

Truth is, planetary scientists have known for a long time that Earth is subject to bombardment from space. You need only look at the Moon's battered landscape to realize that we live in a bad neighborhood. Cosmically speaking, Earth and Moon are very close neighbors, and whatever rain of debris has hit our satellite must have had a similar effect on Earth. Were it not for the constant reworking of Earth's surface by erosion and plate tectonics, our planet would be as densely cratered as the Moon.

The real scientific question is not whether Earth has been battered by impacts over its long geologic history. We know that for a fact. What is revolutionary in today's thinking is that our fragile environment can be disrupted by relatively small impacts. And it is not just the immediate effects of an impact

Target: Earth
by David Morrison

we should be concerned with; Comet S-L 9 has helped us get a feel for consequences that can linger long after the initial strike.

All this is a far cry from barely 20 years ago, when few scientists even considered the possibility that asteroid hits could shuffle Earth's biological deck.

Survival of the Toughest

Our thinking about how asteroid impacts influence Earth's environment began to change with a serendipitous discovery in 1978. Luis and Walter Alvarez of the University of California at Berkeley were searching geologic rock layers for a largely extraterrestrial element called iridium. Iridium, they figured, drizzled down on Earth at a nice, steady rate in the form of meteor dust. Measuring the amount of iridium in these layers could help scientists figure out how fast sediments collected on the floors of ancient oceans.

What they found instead was a major infusion of the element in a rock boundary between the Cretaceous and Tertiary eras, which marks when the age of dinosaurs ended some 65 million years ago. There was enough iridium in the boundary, they calculated, to equal a 6-mile (10-kilometer)-diameter asteroid.

Based on this evidence, the team announced that a 6-mile-diameter asteroid, a so-called Great Extinctor, had caused an environmental catastrophe that ended the Cretaceous era. They concluded that the dinosaurs had suffered from the global cooldown caused by dust thrown up into the stratosphere by the impact. Today we know that this dust cloud is one of several environmental catastrophes associated with an impact of this size.

As if this astounding discovery were not enough, the Alvarezes went on to propose a sweeping generalization. If a hit of this magnitude killed the dinosaurs, perhaps similar impacts might have happened at other times in Earth's history and played a major—perhaps dominant—role in biological evolution. With asteroid impacts in the picture, suddenly one of a species' most important traits is its ability to survive random cosmic assaults. The winner of an evolutionary race is not a species that is stronger or faster or even smarter, but one that survives an occasional intruding asteroid or two. Natural selection still works, but the context has changed. Asteroids enter the evolutionary game as wild cards.

Over the past 15 years, evidence for a dinosaur-snuffing catastrophe has continued to accumulate, culminating with the 1991 discovery of a telltale crater (named Chicxulub) under Mexico's Yucatán Peninsula. Scientists now estimate from the crater's size that the progenitor asteroid hit with an astounding energy equivalent to more than 5 billion Hiroshima atomic bombs (100 million megatons).

With that sobering statistic in mind, geologists today sift the geologic record for hidden evidence that Earth might have suffered other impact-induced mass extinctions. Although that evidence is slow in coming, the events that killed the dinosaurs 65 million years ago are no longer a matter of serious dispute.

Skies of Fire, Tracts of Ice

Aside from the immediate effects of heat and concussion, two major postimpact events

Some Risky Asteroid Sizes

Small Extinctor
- 100 gigatons (100,000 megatons)
- 0.6-mile diameter
- Affects a hemisphere

Regional Bludgeon
- 100 megatons
- 100-yard diameter
- Destroys a continent

Tunguska
- 10 megatons
- 50-yard diameter
- Wipes out a city

Great Extinctor
- 1,000 gigatons (100,000,000 megatons)
- 6-mile diameter
- Global disaster

ASTRONOMY AND SPACE SCIENCE   95

did the dinosaurs in: a firestorm followed by excessive cold. What we saw on Jupiter during Shoemaker-Levy's 1994 impact was a window to this lethal combination of fire and ice.

After fragments of Comet S-L 9 exploded in Jupiter's atmosphere, they produced huge fireballs of hot gas and dust that rose high above the planet's clouds.

Because Jupiter's atmosphere was literally blown away from above the impact site, hot gas and dust were funneled up to altitudes of 2,200 miles (3,500 kilometers). But in Jupiter's strong gravity, whatever went up had to come down. So for about 20 minutes after each impact, the dust-laden plumes fell back into the atmosphere, reentering with a horrendous release of energy. The heat from this reentry was so intense it was easily detected from Earth.

The same thing happened on Earth 65 million years earlier with the Chicxulub impact. But in our case, the plume included large quantities of rock and dust blown out from the crater. When this ejecta rained down about 30 minutes after the impact, it produced a meteor shower of almost unbelievable proportions. Meteors turned the sky red-hot and ignited terrestrial forests and grasslands. Telltale soot from this firestorm is found in sediments from the Cretaceous/Tertiary boundary at sites all over the world. Most land animals probably perished by fire.

Then came the cold.

On Jupiter, large black dust clouds remained clearly visible at each impact site for weeks, suspended in the stratosphere. Even more than a year after the impacts, not all the dust has settled back into the deep atmosphere. These bruises have substantially reduced the amount of sunlight that reaches Jupiter's lower cloud tops.

In Earth's case, the Great Extinctor hit in shallow water and excavated several times its mass in finely fractured target rock. Indeed, the total amount of material measured in the worldwide Cretaceous/Tertiary boundary layer is about 100 times greater than the mass of the 6-mile asteroid that actually hit. Most of the layer has large grains that were blown horizontally by the impact and plume particles that were heated during atmospheric reentry. But a sizable portion of the layer is in the form of fine dust that remained suspended in the atmosphere for months, blocking photosynthesis and plummeting temperatures on the dark surface beneath the clouds.

Suppression of photosynthesis led to a breakdown in the ocean food chain that killed most marine creatures. Survivors had to hunker down to a global drop in temperature not unlike the "nuclear winter" scenario postulated during the Cold War.

Geologic evidence strongly suggests that global cooling and other environmental consequences of an enormous meteor impact (left) caused the mass extinction of the dinosaurs some 65 million years ago.

Still other environmental catastrophes drove the mass extinction 65 million years ago, including global acid rain and destruction of the ozone layer. Species that survived one stress may have succumbed to another. But it was the thermal pulse from reentering ejecta and the subsequent darkness caused by stratospheric dust that did most of the killing.

Death by the Megaton

It is hard to imagine that all this carnage came from a relatively small asteroid no more than 6 miles wide. But what the Great Extinctor lacked in size, it made up for in kinetic energy, measured the same way we measure the yield of thermonuclear weapons.

The energy of the Chicxulub impact was about 100 million megatons—that is, about 5 billion times the size of the atom bomb dropped on Hiroshima in 1945. Fortunately, events like Chicxulub are quite rare, occurring at intervals of 100 million years or more. But what about smaller impacts? Could these produce similar environmental effects?

Brian Toon and Kevin Zahnle at the National Aeronautics and Space Administration's (NASA's) Ames Research Center have calculated the effects of smaller impacts. According to their work, a 2-mile (3-kilometer)-diameter impactor, only 2 percent the mass of a Great Extinctor, would kick up enough dust to block photosynthesis. And given what we know about the number of small, near-Earth objects, such small hits are statistically possible during our lifetimes.

Earth accumulates about 100 tons of extraterrestrial material every day under a constant rain of interplanetary debris. Most meteoroids enter the atmosphere and burn up unnoticed. Some survive the fiery heat of entry and are slowed down by air friction to a speed of about 200 miles (320 kilometers) per hour. What's left hits the ground as a meteorite.

BIG ONES DIG DEEP

It is quite a different story when Earth's atmosphere cannot slow down large meteoroids. Striking at their space velocity of 9 to 12 miles (15 to 20 kilometers) per second, they explode on impact with an energy 100 times their mass in dynamite. Since such an event would be much more destructive than a simple meteorite fall, it is natural to wonder what size meteoroid could devastate Earth's surface.

Calculations by Christopher Chyba and Kevin Zahnle show that large objects entering Earth's atmosphere undergo tremendous stress due to air resistance. As the object fragments and covers more area along its line of travel, it creates more atmospheric resistance. Soon this process leads to a catastrophic breakup that causes the asteroid to decelerate explosively.

Visitors to Wyoming's Grand Teton National Park witnessed a near miss on August 10, 1972, as a meteorite crossed the daytime sky. Scientists estimate that such an object would pack as much energy as a large nuclear bomb.

If such an explosion takes place at altitudes above 9 miles (15 kilometers), there is little harm done near Earth's surface. The outcome is not so rosy if a larger object digs deeper into our atmosphere.

Ordinary stony asteroids, which have a rather crumbly composition, must be larger than 165 feet (50 meters) across to do any damage at the ground. Such a projectile packs about 10 megatons of energy, comparable to the largest nuclear bomb. Thus, Earth's atmosphere protects us from smaller impacts, but not from those with 10 megatons or more of energy. We can expect a 10-megaton impact about once per century.

To find one of the largest atmospheric explosions in recorded history, we need not look to hieroglyphs or deduce hidden imagery in ancient mythology. Earth reeled under a tremendous assault as recently as 1908, with the great Tunguska bolide.

On June 30 of that year, a 15-megaton explosion took place in the atmosphere above a desolate region of central Siberia, devastating more than 775 square miles (2,000 square kilometers) of forest. The Tunguska region is so remote that it was not until 17 years after the explosion that a scientific expedition visited the site. When the team arrived, they found no craters or evidence of the meteoritic fragments.

Many astronomy textbooks state that the Tunguska impactor was probably a comet. If the Tunguska impactor had been cometary, however, it would have exploded at a higher altitude and done no damage. Had it been composed of denser iron, it would have reached the ground and made a crater. Meteor Crater in Arizona was formed by just such a metallic meteorite, with the same 15-megaton energy as Tunguska.

We now recognize the culprit as a rocky asteroid, about the size of a city office building, that decelerated and exploded at an altitude of 5 miles (8 kilometers). Last year, scientists apparently even found tiny fragments of this rocky object embedded in tree resin at the impact site.

Playing Chicken with Oblivion

If "ground zero" for Tunguska (or Meteor Crater, for that matter) had been a city, the destruction would have been terrible—equivalent to that of a large nuclear bomb. Such impacts happen about once every 300 years anywhere on Earth's surface, or about once per millennium on the land. Even if 10 percent of Earth's land were densely populated, this amounts to a hit on a city only once in 10,000 years.

The gaping Meteor Crater (below) in northern Arizona was excavated about 25,000 years ago by the high-velocity impact of a large asteroid. Nearly all of the 60,000-ton meteor was vaporized in the explosive collision.

There are other ways cities can suffer the effects of an impact. Were a Tunguska-class object to land in the ocean, it would produce tsunamis with tremendous potential for destruction. Tsunamis travel much farther from the impact site than does the direct-blast wave, producing widespread destruction in coastal areas thousands of miles away.

For every person killed directly by the impact, about 10 more would be killed by an impact-induced tsunami. Yet even the tsunami risk is less than that of succumbing to an ecological disaster brought about by still-larger impacts, which could trigger crop loss and starvation on a global scale. The risk of dying from such a global event is about 10 times greater than that from a tsunami, or 100 times greater than that of a direct hit.

But a simple comparison of risk between impacts and their secondary effects does not tell the whole story. No other natural disaster we know of has global dimensions. Even the worst earthquake or flood affects only a few percent of Earth's population. An impact-induced ecological catastrophe, however, could kill a billion people and destabilize civilization. It would end the world as we know it.

In other words, larger impacts, ones caused by asteroids as small as 330 feet (100 meters) in diameter right on through the 4-mile (6-kilometer)-diameter-or-larger Great Extinctors, pose the greatest risk to humanity. Even though such massive hits are extremely rare, their global consequences bode worse for humanity than the cumulative effects of smaller and more-frequent hits.

Since the chance of a major impact in our lifetime is so low, impacts are a danger most people ignore. Therein lies the problem with statistical arguments. Being human, the weight we assign to threats in large part determines how "real" they become in our collective mind. Either there is an asteroid up there in an orbit that will lead to a collision with Earth within the next century, or there is not. If not, then we have nothing to worry about. But if there is a big rock up there with our name on it, we had better find it and protect ourselves.

There may be 1,000 near-Earth objects (NEOs)—whose orbits cross that of Earth—large enough to cause a global environmental collapse if they hit. But of these 1,000 objects, we have actually discovered fewer than 100—less than 10 percent. We know from orbital calculations that none of these 100 objects threatens Earth. But that says nothing about the 900 that remain undiscovered. One of them could hit at any time, perhaps without warning.

Earth's Guard Is Down

The sad reality is that we humans are not doing a very good job of policing the inner solar system, where threatening asteroids lurk undetected. Virtually all the discoveries of near-Earth objects are made by just four teams of astronomers, one of which is terminating its search. This leaves only a handful of people to carry out the beginnings of what could be an effort to save humanity from the next mass extinction.

While the search effort continues, albeit haltingly, another facet of the humans-versus-asteroids story is gathering momentum. Former Cold War adversaries have teamed up to explore ways of deflecting or even destroying threatening asteroids.

But the first step in any asteroid defense program is to carry out a census of potentially threatening objects, beginning with the most dangerous—those larger than 320 feet (1 kilometer) in diameter. In 1990, Congress requested that NASA study ways to increase the discovery rate of these near-Earth objects. An international team of astronomers proposed a program called the Spaceguard Survey, designed to obtain a complete census of these larger asteroids.

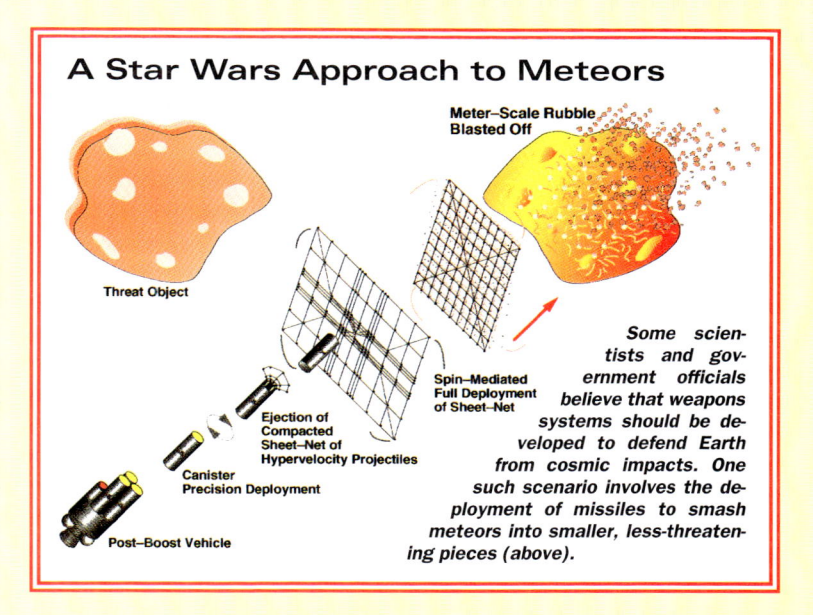

A Star Wars Approach to Meteors

Some scientists and government officials believe that weapons systems should be developed to defend Earth from cosmic impacts. One such scenario involves the deployment of missiles to smash meteors into smaller, less-threatening pieces (above).

The results were reported to Congress in 1992, and NASA made available about $1 million in additional funds to upgrade existing search programs. At the same time, the International Astronomical Union (IAU) appointed a working group on near-Earth objects to promote international cooperation in this search. Since everyone on Earth is at risk from cosmic impacts, it seemed only reasonable to share the cost of the survey.

And even before the dust of Comet Shoemaker-Levy 9's impact completely settles on Jupiter, Congress is asking NASA to act, this time with a greater sense of urgency. They want the agency to work with the U.S. Air Force, which already maintains a network of telescopes used to track faint Earth satellites, to find a way to accomplish an NEO survey within the next decade. A team led by Gene Shoemaker is currently preparing a report on ways to carry out this mandate.

So, in a way, Earth's fragile life-forms are finally coming to grips with a persistent problem. For the first time in our planet's history, through the intellect of its human species, life can avoid a fate that up to now has been inevitable. After observing the geologic evidence for mass extinctions and bearing witness to Jupiter's horrific pounding, we, the people of Earth, are getting positioned for action.

Earth, the planet of life, need never be a target again.

Ask the Scientist

▶ *Are comets ever so bright that they actually lighten the sky to daytime levels during the night? Are there many comets in the sky every night that we just can't see because we have no telescope?*

There have been some very bright comets throughout history, but none so brilliant as to turn night into day. Even the brightest are more than 100 million times fainter than the Sun itself. Comets are chunks of ice that orbit around the Sun, sometimes getting so close that we can see their long, wispy tails with our unaided eyes. Most are so faint, however, that telescopes are needed to see them. Dozens of comets are visible to amateur and professional astronomers each night.

▶ *Did the people in ancient civilizations really believe that, for example, the constellation Orion was actually the outline of one of their gods? Why do so many celestial bodies have names derived from mythology?*

The people of many ancient civilizations believed that their gods lived in the heavens, but they did not necessarily believe the constellation outlines to be the gods themselves. The constellation names were probably created by wandering nomads who told stories in return for food and lodging, and who used the sky as a tapestry on which their listeners could imagine pictures. Because the sky was so important to these people, the names of our constellations, stars, and planets all have their roots in ancient mythologies.

▶ *Is it true that the signs of the zodiac no longer correspond to the days published in the horoscope column? If so, why not? What came first, astronomy or astrology?*

Millennia ago, astrology and astronomy were the same field of study. Sky watchers charted the heavens to understand the will of the gods, and to predict events on Earth. Eventually the science of astronomy broke away to study the physical nature of the universe. Today the original zodiacal signs no longer correspond to the Sun's location at the time of your birth. This is caused by a 25,800-year wobble experienced by Earth. Many of today's zodiacal "signs" are off by at least 30 days.

▶ *Do most astronauts feel ill from the turbulence experienced during reentry into Earth's atmosphere? Do they often have trouble walking once they leave their craft? Do they require special physical therapy?*

The biggest problem astronauts suffer is from the effects of weightlessness. During their first few days of spaceflight, astronauts experience changes in their balance systems, leading to disorientation, dizziness, and motion sickness. This passes as their bodies become accustomed to the new environment. Muscles, however, tend to weaken in this strange "zero-g" setting. Muscle weakness can show up after a long spaceflight as an inability to walk well, although this, too, soon passes as the astronauts use their muscles once again on Earth.

▶ *When was the first planetarium built? Are most associated with museums or with universities?*

The first planetarium theater was built in 1657 by Andreas Bush. He constructed a large globe inside of which 12 people could gaze at the fixed stars overhead. The first modern planetarium opened in 1923, in Munich, Germany. The first in the United States—the Adler Planetarium—opened in Chicago in 1930. Today some planetariums are associated with universities or schools, but most are associated with museums, and nearly every large city has at least one.

▶ *I know that solar eclipses occur only rarely on Earth. But is the Moon always eclipsing the Sun someplace, such that the shadow is just going out in space rather than hitting Earth? Could the space shuttle find such a place?*

Because the Moon is solid, and all solid objects cast a shadow, the Moon is always eclipsing the Sun—from somewhere. Most of the time, the Moon's shadow falls into space, creating an eclipse where no one can experience it. Only when it encounters Earth do people experience a solar eclipse. Satellites do encounter the Moon's shadow when it doesn't fall onto Earth, but, because most satellites travel so close to our planet, this would not occur very often.

▶ *In craters on Earth, are the remnants of the causative meteor easy to detect? Are there many meteor-impact craters on our planet?*

More than 140 impact sites are known on Earth, the largest of which is a 125-mile (200-kilometer)-diameter basin in Sudbury, Ontario, Canada. Only 10 percent retain fragments from the original meteorite, however. This is because, during a crater-forming impact, the kinetic energy of the falling body—which may be several miles across and moving at many miles per second—causes the impactor to vaporize. Tiny fragments of the original meteorite may be found in the vicinity, having been thrown from the explosion.

▶ *What causes a halo to appear around the Moon? Is a lunar halo really a harbinger of bad weather?*

Ice crystals high in the upper atmosphere of Earth can sometimes cause a beautiful halo around the Moon or, in the daytime, around the Sun. These crystals bend the Moon's light in such a way that they form a ring 22 degrees wide. A lunar or solar halo occurs when very thin cirrus clouds are present, and these clouds are usually forerunners of a storm system on its way.

▶ *What kind of telescope would you recommend for a backyard stargazer? My 12-year-old son has developed an avid interest in astronomy, and wants to buy a telescope.*

Before buying a telescope, your son must ask himself a few questions. First, how familiar is he with the sky? If he cannot identify at least 12 constellations, he should subscribe to a popular astronomy magazine and use their star maps regularly.

Second, what are his interests? If the Moon and planets are his passion, a smaller "refractor" (built with lenses) will be fine. If he wants to see fainter stars, star clusters, nebulae, and galaxies, a larger "reflector" (built with mirrors) is a must.

Third, how much can he afford? Quality telescopes usually cannot be purchased for less than $300. If he's ambitious, he may wish to build one himself. In any case, he should join a local amateur astronomy club.

A good beginner's scope is a 4-inch (10.16-centimeter)-diameter reflector. To find good deals, talk with members of your astronomy club or the staff at the local planetarium.

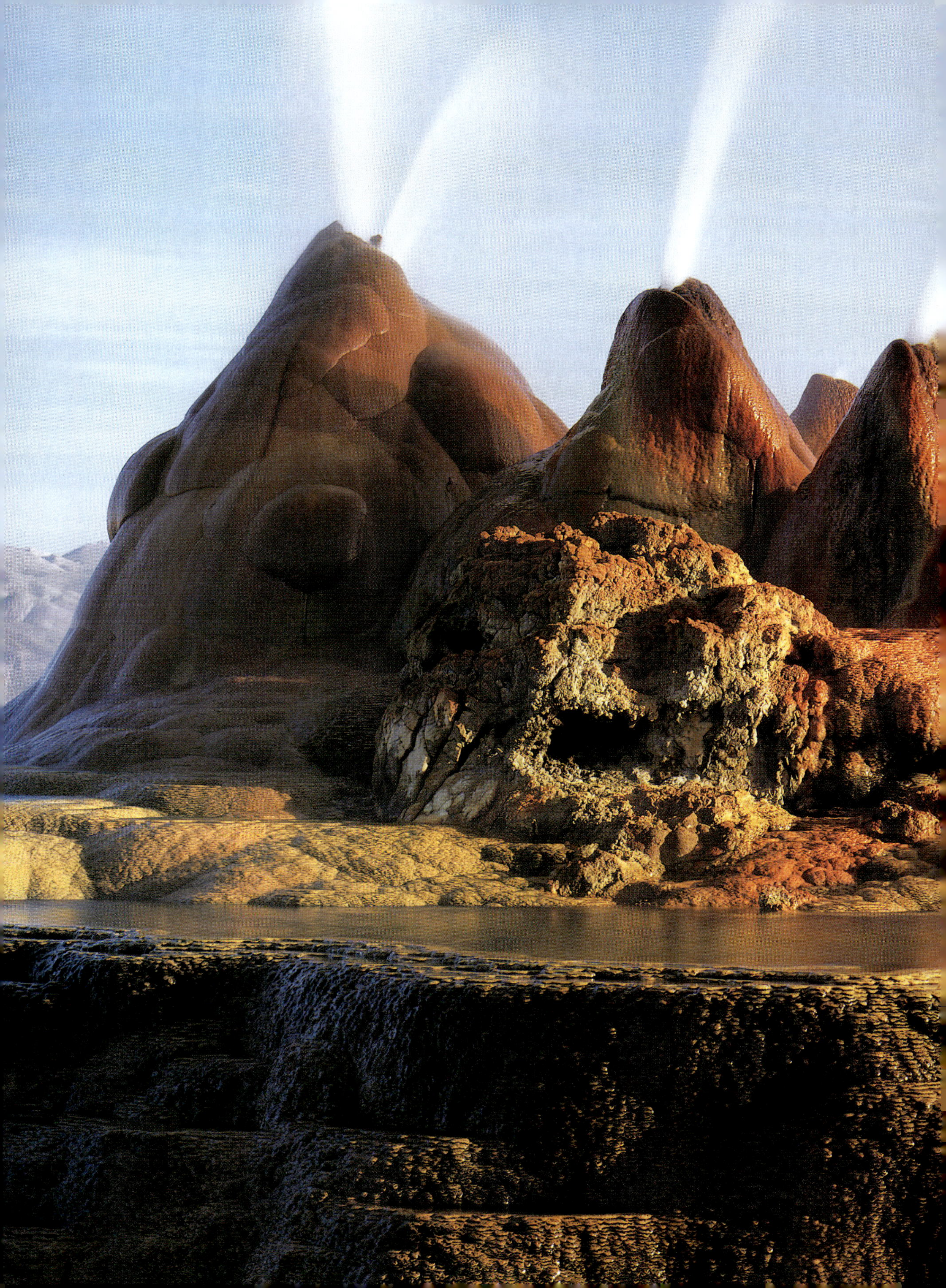

Earth and the Environment

■ Small groupings of geysers occur with some regularity in a region of underground thermal activity that stretches from California northward to Idaho and Montana. Some geysers, like those in Yellowstone National Park, have become popular tourist attractions. Others, like the ones at left in Nevada's Black Rock Desert, remain virtually undiscovered, free to erupt in pristine isolation.

CONTENTS

Lights Above Thunderstorms	104
Mountain Isles, Desert Seas	110
Sky Reading	114
Alien Invaders	120
The Measure of a Mountain	127
Ask the Scientist	134

Strange Lights Above Thunderstorms

by David K. Hill

Forget about the circus. The greatest show on or above Earth is the summer thunderstorm. It climbs the horizon on a sticky afternoon, a piled-up cloud heap taller than Mt. Everest and as white as Antarctic ice. Swallowing blue sky, the storm moves in from the west with a prolonged grumble of thunder. A whoosh of cold air drawn from 40,000 feet (12,200 meters) sets trees in motion, and outsized raindrops pelt the ground. The cloud's blue-black underbelly forms a dramatic backdrop for vivid, branching stabs of lightning.

It's even better at night. Pulses of lightning interrupt the darkness, and crescendos of thunder cut through the rain patter. And when the storm retreats off to the east, its thunder no longer audible, the intricacies of the lightning strokes become visible again as they arc through the clouds, tracing spiderweb trails of electricity.

NEW OPTICAL PHENOMENA

But it may be that the showstopper has been playing out unseen, on a stage high above the cloud tops. In the past few years, scientists have captured images of mysterious new optical phenomena lighting up the night sky miles above thunderstorms. The immense flashing lights extend from above the clouds out to the very edge of the atmosphere. Scientists are simultaneously entranced and mystified. "This is a bona fide baffler of a scientific problem," says John Molitoris, a physicist at the Lawrence Livermore National Laboratory in Livermore, California. "This is one you can really sink your teeth into."

For years, airplane pilots have talked about seeing strangely colored flashes and columns of light above thunderstorms. And for years, scientists greeted these reports with raised-eyebrow skepticism. The flashes were supposedly happening in a part of the atmosphere that researchers had assumed was devoid of dramatic optical phenomena, says physicist Davis Sentman of the University of Alaska in Fairbanks.

FANTASTIC LIGHT DISPLAYS

A case of scientific serendipity in July 1989 overturned that assumption in a hurry. A physicist named John Winckler, of the University of Minnesota in Minneapolis, was testing a special low-light video camera that was going to be sent up on a research rocket. He pointed the camera out the window of his laboratory and recorded the sky above a thunderstorm that was rumbling over Lake Superior, some 155 miles (250 kilometers) north of the lab. In a review of the videotape, a bizarre flash in the clear air above the thunderstorm appeared on the monitor. It didn't resemble ordinary lightning at all, but looked like twin fountains of light licking up into the sky to a height of 12 miles (19 kilometers) above the storm.

This single observation of a strange optical flash sparked intense interest among atmospheric physicists. It caught the attention of the National Aeronautics and Space Administration (NASA) as well. Could this newly discovered phenomenon pose a threat to the space shuttle at takeoff or during its descent to Earth, perhaps by disrupting electrical systems on board? So later in 1989, William Boeck of Niagara University in New York and NASA scientists at the Marshall Space Flight Center in Huntsville, Alabama, began to painstakingly review hours of videotape imagery taken from the space shuttle, looking for similar flashes. The search proved fruitful. "We've turned up about 20 of these events," says Boeck. He also notes that the flashes were always preceded by lightning in the cloud below.

During the summers of 1993 and 1994, scientists took to the sky in research jets to hunt flashes. A team from the University of Alaska, led by Sentman and Eugene Wescott, trolled the skies above the American Midwest with low-light-level video cameras. In 1994, using a color camera, they found that the flashes were blood red and extended even higher than the one Winckler had observed. They reached some 60 miles (100 kilometers) above Earth, into the ionosphere, a part of the outer atmosphere where the air

Unusual optical phenomena now known to be linked to lightning include the blue jet (inset), a flash or column of light shooting up from the top of a thunderhead, and sometimes climbing 20 miles above the clouds.

is extremely thin. Some of the flashes appeared to dangle bluish tendrils beneath them like hovering jellyfish. They lasted for only about 16 milliseconds before disappearing. Hence, the scientists named these flickering specters "red sprites."

The Alaska team captured a second class of optical flash on video, which they named the "blue jet." Unlike red sprites, which seemed to materialize briefly high above the clouds, blue jets sprouted right out of the tops of thunderheads like weird, ephemeral flowers. Some blue jets were narrow columns of light that widened gradually as they reached up to a height of 20 miles (32 kilometers) above the cloud top. Others resembled the burst of spray from a whale's blowhole. The blue jets streaked into the sky at speeds of up to 223,000 miles (360,000 kilometers) per hour.

SEEING IS BELIEVING

How were these fantastic light displays ignored by science for so long? Their brevity is one reason—near the threshold of human visual resolution. Also, the phenomena are often shielded from viewers on the ground by layers of intervening cloud.

Tips for Sprite Watchers

Can a person go out and view red sprites with the naked eye? Yes, says Walter Lyons, a meteorologist with Mission Research Corporation in Santa Barbara, California, if you know where to look. Viewers must dark-adapt their eyes, and have a distant thunderstorm to look at with no intervening clouds, in a part of the sky without city lights. Also, he says, you have to keep yourself from looking where the eyes are naturally drawn—the lightning flickering within and below the distant clouds. "You have to look above [the storm]. But if you stand there long enough, sooner or later you'll start seeing sprites."

Davis Sentman, a physicist at the University of Alaska in Fairbanks, is talking with the National Aeronautics and Space Administration (NASA) about creating a "sprite watchers' guide" with similar viewing tips. He is also planning a "sprite watchers' reporting system" on the World Wide Web, for people to report sightings.

What about blue jets? Eugene Wescott from the University of Alaska says they are much harder to spot than sprites. The best viewing would be from a plane flying at or above 43,000 feet (13,000 meters), Wescott says, at least 60 miles (100 kilometers) away from the storm. And even that could be a dangerous place to be, Wescott adds, since blue jets may pose a hazard to all types of aircraft.

In all thunderstorm viewing, take cover if lightning strikes nearby. Lightning kills about 100 people every year in the United States. The interior of a car is a safe place to enjoy the show.

And, until 1993, scientists were not looking for them. Pilot reports notwithstanding, there was not "supposed" to be much of anything happening that high in the atmosphere. But when Sentman and Wescott aimed their airborne cameras in the right direction at the right time, red sprites and blue jets blossomed on their film.

Meanwhile, back on the ground, a second team of researchers had set up a sprite-observation site in northern Colorado on Yucca Ridge. It is a place where scientists can watch thunderstorms over the Great Plains, an area prone to intense nocturnal storms. "We have an unobstructed view over the Pawnee National Grasslands. It's grass and elk and nothing else," says Walter Lyons, a meteorologist with Mission Research Corporation in

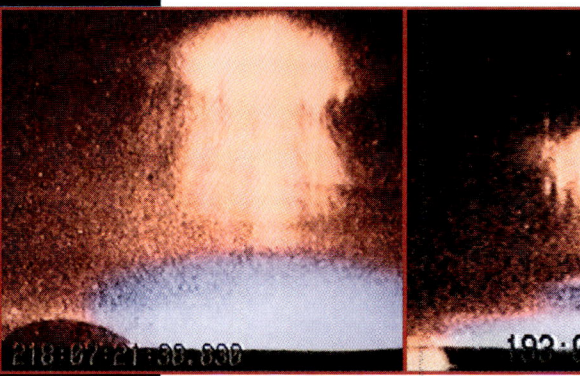

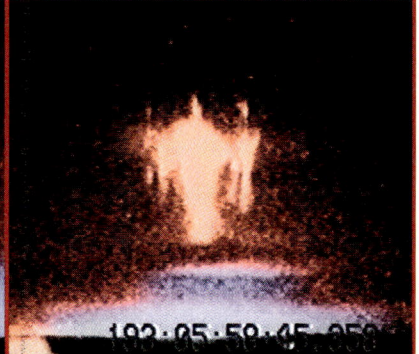

A red sprite can resemble a jellyfish in the ionosphere, especially when tendrils of light develop beneath it (above left). Most sprites last only a few milliseconds, making them very difficult to photograph.

Santa Barbara, California. "It's dark, and it has an absolutely unparalleled view of the prairies to the east."

From their perch over the prairies, Lyons and his team videotaped more than 1,000 sprites dancing over the tops of distant thunderstorms. But they weren't generally seen over ordinary, run-of-the-mill thunderstorms. They seemed to form preferentially over particularly large, long-lasting thunderstorms, the kind that often roam the Great Plains in summertime, bringing torrential rains, damaging winds, and frequent lightning.

The sprites above these giant thunderstorm complexes appeared singly and in groups, lasting from 17 to more than 200 milliseconds. They were up to tens of miles wide and extended to over 40 miles (65 kilometers) above Earth. "And the biggest question is," says Lyons, "What in the world are they?"

Another Layer of Complexity

Lyons and other scientists think the sprites are related to powerful bolts of lightning in the storms below. Ever since Benjamin Franklin risked his life in 1752 by flying a kite underneath a thunderstorm, cloud physicists have struggled to understand exactly how lightning works. They have made progress, building elaborate theories to explain storm electrification, but uncertainties remain. And

Fulgurite—so-called "fossil lightning"—forms when a superheated bolt of lightning bores into the ground or melts sand and rock into the branching shape of the strike. A meteorite impact can also form fulgurite.

now the discovery of red sprites and blue jets adds another layer of complexity.

Scientists know that thunderstorms contain strong vertical winds called updrafts and downdrafts. Updrafts fuel the storm with warm, moist air. These vertical air currents also act to help separate positive and negative electric charges within the cloud, allowing the buildup of electric potential that is relieved by a lightning stroke. Cloud physicists have found that when streams of tiny water droplets and ice crystals jostle with mini-hailstones called graupel, the smaller particles become positively charged, and the graupel negatively charged. The top of a cloud becomes positively charged as updrafts sweep the small ice crystals upward. Graupel falls into lower portions of the cloud, accumulating negative charge there. As this negatively charged region moves over Earth below, it attracts positively charged particles on the ground, like iron filings drawn to a magnet. This "shadow" of positive charge follows the cloud as it moves overhead.

When the charge difference between the cloud and ground, or between different parts of the cloud, becomes large enough, a giant spark—called lightning—flashes between them, attempting to restore the charge balance. Lightning flashes about 100 times per second around the globe. Each stroke heats a 1-inch (2.54-centimeter)-diameter column to an incredible 50,000° F (27,800° C). The air, not accustomed to being five times hotter than the surface of the Sun, expands explosively, sending out shock waves that we hear as thunder.

Lightning can flash entirely within the cloud, between the negative part of the cloud and the positive Earth shadow below, or even between the positive region at the top of the cloud and negatively charged Earth regions away from the storm. This last type of lightning is called a "positive" cloud-to-ground stroke. Such strokes account for only about 5 to 10 percent of all cloud-to-ground strokes, but scientists think these positive lightning strokes may somehow cause sprites.

The Schumann Resonances

One reason they think so is because positive lightning strokes emit a distinctive pulse of energy in extremely low-frequency radio signals. Earle Williams and collaborators at the Massachusetts Institute of Technology (MIT) in Cambridge have been monitoring these radio frequencies in their study of a low-frequency "hum" caused by lightning around the globe. The hum, called the Schumann resonances, is a natural vibration in the huge cavity between Earth and the ionosphere. "The Earth is a good conductor, the ionosphere is a good conductor, and we live in an insulator in between," says Williams. Every lightning bolt, anywhere in the world, leaks a bit of energy into the

Schumann resonances. A big positive cloud-to-ground stroke, says Williams, "rings the entire Earth-ionosphere cavity" and causes a noticeable jump in resonance.

While monitoring the Schumann resonances from a station in Rhode Island, the MIT group would talk to the Yucca Ridge sprite watchers by telephone. When someone in Colorado would yell "Sprite!" the Schumann resonances recorder would jump simultaneously. The bursts in radio waves and sprite locations were also coincident with the location of large positive lightning flashes, as retrieved from the National Lightning Detection Network, a system that uses an array of electromagnetic sensors to map lightning strokes in the continental United States.

Searching for Answers

Williams says that the largest positive lightning flashes might cause a brief, intense increase in the electric field above the thunderstorm, enough to cause a discharge of electricity and set the upper atmosphere glowing for an instant. Sprites, therefore, might be the upper-atmosphere counterpart to the lightning found near Earth's surface. The different responses to electric-field breakdown are due to the huge difference in the thickness of the atmosphere in the two regions. "The air density [where sprites occur] is 10,000 times less than the density where ordinary lightning takes place," says Williams. So instead of the narrow, concentrated channel carrying current in lightning, sprites may be the visual expression of a diffuse electrical discharge.

Other theories have been proposed to explain sprites. Yuri Taranenko of the Los Alamos National Laboratory in New Mexico suggests that a cosmic ray from space could trigger an avalanche of electrons in the enhanced electric field above thunderstorms. This model assumes that a high-energy electron resulting from the collision of a cosmic ray with an air molecule will then smash into other air molecules, liberating additional electrons in an upward-propagating cascade. The electron avalanche would energize air molecules and set them glowing, Taranenko proposes.

Explanations for blue jets have proved more elusive. Theorists are hindered by a lack of observations. While thousands of sprites have been seen, fewer than 100 blue jets are in the books. But airline pilots report that Panama may be a hot spot for blue-jet formation, and Wescott and Sentman hope to mount an expedition there in the near future.

Potential Impact of Sprites

Although sprites were unknown until recently, scientists now wonder if they affect the atmosphere in important ways. For example, they could alter the chemical composition of the upper atmosphere. Although short-lived, sprites could set off lasting changes. "Think of a spritz from a perfume bottle," says Sentman. "It only takes an instant, but the effects last for a long time." For example, researchers speculate that sprites could affect the ozone cycle by changing concentrations of nitrogen oxides in the high atmosphere. Increased levels of nitrogen oxides could increase the concentration of ozone. Since ozone is a potent greenhouse gas, scientists also wonder if sprite-induced changes in ozone could affect the global temperature balance.

But a crucial question related to the potential impacts of sprites is: Are they a worldwide phenomenon? To investigate this question, in February 1995 the University of Alaska team went to South America to search for tropical sprites and jets. Operating out of Lima, Peru, they flew over the Amazon rain forest and its ubiquitous thunderstorms. They found sprites, capturing a couple of dozen on video. This result is important because tropical regions like the Amazon Basin account for most of the world's thunderstorms. If sprites turn out to be important to atmospheric chemistry and/or global climate change, much of their impact will likely come out of equatorial regions.

Sprite Chasers

While in South America, the scientists didn't see any blue jets, but they had an electrifying encounter with a thunderstorm. Returning from a mission one night, they found themselves engulfed by a storm. "We began to see Saint Elmo's fire coming off the wing pods, the taillets, and the tail," says Wescott, describing the luminous manifestation of a strong electric field. Suddenly there was a blinding flash of light—the plane had been struck by lightning. "We really didn't want to be there," says Wescott, but they had the presence of mind to take spectrographic imagery of the Saint Elmo's fire. Luckily, there was no damage to the plane or its scientific cargo.

Lyons and collaborators plan to stock the Yucca Ridge observation site with sophisticated equipment that they hope will help resolve some of the mysteries of sprites. It's time for scientists to move past observations such as "it looks like a huge jellyfish," says John Molitoris of the Lawrence Livermore National Laboratory. "The question is, *Why* does it look like a huge jellyfish?"

To better probe the details of sprites, Molitoris plans to install a computerized remote-sensing platform in Colorado, interfaced with Lyons' equipment. Called IROCS (InfraRed Optical Camera System), it couples a superfast camera with a battery of other sensors. All can be aimed at the same part of the sky, and when a sprite appears, can take simultaneous measurements of light emissions at different wavelengths. This could yield information about chemical changes in the atmosphere caused by sprites. And by operating a faster camera than used in previous sprite observations—imaging at up to 2,000 frames per second instead of 60—Molitoris hopes to capture details of sprites' evolution.

The next few years should bring huge advances in our understanding of sprites and jets, as observers accumulate new information and theorists labor to make sense of it all. Right now, Sentman says, it is still unclear whether these newly discovered phenomena will turn out to be important parts of the atmosphere's internal dynamics, or just an interesting sideshow. "Scientifically, sprites could turn out to be extremely important," he says. "Or they could be in the same category as rainbows [and] halos..."

These things, Sentman says, may not play significant scientific roles, but their beauty provides inspiration for "artists, poets, and lovers." Even if red sprites and blue jets turn out to be mere optical curios, they will still be additional reasons to watch a thunderstorm light up the night sky.

People seeking to view a red sprite are advised to look above a distant nighttime thunderstorm. For all practical purposes, a blue jet (inset) can only be seen from an airplane.

Mountain Isles, Desert Seas

by James Bishop, Jr.

Many worlds converge in the area of southeastern Arizona known as the Madrean Archipelago. There geology, climate, and biology conspire to form "sky islands" populated with myriad plants and animals, many of which are endemic or wildly defiant of what would otherwise be the limits of their ranges.

The names of the mountains making up the archipelago—the Huachucas and Chiricahuas, the Tucsons and Santa Catalinas, the Dragoons and the Rincons, the Santa Ritas, the Pinaleños, the Peloncillos, and the Animas—reverberate with legends spanning the human history of North America. Twelve thousand years ago, the indigenous inhabitants there used fluted spears to hunt giant ground sloths. Recent centuries have seen the Spanish conquistadores, the Apache warriors Cochise and Geronimo, and the outlaws and lawmen of Tombstone. But human history pales beside the region's natural history.

In Arizona's Madrean Archipelago (left), geology, climate, and biology have conspired to form a "sky island" complex that supports an extraordinary diversity of plant and animal life.

A Sculpted Archipelago

Between the Miocene epoch in Earth's geologic history (about 15 million years ago) and the end of the Pliocene epoch (about 2 million years ago), a period of tectonic uplift and violent volcanism built and sculpted the archipelago we see today. The uplifted fault blocks are composed of intermingled metamorphic and sedimentary rocks, intrusive granites, and extrusive volcanic rocks. This complex and varied geology produces a wide range of surface-rock types and soil chemistries, which accounts for much of the biological variation between the region's different mountain ranges.

Millions of years of block faulting, folding, and erosion have broken these mountain ranges into many relatively small and disconnected parts, or islands, and also have filled the valley troughs and basins with alluvial silt, sand, and gravel. The resulting landscape is a system of elevated islands linking two major mountain chains—the Rocky Mountains and the Sierra Madre Occidental—in a sea of low-lying desert and grassland basins.

Stepping-Stones

Because of their peculiar geologic and climatic properties, these islands act as "stepping-stones," allowing many southern Neotropical species of plants and animals to extend their ranges northward from the Sierra Madre Occidental. Northern species may travel southward from the Rockies.

About 20 sky-island complexes are known to exist worldwide, most of them in Africa and South America. They all share many characteristics, most notably their large number of endemic species. But only the Madrean sky islands contain a variety of plants and animals that would normally range from subtropical through temperate latitudes. Only there is a desert porcupine likely to meet a forest-dwelling coati.

Formation of Biotic Communities

During the Tertiary period in Earth's geologic history, North America had three very broadly defined plant associations. The northern half of the continent was covered with cold, temperate coniferous and mixed-deciduous plants (the Arcto-Tertiary Geoflora). The south, on the other hand, supported broad-leaved evergreen plants adapted to warm and wet conditions (Neotropical-Tertiary Geoflora). Finally, in more-arid areas within the southern portion was a more dry- and heat-adapted plant association called the Madro-Tertiary Geoflora.

North meets south in the Madrean Archipelago. The Mount Graham red squirrel (above left) belongs to a species more typical of the Rocky Mountains to the north. By contrast, the elegant trogon (left) and the twin-spotted rattlesnake (above) are common farther to the south.

By the end of the Tertiary period about 2 million years ago, the northern temperate plants and the southern warm-wet plants were retreating north and south, respectively,

EARTH AND THE ENVIRONMENT 111

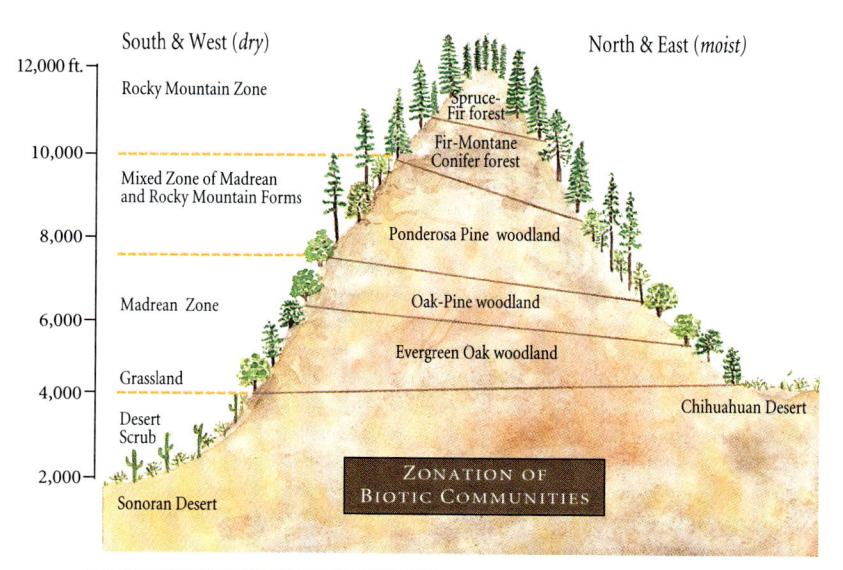

South & West (*dry*)   **North & East** (*moist*)

- 12,000 ft.
- Rocky Mountain Zone
- Spruce-Fir forest
- Fir-Montane Conifer forest
- 10,000
- Mixed Zone of Madrean and Rocky Mountain Forms
- 8,000
- Ponderosa Pine woodland
- Madrean Zone
- 6,000
- Oak-Pine woodland
- Evergreen Oak woodland
- 4,000
- Grassland
- Chihuahuan Desert
- Desert Scrub
- 2,000
- Sonoran Desert

ZONATION OF BIOTIC COMMUNITIES

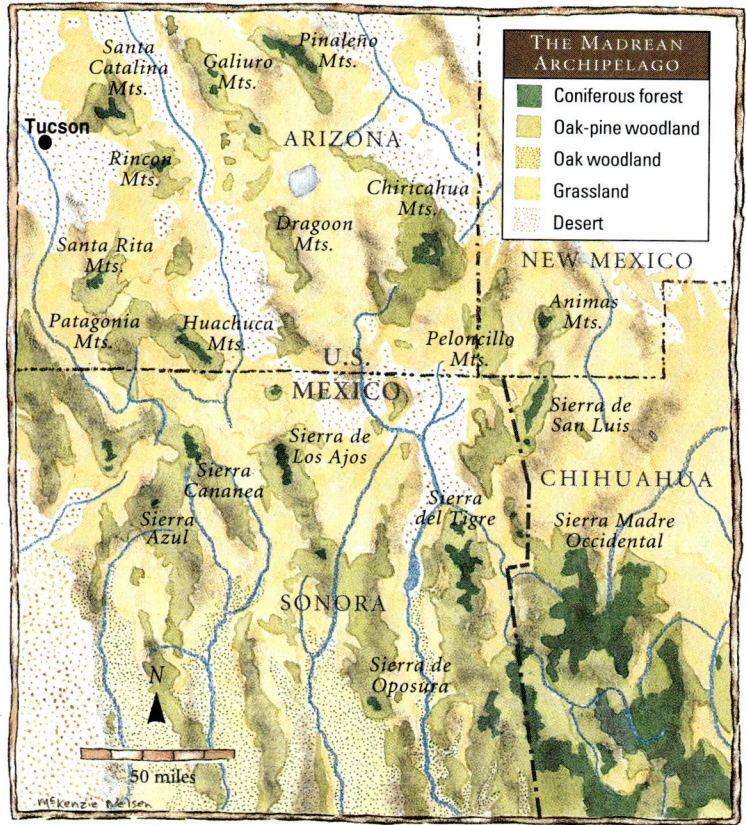

THE MADREAN ARCHIPELAGO

- Coniferous forest
- Oak-pine woodland
- Oak woodland
- Grassland
- Desert

Tucson • ARIZONA

Santa Catalina Mts., Galiuro Mts., Pinaleño Mts., Rincon Mts., Chiricahua Mts., Dragoon Mts., Santa Rita Mts., Patagonia Mts., Huachuca Mts., Peloncillo Mts., Animas Mts., NEW MEXICO, U.S., MEXICO, Sierra de San Luis, Sierra de Los Ajos, CHIHUAHUA, Sierra Cananea, Sierra del Tigre, Sierra Madre Occidental, Sierra Azul, SONORA, Sierra de Oposura

N — 50 miles

McKenzie Nelson

---

the piñon-juniper communities moved onto the lower flanks, and the basins were invaded by deserts and grasslands.

So today a descent from the mountaintops passes through a series of stacked biotic communities: montane coniferous forests of Douglas fir and Engelmann spruce; woodlands of coniferous oak and Chihuahuan and Apache pine; deciduous oak savanna with blue oak surrounded by short-grass prairie and grama grass; tropical deciduous forest with tree morning glory; subtropical thorn scrub dominated by saguaro cactus; and finally the creosote and mesquite of the subtropical desert. As a result of this diverse environment, bears can feed on prickly-pear cactus in the morning and grass roots in semi-alpine meadows in the afternoon, having traversed five biotic communities in just a few hours.

COMPLEX DIVERSITY

"The islands share many of the same habitat types, life zones, and the same environmental forces affecting the distribution of plants and animals," says Mark Fishbein of the Department of Ecology and Evolutionary Biology, University of Arizona in Tucson. "On the other hand, surprising disparities exist between them. In that respect, these mountain islands, adrift in desert seas, are unbelievably complex."

Each island's latitude, elevation (which influences both temperature and moisture), exposure, soil type, and slope gradient determine which combination of species it can

and the dry-loving Madro-Tertiary Geoflora began to dominate the U.S. Southwest. At the same time, remnants of the retreating biota moved upslope into moister mountain refugia in the Madrean Archipelago. As conditions became hotter and drier, the evergreen woodlands were forced higher into the mountains,

A light snow cover in the Santa Rita Mountains (left) is no surprise, especially given the fact that the formations of the Madrean Archipelago support everything from spruce forests to parched deserts (illustrations, facing page).

host from the Chihuahuan Desert to the east, the Sonoran Desert to the west, the Rockies to the north, and the Sierra Madre Occidental to the south.

Ten thousand years ago, as the plants and animals of the area were forced to higher elevations on the Madrean islands in order to survive, populations became separated from one another by impenetrable stretches of desert. The inability of isolated populations to interbreed, and the effects of genetic drift, produced the great number of plant and animal forms found only there.

Arizona's sky-island mountains, like the Galápagos or the Hawaiian Islands, act as "isolated cradles of evolution," says Peter Warshall, consultant to the University of Arizona. The tiered order of biotic communities allows for vertical migration up- and downslope, but the broad valleys between mountain islands act as barriers to horizontal migration as effectively as oceans do.

Because they have been isolated a relatively short time—only about 10,000 years—most of the islands' endemics are subspecies, such as the Mount Graham red squirrel *(Tamiasciurus hudsonicus grahamensis)*, found only atop the 10,720-foot (3,300-meter) Mount Graham, and the Arizona ridge-nosed rattlesnake *(Crotalus willardi willardi)*, found only in the Huachucas and in a few other nearby mountain ranges. Right next door, in the Animas and the Peloncillo mountains, is another subspecies *(C. w. obscurus)* that is separated from a third subspecies in Mexico. But some have evolved into full species, such as the Arizona shrew *(Sorex arizonae)*, first described in 1977 and living on four of the archipelago's islands, and the Ramsey Canyon leopard frog *(Rana subaquavocalis)*, found only in one 5-mile (8-kilometer) area in Ramsey Canyon in the Huachuca Mountains.

Altogether, Arizona's sky-island complex harbors 2,000 known plant species, more than 75 reptile species, and 265 bird species, 30 of which—like the elegant, or coppery-tailed, trogon *(Trogon elegans)*—have their northern limits within the sky islands. The Madrean Archipelago also hosts about 90 native mammal species, the most diverse array found anywhere in the United States.

A Collision Course

While research is just beginning to illuminate the many biological mysteries of Arizona's sky islands, other forms of human activity are on a collision course with the region's biodiversity. Ranching, skiing, logging, and mining have brought more change to the sky islands in a century than nature managed in millennia.

But, led by the Sky Alliance and The Nature Conservancy, there are stirrings of a conservation effort. By any measure, the greatest challenge facing the islands is the reconciliation of humans and nature. And there, where an entire species may consist of only a few hundred individuals, rapprochement cannot come too soon.

Sky Reading

by Scott Weidensaul

A humid summer Pennsylvania day was coming to a close, an eternity of hazy, blistering sun. But when I glanced out the windows, there was a gray smudginess to the western horizon. A cold front, which had been sulking over the Ohio Valley for days, was finally pushing across the state, roughhousing a line of storms before it.

A Yardful of Clues

I was hardly surprised. Even if I had not checked the barometer hung on my office wall, I would have known that the air pressure was dropping, thanks to my aching sinuses. But it was more than that. The robins and grackles nesting in the trees outside had been bickering more than usual all afternoon, and the green frogs that hold court in my pond were plunking away with their calls. Old weather proverbs say that when the fowl in the barnyard squabble, or when frogs croak at midday, rain is near at hand.

Just to make sure, I flicked on my battered weather radio, a pocket-sized unit pretuned to a local National Weather Service (NWS) frequency.

"The National Severe Storms Forecast Center has issued a severe-thunderstorm watch for much of eastern and central Pennsylvania, effective until 9:00 P.M.," the radio said. "A strong cold front moving across Pennsylvania is triggering powerful thunderstorms in the warm, moist air ahead of it. Some of these thunderstorms may produce strong winds and hail. Persons involved in outdoor activities should pay close attention to weather conditions around them."

Long before the first thunder reached me, it was clear a storm was coming my way. The cows in my neighbor's pasture were all huddled together, and the bees had vanished from the garden. The silvery bottoms of the maple leaves showed as the strengthening wind turned them over, and it was then I remembered an old ditty I had learned in childhood: "When leaves show their undersides, be very sure that rain betides."

A few minutes later, as the sky grew an ominous black, I turned from folklore to science for a more precise forecast—and was glad I did. The weather radio informed me that a tornado had been detected about 24 miles (40 kilometers) west of my home. As sudden, violent winds pummeled the house, I kept an eye peeled for a funnel cloud, and one ear to the radio, prepared to take refuge in the basement if either gave warning of a tornado.

The storm passed in an hour with only a few broken branches in the yard, but three other twisters touched down within an hour's drive of my home. Remarkably, thanks to a powerful new sensing device more than 100 miles (160 kilometers) away, the tornadoes were pinpointed even before they emerged from the storm, giving vital warning to those in their paths.

High-Tech Storm Forecasts

While weather lore like watching birds and clouds is as old as humanity, the science of meteorology is undergoing some exciting advances. To learn what is on the horizon (so to speak), I talked to Tom Dunham, a meteorologist at the National Weather Service office in State College, Pennsylvania, the same center that tracks the storms that cross my area.

Dunham explains that the NWS is in the middle of a major overhaul of its operations, from installing new equipment to streamlining its offices. "The weather service is getting new observation systems, satellites, supercomputers, new weather-radio systems—a complete [upgrade] of equipment and technology over the next 10 years. But the most significant advance, the keystone to this whole modernization, is the Doppler weather radar," Dunham says.

Government meteorologists have been using radar since the late 1950s to detect storms; in fact, the last of those rugged original units was scheduled to be replaced in 1996. Radio waves are broadcast into the atmosphere, where they bounce off precipitation; the echo forms a fairly crude picture of what is happening in the air. Doppler radar can detect much finer details, Dunham says, "but the thing that is special about it is that Doppler radar sees the wind, especially tornado winds in thunderstorms."

This state-of-the-art radar relies on the Doppler effect—the same principle that makes the sound of a car horn rise in frequency as the vehicle approaches, then fall sharply as the car passes—and is named

Until relatively recently, weather forecasting meant simply looking at the sky and recalling the appropriate proverb. Today, Tom Dunham (above) and other meteorologists use high-tech methods to develop far more precise predictions.

after its discoverer, 19th-century Austrian physicist Christian Doppler. Wind blowing toward the radar unit increases the frequency of the reflected waves, while wind blowing away lowers the frequency. Measuring the difference, computers working with the Doppler radar can create a strikingly detailed image of what is happening inside a storm.

Such images are critical when it comes to forecasting the most-violent weather, such as severe thunderstorms that may spawn tornadoes. "Tornadoes almost always form

EARTH AND THE ENVIRONMENT

in the mid-level of a thunderstorm and work down," Dunham says. "The beauty of this is that the radar can see the tornado before it emerges, before it's visible. It used to be that in almost all cases, we had to have a visual sighting before we could issue a tornado warning." Now forecasters can save critical minutes—and perhaps lives.

Although the National Weather Service is cutting back from 250 forecasting offices to about 118, all of the remaining offices are expected to have their own Doppler array by the summer of 1996—a change from the past, when even some urban sites lacked their own warning radar, Dunham says. That lack led to forecasters missing a number of fast-developing weather disasters, like the Big Thompson flash flood in Colorado in the 1970s, or the Plainfield, Illinois, tornadoes that hit near Chicago in the early 1990s.

Because the Doppler units have a functional range of more than 100 miles (160 kilometers), the overlap between offices

Meteorologists can predict the formation of a tornado with remarkable accuracy thanks to Doppler radar, which works with computers to create a detailed image of the dynamics inside a storm.

means that only a few areas of the country (mostly in remote, mountainous areas of the West) will not be covered by the new technology, Dunham says.

SATELLITES AND SENSING SYSTEMS

The changes have been not only in radar. In the late 1980s, one of the two satellites that hold a stationary orbit above the United States failed, leaving weather forecasters in the dicey position of relying on just one satellite with no backup.

In 1994, the satellite known as GOES-8 (GOES stands for Geostationary Operation Environmental Satellite) was launched, and GOES-9 was launched in May 1995.

"Having two new ones up there was a big relief," Dunham admits. "We're probably as comfortable with our satellites as we've been in a long time."

Weather Folklore and Proverbs

You do not need to be a meteorologist to predict the local weather, at least in the short term. Much of the old weather folklore of our ancestors (often framed as rhymes, to make them easier to remember) is rooted in fact. These rules of thumb can provide the satisfaction of knowing what the sky holds—without turning on the television or radio.

- **West wind dry, east wind wet.** In North America, the prevailing winds are from the north and west, ushered in by dry high-pressure systems. Storms, on the other hand, often bring rain on east winds. (Exceptions are the Pacific coast and Florida, where west winds bring moisture off the ocean.)
- **If the new moon holds the old moon in her lap, expect fair weather.** Only when the sky is clear and the air stable is it possible to make out the dark portion of a crescent moon—conditions that occur in the middle of a high-pressure system. Turbulent, unstable air, on the other hand, makes the stars twinkle, a sign of impending foul weather.
- **When the night has a fever, it dries in the morning.** This saying, attributed to Native American peoples in New England, is highly reliable—rising temperatures overnight almost always bring rain within a few hours.
- **A ring around the Moon means snow or rain.** As storms approach, the first clouds to overspread the sky are often the highest bands of cirrostratus—sheets of minute ice crystals that fog the night sky, creating a halo around the Moon.
- **Mare's tails and mackerel scales make tall ships carry low sail.** The "tails" of this old saying are cirrus clouds, the extremely high clouds that often take the shape of long streamers, and altocumulus clouds, which can look like wide expanses of fish scales. These patterns often show up in advance of a warm front, anywhere from 12 to 36 hours before bad weather—when sailing craft would furl most of their sails.
- **If the Moon's face is red, of water she speaks.** Attributed to the Zuni people of the American Southwest, this is a poetic rendering of a meteorologic truth—an orangish or reddish Moon often presages rain, because a moist atmosphere causes the color.
- **When dew is on the grass, rain will never come to pass.** A clear sky, dry air, and falling temperatures on summer nights usually bring heavy dew, which is a fairly reliable sign of clear weather for the next 24 hours or so.
- **Falling smoke means rain.** On a clear day with high pressure, smoke usually rises high up into the sky. As low pressure—and rain—approach, however, smoke hugs the ground.
- **Red sky at night, sailors' delight; red sky at morning, sailors take warning.** This old example of English seafaring lore (which is widely thought to have biblical origins) is one of the best-known bits of weather wisdom—and the most frequently misunderstood. A clear sunrise or sunset is usually reddish, and a sign of fair weather. If high or medium-altitude clouds are moving in, however, the low Sun may turn the entire sky red. Here is another version of the same proverb:

> Evening red, morning gray
> sends the traveler on his way.
> Evening gray, morning red
> brings the rain down on his head.

An amateur meteorologist can develop surprisingly accurate short-term weather forecasts simply by observing the appearance of the Moon.

Home Weather Instruments

Keeping track of your local weather, and doing your own short-term weather forecasting, are much more enjoyable and accurate if you have several simple instruments.

• **Thermometer.** Buy a quality mercury thermometer, and place it out of the sun and wind. Do not mount the thermometer against the outside of a window, because heat from the house may affect the reading in winter.

Proper instrumentation is a must for record-keeping buffs. Some devices measure several weather parameters simultaneously.

• **Barometer.** A barometer measures air pressure in inches (or millibars) of mercury, usually with a needle that rises in high pressure and falls in low pressure before a storm. Because altitude also affects barometric pressure, a barometer must be set for any elevation other than sea level.

• **Wind indicator.** An ornate weather vane works well, but so does a simple streamer of nylon cloth mounted on a tall pole, with crossbars indicating the four directions (use a compass to calibrate them). Be sure to mount the indicator away from trees or buildings, which will interfere with the true wind direction.

• **Hygrometer (humidity meter).** A hygrometer measures the amount of moisture in the air, which is usually expressed as *relative humidity*—the percentage of water vapor the air can hold, which increases as the temperature rises. When the relative humidity is 60 percent, it means the air contains 60 percent of the water vapor it is capable of holding at that temperature.

The simplest hygrometer is a wet- and dry-bulb thermometer set—two mercury thermometers with a cloth wick (such as a shoelace) covering one bulb and leading to a reservoir of water. The drier the air, the faster the water evaporates, cooling the wet bulb; you will need a conversion chart to determine the humidity percentage.

• **Rain gauge.** Most garden stores sell inexpensive plastic rain gauges, which can be mounted on a fence post or stake. Place the gauge somewhere out of the wind, but not underneath trees that will deflect rain.

• **Weather radio.** Whether in pocket-sized or tabletop models, these radios are pre-tuned to three frequencies used by the National Weather Service (NWS) for forecasts and severe-storm warnings; one or more frequencies can be picked up in most areas of the country. Some units can be automatically activated in the event of a severe-storm forecast.

Among the other technological advances being installed by the National Weather Service are:

• The Automated Surface Observing System (ASOS), which consists of an array of automatic sensing devices to measure rainfall, temperature, wind speed, cloud height, and other atmospheric conditions. The ASOS "platforms," as they are called, are being installed at airports nationwide, and have often replaced local, staffed weather-service offices.

Not everyone is happy about that change, Dunham says, in part because the ASOS has difficulty detecting freezing rain and cannot observe thunderstorms. But because any airport with air-traffic control has human observers to watch out for storms, and because the ASOS will provide 24-hour data from more locations than is now the case, the system is expected to be an overall improvement, he says.

• The Advanced Weather Interactive Processing System (AWIPS), a sophisticated network that will handle virtually all the computer operations in a weather office, replaces four separate systems now being used. AWIPS should be especially helpful when newer computerized forecasting models come into use in the not-too-distant future,

A ring or halo around the Moon indicates the presence of cirrus clouds in the atmosphere, an often-accurate indicator of stormy weather just around the corner.

Dunham says, allowing meteorologists to predict local weather events like thunderstorms with unprecedented precision.

PROVERBIAL STORMS
Such technological wonders are fine, but what of the ancient weather wisdom that served our ancestors through the centuries? Many of the short-term proverbs—those that predict the weather a day or two in advance—have solid roots in science, Dunham says.

The best known of all, perhaps, is the sailor's rhyme *Red sky at morning, sailors take warning*. "You usually get high-level cirrus [clouds] in advance of a storm. Those clouds can turn the sky red when the Sun comes up and shines on them," Dunham said. The same high, thin clouds can create a halo or ring around the Moon at night—a time-honored sign of coming storms.

Weather folklore is far less accurate when it seeks to predict long-term trends, such as the severity of the coming winter. "Every year, you get the woolly-bear forecasts," Dunham says, referring to the brown-and-black caterpillars whose bands are supposed to presage the winter snows.

"I spent 15 years working in Buffalo, and the woolly-bear forecast I remember best was in 1977. That fall, one of the old-timers said the color of the woolly bears was so mild, there would probably be no winter at all," Dunham says. "Instead, 1977 had one of the worst blizzards in memory."

Contrary to folklore, scientists insist that absolutely no relationship exists between the appearance of woolly-bear caterpillars and the severity of the upcoming winter.

(Interestingly, there is little consistency even among those who swear by woolly bears. In some regions, a thick black band is supposed to mean harsh weather; in others, a thin black band. In still others, it is the richness of the rusty-brown color that is supposed to be the clue. Entomologists, however, point out that the black stripes on a woolly bear get progressively thinner through the fall, as the caterpillar ages and molts, and have nothing to do with weather.)

Most of the long-term weather folklore—that a heavy nut crop or thick onion skin means a bad winter, for instance—has more to do with conditions in the past growing season than the months ahead, Dunham explains.

After the storm had passed with its winds and tornadoes, and I was outside cleaning up the scattered branches and leaves left in its wake, I realized that my headache had gone away with the rising air pressure. The bees were once again at work among the flowers, the robins and grackles had settled down to a truce, and the cows were scattered in the pasture, grazing.

I did not need sophisticated satellites or radar to tell me that there was nothing ahead but fine weather.

EARTH AND THE ENVIRONMENT

ALIEN INVADERS

by Jenny Tesar

Asian cockroaches in Europe, Brazilian pepper trees in Florida, Kenyan lizards in Hawaii, African killer bees in Mexico, American box elders in China, Norway rats everywhere. These world travelers should have stayed home. Instead, they have found new homes and are wreaking environmental and economic havoc.

Unwelcome Exotics

Nonindigenous species—that is, species living where nature did not intend them to be—are often referred to as aliens, or exotics. These animals and plants seldom slip quietly into their new homes. Arriving in areas lacking the predators that limit their populations back home, aliens often multiply rapidly. Soon they are battling native species for food and living space. The natives seldom come out on top. Rather, populations of native species die out, leading to a loss of biodiversity. Instead of habitats filled with a great variety of species, the alien populations take over, creating monocultures suitable for only a few cosmopolitan species.

In some cases, organisms depend on natural mechanisms to move into new territories. For instance, seeds and insects may be carried to a distant region by the wind or hitch a ride on birds. In the great majority of cases, though, humans are responsible for introducing species into new habitats. Some dispersal is inadvertent. For example, plankton and mollusks may be transported from one body of water to another in seawater carried as ballast by cargo ships. As the ships unload their ballast after arriving at their destinations, organisms in the water find new homes, especially if salinity and other conditions are favorable. Many introductions of exotics are deliberate, however. European settlers in the New World, wanting to feel at home, imported English sparrows and starlings. Gypsy moths were introduced to the United States in the 1860s by a French biologist who hoped to establish a silk-producing industry in Massachusetts. Eucalyptus trees were imported from Australia to California in 1853 by timber companies that wanted to start tree plantations.

Without natural enemies, introduced species often multiply rapidly. For example, in their native environment, Australian paper-

The introduction of gypsy moths (laying eggs, left) to U.S. soil by a French biologist has proved to be an ecological disaster. Each year, voracious gypsy-moth larvae defoliate and kill innumerable trees.

Alien invaders are nonindigenous species that colonize new habitats—and then wreak havoc. The aggressive starling, for example, was first brought to the United States in 1890; it has since multiplied into the millions, damaged valuable crops, and displaced native songbirds.

bark (cajeput) trees are kept in check by more than 400 species of insect predators. In Florida, however, where paperbarks were introduced for landscaping purposes, the species has no enemies. The paperbarks have infested the Everglades, where in 1995 they were expanding their range at an estimated rate of 50 acres (20 hectares) a day. As the trees dominate an area, they replace a wet-prairie community containing 60 to 80 species of plants with a community containing only three or four plant species.

Similar problems emerged after yellow jackets were accidentally imported into Hawaii in the late 1970s, in a shipment of Christmas trees. Among the insects that these wasps eat in their new home are caterpillars that feed on the pollen of ohia (lehua)-tree blossoms. Fewer caterpillars means fewer adult moths, which means less food for several species of endangered songbirds.

Alien species also cause other types of environmental damage. Goats, imported to Hawaii by Europeans, eat plant shoots and trample the ground, opening the way to erosion. An infestation of goats and burros in the Galápagos Islands has left many native plants destroyed and tortoise nests trampled. Crystalline, a species of ice plant introduced into California, removes salt from the soil, leaving the soil poisoned after the plant dies. Tamarisk trees remove so much water from the soil—as much as 200 gallons (750 liters) per day—that they can transform southwestern wetlands into deserts.

Such interlopers come with a high economic cost. In a 1993 report, the U.S. Office of Technology Assessment (OTA), a congressional research group, estimated that between 1906 and 1991, the 79 most harmful nonindigenous species in the United States caused $97 billion in crop damage and pest-control efforts. The report also estimated that 15 additional species could cause another $134 billion in future economic losses.

The first line of defense against the invaders is prevention. Preventative efforts include educating people on the dangers of transporting species to new lands, and establishing screening procedures at airports and other points of entry. Once a species becomes established in a new environment, a variety of eradication methods may be tried. But these techniques are often ineffective and can be extremely costly.

Water crises in the arid U.S. Southwest have deepened due to the spread of tamarisk trees. These Egyptian imports can absorb up to 200 gallons of water per day.

Power-Happy Snakes

One of the most frightful aliens is the brown tree snake, a native of coastal Australia, New Guinea, and northwestern Melanesia. Sometime in the late 1940s, it found its way to the

The brown tree snake is an unwelcome immigrant to the island of Guam; it is blamed both for the extinction of nine bird species and for frequent power outages.

Pacific island of Guam. Today about 1 million of these snakes populate the island.

A slender snake that grows to more than 8 feet (2.4 meters) in length, the brown tree snake is active at night and hides during the day. It has a definite fondness for slithering up trees in search of birds and bird eggs. The snake has been blamed for wiping out at least nine species of birds native to Guam. It is not averse to eating newborn puppies and kittens, nor to biting sleeping infants (although its venom is not fatal). The snakes also climb utility poles, shorting out electrical lines; every few days, Guam experiences a power outage caused by these snakes.

It is probably too late to exterminate the brown tree snake on Guam, but environmentalists and public officials are hoping to keep it out of the Hawaiian islands, which so far are free of snakes and already have many bird species endangered by other invaders. Planes fly regularly from Guam to Hawaii, and on several occasions in recent years, stowaway brown tree snakes have been found at airfields on the Hawaiian island of Oahu. Hawaiian inspectors and volunteers have formed an organization known as the Snake Watch Alert Team (SWAT) that is constantly on the lookout for brown tree snakes; SWAT members even use snake-sniffing dogs to check incoming cargo and luggage.

Smothering the South

Kudzu is known as the vine that ate the South. It grows rapidly, covering everything from farmland to utility poles and abandoned cars. A native of Asia, it was introduced into the United States by members of a Japanese delegation who planted a kudzu vine to decorate their exhibit at the 1876 Centennial Exposition in Philadelphia. Americans admired the plant, with its large grassy-green leaves and clusters of reddish-purple flowers. They asked the Japanese exhibitors for seeds and seedlings, and planted the vines in their gardens. Then, in the 1930s, kudzu received its most important boost when soil conservationists began using it to control soil erosion in the South. By the 1950s, more than 70 million kudzu seedlings had been planted in the United States.

As promised, the plants stopped erosion and enriched the soil. But without natural enemies, the vines began to grow out of control—as much as 1 foot (30 centimeters) per day. And getting rid of even one plant is not easy. The plant has a monstrous carrotlike root that can weigh 400 pounds (180 kilograms). As many as 50 vines—each of which may be 100 feet (30 meters) long—can grow from one root.

U.S. farmers have been fighting the spread of kudzu for decades. One effective tactic is to use a herd of goats. Researchers at the University of Georgia in Athens found that four goats can consume 1 acre (0.4 hectare) of kudzu in two years. Another tactic might be to make kudzu roots as popular a foodstuff in the United States as they are in Japan and China, where they are used to make breads, cakes, soft drinks, and herbal medicines.

Lake Denizens

Among the organisms that probably reached North America in water from ballast tanks, few have caused bigger problems than the diminutive zebra mussel. Little more than 1 inch (2.5 centimeters) across, this mollusk is so named because of the dark-brown stripes on its light-tan shell. A native of the Caspian Sea, the mussel first appeared in the Great Lakes in the mid-1980s. With few enemies, it quickly spread, not only throughout the Great Lakes, but into rivers such as the Mississippi, Illinois, Susquehanna, and Hudson.

The zebra mussel's rate of reproduction is tremendous. A female, which reaches maturity in less than a year, may produce more than 1 million eggs annually. The larvae that hatch are free-swimming and easily carried by moving water into new environments. As the larvae metamorphose into adults, they develop small tuft fibers called byssal threads. These threads enable the mussels to attach to any hard surface, including rocks, pipes, buoys, plants, piers, and even other mussels. Zebra mussels feed on plankton and algae, depriving native species of these foods. In parts of the Great Lakes where zebra mussels and native clams began to share territory and compete for food, the native clams are now almost gone.

Initially, it seemed that the zebra mussels' voracious feeding habits would be beneficial. By filtering plankton and other suspended particles from water, they improved water clarity; once-murky lakes sparkled and looked attractive. People's enthusiasm soon waned, however. In the clear water, large pike could easily see prey. They gobbled up game fish, including young fry, leaving fewer fish for sport-fishing enthusiasts. Then a more serious problem developed. Clear water meant that sunlight could penetrate to lake bottoms. Algae flourished, creating foul-smelling masses of rotting material that floated to the lake surfaces and onto nearby beaches. In 1994, such a situation developed on Lake St. Clair, located between Lake Erie and Lake Huron, leading Michigan to become the first U.S. state to call out the National Guard to help combat an alien species.

Large colonies of zebra mussels are often found in reservoirs and on intake pipes at electricity-generating plants and water-treatment facilities. In 1991, a Detroit Edison power plant in Monroe, Michigan, reported mussel densities as high as 750,000 animals per square meter (about 11 square feet). Two years earlier, the mussels had clogged the only intake pipe of the water-treatment plant in Monroe, which withdraws water from Lake Erie; as a consequence, the town's industries, businesses, and schools were forced to close for two days. Zebra mussels also colonize on boat hulls, foul the engines of powerboats, and sink buoys.

Tiny zebra mussels have caused massive damage to many North American lakes and waterways. These prolific creatures—a single female can produce 1 million eggs annually—have decimated populations of native clams, destroyed fish spawning grounds, and clogged water-supply systems.

Large sums of money have been spent or budgeted to control the mussels. An estimated $5 billion will be spent to battle these mollusks just during the 1990s. At utilities and water-treatment plants, control measures include using chemicals that are lethal to the mussels, such as chlorine and potassium permanganate. Because zebra mussels are very sensitive to high temperatures, some electric plants are experimenting with the diversion of waste heat into intake structures to kill the mussels or prevent their settlement. Scientists are also studying the possibility of controlling populations by disrupting their reproductive process.

To make matters worse, a second species of zebra mussel has arrived in the United States, and it appears to thrive in warmer waters than its predecessor. Fortunately, some native birds, including mallard and canvasback ducks, feed on zebra mussels. So do gobies, alien fish from Southeast Asian waters. However, many scientists doubt that these predators will be able to control the population of zebra mussels.

Fish Tales

A 1991 OTA report noted that more than half of the fish listed as endangered in the United States were threatened by introduced species. Nearly one-third were threatened by species introduced in conjunction with sport fishing. An alien species is not necessarily an immigrant from a foreign country. Often people introduce a species from one habitat to another habitat within the same country. In the United States, such transplantation has been done extensively with game fish to provide anglers with more-challenging sport. As a result, there has been a major loss of biodiversity in the nation's rivers and lakes.

On rainy days, Floridians may find walking catfish traversing their lawns. These bizarre predators are descendants of Asian fish imported for use in aquariums.

In some cases, the aliens are large predators that devour native fish. In other cases, the aliens eat the eggs and food of native fish. Largemouth and smallmouth bass introduced into Maine's waterways have threatened native Atlantic salmon. The introduction of Pacific salmon into the Great Lakes has harmed native lake trout. In California, European brown trout have replaced native golden trout; in Arizona, several species of trout and catfish have replaced native Yaqui catfish and Apache trout. Such displacements have given at least some state fisheries programs second thoughts about introducing more species to their waters. For example, plans to introduce Pacific salmon into the Delaware River system have been canceled.

Disenchanted hobbyists also create problems when they dump unwanted guppies, goldfish, and other home-aquarium fish into nearby lakes, rivers, and streams. This is a particularly common problem in Florida, where the consistently mild climate and warm freshwater approximate the natural environment of many aquarium favorites. Even piranhas have been pulled from ponds by fishermen!

Then there is the problem of the walking catfish, a native of Thailand. Small albino forms were originally imported to Florida for use in aquariums. In 1965, some of these fish ended up in a canal. Using their strong pectoral fins, they "walked" overland to infest additional bodies of water, breathing atmospheric air through a specialized respiratory organ attached to the gills. The catfish reverted to their original gray coloration; they now grow to lengths of 1 foot (30 centimeters) or more. Today it is not uncommon to see walking catfish wandering through Florida backyards whenever there is a rain shower.

Rampaging Rodents

In Louisiana's numerous swamps, the most-troublesome invaders are nutrias, large rat-like rodents with formidable orange teeth; webbed hind feet; soft, slate-colored fur—and enormous appetites. Millions of these critters, which are native to Argentina, are chewing their way through Louisiana's wetlands, eating just about any plant they can find and destroying major breeding grounds

Nutrias—large, furry rodents noted for their orange front teeth—have invaded Louisiana swamps. These Argentine imports devour delicate aquatic plants and damage valuable cash crops.

for fish and shellfish. Their only predators are alligators and people.

Most of Louisiana's nutria population can be traced back to 13 animals imported in 1938 by E.A. McIlhenny, the Tabasco-sauce magnate, who envisioned a profitable fur business from the animal's pelt. McIlhenny kept his nutrias in cages, and they bred rapidly. Unfortunately, two years later, several dozen of the animals escaped during a hurricane. The rest, as they say, is history.

In the 1970s, Louisiana trappers were taking nearly 2 million nutrias annually. Then the market for nutria furs collapsed, and the rodents' population soared even higher. Today there are as many as 6,000 nutrias per square mile (2.6 square kilometers) in some parts of Louisiana. Efforts are being made to revive the fur trade and to find other uses for the animals. The Louisiana Nature and Science Center has even held "nutriafests," with cook-offs where people prepare dishes using nutria meat. The idea is not as far-fetched as it might sound; in some parts of the world, nutria meat is considered a delicacy.

The beautiful magenta blossoms of purple loosestrife (above) belie the plant's ability to choke out all other vegetation and clog waterways in fragile wetlands. At least 13 U.S. states ban its importation and distribution.

Summer's Purple Bully

Purple loosestrife is a tall, attractive wetlands plant that bears long spikes covered with hundreds of magenta flowers. It grows in a variety of places—in sunny meadows, along stream banks and ponds, and near swamps. A native of Europe, it arrived in North America early in the 19th century,

EARTH AND THE ENVIRONMENT 125

perhaps as its seeds were carried in bales of hay or in a ship's ballast water. The plant was also imported for medicinal purposes; purple loosestrife was considered useful in treating dysentery, diarrhea, ulcers, sores, and wounds. It once was widely used in beautification projects in southern Canada and the northern United States.

But purple loosestrife's beauty conceals a pushy, aggressive personality. A single plant can produce more than 2.5 million seeds annually, enabling the species to spread rapidly. The roots are strong and serve as a storage organ for food; if the aboveground parts are cut or killed by foliage herbicides, new stems grow from the roots. Plant growth is rapid, especially in the seedling stage. As a result, native marsh grasses are crowded out, and diverse habitats are transformed into monocultures.

Although purple loosestrife provides nectar and pollen for bees and attracts red-winged blackbirds that nest among the densely clustered stems, the plant's spread

The cane toad, imported to Australia to eat grayback beetles, has instead shown an appetite for beneficial insects—and a tendency to kill pets with its venom.

has eliminated the natural foods and cover needed by waterfowl and other native wetland inhabitants. On farms, it slows the flow of water in irrigation systems and reduces hay meadows that provide food for livestock. At least 13 states prohibit the importation and distribution of purple loosestrife.

Current efforts to control this purple invader center on cutting, burning, herbicides, and—in small localized stands—uprooting the plant by hand. These methods are costly, time-consuming, and require continued maintenance; in addition, herbicides often kill desirable plants. In some places, a natural enemy from Europe, the leaf-eating beetle, has been released. But there are concerns that the beetle may itself become an undesirable invader.

Creating More Problems

Sometimes people import an organism to solve a problem, but instead create a new problem. One such creature, the mongoose, was imported into Hawaii in the 1880s to fight rats that were damaging sugarcane fields. Unfortunately, the two species seldom came in contact because rats are active mainly at night, while mongooses are active during the day. Instead of feasting on rats, the mongooses preyed on native, ground-nesting birds, many of which—like the nene, the state bird—are now endangered.

Australia has a similar problem on its hands. In 1935, officials imported 101 toads from Hawaii to exterminate grayback beetles that were decimating sugarcane crops. The toads, actually natives of Central and South America, are variously known as cane toads, marine toads, and buffalo toads. They like moist environments, unlike the beetle grubs, which are found in the ground during the dry season. The toads ate some of the grubs, but they also began eating many other foods, including honeybees and other beneficial insects, pet food, and garbage.

A large gland at the back of a cane toad's neck secretes a poisonous fluid that is deadly to would-be predators. At least part of Australians' acquired hatred of the toads stems from the deaths of pet dogs and cats that had attacked toads. There are indications that some native animals have learned how to fight back. For example, birds known as shrikes, or butcher-birds, flip toads on their back, peck at the vulnerable stomach, and eat the toads' innards. But such predation is unlikely to stem the spread of this toad. A female cane toad deposits some 20,000 eggs at a time, so it is probable that—like many alien invaders—this species will remain in its new environment long into the future.

The Measure of a Mountain

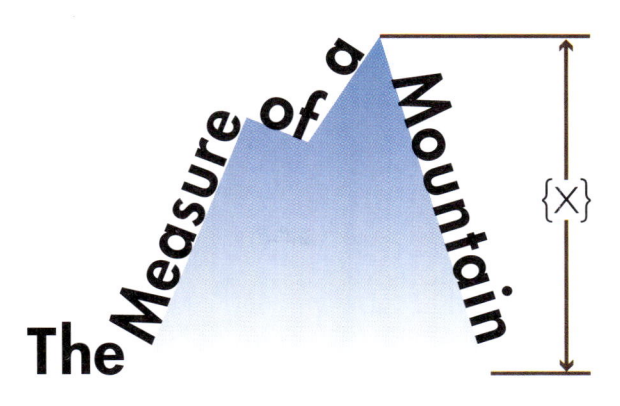

by Larry O'Hanlon

There are certain superlatives about the planet that people are confident in. For instance, most schoolchildren can name the highest mountain in the world: Everest, of course. But a quick check in some reference books reveals a surprising amount of uncertainty about the world's apex.

A QUESTION OF ELEVATION

Look in the 1995 *Information Please Almanac*, for instance, and you will see the ambivalence. In a table of mountain heights, the entries for Everest and K2—the latter a mountain on the border of China and Pakistan—are marked with a footnote that reads, "India gives the height [of Everest] as 29,028 feet [8,854 meters]; however, in 1987, an Italian expedition recalculated its height to be 29,108 feet [8,878 meters] and K2 to be 29,064 feet [8,865 meters]."

This footnote traces back to a heated elevation skirmish a decade ago between Everest and K2. The controversy was triggered by a single, shaky measurement made with the help of radio beacons from naval satellites. It created the slight possibility that Everest could be dethroned from its century-long reign at the top of the world.

It was not. The 1987 Italian expedition—the one immortalized in that footnote—used the latest satellite-location technology to confirm that Everest is indeed higher than K2. But just how high? That was tough to say. Because of technical problems, the actual elevations the Italian team calculated were considerably inaccurate, even though they could still be used to make a relative comparison of one mountain to the other.

The Italian team had not made some careless blunder. As straightforward as it may seem, calculating the elevation of a mountain above sea level is incredibly complex and fraught with uncertainties. We know the distance from Earth to the Moon down to an accuracy of less than 1 inch (2.5 centimeters). Yet despite all of our modern technology, we know the elevation of Everest above sea level only to within about 1 foot (0.3 meter). The exact height of Everest—and of all of the highest mountain peaks on the planet, for that matter—still eludes us.

Scientists have found that determining the precise elevation of Mount Everest (above) and other lofty peaks is fraught with difficulty and uncertainty.

Propelled by our fascination with record-making statistics and the mapmaker's hunger for ever-more-accurate elevations, surveyors have employed sweat and dogged determination to lug their equipment up onto the

EARTH AND THE ENVIRONMENT 127

shoulders of lofty mountain ranges. The British surveyors who first gauged the peaks of the high Himalayas began their work literally at the seashore and marched for thousands of miles across India, making precise measurements all along the way. Today surveying is still a sweaty job. Remote valleys must be traversed and frigid slopes scaled. But the instruments used now include satellites, software, and supersensitive gravity meters.

Scientists, though probably just as fascinated by records as other people, are using the new technology to do more than calculate precise elevations above sea level for the world's highest mountains. They want to know how locations throughout high mountain ranges are moving relative to each other. By measuring minute changes in the motion of Everest and the rest of the Himalayas, for example, scientists hope to resolve some monumental questions about how great mountain ranges are raised and what earthquake hazards those motions will pose to people.

THE DIFFERENCE A FOOT MAKES
Among those working in the thickets of these efforts is Bradford Washburn, honorary director of the Museum of Science in Boston and a patriarch of mountain cartography. In the early 1950s, Washburn's surveying established the elevation of Alaska's Mount McKinley, the highest peak in North America. Now 85, Washburn still remains involved in cartography, most recently by overseeing climbs at McKinley and Everest to gather new elevation data.

One of Washburn's cohorts in the measuring of great mountains is geophysicist Roger Bilham of the University of Colorado in Boulder. Like Washburn, he has no doubts about which mountain, Everest or K2, is king. The controversy of a decade ago, Bilham says, was "a load of rubbish"—partly a result of uncertainties over the new measurement of K2, and partly a result of media fascination with the chance that Everest could be dethroned.

After all, scientists already had ample evidence that Everest was the highest mountain from two surveys earlier in this century by Indian and Chinese teams. The Indian survey, carried out from 1953 to 1954, put Ever-

Measuring the height of the Himalayas is complicated by variations in snow cover and the refraction of light in the rarefied air—a phenomenon that can give distant observers mirages of false summits.

est's elevation at 29,028 feet. The Chinese surveyors did theirs from 1966 to 1968 and then again in 1975, calculating the height at 29,029 feet (8,854 meters).

These measurements were an improvement in accuracy over the elevation of 29,002 feet (8,840 meters) calculated by the British during their Great Trigonometric Survey of India. In the 19th century, survey parties slogged across India, painstakingly measuring the British colony. During surveys at the foothills of the Himalayas, teams often had to traverse malaria-infested jungles and marshes. Many did not make it out alive.

PAINSTAKING CALCULATIONS
The British surveyors relied on the age-old technique of "carrying in" sea level. Beginning at the edge of the sea, the surveyors marched foot by foot, yard by yard, for thousands of miles, measuring their position and elevation at increments of as little as a few feet. Each increment was marked by two poles. To measure the difference in elevation between them, the surveyors used an optical level: a telescope mounted on a swiveling, leveled base and placed halfway between the two poles. The surveyor would sight the poles individually, reading off measurements that were then used to calculate the change in elevation over the distance marked by the two poles.

When the surveyors reached the southern margin of the Himalayas, they had enough measurements to calculate both their distance from the sea horizontally and their elevation above sea level. Now in sight of the peaks, some 100 miles (160 kilometers) north, the surveyors used theodolites—telescopes for precisely measuring vertical and horizontal angles—to determine the elevation of each summit. The two theodolite sightings of the summit and the horizontal baseline between them traced out an imaginary triangle. This and some trigonometry allowed the surveyors to calculate the height of the summit above the theodolites. And given that the elevations of the theodolite stations were known (they had been carried in from the sea), the surveyors could then figure out the elevation of Everest's summit above sea level. In May 1852, the British identified Everest as the highest mountain in the world.

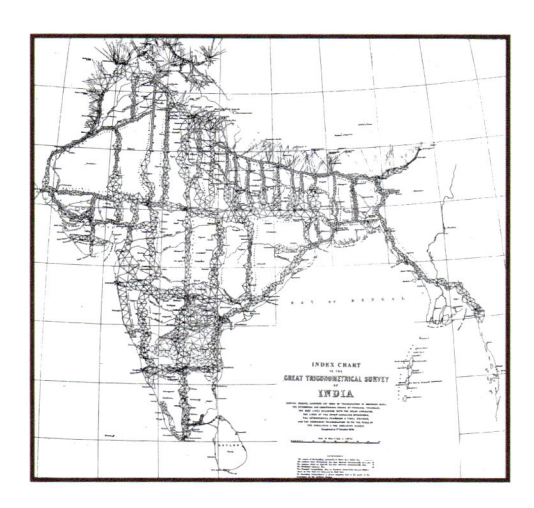

Afterward, however, the question lingered: Just how high, *exactly*, was the roof of the world? Over the next half century, the official height of Everest was changed several times as new information became available. Scientists later determined that the British measurement contained errors amounting to a total of 120 feet (37 meters). Certain errors counterbalanced others, leaving the British, by dumb luck, only 26 feet (8 meters) shy of the currently accepted elevation of 29,028 feet.

Mount Everest's summit was first identified as the top of the world in 1852 by a British team that completed a grueling trigonometric survey of India.

CORRECTING PAST ERRORS

This century, Indian and Chinese survey teams remeasured Everest. Like the British before them, they still carried in sea level—the Indians from Bombay, some 1,000 miles (1,600 kilometers) away, and the Chinese from the Yellow Sea, some 2,500 miles (4,000 kilometers) distant. But both teams went to great lengths to correct for the errors that plagued the British.

One major source of error is refraction, the bending of light beams that occurs when they pass through air layers of different temperature and pressure. As light travels down from the summit of a mountain, it passes through many such layers. As a result of refraction, the surveyor could actually sight what is a mirage, not the peak itself. To reduce refraction errors, the Chinese team carried in sea level to within 5 to 12 miles (8 to 19 kilometers) of Everest's summit. This reduced the amount of air the light had to pass through on its way to the theodolites. The Chinese also launched weather balloons near their theodolite stations to measure the temperature and pressure changes in the atmosphere. This allowed them to better estimate the refraction errors they would have to correct for.

The surveyor's next hurdle in measuring a mountain peak is the shape of the peak itself. When surveyors set the crosshairs of their theodolites on the summit, there is a risk that the instruments might not all measure to exactly the same point. In 1975, the Chinese installed the first survey "beacon" on Everest, a red reflector that could be seen through a theodolite for 10 miles (16 kilometers). By reshooting their angles to this reference point, the Chinese were able to minimize errors.

Although refraction errors and inconsistent theodolite sightings are a problem, they can be reduced. A more difficult problem for surveyors is the effect Earth's gravitational field has on surveying instruments.

An image taken from a U.S. space shuttle provides an unusual view of Mount Everest (center) and many of the towering Himalayan peaks that surround it.

Surveyors level their scopes and poles before taking a measurement. One way to do this is with a plumb bob, a metal weight on a string. Ideally, the plumb bob would point toward Earth's center. But in mountainous areas, this is not the case: the plumb bob is often pulled to the side by local gravity. Mountain ranges are massive enough to exert a measurable gravitational tug sideways on optical levels and theodolites. At sites some 18 miles (29 kilometers) from Everest's peak, the gravity of the Himalayas can knock theodolite sighting off enough to add about 20 feet (6 meters) of error, Washburn notes. To minimize these errors, the Chinese made careful "terrain corrections" to their measurements. They kept track of the topography on either side of their path, estimating the gravitational effects of various features.

There is another source of error linked to gravity that the Chinese and Indians labored to correct. It continues to plague mountain surveyors today: the unevenness of sea level. The British assumed they could use leveling to extend an imaginary line from the shore all the way inland along Earth's gentle curve to a point beneath the Himalayan peaks.

There the line would lie at a depth corresponding to where the sea would be if the continent were not in the way. This carried-in sea level could be used to measure the elevation of the theodolite stations and, by extension, Everest's summit.

In reality, sea level varies depending on location. If the ocean could flow beneath the continents, it would undulate like a wrinkled blanket. This is so because the interior of the planet is lumpy and irregular. Because Earth's interior is lumpy, so is its gravity field. Sea level simply conforms to the warps.

Geodesists, scientists who measure the physical features of Earth and its gravitational field, have a mathematical way to describe the changing sea level. They call it the geoid, an imaginary surface representing the elevation of sea level worldwide with respect to gravity. Think of it as an elastic sheet: the gravity of high mountain ranges pulls the sheet up, and deeply buried masses pull it down. So carrying in sea level doesn't always provide the right reference point for measuring mountains. Because of gravity changes, sea level under a mountain could be very different from sea level at the nearest ocean.

The 19th-century British surveyors did not know enough about the geoid beneath the Himalayas to account fully for its unevenness. But the Indians and the Chinese surveyors of this century did, and they took great pains to correct for it in their measurements. The Chinese, for instance, took measurements of local gravity about every half mile (0.8 kilometer) with a gravity meter. The core of this device is a weight hanging on a fine spring; the weight responds to changes in the force of gravity tugging downward. The best meters can measure a change in gravity across the thickness of a cracker. Such gravity measurements were used to correct more precisely for local deviations in sea level in the Himalayas. Corrections like these enabled the Chinese and the Indians to better keep track of local sea level.

Adjusting for warps in the geoid and other sources of error, the Chinese calculated that Everest is 29,029 feet above sea level, with a possible error of plus or minus 1.2

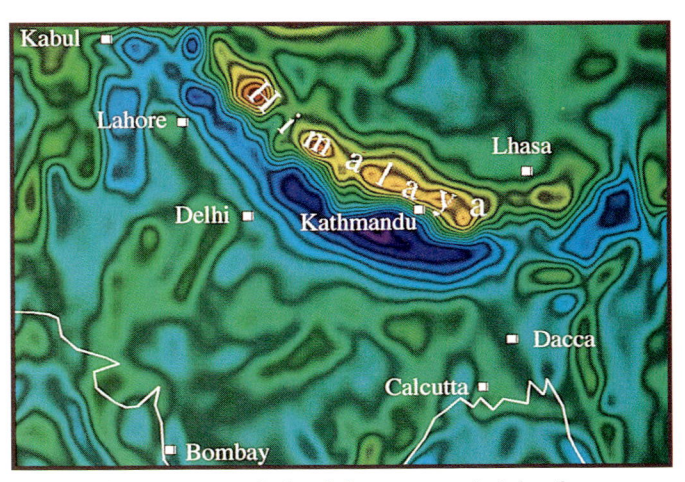

The strong gravitational forces generated by the Himalayas, indicated by contours on a color-enhanced image (above), actually affect the sea level—and thus the true measure of a mountain—in the region.

feet (0.36 meter). This was remarkably close to the earlier Indian measurement of 29,028 (plus or minus 10 feet [3 meters]). That kind of agreement, Bilham writes, is like driving two tunnels from opposite ends of a 4,000-mile (6,400-kilometer)-wide mountain and meeting in the middle with only 14 inches (36 centimeters) of error. The Chinese elevation for Everest is likely the most accurate measure of a mountain ever made.

Space Geodesy

Since the Chinese and Indian surveys, there have been great leaps in the science of surveying the planet from orbit, a specialty called space geodesy. Using the 24 satellites of the Global Positioning System (GPS), the precise location of a solitary ant in the Arabian Desert can be pinpointed. Space geodesy is becoming important to scientists who have relied on manual surveys—geologists, for example, who want to keep track of subtle changes along earthquake faults.

Today GPS is the reigning satellite-location technology. Its satellites broadcast specially coded radio beacons that a person with a receiver can use to fix his or her distance from the satellite. GPS receivers can determine the location of any point on Earth to a fraction of an inch, including the elevation of a mountain.

There is just one problem: GPS receivers measure elevations with respect to Earth's center of mass, which is the center of the satellites' orbits. Most people, however, want to know how high mountains are with respect to *sea level*, not the center of the planet. But sea level is invisible to GPS satellites.

The software in the receiver, however, can convert the satellite measurements into elevations from various other reference points. You could, of course, find out your distance from Earth's center of mass. Or you could find out your elevation above a rough estimate of the geoid surface beneath your particular location. Or you could get your elevation above yet another imaginary surface—the ellipsoid, a smoothed-out approximation of Earth's shape adopted by international agreement.

In great mountain ranges like the Himalayas, all bets are off because the geoid is so poorly known. To convert a GPS elevation to a sea-level elevation, you must map the local geoid, the surface of which represents sea level. This is the irony of GPS: although it has the potential to measure mountains with remarkable precision, its accuracy is limited by a problem that has plagued surveyors since the Great Trigonometric Survey: ignorance of the geoid. The most advanced GPS receivers measure down to an accuracy given in millimeters. Therefore, to really exploit the potential of GPS, the geoid must be mapped with similar precision.

"I once thought that with GPS, we could just stick all this data into a computer, and the number would come right out," Washburn says. "But there's a whole lot of garbage you've got to do in the field. That's where we're all wallowing on McKinley and Everest."

That "garbage in the field" is mapping the geoid beneath a mountain range—no small feat. To map the local geoid accurately, Washburn says, a cartographer has to measure gravity and elevation at each point in a neat grid over the range. In the field, this is often "utterly impossible" because many of the points on that grid could be on nearly vertical slopes or in unreachable areas. So for the moment, using GPS data to measure the true elevation of a mountain is far from simple.

Although scientists have not pinpointed Everest's precise elevation, adventurous climbers can rest assured that they are tackling the world's highest mountain.

Despite these geoid woes, there may be no reason for mountain cartographers to stow their GPS receivers in the closet just yet. In 1996, the U.S. Defense Mapping Agency will release a new global geoid model. It will describe the contours of the geoid with much greater accuracy than the geoid currently in use. That means geodesists will be able to measure things with less uncertainty about just where sea level lies.

The new geoid will not be a cure-all, however. Its large-scale accuracy is quite good, but it suffers on the fine scale—say, the width of a mountain range. There is talk of a new airborne gravity-measuring system in the works, which may make mapping the geoid in rough terrain more practical. But in the meantime, to obtain the best possible results, surveyors will still have to lug gravity meters as well as GPS receivers up the slopes of the world's highest mountains.

A Scientific Contribution

If it is so hard to measure mountains with satellites, why do it at all? In part because people are fascinated by record-making statistics—the highest, the longest, the best, and the oldest. But for most scientists, it is not really about some elevation horse race between the world's highest mountains; it's about science. Many geologists are using GPS to measure minute changes in Earth's crust. Because mountain ranges often mark places where the crust is highly active, GPS users are drawn to them.

Bilham leads a team of scientists whose goal is to understand the mechanics of how the Himalayas have been lifted from the crust. At the moment, Bilham says, Everest is rising at about 0.25 inch (0.6 centimeter) per year, due to the ongoing collision of the tectonic plates carrying India and Tibet. There are a few theories explaining the forces that continue to raise the Himalayas, but it is not clear which is correct.

Bilham and his colleagues have journeyed to the Himalayas several times since 1990, taking GPS elevations from the southern tip of India, across Mount Everest itself, and into central China. Bilham will use all this information to model the motions of the crust across the Himalayas and perhaps figure out their inner mechanics.

But of more immediate concern to Bilham and his colleagues is the quake risk. "There is a major earthquake overdue in the Himalaya, we believe," he says. The last great quake in the region was in 1934, and it killed some 20,000 people. Measuring the ground motions across the region precisely may help Bilham and his colleagues predict where the next one may occur. Since they measure the motions of their survey points relative to each other—not relative to sea level—the geoid is thankfully not an issue.

Using his mountaineering experience and contacts, Washburn is also trying to use GPS to make a contribution to science (although no doubt the cartographer's obsession with better elevation data plays a role, too). At his age, Washburn says, he is not going to be climbing Everest. But recently he has recruited others, chiefly recreational climbers, to do fieldwork, and has supplied them with guidance and equipment. So far, he and his colleagues have collected extensive GPS measurements from both McKinley and Everest.

Washburn hopes that eventually this GPS data may prove useful to geologists like Bilham. "All we're doing is adding more reliable, factual data to a pile of fundamental data," he says. In the meantime, he is happy just to stay involved in mountain cartography. As for coming up with new elevations for Everest and McKinley that mapmakers will accept, it may take a while. The reason? There is still that ornery geoid to contend with.

"I'm not going to announce any altitude until we have an array of data that the National Geodetic Survey will look at and say, 'That's good,'" Washburn says.

Anyone trying for a new elevation of Everest with GPS will be hard-pressed to beat the extraordinarily accurate Chinese and Indian measurements. Is a gain in accuracy of inches worth all the geoid mapping and effort that will be required to make GPS work in a place like the Himalayas? Washburn's answer calls to mind mountaineer George Leigh Mallory's response when someone asked why anyone would want to climb Mount Everest: because it's there, of course. "Every time there's a new method of surveying, somebody wants to use it," Washburn says. "The value of it is simply additional knowledge. As to whether it would pay for lunch, the answer would have to be no."

Perhaps, in the near future, the GPS mess will be solved to everyone's satisfaction. Maybe then our almanacs will be less ambivalent and read something like, "Mt. Everest is 29,029.57 feet [8,854.02 meters] above mean sea level, plus 0.25 inch [0.6 centimeter] per year of uplift since the survey in 2005." Until then, we may not know the exact elevation of Earth's highest peak—but we do know that Everest is still king.

Ask the Scientist

▶ *Do scientists recognize a difference between a pond and a lake? Do the terms* stream, brook, creek, *and* river *have official definitions that distinguish each from the others?*

According to William A. Niering, the author of *Wetlands*, one of several Audubon Society Nature Guides, the primary difference between a lake and a pond is size: lakes are usually deep, large bodies of water; ponds are smaller and shallower. Ponds usually have water of uniform temperature, whereas the water temperature of a lake changes with depth. Ponds may or may not, depending on their location, be permanent the entire year; lakes are less transient. A pond may, in fact, be a lake's remnants, its last stage.

Perhaps muddying the water, Robert E. Coker, Ph.D., Kenan Professor of Zoology at the University of North Carolina in Chapel Hill, observes that human-made reservoirs and ponds are increasing at a phenomenal rate; he considers them to be different from naturally formed lakes. "Many bodies of water that we often call 'lakes' are not lakes—at least, not if we think of a lake as 'a substantial body of standing water in a depression of land,'" he argues.

The terms *stream*, *brook*, *creek*, and *river* do not have widely accepted distinctions. For instance, while the middle section of a river—away from the headwaters—is often called a stream, many "streams" are actually small rivers.

▶ *I recently saw the film* Twister. *Did the special effects shown in that movie realistically portray what could (or does) happen during a tornado?*

"For the most part, Hollywood did its homework," says Vincent Miller, an Oklahoma City, Oklahoma, meteorologist and avid storm chaser who was a weather consultant for the film *Twister*. Inevitably, some artistic license might have been involved in certain sequences—for instance, in the biggest, or very last, tornado shown in the movie, it is unlikely that the people depicted in the scene would have been able to maintain their grip and survive. Yet that storm looked real enough to even professional meteorologists. Many who saw it assumed it to be a facsimile of an actual historic tornado.

In another scene notable for its wild sequence of exploding power lines, a tornadic condensation funnel lowers very rapidly from cloud base to the ground. Although such a manifestation is unusual, it has been known to occur.

According to Miller, the computer-animated tornadoes shown in *Twister* were generally attached to real clouds, except when the filming was necessarily conducted in sunny weather and video sequences of clouds had to be substituted.

▶ *Is it true that most of an iceberg is actually underwater, and that the visible section represents only perhaps 10 percent of the object? Is the* Titanic *the last known ship to have been wrecked by a collision with an iceberg?*

No to both questions. Even when the ends of glaciers are first "calved" (break off into the sea), icebergs can tower more than 300 feet (90 meters) above the water and would never be mostly submerged. Furthermore, by the time an iceberg drifts into warmer waters, much of it has melted into a smaller, irregular "splinter" (such as the berg that sank the *Titanic* in 1912).

Although the *Titanic* is the most infamous collision, it was not the last. Writes James L. Dyson, author of *The World of Ice*: "Few ships collide with icebergs. From the *Titanic* disaster until 1959, the only loss of life from such a mishap occurred during World War II, when a merchant ship sank after colliding with a berg in the North Atlantic."

But the fate of the Danish passenger-cargo ship *Hans Hedtoft* on January 30, 1959, serves as a grim reminder that the danger of icebergs has been far from eliminated. The *Hans Hedtoft*, an ultramodern vessel built especially for winter use in northern waters, had a double steel bottom, an armored bow and stern, and seven watertight compartments. The ship was equipped with radar and the most modern electronic aids for navigation then available. Its builders, like those who had designed and built the *Titanic*, regarded the *Hans Hedtoft* as essentially unsinkable.

On that fateful day, the Danish vessel rounded Cape Farewell at the southern tip of Greenland and collided in heavy seas with an iceberg; its crew immediately radioed that the ship was sinking. Despite a prompt and extensive search effort, not a trace of the *Hans Hedtoft* or any of its 95 passengers and crew was ever found.

▶ *Some years back, a great deal of publicity was given to using gasohol as an automotive fuel. What is gasohol? Is it still produced for sale in the United States?*

Automotive writer James E. Duffy, in the 1992 tome *Auto Fuel Systems: Technology*, defined gasohol as a mixture commonly comprised of about 90 percent unleaded gasoline and 10 percent grain alcohol (ethanol). The chief advantage of gasohol over gasoline is that the ethanol can be produced from readily available organic matter, whereas gasoline is manufactured from petroleum, a more expensive resource. Although a federal appeals court ruled on April 28, 1995, that the Environmental Protection Agency (EPA) may not mandate the sale of gasoline mixed with ethanol, gasohol is still produced for sale in the United States by Methanex Corporation, based in Akron, Ohio.

▶ *To the uneducated eye, would a diamond brought up from a mine bear any resemblance to the diamond gem we see used in jewelry? What about a ruby or an emerald?*

"Gemstones brought up from the depths of a mine tend to be combined with other rocky materials and so often bear little resemblance to a polished gem," writes Russell P. MacFall, author of *Gem Hunter's Guide*.

For instance, mined diamonds often occur in so-called "yellow ground"—the uppermost layer of deep, funnel-shaped pipes of diamond-bearing igneous rock that have been injected into Earth's crust by ancient volcanoes.

Similarly, rubies are often mined from irregular deposits of white calcite, and emerald crystals from quartz or black tourmaline—among a host of other igneous composites.

▶ *What material gives England's White Cliffs of Dover their white color? Are the cliffs subject to erosion from the surrounding waters?*

Running like bulwarks along most of the Sussex coast, the famous cliffs at the end of the English Channel are laced with dark horizontal lines of flint, but their white color is derived from the flaky limestone commonly referred to as "chalk."

The erosion is substantial. As Nigel Calder poetically writes in *The English Channel*, "The cliffs are ragged because of rapid erosion, and the white lighthouse [at Dover] looks as if it will soon be due to fall into the sea."

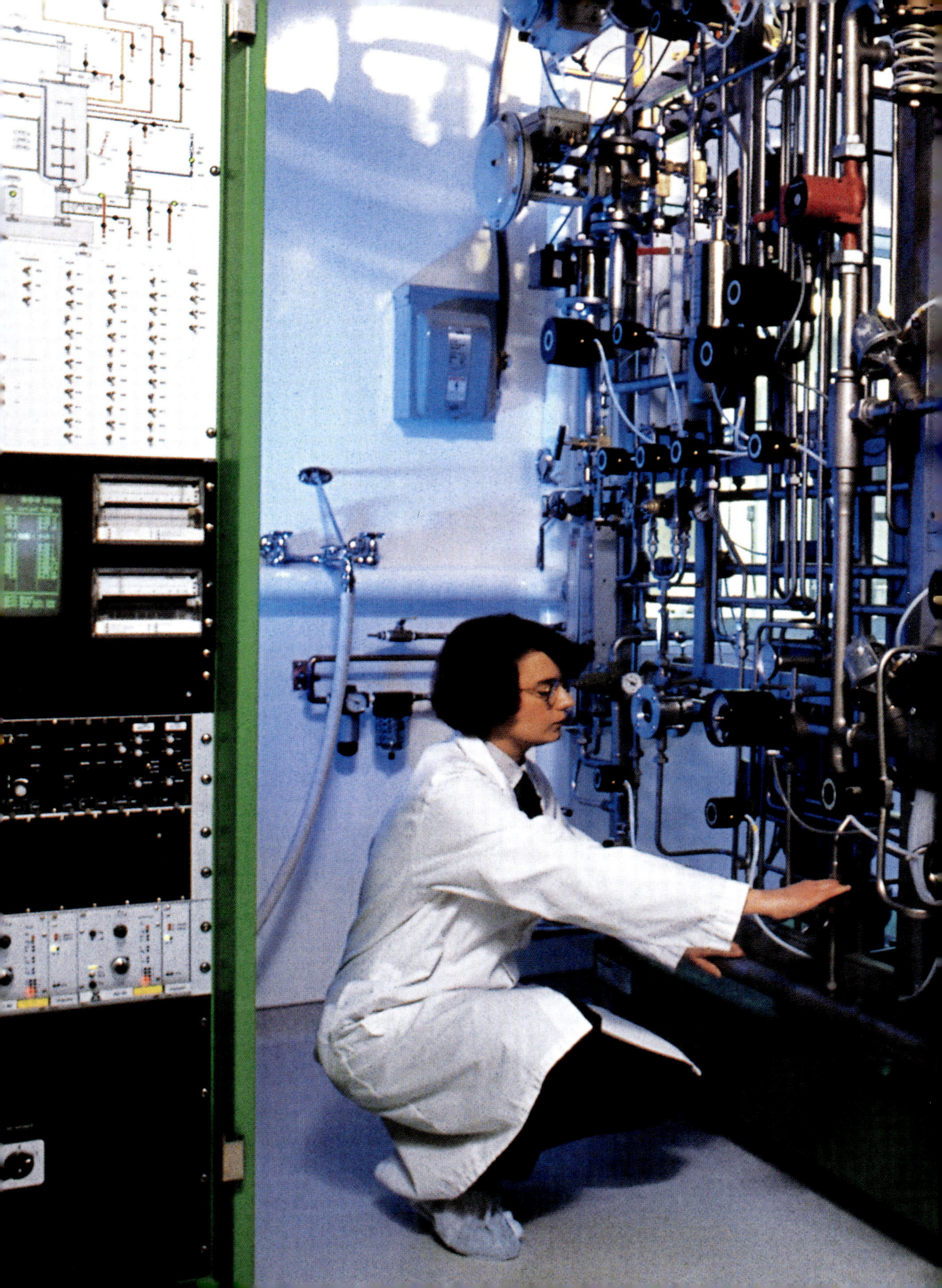

HUMAN SCIENCES

▪ RESEARCHERS AROUND THE WORLD ARE WORKING FURIOUSLY TO DEVELOP A VACCINE AGAINST THE HUMAN IMMUNODEFICIENCY VIRUS (HIV), WHICH CAUSES AIDS. ALREADY, SEVERAL LEADING PHARMACOLOGICAL LABORATORIES HAVE DEVELOPED EXPERIMENTAL VACCINES, SOME OF WHICH ARE UNDERGOING RANDOMIZED TRIALS. SO FAR, NO AIDS VACCINE HAS BEEN APPROVED FOR LARGE-SCALE DISTRIBUTION.

CONTENTS

Beating the Odds	138
Hyperbaric Medicine	145
Tissue in a Test Tube	151
Vegetarian Diets	156
What Are My Chances, Doc?	162
Darwinian Medicine	165
Ask the Scientist	172

Magic Johnson's triumphant return to professional basketball has cast a spotlight on medical advances and changes in public opinion that bring new hope to those battling the human immunodeficiency virus (HIV) that causes AIDS.

Beating the Odds

by Susan Brink

Back in the Los Angeles Laker lineup with a jazzy new left-handed hook shot added to his considerable repertoire, Earvin Johnson, known to an adoring public as Magic and to his teammates as Buck, bounced once again into the public eye. His fighting words of 1991, when he announced his retirement from basketball, rang triumphant. "I am going to go on, and I'm going to beat it," he had said, at a time when few believed *anyone* could beat "it"—infection with the human immunodeficiency virus (HIV) that causes AIDS.

Back then, fears about the disease ran rampant. Johnson triggered a storm of criticism when he tried to stage a comeback in 1992, for possibly endangering the health of his teammates and other players. His second comeback unleashed bear hugs from his teammates, an avalanche of support from across the country—the 4.4 Nielsen rating was the highest ever for a late-night regular-season game on cable television—and long-sought evidence for 3-year-old Earvin Johnson III that Daddy was still a pro.

SEA CHANGE

In little more than four years, Johnson's very public saga speaks both to the accumulation of findings on the medical front and to growing compassion and acceptance on the social front. Where experts once believed that HIV infection charts a path inevitably leading to AIDS and death, a small percentage of infected people force a qualifier: now make that *almost* inevitably. Where once little could be done as the infection progressed to AIDS, now an arsenal of antiviral drugs keeps the disease at bay, and some of the major infections that once meant sure death, such as pneumocystic pneumonia, can be battled with other drugs.

And after living with AIDS for some 15 years, society itself has seen its worst fears about casual transmission fail to materialize. Most people now accept that the virus cannot be transmitted in a restaurant by an infected cook, or on a bus through a sneeze, or through a mosquito bite, or via a toilet seat. And with the toll of individuals infected, sick, or dead well over a million, the universe of healthy people who have known or loved someone with AIDS has expanded well beyond previous stereotypes.

The newfound compassion, moreover, is backed by a federal law with teeth: the Americans with Disabilities Act (ADA) protects against discrimination those who are HIV-positive as well as those with AIDS. Not only in very public places like the professional sports arena, but in everyday places

138 HUMAN SCIENCES

like schoolyards and gyms, HIV-infected people are often both healthy enough and accepted enough to play.

Magic Johnson's energy and well-defined muscles may have surprised some viewers. They shouldn't have. People carrying the virus can stay healthy for 8 to 10 years after infection. During that time, they can build houses, play basketball, and otherwise do whatever they feel like doing, although the risk that intense exercise poses to their health—or the benefit of it—is still controversial.

"The beneficial effects of feeling good about how he performs are going to outweigh any negative effects," maintains Steve Miles, M.D., an AIDS physician at the University of California at Los Angeles (UCLA). But some research has shown that intense exercise, especially in a competitive environment, can take a toll on an already weak immune system. "I'm not too sure he's doing the right thing," says DeSales Lawless, M.D., an immunologist at Rockefeller University in New York City, of Johnson's return to basketball. In any case, close monitoring by a physician is essential. And HIV-positive individuals are likely to be active much longer than early fears predicted. Since the epidemic was identified in 1981, survival time after progressing to the disease has roughly doubled.

How AIDS Spreads
Cases due to heterosexual sex have risen more than 150 percent since 1990.

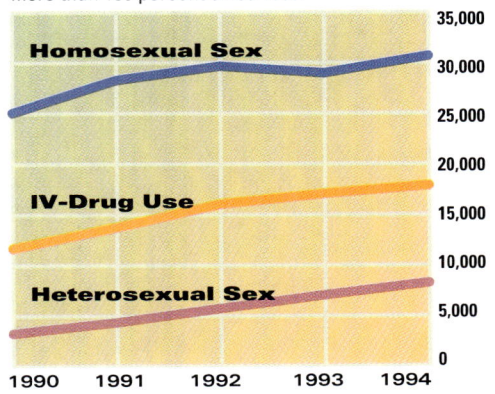

Holding the AIDS Virus at Bay
HIV turns infected cells into virus factories. Drugs that block specific reactions can foil the process, at least temporarily.

1 Virus attaches to receptors on a host cell, releasing its genetic material as RNA.

2 An enzyme converts the viral RNA into DNA. Drugs called nucleoside analogues, such as AZT, can interrupt this process.

3 The viral DNA is integrated into the host cell's chromosomes.

4 The infected cell produces new viral RNA, which generates proteins and other constituents of whole viruses.

5 The protease enzyme creates more proteins by cutting them into shorter pieces. Protease inhibitors fight replication by neutralizing the enzyme.

6 The newly milled proteins fold together to form HIV capsules.

7 Completed HIV capsules break away to infect other cells.

WARM RECEPTION
The extra time is no mere Job-like prolongation of suffering. Quality of life has improved, too, even for those whose infection has moved into the AIDS realm. Judith Billings of Puyallup, Washington, revealed in January 1996 that she has AIDS—and expects to announce that she is a candidate for Congress. A year ago, Billings, 56, now in her second term as state superintendent of schools, tested positive for HIV and for pneumocystic pneumonia—one of the "opportunistic infections" that define AIDS—after suffering respiratory problems for months. But she

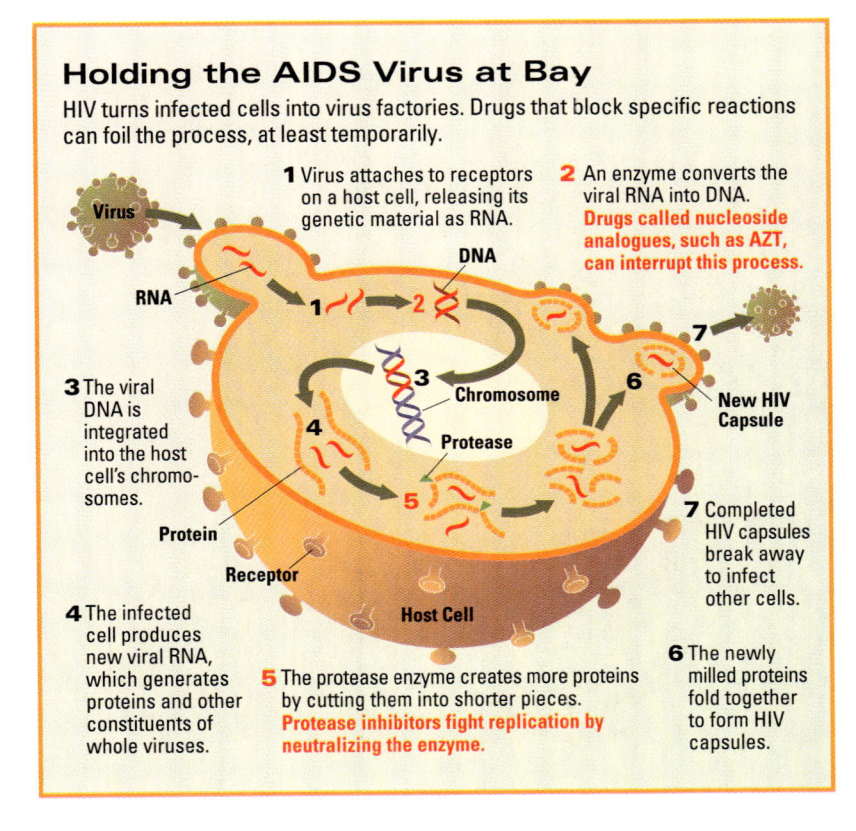

HUMAN SCIENCES 139

believes she may have been infected as early as the late 1970s or early 1980s when she underwent multiple attempts, which failed, to get pregnant with donor sperm. Many such efforts occurred before sperm banks screened donors for HIV. Despite having already progressed to AIDS, she says a combination of the drugs AZT and ddC and a protease inhibitor has restored her energy, and the public's response to her news has restored her faith. "Reaction has been much more positive than negative," she says.

Judith Billings (above) has not let AIDS deter her dedication to public service. She has vowed to continue her work as the top educator in Washington State—and has even considered a run for the U.S. Congress.

Some of the thanks for the extension of high-quality years go to ordinary physicians. With a decade's experience behind them, many have developed an eye for danger signs. A headache may be just a headache, for example, in an HIV-positive person who still has at least 500 CD4-type white blood cells per milliliter of blood. These are the cells that the AIDS virus seeks out and kills—and in someone with a CD4 count that has dipped to 100, a headache could signal toxoplasmosis, a potentially deadly parasitic infection.

A seasoned clinician can be lifesaving. At the third Conference on Retroviruses and Opportunistic Infections, held in February 1996 in Washington, D.C., University of Washington AIDS researcher Mari Kitahata announced findings that AIDS patients with doctors most experienced in treating HIV had a 31 percent lower risk of death at any given time than did patients with the least experienced doctors. The veteran physicians, who had had more AIDS patients or better training in AIDS, checked CD4 counts more often and were quicker to prescribe drugs that prevent a dangerous infection. But Kitahata also found surprisingly little to separate an experienced AIDS clinician from a greenhorn. Once doctors had five AIDS cases under their belts, the rest of their AIDS patients' survival rates rose to match those of the original doctors with more experience.

A Little Help

Clinicians also get credit for longer and better lives. AZT (also known as azidothymidine or zidovudine, the oldest antiviral drug) has never lived up to its initial hype as a miracle pill, at least not when used alone. Its benefits fade quickly as the wily AIDS virus engineers itself to resist the drug. But recently, several studies have shown that patients live longer if they take AZT along with another antiviral medication.

That validates the high hopes many scientists had held for years in such "combination therapy." And now doctors have the hard numbers they need before asking patients to mix a cocktail of several toxic drugs. The recent raft of studies found, for example, that patients who had never taken antiviral drugs before and who took AZT

A Rise Among Women
Men are stricken in greater numbers, but the rate is rising faster among women.

Where Did the Virus Go?

Rarely do scientists slaving away on a decades-long epidemic wax effusive. But at the third Conference on Retroviruses and Opportunistic Infections, held in February 1996 in Washington, D.C., researchers were gushing over a new breed of anti-AIDS drugs. "We're in a new era," said one AIDS specialist. "It's a time of hope again," exulted another.

The source of their enthusiasm was a class of drugs called protease inhibitors. They work by snipping an enzyme the virus needs in order to be able to infect human blood cells. Saquinavir, produced by Hoffmann-La Roche as Invirase, is already on the market, and others are likely to gain U.S. Food and Drug Administration (FDA) approval in 1996.

Whole bevies of scientific teams are bringing in glowing results. One team evaluated a clinical regimen composed of old standby AZT, the similar drug 3TC, and a protease inhibitor from Merck called indinavir. Over 85 percent of patients taking the three-drug combination saw their blood levels of HIV drop literally to undetectable levels—a stunning effect and one that lasted for many months.

Another group of researchers gave the protease inhibitor ritonavir, made by Abbott Laboratories, to sick patients who were already taking various kinds of anti-HIV drugs. Those who added ritonavir to their regimen suffered 43 percent fewer deaths than did those who took a placebo. On top of all the good news, protease inhibitors do not seem to cause the violent side effects that are seen with many other AIDS drugs.

Protease inhibitors are not being touted as miracle pills. They do not cure the disease, and the ever-resourceful AIDS virus may learn to resist them, says Anthony Fauci, head of the National Institute of Allergy and Infectious Diseases. But Fauci notes that "caveats aside, this is unequivocally better than anything we've seen before."

Traci Watson

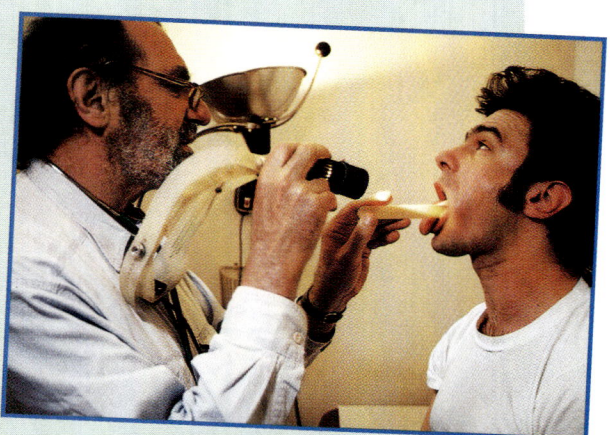

In recent clinical tests, a new class of anti-AIDS drugs called protease inhibitors appears to improve the health of many patients with the disease.

with one of two other drugs—ddI or ddC—had a 38 percent lower risk of dying than those who took AZT alone.

Researchers also have discovered that, while most people still begin to sicken seven to nine years after infection and die within 11 years, others remain disease- and symptom-free for a decade or more. They are a fortunate group of people—some scientists call them long-term nonprogressors—who are HIV-positive but show no measurable sign of disease for 10 or more years and have a stable, near-normal count of CD4 cells. The "viral load"—the amount of HIV in their system—may be negligible. These long-term nonprogressors are as noteworthy for their scarcity as for their good fortune. Only about 5 percent of HIV-infected people fit the description. They are a mixed bunch. "We have women, men, all different ethnic backgrounds; we have people who got it from IV drugs, from homosexual sex, from heterosexual sex, from transfusions," says Jay Levy, a researcher at the University of California at San Francisco who studies such patients.

Nor do they share a common lifestyle. Compare Newton Butler, 42, of Oakland, California, who tested positive in 1984, with Charles Baranosky, 44, of San Francisco, who tested positive in 1986 (a frozen blood sample from a previous hepatitis study indicates he may have been HIV-positive as early as 1978). Butler works out regularly, takes hikes, eats well, and avoids "pure toxins, like tobacco." Baranosky is 30 pounds overweight, has a nightly cocktail or two, eats rich and fattening foods, and knows he should quit smoking. "The cigarettes are going to kill me before anything," he says.

COMMON SENSE

The two men illustrate an area in which medicine has few hard answers. Just how much benefit do good nutrition, a reasonable amount of exercise, and steering clear of unhealthy habits bring to those carrying HIV? Most experts fall back on plain common sense. "We have to promote a healthy lifestyle across the board," says Bruce Walker, a researcher at Massachusetts General Hospital in Boston who has joined forces with a San Francisco Department of Public Health group that has been studying long-term HIV-infected survivors of the 1978 hepatitis study.

At least Baranosky and others can monitor their health status more precisely than before. At February's conference, scientists announced solid support for a new tech-

Years of intensive study have yielded new insights into the pathology of HIV (seen as dark splotches in infected cells, left). Such fundamental clinical research has helped pharmaceutical companies bring a number of promising antiviral drugs to market.

nique to chart a patient's health: inventorying the viral load. Led by AIDS researcher John Mellors of the University of Pittsburgh, scientists described the link between viral load and an infected but healthy person's chances of getting sick. Their study found that none of the patients with the lowest viral load died of AIDS in the following five years, while 65 percent of those with the highest viral load died of the disease. The findings should further stretch survival because doctors can evaluate a healthy but infected person's risk of getting sick. Those on the threshold of developing AIDS symptoms can start antiviral-drug therapy, which is expected to raise the proportion of HIV-positive people—now at 10 to 50 percent—who take such drugs. The new measure will show whether someone could benefit—and when it is safe to delay taking the drugs. Doctors can also monitor viral load when they prescribe antivirals: if viral load rises steadily, it is time to try a new drug combination.

New hope and new treatments carry a stiff price tag, and with a health-care system driven by cost-cutting pressures, who will pay the bills is far from clear. Bare-bones AZT treatment costs over $2,000 per year. The new protease inhibitors are difficult and expensive to make. The only one now on the market, Hoffmann-La Roche's Invirase, costs at least $6,000 a year. The cost of successful triple-drug combinations could total $12,000 to $18,000 per year.

It is already not unusual for patients to combine a dozen different drugs following conventional treatment standards in an attempt to stay healthy. Most insurance companies cover what has proved effective. It is also not unusual for people to shun all therapies, fearing a litany of side effects like fever, chills, numb fingers and toes, and confusion that they believe may be worse than the disease's normal toll.

Regardless of their choices in coping with the disease, long-surviving HIV and AIDS patients are a hearty, odds-beating bunch. But remnants of fear linger—perhaps especially among people in regions relatively untouched by the disease.

While many Americans may have begun to regard HIV infection as another chronic disease, new legislation could set HIV-infected military personnel apart. On February 10, 1996, President Clinton signed a defense authorization bill containing a provision to force the discharge of all HIV-positive service members—1,049 men and women—within six months. The provision was sponsored by U.S. Representative Robert Dornan (R-Calif.), who says HIV-infected service members have been a burden because they had to be excluded from combat and certain jobs. "This goes against the American grain," says Patricia Fleming, White House AIDS policy director. There is bipartisan support in both houses of Congress to repeal the ban on HIV-positive military personnel.

A Rise Among Minorities
The white share of cases has dropped from 49.3 percent in 1990 to 41.4 percent in 1994.

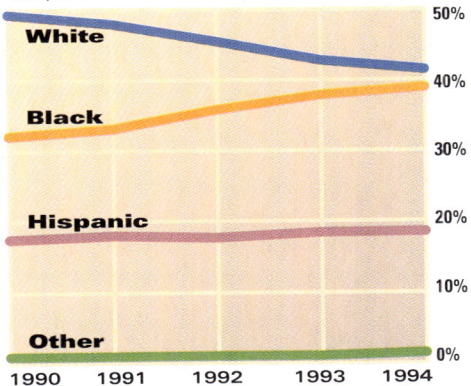

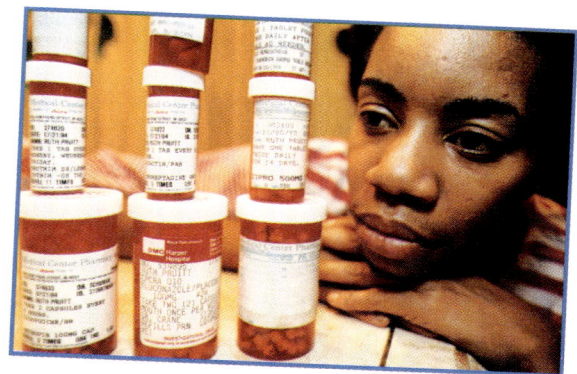

Many HIV-positive patients take 10 or more drugs in an effort to stay healthy. This so-called combination therapy is costly and can produce a litany of side effects in some individuals.

ENFORCEMENT

Ironically, it was the Americans with Disabilities Act, another federal law, that in 1990 gave civilians with HIV or AIDS rights previously denied them. The epidemic's history of discrimination chronicles dentists who have refused to clean the teeth of HIV-infected people, employers who have fired them—or refused to hire them in the first place—or who have cut off their health benefits, and, as a final insult, funeral directors who have refused to embalm their bodies. The U.S. Department of Justice and the Equal Employment Opportunity Commission (EEOC) have aggressively enforced the law. HIV and AIDS cases now make up

HUMAN SCIENCES 143

nearly 21 percent of the cases filed by the EEOC. One of the most common forms of discrimination—denial or limitation of health-care benefits to people with HIV or AIDS—has slowed since the EEOC won several such suits. The biggest was settled in December 1995: a New York City–area laborers union welfare fund, the 11,000-member Mason Tenders District Council Welfare Fund, agreed to pay a total of $1 million to at least three AIDS patients and the survivors of 11 others found to have been denied health benefits illegally. Although the fund paid for coverage of other costly diseases like cancer and heart disease, it refused to cover HIV and AIDS. "We've got to stop talking about fear and start talking about right and wrong. Thanks to the ADA, fewer people are having their freedoms limited because of stereotypes and fears about HIV and AIDS," says Deval Patrick, assistant attorney general of the Justice Department's civil-rights division.

Public fears have quieted. A recent poll, for example, found that the public still rates AIDS as the nation's greatest health problem, but the 37 percent who call it that is down sharply from 70 percent in 1987. The poll also found that an overwhelming 74 percent of Americans approve of Magic Johnson's return to professional basketball. A paper published in the February 19, 1992, *Journal of the American Medical Association* (JAMA) found that in 1985, 39 percent of those polled said children with AIDS should not attend public schools. By 1991, the disapproval figure had dropped to only 9 percent.

Higher Profiles

Another sure indicator of widening acceptance is a higher media profile. Early in 1995, Ric Munoz, an HIV-positive marathon runner, starred in a Nike commercial. In June 1995, on AIDS Compassion Day, the NBC soap opera *Another World* featured Keith Christopher, an openly gay and HIV-positive actor, playing an openly gay and HIV-positive character—probably a television first. And there is even a slick magazine aimed at people with HIV or AIDS. Its founders coined the name *POZ*, which refers to thinking and acting positively as well as to infection status.

New York City Councilman Tom Duane has personally felt a positive shift in public attitude. In 1989, he lost a city-council race, about a year after testing positive for HIV. He kept his HIV status a secret during that campaign. "The world was different, and I was not ready yet," says Duane, 41. In 1991, he went public and went on to become the first U.S. citizen to win elective office after announcing he was HIV-positive. By then, his own attitude had shifted. In the first act of his 1991 campaign, he sent a letter to every voter in his district in which he revealed he was HIV-positive. He won the general election with 75 percent of the vote. Duane's political fortunes have fluctuated—he lost the 1994 Democratic nomination for Congress—but his immune system appears to be holding up. He is still healthy and takes no anti-HIV medication.

Duane was ambivalent about giving up his privacy; he felt, understandably, that his health is his business. But in politics—and the spotlight in general—the public wants to know. And it appeared that compassion and tolerance won out over fear and panic with near-universal agreement that Magic Johnson deserved another shot on the basketball court. His example may inspire other survivors to risk taking their own long shots.

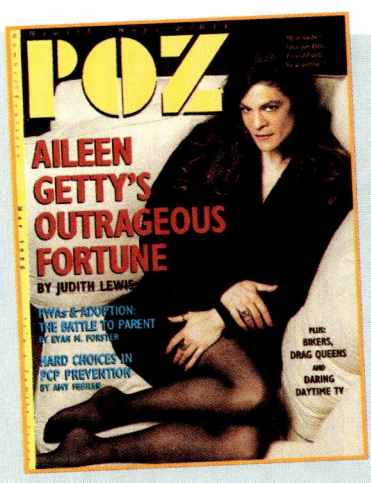

The commercial success of POZ (above), a slick magazine targeted to people infected with HIV, reflects the widening public profile of a disease that was previously shrouded in secrecy and stereotypes.

Hyperbaric Medicine: High-Pressure Therapy

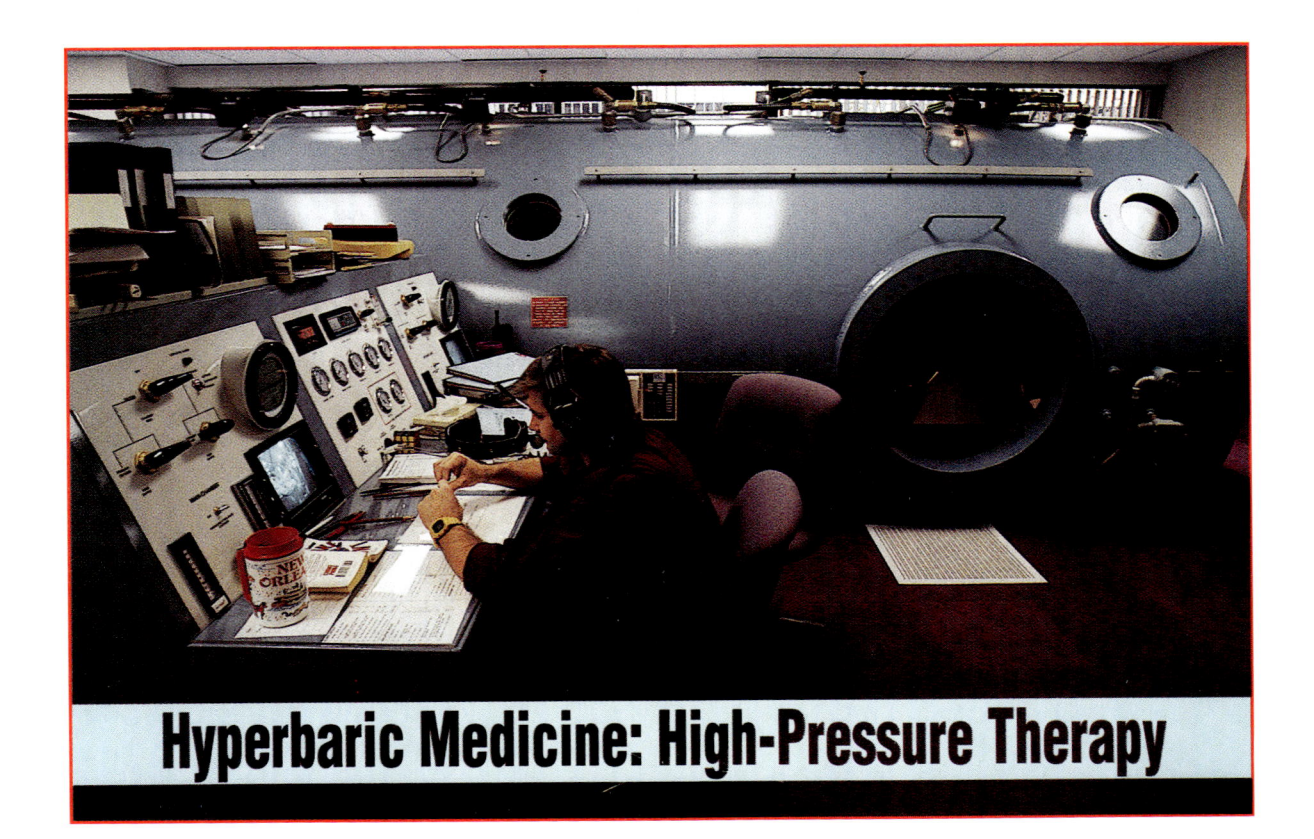

by Zahid B. M. Niazi, M.D., and C. Andrew Salzberg, M.D.

Taking a vicious slam, Martin Gelinas of the Vancouver Canucks hit the wall of the hockey rink at crushing speed. So massive was the bruise to his thigh that he might have been sidelined for the season. Instead, two weeks later, Gelinas amazed both coaches and fans by returning to the ice—fully recovered. Magic?

No, medicine . . . hyperbaric medicine. Canuck trainers credit Gelinas' speedy healing to 90-minute, twice-a-day sessions spent breathing pure oxygen inside a hyperbaric, or pressurized, chamber.

By definition, hyperbaric medicine is the use of greater-than-normal air pressure to treat either injury or disease. It is best known for the treatment of "the bends," or caisson disease—two names for the decompression illness suffered by divers and miners who surface from the deep too quickly. Hyperbaric treatment of sports and other injuries is grounded in the medical understanding that increased pressure boosts the amount of oxygen that dissolves in the body. This extra oxygen literally fuels the healing of injuries. The added pressure also reduces swelling in and around the wound.

But the speedy healing of bruises, sprains, and breaks is just a small part of hyperbaric medicine in the 1990s. In recent years, hyperbaric therapy has been used to treat decompression illness and carbon monoxide poisoning, improve chronic circulation problems, treat life-threatening infections, and even enhance the success of plastic surgery. These and other new uses for hyperbaric therapy mark an exciting renaissance in an old field of medicine.

THE CHAMBER

The centerpiece of hyperbaric medicine is the hyperbaric chamber—in its simplest form, a cylinder just large enough for one

The pressurized, oxygen-rich environment generated inside a hyperbaric chamber (above) provides a painless, noninvasive, and effective means of treating wounds, infections, and various other disorders.

HUMAN SCIENCES 145

Professional athletes—both here and abroad (below)—have availed themselves of hyperbaric medicine. The San Francisco 49ers (left), Vancouver Canucks, and many other teams have their own hyperbaric chambers to help their star players recover from deep bruises and other injuries.

person. Its door and any windows have rubber gaskets for a strong, airtight seal. Attached to the chamber are compressors that pump in extra air until the pressure inside is two to six times greater than atmospheric, or outside, pressure. As the pressure increases, it squeezes more oxygen molecules into each square inch of air, thereby increasing the partial pressure of oxygen being inhaled, or breathed in. A "multiplace" chamber is larger, and more than one patient can be treated at the same time with nurses and physicians in attendance if needed.

Today's hyperbaric chambers also come equipped with head tents or face masks through which patients can breathe pure, or 100-percent, oxygen. (Normal air contains roughly 21 percent oxygen and 78 percent nitrogen, with a smattering of carbon dioxide and other gases.) Within an hour, a person breathing pure oxygen at twice normal air pressure increases the amount of oxygen in his or her blood by as much as 15 times.

Hyperbaric History

The one-person or "monoplace" hyperbaric chambers used today resemble slender space capsules. Inside, the patient rests comfortably in a padded chair, dons an oxygen mask, and has the option of watching television or films, listening to music, or even playing video games. This is a far cry from the first hyperbaric chamber ever used in medicine—the British "Domicilium" of the mid-17th century.

The Domicilium was little more than a crude box with hand-operated bellows for pumping in extra air. Still, it helped ease the labored breathing of patients with "consumption," or tuberculosis. In the mid-1800s, developers added steam pumps to increase the amount of air that could be squeezed into a hyperbaric chamber. Later, electrical pumps were introduced to reduce noise.

An important pioneer of 20th-century hyperbaric therapy was the American physician O.J. Cunningham, who in 1928 helped build an institute of hyperbaric medicine in Cleveland, Ohio. The clinic came to be known as the "steel-ball hospital." It was 64 feet (19.5 meters) wide, five stories high, and featured a large, airtight steel chamber that could house 12 patients. This oversized hyperbaric chamber came furnished with beds, toilets, and even showers so that patients could stay for days at a time. Cunningham fervently believed that most any kind of disorder—from the flu to brain tumors—could be cured with hyperbaric therapy. But he had little scientific basis for his claims, and a questionable success rate. So, in 1930, the American Medical Associa-

In 1928, American physician O.J. Cunningham built a controversial hyperbaric institute in Cleveland, Ohio. The facility's enormous steel sphere (right) was large enough to house a dozen full-time patients.

Hyperbaric chambers are best known for their success at treating decompression sickness. This illness, also called the bends, afflicts divers (bottom) or miners who return to the surface from the depths too quickly.

tion (AMA) and the Cleveland Medical Society closed Cunningham's unusual hospital. The reputation of hyperbaric medicine was tarnished further in the 1950s, when it was used irresponsibly to treat senility, impotence, and other problems for which it had no proven benefit.

It was in the mid-1940s that doctors first began using hyperbaric chambers to successfully treat decompression illness, or "the bends"; this is still the application for which hyperbaric chambers are best known. When an underwater diver or miner returns to the surface from a great depth, he or she has passed from an environment of high pressure to one of much lower pressure. As a result of this drop in pressure, oxygen, nitrogen, and other gases inside the body begin to expand. If the change from high to low pressure is slow enough, the expanding gas can safely pass out of the lungs. But if the change comes too quickly, as in a diver swimming rapidly to the surface, the trapped gases form painful and dangerous bubbles within the body. The result is decompression sickness, which can be fatal if left untreated. Treatment involves returning the affected person to a high-pressure environment, such as that found in a hyperbaric chamber. This allows the air bubbles trapped inside the body to slowly redissolve. The gas then leaves the body normally through the lungs as pressure is gradually returned to normal.

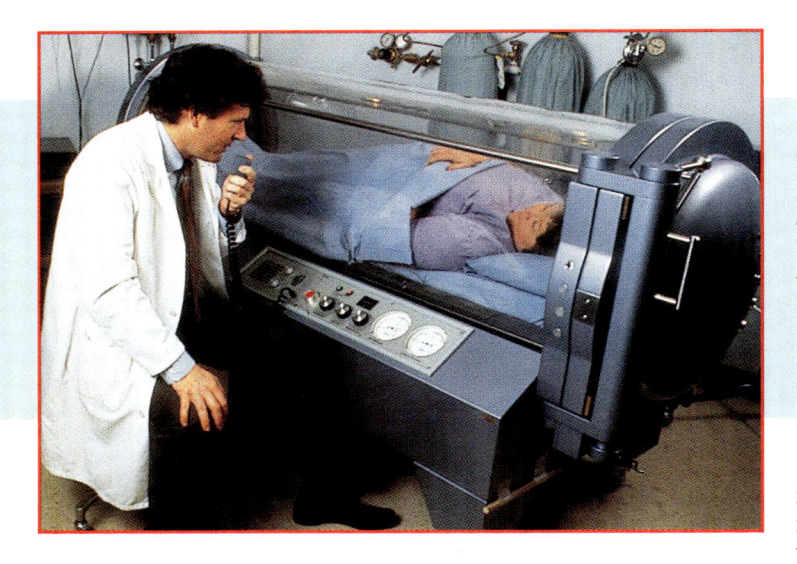

More than 200 U.S. hospitals have installed hyperbaric chambers to help individuals recover from smoke inhalation, carbon monoxide poisoning, plastic and reconstructive surgery, and gangrenous infections.

A portable hyperbaric chamber (bottom) has been developed for use by climbers suffering from altitude sickness. This debilitating ailment occurs at high elevations, where air pressure and oxygen content are low.

In the 1950s and 1960s, doctors began treating certain infections by placing their patients inside hyperbaric chambers. As mentioned, a person inside such a chamber takes in many times the normal amount of oxygen with each breath. This process hyperoxygenates the blood and tissues, and so helps kill certain disease organisms that cannot grow in the presence of oxygen. Extra oxygen in the blood also stimulates the production of white blood cells, which help clear all types of bacteria from the body.

Around the same time, physicians began using hyperbaric therapy to save the lives of people suffering smoke inhalation or carbon monoxide poisoning. Particles in smoke block the lungs, preventing oxygen from reaching the blood. Carbon monoxide in turn interferes with the blood's ability to carry oxygen. In either case, the affected person can literally suffocate without treatment. Hyperbaric treatment is ideal because it quickly increases oxygen levels in the patient's body.

NEW USES

Hyperbaric therapy has likewise helped save the lives of so-called "blue babies," infants who are born with heart defects. Without treatment, the babies usually die because their weakened hearts cannot supply their brains and bodies with enough oxygen. By superoxygenating the infants' blood, hyper-

Step Inside

What is it like to sit inside a hyperbaric chamber? The sleek, one-person hyperbaric chambers used by sports teams today are designed for comfort and relaxation. So step inside, lean back in the padded recliner, and slip on your oxygen mask. As an attendant slowly increases the pressure of the air inside the chamber, you may feel a popping in your ears, rather like that felt when descending in an airplane. You may likewise experience a temporary distortion in your vision.

Now relax—watch a video, listen to music, or just nap. Your chamber also features several windows and portholes to ease any feelings of claustrophobia. A nurse or technician remains just outside.

How long you spend in the chamber, and at what pressure, depends on the reason for your treatment. A diver with the bends, for instance, might spend several hours at a time over several days—say, six hours the first day, four hours the next, and so on. On the first day, the diver might return to the undersea pressure from which he or she surfaced. Then, gradually, technicians would decrease the pressure within the chamber until it reached normal levels.

The interior of a hyperbaric chamber (above) is designed with the patient's comfort in mind. Most units contain a padded recliner, windows, and equipment for watching movies or listening to music.

By contrast, someone receiving treatment for a sports injury would receive pressure about twice that of normal air. His or her therapy sessions would typically last 60 to 90 minutes and be repeated once or twice daily over two to three weeks.

---

baric therapy increases their chance for survival until surgeons can repair their defective hearts.

Studies suggest that hyperbaric therapy may help adult patients recover from heart problems as well. People who have suffered heart attacks, for example, seem to benefit from the extra oxygen that hyperbaric therapy supplies to their weakened hearts. Cardiac surgeons were the first to use hyperbaric operating rooms. Superoxygenating the blood of someone during heart surgery may help reduce the injury that occurs when surgeons must temporarily stop circulation to the heart and brain. After surgery, hyperbaric therapy can speed the healing of the surgical wounds.

Chronic, or long-term, circulation problems can also be treated with hyperbaric medicine. A good example is seen with diabetes, a disorder that often leads to poor circulation in the limbs. As a result, diabetics often develop painful, slow-healing ulcers on their legs, feet, and toes. By pumping the blood with extra oxygen, hyperbaric therapy helps heal these stubborn sores. And the increased pressure helps restore circulation by encouraging the growth of tiny blood vessels called capillaries.

People with vasculitis, a painful inflammation of the blood vessels, are also spending time in hyperbaric chambers. The increased pressure reduces their agonizing swelling, and the extra oxygen helps heal sores associated with the disorder.

Some innovative plastic surgeons are using hyperbaric therapy as well. They have found that patients who spend time in a high-pressure chamber after receiving a skin graft heal better and faster. The treatment has similar benefits for burn patients, who do not become as dehydrated when they are prescribed time in the chamber; their burns heal more rapidly also.

Hyperbaric therapy seems to be especially useful in the case of complicated breaks and other wounds that are cut off from their normal blood supply. This sometimes happens when the injury destroys or damages vital blood vessels. Hyperbaric treatment can overcome the problem by literally forcing oxygen out of the blood and diffusing it through surrounding tissues.

Hyperbaric Hazards

As wondrous as hyperbaric therapy may sound, it is not without risks. The oxygen-rich environment of the hyperbaric chamber can quickly transform the smallest spark into a raging fire. So before entering the chamber, a patient must remove metal jewelry or any other object that might produce a spark. The patient must also be sure that his or her hair and body are free of any flammable substance such as hair spray, perfume, or even a lotion containing alcohol or mineral oil. Although such precautions are vitally important, the risk of fire within a properly maintained chamber is very small. Indeed, there has never been a fire in a one-person hyperbaric chamber in the United States.

Yet another exciting use for hyperbaric therapy is in radiation treatment for cancer. In the 1950s, radiotherapists found that the high-oxygen environment of a hyperbaric chamber somehow helped their radiation beams kill cancer cells. More recently, researchers have found that hyperbaric therapy also helps heal the radiation burns left behind by such treatments.

Still other researchers are eager to see if hyperbaric therapy can help patients infected with the human immunodeficiency virus (HIV), the virus that causes AIDS. Super-oxygenating their blood may help these patients resist the infection's spread—and so stay healthy longer. Such therapy may also help those already weakened by AIDS to recover from the many secondary infections that accompany the disease.

INCREASING POPULARITY

It is estimated that there are some 300 hyperbaric chambers in use in the United States. Most are in hospitals, reserved for lifesaving emergencies such as decompression illness, smoke inhalation, and carbon monoxide poisoning, and for the treatment of serious, hard-to-heal wounds. A growing number of one-person chambers are being bought by sports teams with enough money to afford their own. The payback is the quickened healing of their superstar players. Some players even use time in the chamber as a high-tech remedy for fatigue, even jet lag.

It may be only a matter of time before hyperbaric chambers appear in physical-therapy offices, alongside other high-tech treatments such as ultrasound and magnetic stimulation. After all, if hyperbaric chambers can get a professional player back to work faster, this therapy can do the same for those injured in the workplace or during recreational sports.

At present, health-insurance companies pay for hyperbaric therapy only for certain well-established uses, such as decompression illness, carbon monoxide poisoning, or gangrenous infections. (In a hospital, a stint in a hyperbaric chamber may cost $100 to $400 an hour.) But experts say that the newer uses of hyperbaric therapy are rapidly becoming more accepted.

TISSUE IN A TEST TUBE

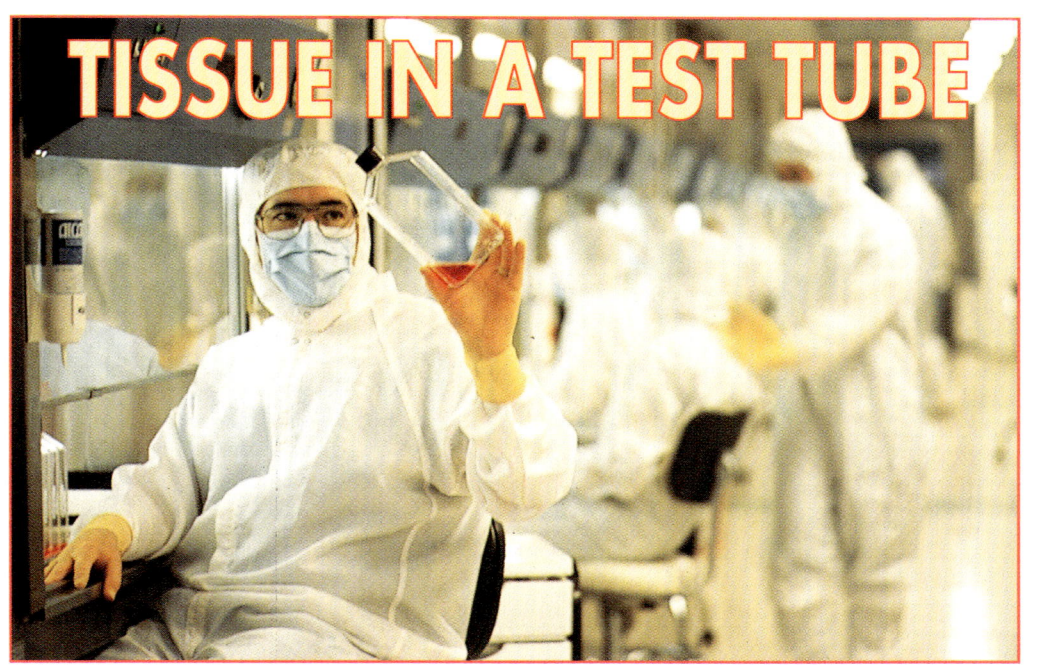

by Edward Edelson

Thanks to recent developments in the field of tissue engineering, scientists can grow replacement skin, cartilage, and organs in the laboratory.

She was a 35-year-old mother of two who suffered searing burns over 75 percent of her body when a gas heater exploded. When the patient was brought to Indiana University Hospital in Indianapolis, Raj Sood, M.D., director of the burn unit, shaved off a small piece of her uninjured skin and sent it to Genzyme Tissue Repair, a small biotechnology company in Boston.

Within 30 days, Genzyme had grown enough skin from that small shaving to cover all of the patient's wounds. She left the hospital after 46 days—about one-third of the time normally required for treating such burns.

A 16-year-old boy suffered severe burns over 60 percent of his body because he was fooling around with a bottle of gasoline and a match. When the boy arrived at the University of California at San Diego Hospital, John Hansborough, M.D., director of the regional burn unit, covered the wounds with skin that had been grown at Advanced Tissue Sciences, a biotech company in nearby La Jolla. The boy was out of the hospital in 47 days.

REPLACEMENT SKIN . . . AND MORE

Both of these patients healed quickly because of recent advances in the field of tissue engineering. Scientists working at small, pioneering biotechnology companies are growing not only replacement skin but heart valves, breasts, ears, cartilage, and other body tissues. Most products are several years down the road, but a few are already available on a limited basis.

The skin used on the San Diego boy is the product of an intellectual partnership formed in 1984 between Joseph Vacanti, M.D., a surgeon at Harvard Medical School in Boston, and Robert Langer, a chemical engineer at the Massachusetts Institute of Technology (MIT) in Cambridge. It occurred to Vacanti, when confronted with a frustrating shortage of livers to transplant, that "maybe we could make new livers." He took his problem across town to Langer, a close friend.

"My idea was to take degradable polymers, make scaffolds out of them, and grow cells on them," Langer says. "The idea was that if we made the scaffolds the right way, the cells would be able to reorganize themselves and make new tissues. The scaffold polymers would then degrade and disappear, leaving behind only new tissue."

Langer and Vacanti are now on the scientific advisory board of Advanced Tissue Sciences in La Jolla. It is there that two engineered skin products are being developed.

Both are made from the living skin cells of infant foreskins—by-products of circumcisions.

To create new skin, researchers extract cells called fibroblasts from the foreskins. These cells are seeded onto a meshlike scaffold and supplied with nutrients and oxygen. "We trick the fibroblasts into thinking they're in the body," says company president

Something to Build On

A molecular scaffold is a microscopically small structure, made of polymer fibers, that guides the growth of cells. The polymer used in the scaffolding developed by Robert Langer of the Massachusetts Institute of Technology (MIT) in Cambridge consists of molecules of lactic acid and glycolic acid, materials that have U.S. Food and Drug Administration (FDA) approval for use as sutures. Langer's first scaffolds were manufactured by dissecting individual polymer fibers from sutures and bonding them into the shape desired for the engineered tissue. Now scaffolds are being used to grow a variety of body tissues, ranging from sheets of skin manufactured by Advanced Tissue Sciences to complex structures such as heart valves.

Polymer scaffolds have major advantages. They allow cells to adhere to them, and they permit those cells to form the desired tissue. The polymers are also easy to manufacture and are compatible enough with body tissue to avoid rejection. Most important, they disappear over time as they are absorbed into the body or nutrient medium—leaving behind only the engineered tissue.

There have been several variations on the theme. One modification is to include crystals of sodium chloride (salt) in the polymer scaffold. When the scaffold degrades, the salt crystals are leached out, leaving a porous polymer lattice that allows nutrients to reach the cells.

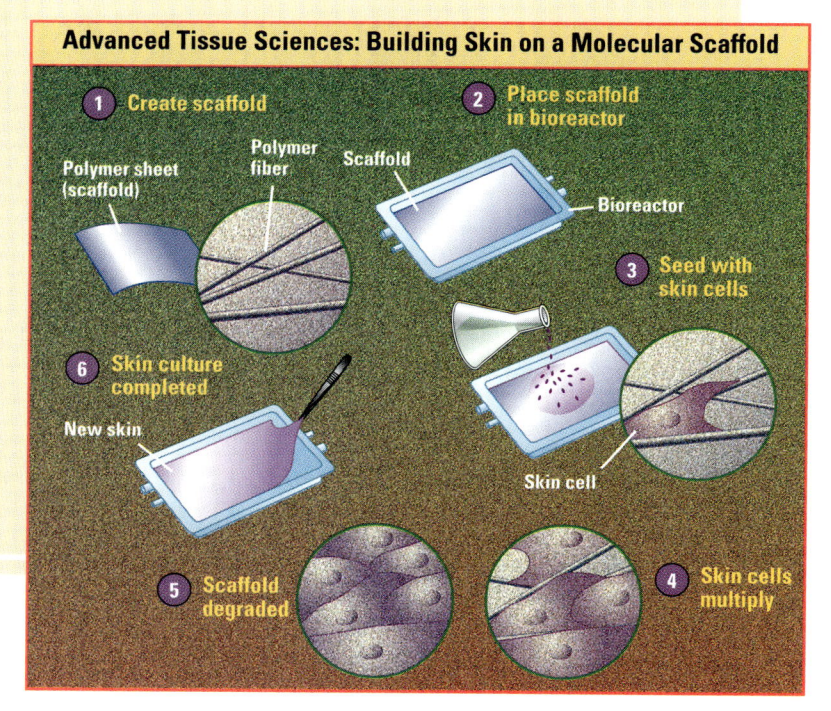

Joseph Vacanti, M.D. (left), and Robert Langer, Ph.D. (right), the founders of Advanced Tissue Sciences, have developed an apparatus for cultivating replacement heart valves and other body tissues.

Advanced Tissue Sciences: Building Skin on a Molecular Scaffold

1. Create scaffold — Polymer sheet (scaffold), Polymer fiber, Scaffold
2. Place scaffold in bioreactor — Bioreactor
3. Seed with skin cells — Skin cell
4. Skin cells multiply
5. Scaffold degraded
6. Skin culture completed — New skin

Gail K. Naughton. The fibroblasts cover the scaffold as they grow, forming a thin sheet.

This sheet resembles the inner layer of skin, called the dermis. The outer layer, the epidermis, is much harder to duplicate. Because the epidermis is the body's first line of defense against the outside world, it has strong immune responses and is easily rejected when transplanted. So Advanced Tissue Sciences adds a synthetic epidermal layer to its dermal sheets to create a product called Dermagraft-TC.

Dermagraft-TC is a "transitional covering" for severe burns. It temporarily replaces the patient's own dermis to prevent dehydration, but is later removed. In clinical tests, it appears to be less prone to rejection than are dressings of cadaver skin—the standard treatment.

the raw material for its manufactured skin. "We're able to manufacture 4 acres [1.6 hectares] of skin from a single foreskin," says senior vice president Michael L. Sabolinski, M.D. He keeps on his desk a sample of cultured skin about 3 inches (7.5 centimeters) in diameter. To prove the toughness of the skin to a visitor, Sabolinski lifts the thin, transparent sample from its dish and tugs on it a couple of times.

Unlike Dermagraft, the Graftskin made by Organogenesis boasts both a dermis and an epidermis. Researchers start by adding fibroblasts to collagen (a fibrous protein found in connective tissue) to create a dermal layer. They then add keratinocytes, cells that are the building blocks for the epidermal layer. Organogenesis claims its culturing technique eliminates rejection.

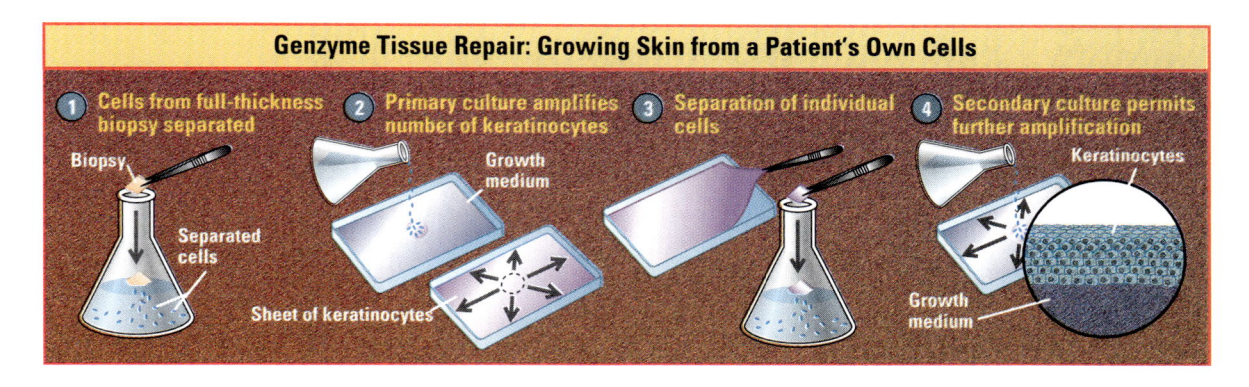

Genzyme Tissue Repair: Growing Skin from a Patient's Own Cells

1. Cells from full-thickness biopsy separated — Biopsy, Separated cells
2. Primary culture amplifies number of keratinocytes — Growth medium, Sheet of keratinocytes
3. Separation of individual cells
4. Secondary culture permits further amplification — Keratinocytes, Growth medium

The company makes a second product, simple Dermagraft, for the treatment of diabetic ulcers. Each year, these gaping foot wounds that do not heal properly force as many as 75,000 diabetics in the United States to undergo amputations. In clinical trials, Dermagraft transplants appear to help the wounds mend. Advanced Tissue Sciences expects to file applications for approval by the U.S. Food and Drug Administration (FDA) of Dermagraft and Dermagraft-TC later this year; the FDA has promised expedited reviews for both.

REDUCING REJECTION RISKS

A Boston-area company called Organogenesis also uses living human foreskin cells as

The first application foreseen for Graftskin is the treatment of venous ulcers, the most difficult type of chronic wound to heal. Approximately 1 million Americans suffer from venous ulcers. Conventional treatment costs $3,000 to $4,000, and takes an average of six months. Because healing is faster with Graftskin, the product could be priced as high as $1,000 and still save medical costs, the company claims. Graftskin has already been submitted to the FDA for approval, and Organogenesis has signed a marketing deal with Sandoz, the drug company.

Genzyme Tissue Repair, another Boston-area company, takes a different approach to engineered skin. Whereas the engineered skin made by Advanced Tissue Sciences the cartilage cells go back into the knee, explains Ross Tubo, scientific director at Genzyme, "they say, 'I recognize the neighborhood, and now I'm going to build the house I used to live in.'"

UNIQUE USES
Applications for engineered tissue seem endless. Take a condition called retrouretal reflux. This bladder abnormality, in which urine flows the wrong way, affects a significant number of children, according to Anthony Atala, M.D., of Boston Children's Hospital. Currently most patients have surgery, which usually means a seven-day hospital stay. Atala's engineered body part could enable patients to go home immediately.

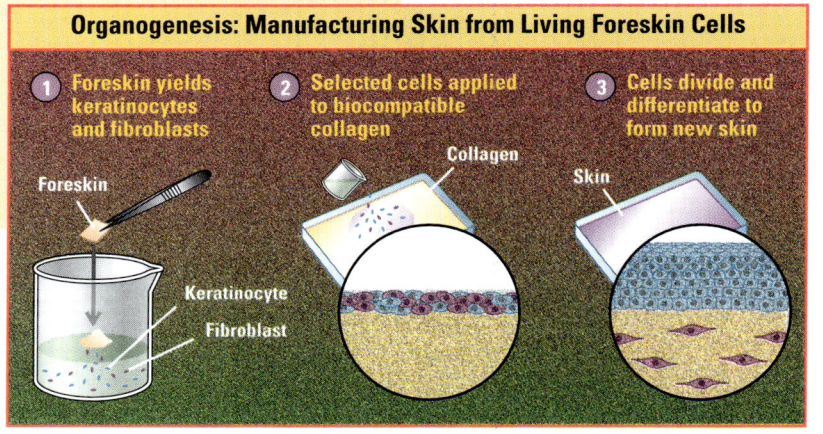

Using the scaffold method devised by Langer and Vacanti, Atala is rebuilding the section of the bladder where it meets the ureter, the vessel through which urine flows. "It takes about four weeks to grow the organ," Atala says. "It's been tested on dogs and pigs—the pig's urinary system is closest to a human's." Human clinical trials were slated to begin as early as May 1996.

In a more challenging effort, David Mooney, a chemical engineer at the University of Michigan in Ann Arbor, is collaborating with James Martin, M.D., of the

Organogenesis executives have enthusiastically promoted the toughness of their Graftskin replacement-skin product (above). The firm has sought federal approval to use the tissue for treating venous ulcers.

and Organogenesis can be used on any patient, Genzyme's manufactured skin is grown with cells from a patient's own skin. Researchers extract keratinocytes from a small biopsy of healthy skin, and culture the cells to manufacture sheets of epidermis that can then be grafted back onto the patient. Obviously, rejection is not a problem.

Using a technique developed in Sweden, Genzyme is also making cartilage that can be used to repair injuries to knees and other joints. Again the starting material is the patient's own cartilage cells, which are engineered to multiply outside the body.

Why do researchers have to grow tissues in the laboratory, rather than in the body? Because in the body, cells are continually sent a mélange of molecular messages that inhibit their growth. Overall, that's a good thing—when cells do not respond to those inhibitory signals, it is called cancer. But outside the body, cells can be put in an environment that allows healthy growth. When

Carolina Medical Center in Charlotte, North Carolina, to tissue-engineer a human-breast replacement for women who undergo mastectomy. Their hope is to start growing a woman's own tissue on a polymer scaffold molded into a realistic shape. Once the tissue has started growing, the scaffold would be implanted in the body, then seeded with cells. It would finish growing in the woman's body. Such a treatment could provide a safe alternative to silicone or saline implants for the 250,000 women who undergo mastectomy each year.

But generating a complex three-dimensional structure like a breast is not easy, Mooney admits. One complication is that a woman's breast consists mainly of fatty tissue, the growth of which is not well understood. Mooney is working with a variety of cell types to learn which will best grow into a realistic breast. "We haven't settled on a specific configuration," he says. "We're just at the stage of doing initial animal trials. We hope to know within three years whether it's a reasonable approach."

If you think that growing a breast is a problem, think about a heart valve, whose regular beating controls the flow of blood in the body. Some 60,000 people a year have faulty heart valves replaced with plastic valves, pig valves, or occasionally with human valves (which are scarce). Christopher Breuer, M.D., who works with Vacanti, is crafting tissue-engineered heart valves.

To cultivate a valve, you take some heart tissue, place it on a polymer scaffold, and pulse nutrients through it. After six weeks or so, the polymer has been absorbed and a valve is beating away. Breuer has made valves from animal cells and implanted them in lambs. So far, the results look promising.

A tissue-engineered valve would be especially useful for patients who undergo valve replacement in their 40s or 50s. Currently such patients are limited to two less-than-ideal options: get an artificial mechanical valve and take an anticoagulant drug for the rest of their lives; or receive a replacement valve from a pig and undergo open-heart surgery again in 10 years when the valve wears out. The hope is that a tissue-engineered valve would solve both problems.

Could scientists grow an entire heart for someone who needs a transplant? The demand is certainly there: of the 40,000 American patients who require heart transplants each year, only about 2,300 get them, because more transplantable hearts are not available. But a full-fledged heart is a major undertaking, partly because of its complex structure and partly because heart-muscle cells so far have not grown well in culture. "It's an open question whether a whole heart can be grown," says Breuer.

How about a liver? Doctors at Cedars-Sinai Medical Center in Los Angeles have forged a liver by sticking cells onto beads coated with collagen. They have tested it outside the body, running blood from patients with failed livers through the artificial liver as a temporary measure until a transplant could be done. Some engineered livers have been implanted in animals, but human trials remain a few years off.

Financial and Regulatory Questions

Whether the new products being tested ever achieve widespread use depends on how well they meet FDA requirements. The potential pitfalls are best illustrated by an episode involving Genzyme's laboratory-grown cartilage cells. Since these cartilage cells are injected into the people who donated them, Genzyme did not believe it necessary to apply for FDA permission to test the Swedish technology in the United States (as is required for any medical treatment that puts foreign substances into the body). But two divisions of the FDA got into an argument about jurisdiction over the procedure, and U.S. clinical tests were forced to a halt after just one patient had been treated.

Beyond regulatory issues, there are financial considerations. With fast-rising medical costs at the center of public attention, much depends on whether insurance companies will reimburse patients for the new engineered-tissue products as they become available for general use. Financial and regulatory questions, which have yet to be answered, can make the complicated biomedical issues look easy.

Varying Views on Vegetarian Diets

by Dixie Farley

Although all vegetarians avoid red meat, individuals approach their diets in different ways. A "semi-vegetarian" takes a comparatively liberal approach, with a diet that includes dairy foods, eggs, poultry, and fish. A "pesco-vegetarian" avoids poultry as well.

Many people are attracted to vegetarian diets. It is no wonder. Health experts for years have been telling us to eat more plant foods and less fat, especially saturated fat, which is found in larger amounts in animal foods than in plant foods.

C. Everett Koop, M.D., the former surgeon general of the Public Health Service, for example, has expressed major concern about Americans' "disproportionate consumption of foods high in fats, often at the expense of foods high in complex carbohydrates and fiber—that may be more conducive to health."

And, while guidelines released by the U.S. Department of Agriculture and the Department of Health and Human Services early in 1996 advise 2 to 3 daily servings of milk—and the same of foods such as dried peas and beans, eggs, meat, poultry, and fish—they recommend 3 to 5 servings of vegetables, 2 to 4 servings of fruits, and 6 to 11 servings of bread, cereal, rice, and pasta: in other words, 11 to 20 plant foods, but only 4 to 6 animal foods. In addition, the revised 1996 guidelines included a positive statement about the healthfulness of vegetarian diets.

It is wise to take precautions, however, when adopting diets that entirely exclude animal flesh or dairy products. "The more you restrict your diet, the more difficult it is to get all the nutrients you need," says Marilyn Stephenson, R.D., of the U.S. Food and Drug Administration's (FDA's) Center for Food Safety and Applied Nutrition. "To be healthful, vegetarian diets require very careful, proper planning. Nutrition counseling can help you get started on a diet that is nutritionally adequate."

Certain people, such as Seventh-Day Adventists, choose a vegetarian diet because of religious beliefs. Others give up meat because they feel that eating animals is unkind. Some people believe it is a better use of Earth's resources to eat low on the food chain; the North American Vegetarian Society notes that 1.3 billion people could be fed with the grain and soybeans eaten by U.S. livestock. On the practical side, many people eat plant foods because animal foods are more expensive.

"I'm a vegetarian because I just plain enjoy the taste of vegetables and pasta," says Judy Folkenberg of Bethesda, Maryland. Reared on a vegetarian diet that included eggs and dairy products, Folkenberg has added fish to her diet. "I love crab cakes and shrimp," she says.

Just as vegetarians differ in their motivation, their diets differ as well. In light of these variations, it is not surprising that the exact number of vegetarians is unknown. In a

National Restaurant Association Gallup Survey in June 1991, 5 percent of respondents said they were vegetarians, yet 2 percent said they never ate milk or cheese products, 3 percent never ate red meat, and 10 percent never ate eggs.

Risks

Vegetarians who abstain from dairy products or animal flesh face the greatest nutritional risks because some nutrients naturally occur mainly or almost exclusively in animal foods.

Vegans, who eat no animal foods (and, rarely, vegetarians who eat no animal flesh, but do eat eggs or dairy products), risk vitamin B_{12} deficiency, which can result in irreversible nerve deterioration. The need for vitamin B_{12} increases during pregnancy, breast-feeding, and periods of growth, according to Johanna Dwyer, D.Sc., R.D., of Tufts University Medical School and the New England Medical Center Hospital in Boston. Writing in the *American Journal of Clinical Nutrition*, Dwyer reviewed studies published between 1983 and 1988, and concluded that elderly people also should be especially cautious about adopting vegetarian diets because their bodies may absorb vitamin B_{12} poorly.

Ovo-vegetarians, who eat eggs but no dairy foods or animal flesh, and vegans may have inadequate vitamin D and calcium. Inadequate vitamin D may cause rickets in children, while inadequate calcium can contribute to the risk of osteoporosis later on.

These vegetarians are susceptible to iron-deficiency anemia because they are not only missing the more readily absorbed iron from animal flesh, they are also likely to be eating many foods with constituents that inhibit iron absorption—soy protein, bran, and fiber, for instance. Vegans must guard against inadequate calorie intake, which during pregnancy can lead to low birth weight; and against protein deficiency, which in children can impair growth, and which in adults can cause loss of hair and muscle mass and abnormal accumulation of fluid.

According to the Institute of Food Technologists and the American Dietetic Association, vegan diets, if appropriately planned, can provide adequate nutrition even for children. Some experts disagree.

Gretchen Hill, Ph.D., associate professor of food science and human nutrition at the University of Missouri, Columbia, believes that it is unhealthy for children to eat no red meat. "My bet is those kids will have health problems when they reach 40, 50, or 60 years of age," she says, "mostly because of imbalances with micronutrients [nutrients required only in small amounts], particularly iron, zinc, and copper." While meat is well known as an important source of iron, Hill says it may be even more valuable for copper and zinc. Copper not only helps build the body's immunity, but also builds red blood cells and strengthens blood vessels. "A lot of Americans are marginal in this micronutri-

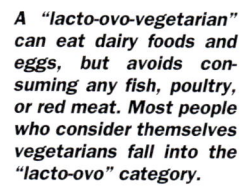

A "lacto-ovo-vegetarian" can eat dairy foods and eggs, but avoids consuming any fish, poultry, or red meat. Most people who consider themselves vegetarians fall into the "lacto-ovo" category.

A "lacto-vegetarian" can eat dairy products, but avoids eggs, fish, poultry, and red meat. Nutritionists suggest that such a person opt for low-fat varieties of milk products.

ent," she says, "and, as a result, are more susceptible to diseases. Children can't meet their zinc needs without eating meat."

Also, vegetarian women of childbearing age have an increased chance of menstrual irregularities, Ann Pedersen and others report in the *American Journal of Clinical Nutrition*. Nine of the study's 34 vegetarians (who ate eggs or dairy foods) missed menstrual periods, but only 2 of the 41 nonvegetarians did. The groups were indistinguishable when it came to factors such as height, weight, and age at the beginning of menstruation.

Can Veggies Prevent Cancer?

The National Cancer Institute (NCI) states in its booklet *Diet, Nutrition & Cancer Prevention: The Good News* that one-third of cancer deaths may be related to diet. The booklet's "Good News" is: vegetables from the cabbage family (cruciferous vegetables) may reduce cancer risk; diets low in fat and high in fiber-rich foods may reduce the risk of cancers of the colon and rectum; and diets rich in foods containing vitamin A, vitamin C, and beta-carotene may reduce the risk of certain cancers.

Part of the FDA's proposed food-labeling regulations states: "The scientific evidence shows that diets high in whole grains, fruits, and vegetables, which are low in fat and rich sources of fiber and certain other nutrients, are associated with a reduced risk of some types of cancer. The available evidence does not, however, demonstrate that it is total fiber, or a specific fiber component, that is related to the reduction of risk of cancer."

As for increasing fiber in the diet, Joanne Slavin, Ph.D., R.D., of the University of Minnesota, in 1990 in *Nutrition Today* gives this advice: "Animal studies show that soluble fibers are associated with the highest levels of cell proliferation, a precancerous event. The current interest in dietary fiber has allowed recommendations for fiber supplementation to outdistance the scientific research base. Until we have a better understanding of how fiber works its magic, we should recommend to American consumers only a gradual increase in dietary fiber from a variety of sources."

The FDA acknowledges that high intakes of fruits and vegetables rich in beta-carotene or in vitamin C have been associated with reduced cancer risk. But the agency believes the data are not sufficiently convincing that either nutrient by itself is responsible for this association.

Pointing out that plant foods' low fat content also confers health benefits, the FDA states in its proposed rule that diets low in fat have been shown to give protection against coronary heart disease, and that it has tentatively determined that "Diets low in fat are associated with the reduced risk of cancer." The FDA notes that diets high in saturated fats and cholesterol increase levels of both total and LDL (low-density lipoprotein) cholesterol, and thus also increase the risk for coronary heart disease, and that high-fat foods contribute to obesity, a further risk factor for heart disease. (The National Cholesterol Education Program recommends a diet with no more than 30 percent fat, of which no more than 10 percent comes from saturated fat.)

For those reasons, the agency would allow some foods to be labeled with health claims relating diets low in saturated fat and cholesterol to decreased risk of coronary heart disease, and relating diets low in fat to reduced risk of breast, colon, and prostate cancer. "Examples of foods qualifying for a health claim include most fruits and vegetables; skim-milk products; sherbets; most flours, grains, meals, and pastas (except for egg pastas); and many breakfast cereals," the proposed rule states.

Dwyer, in her article, summarizes these plant-food benefits: "Data are strong that vegetarians are at lesser risk for obesity, platonic [reduced muscle tone] constipation, lung cancer, and alcoholism. Evidence is good that risks for hypertension, coronary artery disease, type II diabetes, and gallstones are lower. Data are only fair to poor that risks of breast cancer, diverticular disease of the colon, colonic cancer, calcium kidney stones, osteoporosis, dental erosion, and dental caries are lower among vegetarians."

Death rates for vegetarians are similar to or lower than rates for nonvegetarians, Dwyer reports, but are influenced in Western countries by vegetarians' "adoption of many healthy lifestyle habits in addition to diet, such as not smoking, abstinence or moderation in the use of alcohol, being physically active, resting adequately, seeking ongoing health surveillance, and seeking . . . guidance when health problems arise."

SLOW SWITCHING
It is generally agreed that to avoid intestinal discomfort from increased bulk, a person should not switch all at once to foods with large amounts of fiber. A sensible approach to vegetarian diets is to first cut down on the fattiest meats, replacing them with cereals, fruits, and vegetables, recommends Jack Zeev Yetiv, M.D., Ph.D., in his book *Popular Nutritional Practices: A Scientific Appraisal*. "Some may choose to eliminate red meat but continue to eat fish and poultry occasionally, and such a diet is also to be encouraged."

An "ovo-vegetarian" can eat eggs, but avoids all dairy foods and all forms of animal flesh. A person following such a diet should be absolutely certain to obtain adequate calcium and vitamin D.

Replacing Animal Sources of Nutrients

Vegetarians who eat no meat, fish, poultry, or dairy foods face the greatest risk of nutritional deficiency. Nutrients most likely to be lacking, and some nonanimal sources, are:

- *Vitamin B_{12}*—fortified soy milk and cereals;
- *Vitamin D*—sunshine and fortified margarine;
- *Calcium*—tofu, broccoli, seeds, nuts, spinach, kale, bok choy, legumes (peas and beans), greens, and calcium-enriched grain products;
- *Iron*—legumes, tofu, bean sprouts, green leafy vegetables such as spinach, dried fruit, enriched white rice, whole grains, and iron-fortified cereals and breads, especially whole wheat (absorption is improved by vitamin C, found in citrus fruits and juices, tomatoes, strawberries, broccoli, peppers, dark-green leafy vegetables, and potatoes with skins);
- *Zinc*—whole grains, whole-wheat bread, legumes, nuts, popcorn, applesauce, mangoes, and tofu.

Since all plant foods—including fruit—contain some protein, even vegans probably can get enough of this nutrient by eating a variety of fruits, vegetables, and grains every day. To improve the quality of protein and ensure getting enough:

Combine
legumes such as black-eyed peas; chickpeas; peas; peanuts; lentils; sprouts; and black, broad, kidney, lima, mung, navy, and pea beans

with
grains such as rice, wheat, corn, rye, oats, millet, barley, and buckwheat.

There are also foods called protein analogues, made of products such as soy that are made to look like such meats as hot dogs and bacon.

Slowly changing to the vegetarian kitchen also may increase the chances of success. "If you suddenly cut out all animal entrées from your diet, it's easy to get discouraged and think there's nothing to eat," says lifelong veggie eater Folkenberg. "I build my meals around a starchy carbohydrate such as pasta or potatoes. Even when I occasionally cook seafood, I center on the carbohydrate, making that the larger portion. Shifting the emphasis from animal to plant foods is easier after you've found recipes you really enjoy."

Because vegans and ovo-vegetarians face the greatest potential nutritional risk, the Institute of Food Technologists recommends careful diet planning to include enough calcium, riboflavin, iron, and vitamin D, perhaps with a vitamin D supplement if sunlight exposure is low. (Sunlight activates a substance in the skin and converts it into vitamin D.)

For these two vegetarian groups, the institute recommends calcium supplements during pregnancy, infancy, childhood, and breast-feeding. Vegans need to take a vitamin B_{12} supplement because that vitamin is found only in animal-food sources. Unless advised otherwise by a doctor, those taking supplements should limit the dose to 100 percent of the National Academy of Science's (NAS') Recommended Dietary Allowances (RDAs). Vegans, especially children, also must be sure to consume adequate calories and protein. For other vegetarians, it is not difficult to get adequate protein, although care is needed in small children's diets.

Nearly every animal food, including egg whites and milk, provides all eight of the essential amino acids in the balance needed by humans, and therefore constitutes "complete" protein. Plant foods contain fewer of these amino acids than do animal foods.

The American Dietetic Association's (ADA's) position paper on vegetarian diets, published in its journal in 1988 (and updated in 1993) and coauthored by Dwyer and Suzanne Havala, R.D., states that a plant-based diet provides adequate amounts of amino acids when a varied diet is eaten on a daily basis. The mixture of proteins included in foods like grains, legumes, seeds, and vegetables provides a complement of amino acids

In avoiding all forms of animal products, a "vegan" represents the most extreme type of vegetarian. Vegans must monitor their nutritional intake carefully to avoid vitamin deficiencies.

so that deficits in one food are made up by another. Not all types of plant foods need to be eaten at the same meal, since the amino acids are combined in the body's protein pool.

Frances Lappe, in *Diet for a Small Planet*, writes that it is best to consume complementary proteins within three to four hours of each other. High amounts of complete proteins can be gained by combining legumes with grains, seeds, or nuts.

Also available are protein analogues. These substitute "meats"—usually made from soybeans—are formed to look like meat foods such as hot dogs, ground beef, or bacon. Many are fortified with vitamin B_{12}.

The chart on page 160 lists sources of the nutrients of greatest concern for vegetarians who do not eat animal foods. As with any diet, it is important for the vegetarian diet to include many different foods, since no one food contains all the nutrients required for good health. "The wider the variety, the greater the chance of getting the nutrients you need," says the FDA's Stephenson.

The American Dietetic Association recommends:
- Minimizing intake of less-nutritious foods such as sweets and fatty foods;
- Choosing whole- or unrefined-grain products instead of refined products;
- Choosing a variety of nuts, seeds, legumes, fruits, and vegetables (particularly dark-green leafy vegetables), including good sources of vitamin C to improve iron absorption;
- Choosing low-fat varieties of milk products, if they are included in the diet;
- Avoiding excessive cholesterol intake by limiting eggs to two or three yolks a week;
- For vegans, using properly fortified food sources of vitamin B_{12}, such as fortified soy milks or cereals, or taking a supplement;
- For infants, children, and teenagers, ensuring adequate intake of calories and iron and vitamin D, taking supplements if needed;
- Consulting a registered dietitian or other qualified nutrition professional, especially during periods of growth, breast-feeding, pregnancy, or recovery from illness;
- If exclusively breast-feeding premature infants or babies beyond four to six months of age, giving vitamin D and iron supplements to the child from birth or at least by four to six months, as your doctor suggests;
- Usually, taking iron and folate (folic acid) supplements during pregnancy.

With the array of fruits, vegetables, grains, and herbs available in U.S. grocery stores and the availability of vegetarian cookbooks, it is easy to devise tasty vegetarian dishes. People who like their entrées on the hoof also can benefit from adding more plant foods to their diets. You don't have to be a vegetarian to enjoy the wide variety of tasty dishes from a vegetarian menu.

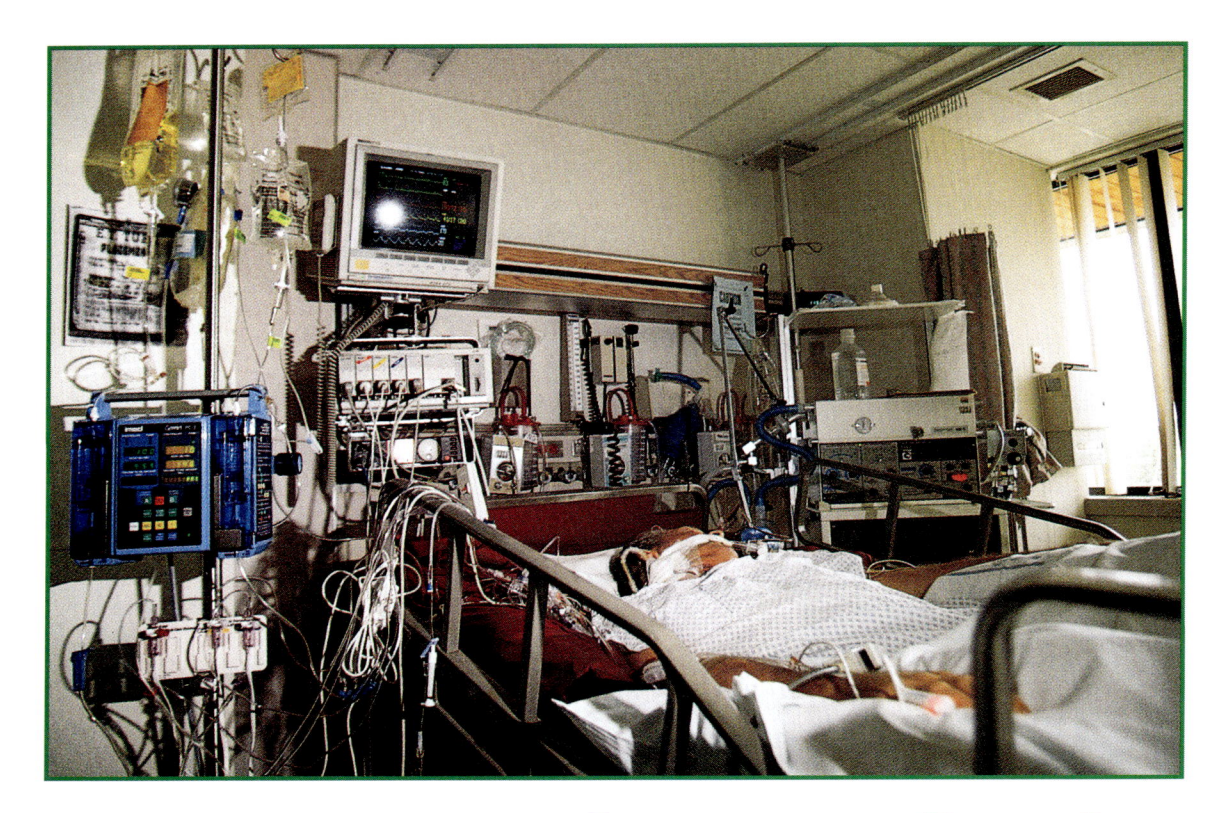

What Are My Chances, Doc?

by Suzanne Oliver

At Dartmouth-Hitchcock Medical Center in Lebanon, New Hampshire, a 68-year-old man sits down at an Apple computer with a nurse. He has coronary heart disease—a catheterization confirmed it. Should he submit to bypass surgery?

The computer already knows that all three of the patient's coronary arteries are severely clogged. It also knows his sex, weight, height, heart function, and other things. Now the patient answers the computer's questions. Would you accept a 25 percent chance of dying in order to alleviate your symptoms? A 20 percent chance?

Out pops a sheet. It informs the tensely waiting patient that his chance of surviving bypass surgery is 97 percent. He chooses the surgery.

This cardiovascular database, developed at Dartmouth and four other hospitals, will be sold by Apache Medical Systems in McLean, Virginia. It's just one of many Apache products [designed to] predict patient outcomes. All told, Apache has on computer the outcomes of 400,000 hospital admittances covering 100-odd diseases. From these statistics, Apache's software can predict patient survival with an accuracy that can sometimes beat that of doctors' hunches. The system's margin of error is plus or minus 3 percent.

To what end? Such predictions help doctors choose the best treatments for patients. They help patients decide whether they want to chance surgery. They assist doctors and families in making decisions about withdrawing life support.

At the University of Michigan Medical Center in Ann Arbor, a 77-year-old woman lies in bed, kept alive by a ventilator. Her

A powerful computer database that predicts a patient's chances of survival is helping doctors solve the difficult treatment dilemmas that frequently confront them in the intensive-care ward or the operating room.

husband doesn't want the ventilator unplugged. Dr. Mark Cowan invites him to glance at a personal-computer screen. It displays a graph with daily estimates of the likelihood that the woman will die during her hospital stay. The likelihood has been 99 percent for seven days. Hardly hopeful odds. The husband agrees to take the woman off the ventilator. She dies almost instantly.

Apache is the brainchild of Dr. William Knaus, an intensive-care [physician] at George Washington University Medical Center in Washington, D.C. After receiving a grant from the Health Care Financing Administration in 1978, he and three colleagues began collecting data on patients in the hospital's intensive-care unit. That data collection eventually expanded to more than 200 hospitals.

Knaus' idea was to create a database to help him make life-and-death decisions about treatments. What better place to look than intensive-care units, where thousands of people lie at the point of death? Long experience with such patients had convinced Knaus that doctors needed help in making life-and-death decisions. "I wasn't smart enough to figure out what to do in each situation," says Knaus.

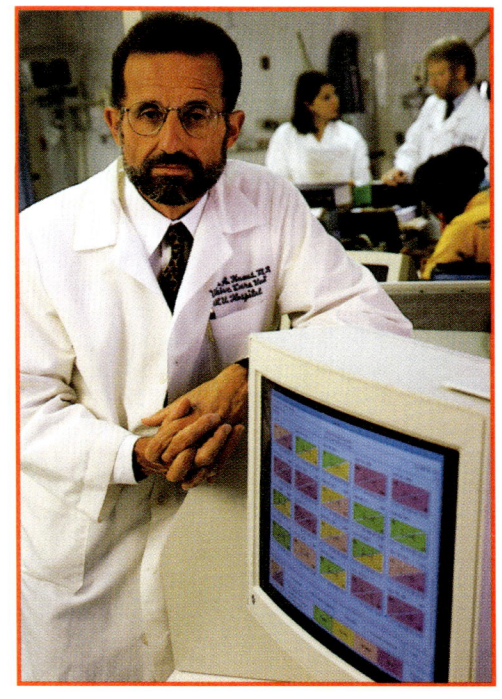

The system's mortality predictions are based on 17 physiological variables, including blood pressure, respiratory rate, temperature, pulse, white blood count, and urine output. [In addition to] those 17, two other variables are age

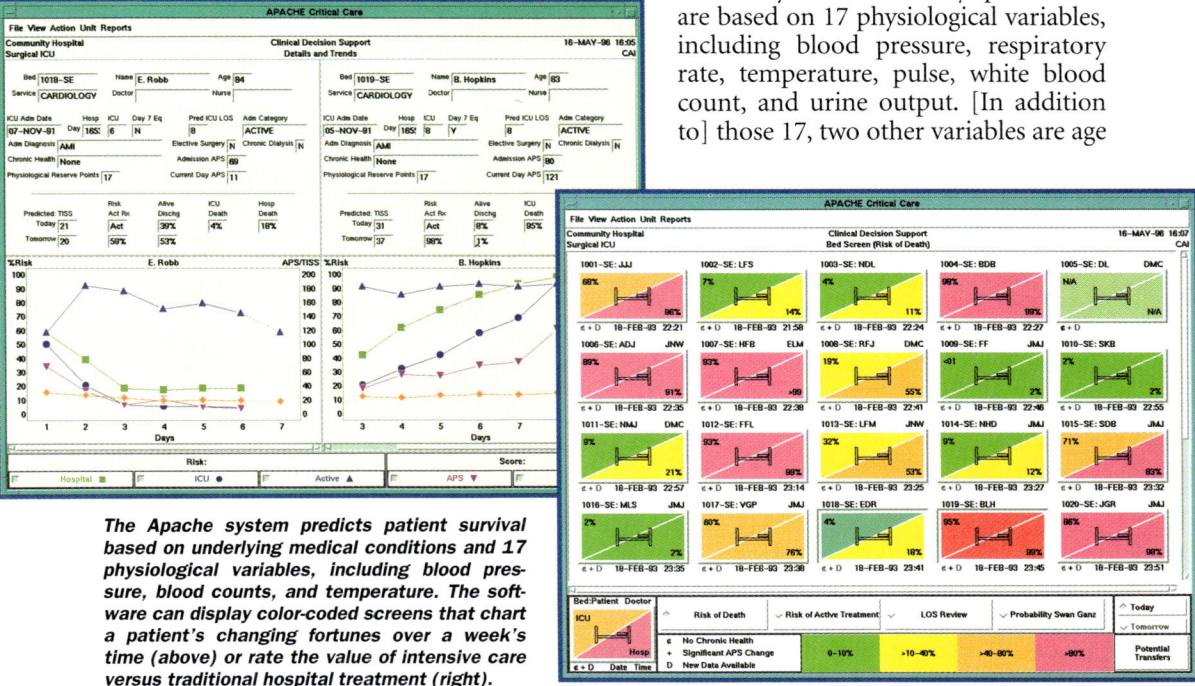

William Knaus, M.D. (above), founded Apache Medical Systems in 1990 after conducting statistical research on patient mortality for more than a decade.

The Apache system predicts patient survival based on underlying medical conditions and 17 physiological variables, including blood pressure, blood counts, and temperature. The software can display color-coded screens that chart a patient's changing fortunes over a week's time (above) or rate the value of intensive care versus traditional hospital treatment (right).

HUMAN SCIENCES   163

and underlying medical conditions. Surprisingly, age has far less to do with patient survival than most doctors think. Another surprise: most preexisting medical conditions do not reduce survival odds, the main exceptions being diseases that suppress the immune system, such as AIDS and cancer.

Unable to get enough grant money to develop Apache further, Knaus in 1990 turned his research into a business—Apache Medical Systems. He raised funds from venture capital investors and Cerner, a hospital software company. Apache's chairman is Gary Bisbee, formerly an investment banker at Kidder, Peabody and founder of Hanger Orthopedic Group. Knaus, 48, still works at George Washington University Medical Center and is Apache's chief scientific adviser. He owns 10 percent of the company, with a probable value of $5 million.

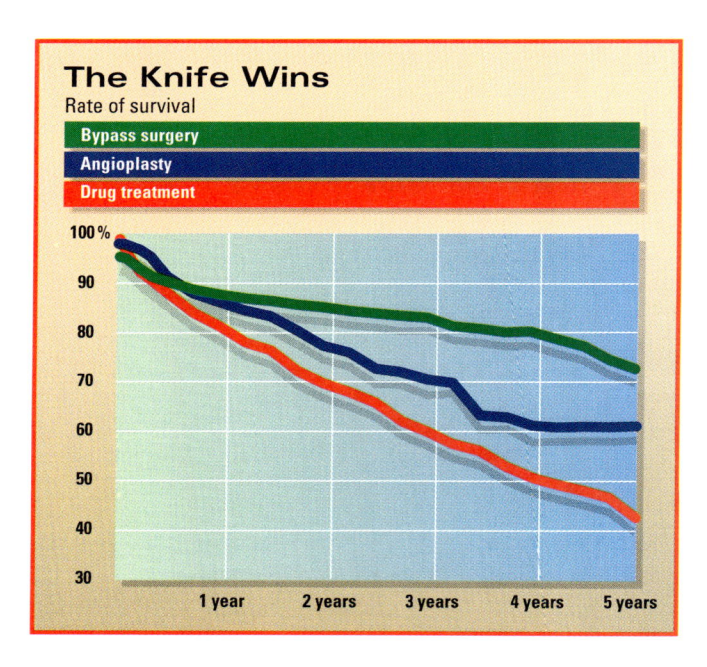

Many hospitals adopted the Apache System to cut costs and measure quality in intensive-care units. Dr. Charles Watts, head of the critical-care unit at the University of Michigan Medical Center, says he saved $2.5 million last year in his intensive-care unit because the Apache System's predictions enabled him to move some recovering patients more quickly out of intensive care, where costs run $2,500 a day and the risk of infection is high. When a patient was demonstrably beyond recovery, it was easier for the family to decide against intravenous feeding that artificially extended his [or her] life.

Apache has also saved many lives. As [often] as three times a day it flashes an unbiased measure of whether a given treatment is working. If mortality risk rises from 50 [percent] to 60 [percent], the doctor knows he must quickly consider another method of treatment. "This is objective material that emboldens the nurse to contact the doctor. It is not just a nurse's opinion against a doctor's," says Watts.

The Apache software for heart disease, developed with the five hospitals, plots five-year survival curves for the patient for three different treatments: drugs, angioplasty (expanding artery channels), and bypass surgery (see chart).

Knaus hopes to push Apache beyond simple survival measurements. He wants Apache to be able to tell patients how fit they will be after leaving the hospital. For instance, will they be able to play golf? To that end, Knaus has begun collecting data on patients who leave the hospital.

Use of Apache, of course, raises ethical questions. Should a machine decide who will get additional treatment and who will be allowed to die? Someone must make these choices, and, Knaus argues, his machine is more objective than a human doctor.

"We have learned that we can't afford to do everything for everyone," says Knaus. "If I were [the patient], I would want to be judged on Apache. It knows only those facts that are relevant to my condition, not race or insurance coverage, which have been used to allocate care in the past."

At any rate, no hospital is going to let a computer play God. The Apache system sometimes out-predicts doctors in a one-on-one match, but the best results still come from teamwork: doctors using Apache to supplement their own instincts.

A Darwinian View of Medicine

by Lori Oliwenstein

Paul Ewald knew from the start that the Ebola-virus outbreak in Zaïre would fizzle out. On May 26, 1995, after eight days in which only six new cases were reported, that fizzle became official. The World Health Organization (WHO) announced it no longer needed to update the Ebola figures daily (although sporadic cases were reported until June 20).

The virus had held Zaïre's Bandundu Province in its deadly grip for weeks, infecting some 300 people and killing 80 percent of them. Most of those infected hailed from the town of Kikwit. It was all just as Ewald had predicted. "When the Ebola outbreak occurred," he recalls, "I said, as I have before, these things are going to pop up, they're going to smolder, you'll have a bad outbreak of maybe 100 or 200 people in a hospital, maybe you'll have the outbreak slip into another isolated community, but then it will peter out on its own."

A DARWINIAN VIEW

Ewald is no soothsayer. He is an evolutionary biologist at Amherst College in Amherst, Massachusetts, and perhaps the world's leading expert on how infectious diseases—and the organisms that cause them—evolve. He is also a force behind what some are touting as the next great medical revolution: the application of Darwin's theory of natural selection to the understanding of human diseases.

Researchers in the field of Darwinian medicine—dedicated to studying how diseases evolve—correctly predicted that a 1995 Ebola outbreak in Zaïre would end quickly due to the rapid progression of the virus.

English naturalist Charles Darwin published his theory of evolution and natural selection in 1859. Researchers are now applying these ideas to gain a better understanding of health and disease.

A Darwinian view can shed some light on how Ebola moves from human to human. (Between human outbreaks, the virus resides in some as-yet-unknown living reservoir.) A pathogen can survive in a population, explains Ewald, only if it can easily transmit its progeny from one host to another. One way to do this is to take a long time to disable a host, giving him or her time to come into contact with other potential victims. Ebola, however, kills quickly, usually in less than one week. Another way is to survive for a long time outside the human body, so that the pathogen can wait for new hosts to find it. But the Ebola strains encountered thus far are destroyed almost at once by sunlight, and even if no rays reach them, they tend to lose their infectiousness outside the human body within one day. "If you look at it from an evolutionary point of view, you can sort out the 95 percent of disease organisms that aren't a major threat from the 5 percent that are," says Ewald. "Ebola really isn't one of those 5 percent."

The earliest suggestion of a Darwinian approach to medicine came in 1980, when George Williams, Ph.D., an evolutionary biologist at the State University of New York at Stony Brook, read an article in which Ewald discussed using Darwinian theory to illuminate the origins of certain symptoms of infectious disease—things like fever, low iron counts, and diarrhea. Ewald's approach struck a chord in Williams. Twenty-three years earlier, he had written a paper proposing an evolutionary framework for senescence, or aging. "Way back in the 1950s, I didn't worry about the practical aspects of senescence—the medical aspects," Williams notes. "I was pretty young then." Now, however, he sits up and takes notice.

While Williams was discovering Darwinian theory in Ewald's work, Randolph Nesse, M.D., was discovering Williams' work. Nesse, a psychiatrist and a founder of the Evolution and Human Behavior Program at the University of Michigan in Ann Arbor, was exploring his own interest in the aging process, and he and Williams soon got together. "He had wanted to find a physician to work with on medical problems," says Nesse, "and I had long wanted to find an evolutionary biologist, so it was a very natural match for us." Their collaboration led to a 1991 article that most researchers say signaled the real birth of the field.

According to Darwinian medicine, a fever may be a defense mechanism evolved by the human body to fight off infections by creating an internal environment that is inhospitable to the invading microbes.

INTERPRETING THE BODY'S DEFENSES
Nesse and Williams define Darwinian medicine as the hunt for evolutionary explanations of vulnerabilities to disease. It can, as Ewald notes, be a way to interpret the body's defenses, to try to figure out, say, the reasons we feel pain or get runny noses when we have a cold, and to determine what we should—or should not—be doing about those defenses. For instance, Darwinian researchers like physiologist Matthew Kluger, Ph.D., of the Lovelace Institute in Albuquerque, New Mexico, now say that a moderate rise in body temperature is more than just a symptom of disease; it is an evolutionary adaptation that the body uses to fight infection by making itself inhospitable to invading microbes. It would seem, then, that if you lower the fever, you may prolong the infection. Yet no one is ready to say whether we should toss out our aspirin bottles. "I would love to see a dozen proper studies of whether it's wise to bring fever down when someone has influenza," says Nesse. "It's never been done, and it's just astounding that it's never been done."

Diarrhea is another common symptom of disease, one that is sometimes the result of a pathogen's manipulating your body for its own good purposes, but that may also be a defense mechanism mounted by your body. For example, once cholera bacteria invade the human body, they induce diarrhea by producing toxins that make the intestine's cells leaky. The resultant diarrhea then both flushes competing beneficial bacteria from the gut and gives the cholera bacteria a ride into the world, so that they can find another hapless victim. In the case of cholera, then, it seems clear that stopping the diarrhea can only do good.

But the diarrhea that results from an invasion of shigella bacteria—which cause various forms of dysentery—seems to be more an intestinal defense than a bacterial offense. The infection causes the muscles surrounding the abdomen to contract more frequently, apparently in an attempt to flush out the bacteria as quickly as possible. Studies done more than a decade ago showed that using drugs like Lomotil to decrease the contractions and reduce the diarrheal output actually prolongs infection. On the other hand, the ingredients in such over-the-counter preparations as Pepto-Bismol, which do not affect how frequently the abdomen contracts, can be used to stem the diarrheal flow without prolonging infection.

Younger diners should be heartened to know that their dislike of broccoli and other vegetables may be an evolved aversion to foods containing high levels of plant toxins—substances that can be harmful to children.

Seattle biologist Margie Profet, Ph.D., points to menstruation as another "symptom" that may be more properly viewed as an evolutionary defense. As Profet points out, there must be a good reason for the body to engage in such costly activities as shedding the uterine lining and letting blood flow away. That reason, she claims, is to rid the uterus of any organisms that might arrive with sperm in the seminal fluid. If an egg is fertilized, infection may be worth risking. But if there is no fertilized egg, says Profet, the body defends itself by ejecting the uterine cells, which might have been infected. Similarly, Profet has theorized that morning sickness during pregnancy causes the mother to avoid foods that might contain chemicals harmful to a developing fetus. If she is right, blocking that nausea with drugs could result in higher miscarriage rates or more birth defects.

EVOLUTIONARY CHECKS AND BALANCES
Darwinian medicine is not simply about which symptoms to treat and which to ignore. It is a way to understand microbes—which, because they evolve so much more quickly than humans do, will probably always beat us unless we figure out how to harness their evolutionary power for our own benefit. It is also a way to realize how disease-causing genes that persist in the population are often selected for, not against, in the long run.

Sickle-cell anemia is a classic case of how evolution tallies costs and benefits. Some years ago, researchers discovered that people with one copy of the sickle-cell gene are better able to resist the protozoans that cause malaria than are people with no copies of the gene. People with two copies of the gene may die, but in malaria-plagued regions such as tropical Africa, their numbers will be more than made up for by the offspring left by disease-resistant kin.

Cystic fibrosis may also persist through such genetic logic. Animal studies indicate that individuals with just one copy of the cystic-fibrosis gene may be more resistant to the effects of the cholera bacterium. As is the case with malaria and sickle cell, cholera is much more prevalent than is cystic fibrosis; since there are many more people with a single, resistance-conferring copy of the gene than with a disease-causing double dose, the gene is stably passed from generation to generation.

"With our power to do gene manipulations, there will be temptations to find genes that do things like cause aging, and get rid of them," says Nesse. "If we're sure about everything a gene does, that's fine. But an evolutionary approach cautions us not to go too fast, and to expect that every gene might well have some benefit as well as costs, and maybe some quite unrelated benefit."

Darwinian medicine can also help us understand the problems encountered in the New Age by a body designed for the Stone Age. As evolutionary psychologist Charles Crawford, Ph.D., of Simon Fraser University in Burnaby, British Columbia, puts it: "I used to hunt saber-toothed tigers all the time, thousands of years ago. I got lots of exercise and all that sort of stuff. Now I sit in front of a computer, and all I do is play with a mouse, and I don't get exercise. So I've changed my body biochemistry in all sorts of unknown ways, and it could affect me in all sorts of ways, and we have no idea what they are."

Radiologist Boyd Eaton, M.D., of Emory University in Atlanta, and his colleagues believe such biochemical changes are behind today's breast-cancer epidemic. While it is

Darwinian researchers contend that humans suffer from back pains, nearsightedness, and other ailments because the body is slow to adapt to modern conditions.

impossible to study a Stone Ager's biochemistry, there are still groups of hunter-gatherers around—such as the San of Africa—who make admirable stand-ins. A foraging lifestyle, notes Eaton, also means a lifestyle in which menstruation begins later, the first child is born earlier, there are more children altogether, they are breast-fed for years rather than months, and menopause comes somewhat earlier. Overall, he says, American women today probably experience 3.5 times more menstrual cycles than our ancestors did 10,000 years ago. During each cycle, a woman's body is flooded with the hormone estrogen, and breast cancer, as research has found, is very much estrogen-related. The more frequently the breasts are exposed to the hormone, the greater the chance that a tumor will take seed.

Depending on which data you choose, women today are somewhere between 10 and 100 times more likely to be stricken with breast cancer than were our ancestors. Eaton's solutions are radical, but he hopes people will at least entertain them; they include delaying puberty with hormones and using hormones to create pseudopregnancies, which offer a woman the biochemical advantages of pregnancy at an early age without requiring her to bear a child.

EVOLUTIONARY COMPROMISES

In general, Darwinian medicine tells us that the organs and systems that make up our bodies result not from the pursuit of perfection, but from millions of years of evolutionary compromises designed to get the greatest reproductive benefit at the lowest cost. We walk upright with a spine that evolved while we scampered on four limbs; balancing on two legs leaves our hands free, but we will probably always suffer some back pain as well.

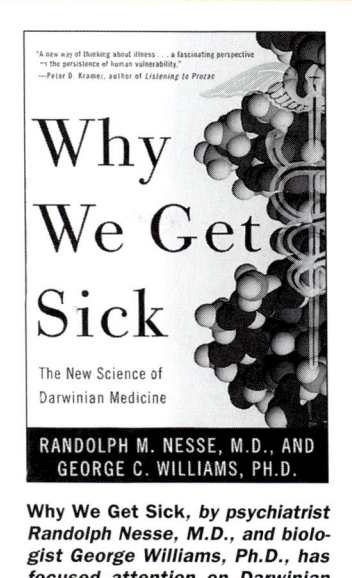

Why We Get Sick, by psychiatrist Randolph Nesse, M.D., and biologist George Williams, Ph.D., has focused attention on Darwinian medicine and its principles.

"What's really different is that up until now, people have used evolutionary theory to try to explain why things work, why they're normal," explains Nesse. "The twist—and I don't know if it's simple or profound—is to say we're trying to understand the abnormal, the vulnerability to disease. We're trying to understand why natural selection has not made the body better, why it has left the body with vulnerabilities. For every disease, there is an answer to that question. And for very few of them is the answer very clear yet."

One reason those answers are not yet clear is that few physicians or medical researchers have done much serious surveying from Darwin's viewpoint. In many cases, that is because evolutionary theories are hard to test. There is no way to watch human evolution in progress—at best, it works on a timescale involving hundreds of thousands of years. "Darwinian medicine is mostly a guessing game about how we think evolution worked in the past on humans, what it designed us for," says evolutionary biologist James Bull, Ph.D., of the University of Texas at Austin. "It's almost impossible to test ideas that we evolved to respond to this or that kind of environment. You can make educated guesses, but no one's going to go out and do an experiment to show that yes, in fact, humans will evolve this way under these environmental conditions."

Yet some say that these experiments can, should, and will be done. Howard Howland, Ph.D., a sensory physiologist at Cornell University in Ithaca, New York, is setting up just such an evolutionary experiment, hoping to interfere with the myopia, or nearsightedness, that afflicts a full quarter of all Americans. Myopia is thought to be the result of a delicate feedback loop that tries to keep images focused on the eye's retina. There is not much room for error: if the length of your eyeball is off by just one-tenth of a millimeter, your vision will be blurry. Research has shown that when the eye perceives an image as fuzzy, the eye compensates by altering its own length.

This loop obviously has a genetic component, notes Howland, but what drives it is the environment. During the Stone Age, when we were chasing buffalo in the field, the images we saw were usually sharp and clear. But with modern civilization came lots of close work. When your eye focuses on something nearby, the lens has to bend, and since bending that lens is hard work, you do as little bending as you can get away with. That is why, whether you are conscious of it or not, near objects tend to be a bit blurry. "Blurry image?" says the eye. "Time to grow." And the more the eye grows, the fuzzier those buffalo get. Myopia seems to be a disease of industrial society.

To prevent that disease, Howland suggests going back to the Stone Age—or at least convincing people's eyes that that's where they are. If you give folks with normal vision glasses that make their eyes think they are looking at an object in the distance when they are really looking at one nearby, he says, you will avoid the whole feedback loop in the first place. "The military academies induct young men and women with 20/20 vision who then go through four years of college and are trained to fly an airplane or do some difficult visual task. But because they do so much reading, they come out the other end nearsighted, no longer eligible to do what they were hired to do," Howland notes. "I think these folks would very much like not to become nearsighted in the course of their studies." He hopes to be putting glasses on them within a year.

"Domesticating" Harmful Organisms
The numbing pace of evolution is a much smaller problem for researchers interested in how the bugs that plague us do their dirty work. Bacteria are present in such large numbers (one person can carry around

Some researchers believe that mild forms of viruses can be harnessed to protect humans from more serious strains—much like early humans used domesticated dogs to stave off their more-threatening cousins, wolves.

more pathogens than there are people on the planet) and evolve so quickly (a single bacterium can reproduce 1 million times in one human lifetime) that experiments we could not imagine in humans can be carried out in microbes in mere weeks. We might even, says Ewald, be able to use evolutionary theory to tame the human immunodeficiency virus (HIV).

"HIV is mutating so quickly that surely we're going to have plenty of sources of mutants that are mild as well as severe," he notes. "So now the question is, Which of the variants will win?" As in the case of Ebola, he says, it will come down to how well the virus travels from person to person.

"If there's a great potential for sexual transmission to new partners, then the viruses that reproduce quickly will spread," Ewald says. "And since they're reproducing in a cell type that's critical for the well-being of the host—the helper T cell—then that cell type will be decimated, and the host is likely to suffer from it." On the other hand, if you lower the rate of transmission—through abstinence, monogamy, or condom use—then the more severe strains might well die out before they have a chance to be passed very far. "The real question," says Ewald, "is, Exactly how mild can you make this virus as a result of reducing the rate at which it could be transmitted to new partners, and how long will it take for this change to occur?" There are already strains of HIV in Senegal with such low virulence, he points out, that most infected people will die of old age. "We don't have all the answers. But I think we're going to be living with this virus for a long time, and if we have to live with it, let's live with a really mild virus instead of a severe virus."

Although condoms and monogamy are not a particularly radical treatment, the idea that they might be used not only to stave off the virus, but to tame it, is a radical notion—and one that some researchers find suspect. "If it becomes too virulent, it will end up cutting off its own transmission by killing its host too quickly," notes James Bull. "But the speculation is that people transmit HIV primarily within one to five months of infection, when they spike a high level of virus in the blood. So with HIV, the main period of transmission occurs a few months into the infection, and yet the virulence—the death from it—occurs years later. The major stage of transmission is decoupled from the virulence." So unless the protective measures are carried out by everyone, all the time, we will not stop most instances of transmission; after all, most people do not even know they are infected when they pass the virus on.

But Ewald thinks these protective measures are worth a shot. After all, he says, pathogen taming has occurred in the past. The forms of dysentery we encounter in the United States are quite mild because our purified water supplies have cut off the main route of transmission for virulent strains of the bacteria. Not only did hygienic changes reduce the number of cases, but they also selected for the milder shigella organisms, those that leave their victims well enough to get out and about. Diphtheria is another case in point. When the diphtheria vaccine was invented, it targeted only the most severe form of diphtheria toxin, though for economic rather than evolutionary reasons.

Over the years, however, that choice has weeded out the most virulent strains of diphtheria, selecting for the ones that cause few or no symptoms. Today those weaker strains act like another level of vaccine to protect us against new, virulent strains.

"You're doing to these organisms what we did to wolves," says Ewald. "Wolves were dangerous to us, we domesticated them into dogs, and then they helped us, they warned us against the wolves that were out there ready to take our babies. And by doing that, we've essentially turned what was a harmful organism into a helpful organism. That's the same thing we did with diphtheria; we took an organism that was causing harm, and without knowing it, we domesticated it into an organism that is protecting us against harmful ones."

A LEGITIMATE FIELD?

Putting together a new scientific discipline—and getting it recognized—is in itself an evolutionary process. Although Williams and Nesse say there are hundreds of researchers working (whether they know it or not) within this newly built framework, the two realize the field is still in its infancy. It may take some time before *Darwinian medicine* is a household term. Nesse tells how the editor of a prominent medical journal, when asked about the field, replied, "Darwinian medicine? I haven't heard of it, so it can't be very important."

But Darwinian medicine's critics do not deny the field's legitimacy; they point mainly to its lack of hard-and-fast answers and its lack of clear clinical guidelines. "I think this idea will eventually establish itself as a basic science for medicine," answers Nesse. "What did people say, for instance, to the biochemists back in 1900 as they were playing out the Krebs cycle? People would say, 'So what does biochemistry really have to do with medicine? What can you cure now that you couldn't before you knew about the Krebs cycle?' And the biochemists could only say, 'Well, gee, we're not sure, but we know that what we're doing is answering important scientific questions, and eventually this will be useful.' And I think exactly the same applies here."

Ask the Scientist

▶ *When someone breaks his or her back, does it mean that the spinal cord has been severed? Are all people who break their backs rendered at least partially paralyzed?*

A broken back usually refers to damage to the bony covering of the spinal cord, called the vertebral column. When the vertebral column is twisted, stretched, or crushed, the vertebrae may be dislodged into misalignment or even broken, causing the fragile spinal cord inside to be stretched or, more seriously, partially or completely severed.

As long as injury to the vertebrae does not result in damage to the spinal cord, no paralysis will occur. However, if the fibers of the spinal cord that carry signals for muscle movement are interrupted, paralysis will occur; the degree of paralysis depends upon the location and the extent of destruction.

▶ *Not so long ago, the expression "nervous breakdown" was frequently used to describe some sort of mental collapse. What were the symptoms of a nervous breakdown? What do they call it now?*

"Nervous breakdown" is a nonspecific term used to describe an individual's temporary inability to function normally because of emotional stress, depression, or other psychological illnesses. Individuals with this problem display a variety of symptoms including poor appetite, insomnia, social withdrawal, and difficulty performing expected responsibilities at home and work.

The problem with the term "nervous breakdown" is that it is very imprecise. Most mental-health professionals avoid this lay expression in favor of more-specific and -descriptive diagnostic labels, such as those used for various anxiety or mood disorders.

▶ *Do doctors understand what makes people yawn? Why do people yawn when they are bored, even if they are not sleepy? Can a yawn ever be truly suppressed?*

People yawn. Dogs and cats yawn. Even fetuses during the second half of pregnancy yawn. Yawning is ubiquitous in the animal kingdom, but no one knows for sure what physiological processes occur to produce a yawn. A number of disparate situations seem to trigger yawning: hunger, sleepiness, boredom, and watching other people yawn.

Some experts believe that yawning is a semi-involuntary arousal reflex, the purpose of which is to reverse reduced levels of oxygen in the blood—and therefore in the brain. Researchers trying to test this hypothesis had subjects breathe higher-than-normal levels of carbon dioxide (to increase yawning) and 100-percent oxygen (to decrease yawning); neither condition affected the rate of yawning. Thus, we still do not really know why people yawn.

Yawning can often be suppressed by increasing mental or physical activity.

▶ *What does it mean when a doctor says that a person's electrolytes need to be balanced? What is an electrolyte? Do electrolytes have something to do with nutrition?*

172 HUMAN SCIENCES

Electrolytes are the positively or negatively charged particles in the blood and other body fluids. Sodium, potassium, calcium, and magnesium ions make up the bulk of positively charged electrolytes. Chloride, phosphate, and bicarbonate comprise most of the negatively charged ones. Each has an important individual and synergistic function. Sodium, for example, is responsible for maintaining the normal volume of blood. Potassium is important for muscle contraction. Bicarbonate is key to preserving the acid-base balance in the body within a narrow range. Calcium is essential for the activity of many enzymes, blood clotting, transmission of nerve impulses, and muscle contraction.

The kidneys, lungs, and gastrointestinal tract are remarkably efficient in assuring the proper levels of electrolytes in the body. If too much of an electrolyte, such as salt (sodium chloride), is consumed and absorbed, the kidneys will excrete the excess in urine. If the calcium level in the blood is high, the intestines turn off further absorption of dietary calcium until the blood level returns to its normal state.

A normal, balanced diet will provide sufficient electrolytes, usually more than are needed (especially salt). The body will then take care of balancing these appropriately. In certain disease conditions, such as hypertension or kidney failure, the doctor may recommend a diet with restricted electrolytes.

▶ *I have read that most heart attacks and strokes occur in the morning. Should that statistic affect one's exercise time? I prefer to do my jogging very early in the day.*

Despite statistics about the most common time for strokes and heart attacks, the real determinant of cardiovascular disease is the status of the blood vessels. If the vessels are wide open, stroke or heart attack will not occur at any time of the day. If the vessels are clogged with fatty deposits, problems could occur day or night. Exertion, stress, and rise in blood pressure—possibly more common in the morning—strain the cardiovascular system and increase the likelihood of serious events.

Jogging at any time of the day (or night) is appropriate and safe as long as the individual warms up and gradually increases the level of exertion. Regular moderate exercise is preferable to intermittent intense exertion. Individuals at high risk for heart attack or stroke should develop an exercise program as part of an overall plan for cardiovascular health under the direction of a physician and an exercise trainer.

▶ *Why does it cost so much to undergo a CAT scan or an MRI? I understand that the equipment is expensive, but a hospital near me had a CAT-scan machine donated to it, and yet it still costs a fortune to have the test.*

There are several reasons why people who have either CAT scans or an MRI are shocked when they see the bills. Both of these procedures, which are used to obtain detailed images of structures within the body, require the use of very expensive equipment. Manufacturers try to recoup the expense of the many years of non-income-generating research that made these machines possible in the first place. Furthermore, new advances in technology require the purchase of updated equipment before the old machines are worn out. Still another reason for the high expense is that hospitals may use these sophisticated technologies to subsidize other, less profitable services.

Finally, there is more to the cost of operating a CAT scanner or MRI machine than just the initial purchase. They are expensive devices to maintain. Technicians must be paid to run them, and physician specialists bill separately to interpret the results of the tests. Despite the cost, these procedures can save money in the long run by providing an accurate diagnosis early and avoiding other unnecessary tests or even surgery.

James A. Blackman, M.D., M.P.H.

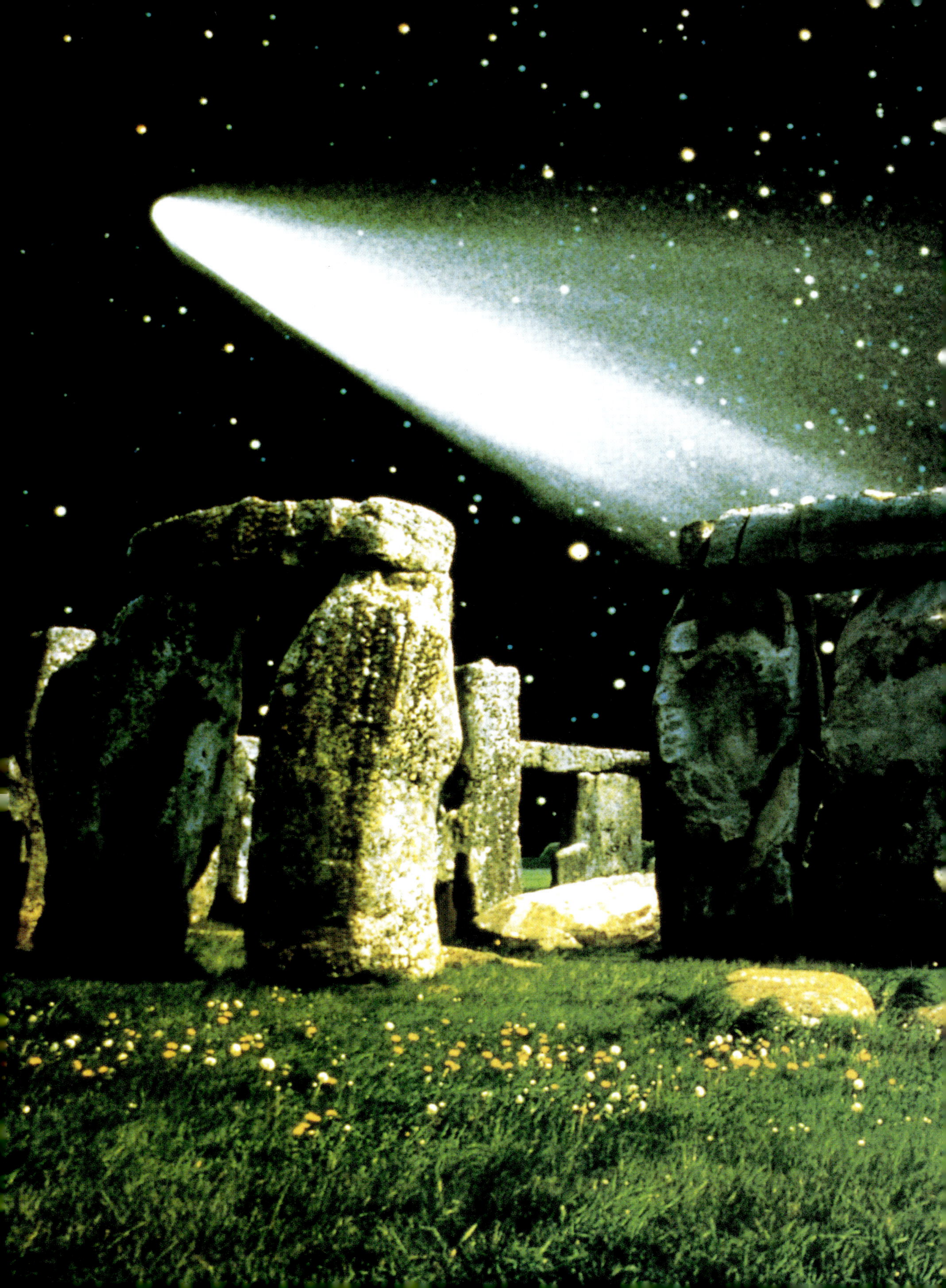

Past, Present, and Future

■ For many thousands of years, details about the construction and function of Stonehenge have been shrouded in mystery. Similarly, the origin of comets millions of years ago is still a matter of conjecture. It will be the task of future generations to definitively answer the questions that stymie the great thinkers of the present.

CONTENTS

The Tomb of the Brothers..........176
Bringing Back the Dinosaurs......184
An Elusive Map of the World......191
Watery Grave of the Azores.......197
The Weather on Trial.................202
Ask the Scientist......................206

The Tomb of the Brothers

by David A. Pendlebury

In May 1995, journalists from news organizations around the world gathered for a press conference at American University in Cairo, Egypt. There they learned of a spectacular discovery in Egypt's Theban necropolis, the sacred burial place of pharaohs who reigned from 1550 to 1050 B.C. Not since 1922, when British archaeologist Howard Carter entered the tomb of Tutankhamen to find, as he said, "wonderful things," had the world's media been as electrified by a discovery from ancient Egypt.

The Cairo press conference was presided over by American archaeologist Kent R. Weeks, Ph.D., chairman and professor of Egyptology and archaeology at the American University. Dr. Weeks described his excavation of an enormous tomb in the Valley of the Kings, some 300 miles (480 kilometers) south of Cairo on the west bank of the Nile. He told those assembled that, in his opinion, the tomb served as the burial place for many—perhaps as many as 50—of the sons of Ramesses II, one of Egypt's greatest rulers. It was the largest, most elaborate tomb ever found in Egypt, he noted.

The exciting news was quickly broadcast on radio and television around the world, and newspapers and newsmagazines trumpeted the discovery. For a moment, the public was transported to the age of Moses.

Intriguing, Albeit Looted
Ironically, the tomb had been under the nose of archaeologists for quite some time. The Valley of the Kings is one of the best-explored places on the planet. The famed tomb of Tutankhamen is, in fact, quite close to the tomb Weeks uncovered. A few explorers of the past century had actually ventured into the tomb's outer chambers and noted its location within the necropolis, but its entrance had thereafter been obscured.

To date, a total of 62 royal tombs have been found in the Valley of the Kings, the last (number 62) being that of Tutankhamen. The tomb Weeks rediscovered is identified on old maps of the area as tomb number 5 (or KV5, for King's Valley tomb number 5). Its entrance lies directly across from the entrance to Ramesses II's own tomb.

Weeks found nothing in KV5 like the precious and delicate objects recovered by Carter from Tutankhamen's tomb: no glittering gold and lapis-lazuli mask, no sleek alabaster vessels, no ebony boxes inlaid with ivory, no chariots, jewelry, fans, or musical instruments. Instead, Weeks and his workers encountered flood debris reaching almost to

Many of ancient Egypt's greatest architectural achievements, including the magnificent temple at Karnak (below), were built millennia ago, during the long reign (1290-1224 B.C.) of Ramesses II.

176 PAST, PRESENT, AND FUTURE

the ceiling of the structure, the result of rare flash floods in the valley. Within this mess, he and his workers found thousands of pieces of broken pottery, scattered beads, fragments of funerary jars and urns, pieces of a stone sarcophagus, broken amulets, and funerary statuettes, as well as animal and human bones, including a knee joint from a mummified human body. From this, it was clear that the tomb had been thoroughly plundered in antiquity. Even such humble objects, however, can prove to be important evidence in understanding the history of a tomb and its occupants.

"Absolutely amazing," observed one Egyptologist. "Very impressive," said another. Other archaeologists gave similar rave reviews. Still, without treasures of the type found by Carter, public curiosity was soon satisfied, and the media frenzy subsided. But Egyptologists continued to marvel at what had been discovered—and at what might still be found within this intriguing, albeit looted, tomb.

"This significant discovery," Weeks says, "will help historians understand more thoroughly the culture, chronology, and history of ancient Egypt during the reign of Ramesses the Great. The structure and size of this tomb will offer insight into the architectural and engineering methods practiced during this time period.

"What makes tomb 5 important," he added, "is that it is a multiple burial of many royal children, a tomb of unique plan and large size, and a tomb whose decoration and artifacts promise to provide information about Egypt's royal family at a crucial period in ancient history. Tomb 5 raises many questions about what else the Valley of the Kings and other areas at Thebes may have to offer. It is an entirely new type of New Kingdom burial structure."

Ancient Egypt developed on the long but very narrow floodplain of the Nile River. Deserts on both sides of the Nile effectively insulated the civilization from outside influences.

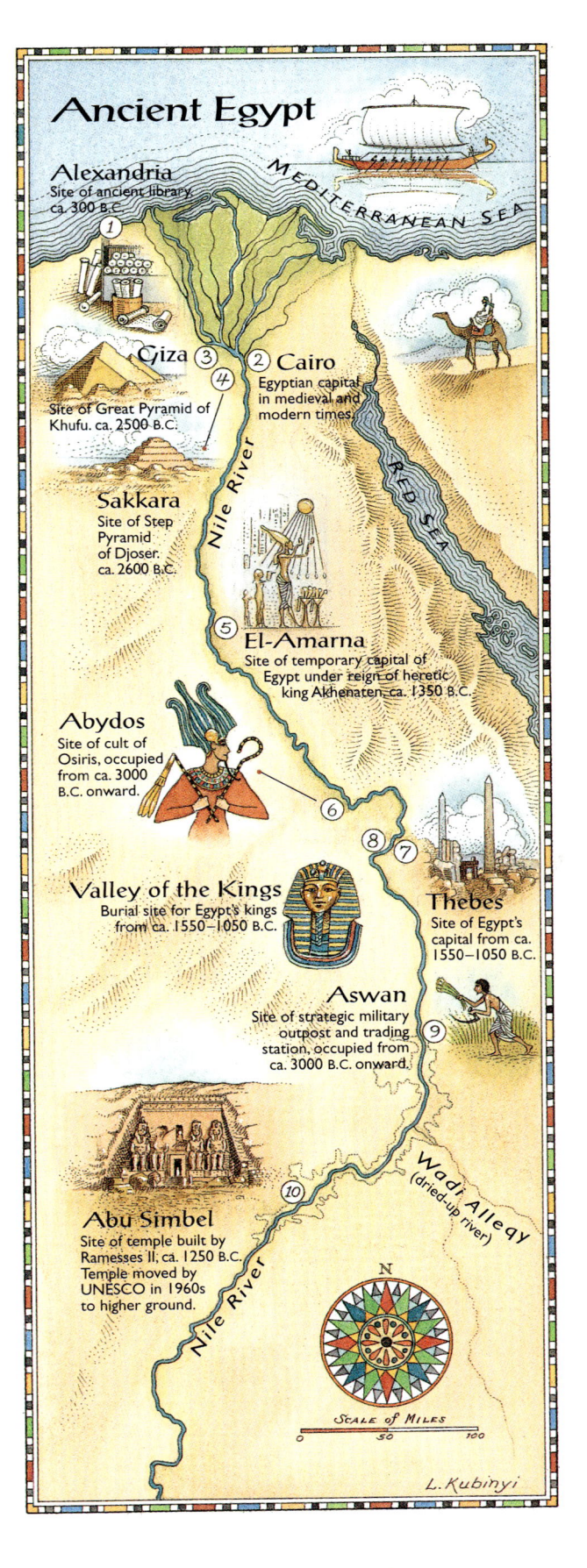

Ancient Egypt

1. **Alexandria** — Site of ancient library, ca. 300 B.C.
2. **Cairo** — Egyptian capital in medieval and modern times.
3. **Giza** — Site of Great Pyramid of Khufu, ca. 2500 B.C.
4. **Sakkara** — Site of Step Pyramid of Djoser, ca. 2600 B.C.
5. **El-Amarna** — Site of temporary capital of Egypt under reign of heretic king Akhenaten, ca. 1350 B.C.
6. **Abydos** — Site of cult of Osiris, occupied from ca. 3000 B.C. onward.
7. **Thebes** — Site of Egypt's capital from ca. 1550–1050 B.C.
8. **Valley of the Kings** — Burial site for Egypt's kings from ca. 1550–1050 B.C.
9. **Aswan** — Site of strategic military outpost and trading station, occupied from ca. 3000 B.C. onward.
10. **Abu Simbel** — Site of temple built by Ramesses II, ca. 1250 B.C. Temple moved by UNESCO in 1960s to higher ground.

L. Kubinyi

Likenesses of Ramesses II are found throughout Egypt. Some are relatively modest (below), while others, like those at the Temple of Amon (above), are truly imposing. The colossal statues of Ramesses II at Abu Simbel (facing page) were moved during the construction of the Aswan Dam to prevent them from being flooded.

THE REIGN OF RAMESSES II

The reign of Ramesses II (1290-1224 B.C.) is prominent in the annals of ancient Egyptian history for several reasons. First and foremost, perhaps, is its mere duration. Ramesses II sat on the throne of Egypt for an incredible 66 years during a time when life expectancy was less than half as long. He outlived many of his children. His 13th son, Merneptah, was well into his 60s when he succeeded his father. Ramesses II himself had reached the ripe old age of 92 by the time he passed into the netherworld.

Second, Ramesses II ruled over a vast territory, one far larger than Egypt's present borders. During his reign, the Egyptian Empire extended from Syria in the north to Nubia in the south. He was a major figure on the world stage, an important player in political and economic matters throughout the Near East. In the fifth year of his reign, Ramesses II led an Egyptian army of 20,000 against the Hittite army in Syria. The outcome of this conflict, known as the Battle of Kadesh, was a draw (although it was proclaimed a glorious victory at home). Sixteen years later, the two enemies became allies when a treaty was concluded between the Egyptian and Hittite empires. Ramesses II married a Hittite princess to further cement the accord, and a lively correspondence between the two royal houses has been preserved.

178   PAST, PRESENT, AND FUTURE

Third, Ramesses II was Egypt's greatest builder. Not only did he erect thousands of monuments to himself, but he also usurped the monuments of others, obliterating the names of his predecessors and replacing them with his own. To this day, his works and his name are found throughout Egypt and even beyond. The forecourt of the Luxor Temple and the great hall of columns of the Karnak Temple, both across the Nile from the Valley of the Kings, were constructed during his reign. He also built a massive mortuary temple, known as the Ramesseum, on the west bank of the river, in the plain below his tomb. The temple of Abu Simbel, which was dismantled and moved to higher ground in the 1960s to preserve it from rising waters caused by the building of the High Dam at Aswan, is also the work of Ramesses II's architects and builders. The great pharaoh even managed to build a new capital city, called Pi-Ramesse, in the northern region of the Nile delta. His burial and that of his sons, however, was in Thebes, in the south.

Tradition holds that Ramesses II is the pharaoh of the Exodus, the one to whom Moses commanded "Let my people go," the one who was besieged by plagues, and the one whose first-born child was killed as punishment for refusing to let the Hebrews depart. It is more likely, however, that the pharaoh referred to in the Book of Exodus was Merneptah, the son and successor of Ramesses II.

The renown of Ramesses II also arises from the scores of sons and daughters he produced—perhaps as many as 100 children in all by eight wives and many concubines. The names of these princes and princesses are recorded on several monuments. The world authority on the reign of Ramesses II, British Egyptologist Kenneth A. Kitchen of the University of Liverpool, has cited a passage from the Book of Psalms (127:4-5) to explain a possible reason why Ramesses II produced so many heirs: "Like arrows in the hands of a warrior are sons born in one's youth. Blessed is the man whose quiver is full of them." It is clear from many sources that Ramesses II lavished attention and praise on his offspring, so the construction of a special tomb for them, one close to the father, is not difficult to imagine.

The Tomb of the Brothers

As mentioned, tomb KV5 was not entirely unknown to Egyptologists—just forgotten. The great archaeological survey of Egypt organized by Napoleon in 1798 recorded a tomb in the Valley of the Kings at the site of KV5. The British explorer and adventurer James Burton (1788-1862) entered the tomb in 1825. He even left a calling card of sorts, using smoke from a candle to imprint his name and date on the ceiling of one of the tomb's outer chambers. In 1902, Carter (1874-1939) cleared a few feet of accumulated debris from the outer chamber of the tomb and noted the cartouche (royal insignia) of Ramesses II, but decided that the tomb was insignificant and abandoned further work there. In fact, Carter later dumped rock and dirt from his subsequent excavations over the entrance to KV5, effectively sealing it until Weeks located the entrance again in 1987.

As director since 1979 of a project to map the Theban necropolis, Weeks has used a hot-air balloon to survey the Valley of the Kings. He has employed sensitive magnetometers to probe the floor of the valley. And he has pored over old notes from early excavators and explorers, such as those of Burton and Sir John Gardner Wilkinson (1797-1875). When Weeks decided to try to relocate and explore KV5, it was as a salvage project: the Egyptian Antiquities Service planned to widen the paved road in the Valley of the Kings to create a turnabout for tourist buses. To find the entrance to KV5, Weeks and his workmen cleared more than 13 feet (4 meters) of rock and debris. They found the entrance to KV5 some 10 feet (3 meters) below the level of the modern road—a discovery that put plans for the bus turnabout back on the shelf.

Weeks spent several seasons excavating the tomb's outer chambers—not a trivial task considering that the tomb, lying well below the modern grade, had become packed with flood

In 1995, archaeologist Kent Weeks (above, at left) discovered a complex tomb in which may lie the mummies of dozens of Ramesses II's sons.

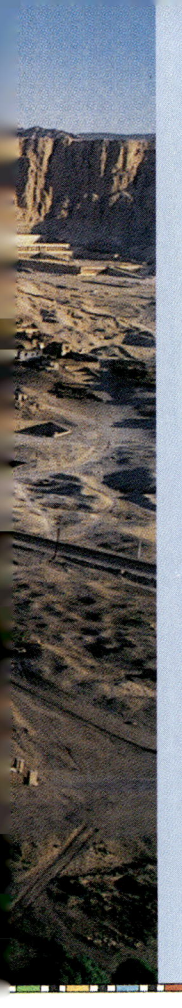

Valley of the Kings

Located 300 miles south of Cairo, the Valley of the Kings is home to 62 tombs.

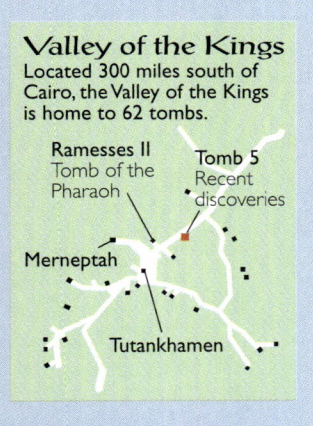

- **Ramesses II** Tomb of the Pharaoh
- **Tomb 5** Recent discoveries
- **Merneptah**
- **Tutankhamen**

By custom, the pharaohs of ancient Egypt began building their tombs as soon as they were crowned. Ramesses II went a step further: besides his own monument—which held his mummified remains—he had an elaborate mausoleum constructed for his sons. In 1995, this so-called "Tomb of the Brothers" (diagrammed at right) was discovered in the Valley of the Kings, not far from the tomb of Tutankhamen (see map above). Today, the Valley is the focus of much archaeological activity (left).

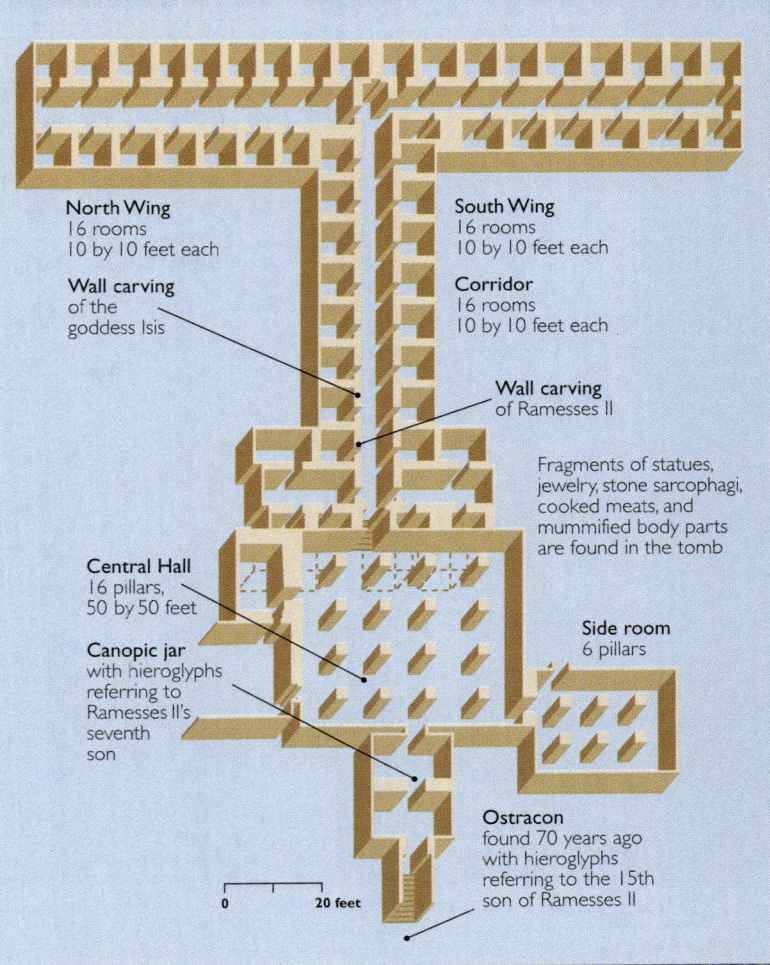

North Wing — 16 rooms, 10 by 10 feet each

Wall carving of the goddess Isis

South Wing — 16 rooms, 10 by 10 feet each

Corridor — 16 rooms, 10 by 10 feet each

Wall carving of Ramesses II

Fragments of statues, jewelry, stone sarcophagi, cooked meats, and mummified body parts are found in the tomb

Central Hall — 16 pillars, 50 by 50 feet

Canopic jar with hieroglyphs referring to Ramesses II's seventh son

Side room — 6 pillars

Ostracon found 70 years ago with hieroglyphs referring to the 15th son of Ramesses II

0   20 feet

---

debris of limestone and gravel, debris that had taken on the consistency of concrete. Directly behind the entrance, Weeks encountered an anteroom, a second room behind it, and then a giant square room measuring 50 feet (15.25 meters) on a side with 16 pillars evenly filling the interior space. An annex, with six square pillars, extended off to the right. Two small rooms were found to the left of the large room. Clearing these spaces and recording the finds were difficult and tedious: the tomb was almost completely choked with debris, and structurally unsound as well, requiring jacks and supports to ensure its stability.

Reliefs and inscriptions on the walls of these chambers, although very badly damaged from flooding and from the effects of a leaky sewage pipe that ran above the tomb, revealed to Weeks that the tomb was associated with the sons of Ramesses II. The king is depicted in these reliefs presenting his sons before various deities. Named on the wall reliefs found in the outermost room of the tomb are Ramesses II's firstborn son, Amenhirkhopsef, and his second born, also named Ramesses. A fragment of a canopic jar, a vessel used to hold one of the internal organs of the deceased that was removed during mummification, was found in the next room; it was inscribed with the name of the seventh son, Seti. A fourth son associated with the tomb is Meryatum, the 15th son of Ramesses II. A pottery fragment inscribed with his name was discovered just outside the tomb's entrance. Of the names of some 50 sons of Ramesses II, known from other monuments and inscriptions, only these four were found in or near the tomb. Since finding these inscriptions in 1987, Weeks has supposed the tomb was built for at least some of the sons of Ramesses II.

On February 2, 1995, Weeks decided to open up a doorway along the back wall of

At the end of a long corridor within the sprawling Tomb of the Brothers stands a relief sculpture of Osiris (above), the ancient Egyptian god of the afterlife. Wall carvings (facing page) in the mausoleum depict Ramesses II presenting his sons to various deities. Other finds may lie hidden beneath the enormous amount of debris.

the large pillared hall. He had assumed that he would find another unremarkable annex, a small room filled with debris like the two rooms to the left of the pillared hall. Instead, after crawling through a small space cut through the top of the debris blocking the door, he and two colleagues found themselves in a huge hallway that extended straight back into the bedrock for as far as they could see. The air was stale. The trio had entered a part of the tomb that had not been visited for more than 3,000 years.

The corridor they found measured more than 100 feet (30 meters) in length. On either side of the hall, they found small rooms, eight on a side and each with its own narrow doorway. At the end of the hallway, they found a statue of Osiris, the god of the underworld, which had been cut in high relief so as to stand out from the wall. More remarkable, they found a second long hallway, which ran perpendicular to the first. Small rooms were also cut from the bedrock along both sides of this hallway—32 in all. The tomb's plan, therefore, was in the shape of the letter *T*.

At each end of the transverse hallway, Weeks encountered steps that seemed to lead downward to a lower level. He imagined that the tomb was designed with a basement, and that burial chambers would be found below each of the small rooms lining the hallways. The small rooms above, he thought, would have served as offering chapels, while the mummified bodies of the deceased offspring would have been entombed beneath these chapels. Further exploration of the tomb would have to wait, however, until the fall season of 1995.

During the fall season, Weeks and his coworkers cleared the debris in both stairways that seemed to lead to the basement, or lower level, of the tomb. Unfortunately, both ended abruptly at a stone wall. Although disappointing, this did not prove that there is no lower level. In fact, the small chapels at the end of the transverse hallway, the rooms closest to the descending stairs, have a fine, smooth plaster floor. When Weeks tapped on this floor, he could tell that a hollow space was directly beneath it. Beginning in June 1996, archaeologists will sink a test trench through this floor to determine whether the hollow space is human-made or a natural formation.

The second revelation from the fall season was the discovery of two new hallways, each lined with more side chambers or chapels. These two corridors, which slope downward at a 30-degree angle, extend from the great pillared hallway in the opposite direction of the Osiris statue. In fact, they extend toward the tomb of Ramesses II, which is a mere 100 feet away. Weeks wonders whether it is possible that the tomb of the brothers might join the tomb of their father. Further clearance and exploration of these new passageways will be a priority for the excavators in the coming season. A French team is currently working in the tomb of Ramesses II. Weeks has joked, "I would dearly love to surprise them.... I'd love to beat the French into their own tomb."

A TREASURE OF INFORMATION
If Weeks and his crew were the first to enter these interior hallways and rooms in KV5 after some three millennia, why, then, have they found no treasure of the type removed from Tutankhamen's tomb? The answer comes from the transcript of an ancient criminal trial, written in ink on papyrus, Egypt's form of paper. The trial was convened around 1150 B.C., and dealt with the prosecution of men who had plundered tombs in the Valley of the Kings. They admitted, in a confession elicited by torture, to robbing the tomb of Ramesses II and, the manuscript says, another tomb "across the path." This second tomb is clearly KV5. Thus, only a century after its completion, the tomb had been violated and ransacked.

Presumably, the high priests made sure the tomb was closed again—and securely. It remained so until Weeks paid a visit in 1995.

The amazing rediscovery of KV5, which is ongoing and quite likely holds more surprises for Egyptologists and the public, has already revealed important new information for historians and archaeologists. The plan, as Weeks says, is unique, and its full extent is yet to be revealed. The family grouping of burials is something new to Egyptologists, who previously have encountered such a scheme only in burials of the sacred Apis bulls in Memphis to the north. The elaborate burial provisions that a king of Egypt made for his sons is also unknown in Egyptian history. If Weeks finds intact burials, treasures not unlike those of Tutankhamen could be found, as could extremely important information on the identity, status, relationships, and health of members of the royal family.

It was the fervent hope of the Egyptians that they would live on in eternity, that their names would not be forgotten. Ramesses II, by constructing buildings and monuments throughout Egypt, ensured that his name would never vanish. He seems also to have achieved the same goal for his beloved sons. The next few seasons of excavation by Weeks and his team may make the names of these ancient Egyptians as familiar to us as that of Tutankhamen.

Bringing Back the Dinosaurs

by Richard Milner

"Large dinosaur bones," says Lowell Dingus, director of the American Museum's fossil-hall renovation, "are extremely heavy, remarkably fragile, and utterly priceless." During the past five years, Dingus' staff has dealt with thousands of them during a massive $34 million overhaul of the American Museum of Natural History's halls of fossil mammals, reptiles, amphibians, and fishes in New York City. Approximately 100 skeletons are featured in the newly refurbished dinosaur displays, making this the most comprehensive exhibition of Mesozoic memorabilia anywhere. Even so, the skeletons represent only 5 percent of the museum's unrivaled collection of 2,000 fossil-dinosaur specimens.

HUMBLE BEGINNINGS

When the imperious Henry Fairfield Osborn founded the American Museum of Natural History's vertebrate paleontology department in 1891, the museum did not have a single dinosaur. He promptly bought up collections, sent field crews out to gather more, and created a new concept for museums by having skeletons of extinct animals mounted in realistic poses. Until then, most museums merely exhibited their slabs of fossils pretty much as they came from the quarry. A bronze bust of Osborn at the museum bears the legend: "For him the dry bones came to life, and giant forms of ages past rejoined the pageant of the living."

Shortly before the turn of the century, a young artist named Charles R. Knight began painting extinct animals for the museum, and Osborn commissioned him to undertake huge murals for the halls of fossil mammals. Knight also created the definitive images of dinosaurs in a series of classic paintings exhibited near the fossil reptiles. So popular were these halls that visitors lined up for hours for a glimpse of Earth's prehistoric past. Soon Knight's images of mammoths and dinosaurs leaped off the museum's walls to enter popular culture. Comic-book artists, toymakers, and filmmakers have copied and recopied his upright tyrannosaurs and sluggish, swamp-dwelling brontosaurs ever since.

Today dinosaurs are more popular than ever, although some of the scientific thinking about them has changed. Many paleontologists now see these animals, which dominated the fossil record for 140 million years, as more active, social, and diverse in their habits than had previously been supposed. New evidence suggests that some were herd animals, that others may have been good parents, and that all our modern birds are descended from small carnivorous dinosaurs. With recent discoveries of new species, along with intensive work on cladistics (the study of branching evolutionary relationships), museum scientists are continuing to puzzle out just how the various dinosaur groups arose and diversified.

TRANSFORMATION TEAMS

Soon after the fossil-hall restoration was announced, Dingus assembled three special groups to undertake the transformation. Phil Fraley, a versatile preparator who worked with Dingus at the California Academy of Sciences, headed a crew that would mount or repair the skeletons; Steven Warsavage directed the installation group; and Jeanne Kelly, of the museum's vertebrate paleontology lab, supervised six preparators who were to clean and conserve the specimens.

"My staff for the mounting and renovation of skeletons was made up of artists and sculptors," says Fraley, "because they tend to be people who have trained themselves in a wide variety of occupations and technologies." Fraley hired five sculptors with diverse skills in metalworking, welding, and fine

As part of the restoration of the fossil halls at the American Museum of Natural History in New York City, dinosaur skeletons were reassembled according to current scientific thinking.

PAST, PRESENT, AND FUTURE   185

arts. Some had worked at art foundries, while others had created sculptures in public parks and even conceptual art (such as the geometric forms that sculptor Dion Kliner once buried in concrete containers "to prevent any meaning from seeping in"). None had ever been especially interested in fossils or dinosaurs; rather, they were chosen for

Artists, sculptors, metalworkers, paleontologists, and curators all contributed their expertise to the dinosaur-skeleton restoration project.

their experience and ingenuity in fabricating durable three-dimensional art. The artists had to learn to work with the curatorial staff. "The designers had their concept," says one crew member, "the curators had the scientific requirements, Phil had his own ideas, and we had to try to satisfy them all. Finished pieces rarely turned out to be what any of them had expected, but they were usually happy with the results."

Before Fraley's group could begin making steel supports for any of the fossil mammals or reptiles, Kelly and her preparators had to clean all 300 specimens, removing dirt and varnish, which often clogged delicate structures. They then coated each bone with plastic diluted in acetone, a mixture that preserves the finest details. Many of the old skeletal mounts were still usable, but Fraley's crew needed to completely dismantle and reposition five specimens, because knowledge about the creatures had changed.

Among those needing major overhauls were the gigantic *Tyrannosaurus rex* and *Apatosaurus* (formerly known as *Brontosaurus*) skeletons. Fraley and his team worked to disarticulate the old king of the tyrant lizards. For decades, it had been standing up on two legs with its tail dragging on the ground—a pose no longer consistent with scientists' understanding of how the creature moved. Since current research holds that birds are the closest living relatives of extinct dinosaurs, the great meat eater had to be posed more like a roadrunner, with its head thrust forward and tail held well above the ground.

HANDLE WITH CARE

"Moving around these tons of dinosaur bones," says Fraley's assistant Paul Zawisha, "one quickly learns to appreciate their extreme fragility. When we first worked on the mammoths in the mammal halls, we removed a metal clamp from one of the tusks, and it exploded like a firecracker. Ribs are crystalline inside and shatter as easily as icicles." Special care was taken when moving the fossils; workers even used special high-tech carts whose wheels don't wobble or vibrate. Remounting large dinosaurs is especially nerve-racking, as it requires hoisting extremely heavy, fragile objects high above unforgiving concrete floors. During the renovation, the crews acquired a great deal of respect for the original preparators of a century ago, who had to confront the same problems without today's technology.

Before any adjustments to the tyrannosaur were attempted, crew member Matt Josephs built a small-scale model in a new, low-riding, running pose. After the miniature

Each and every dismantled dinosaur bone was cleaned and then coated with plastic diluted with acetone. Only then could the skeleton-reassembly process begin. Below, steel braces are connected to an Iguanodon skull. At right, a vise grip is used to hold the armature that will connect the Iguanodon's vertebrae and ribs.

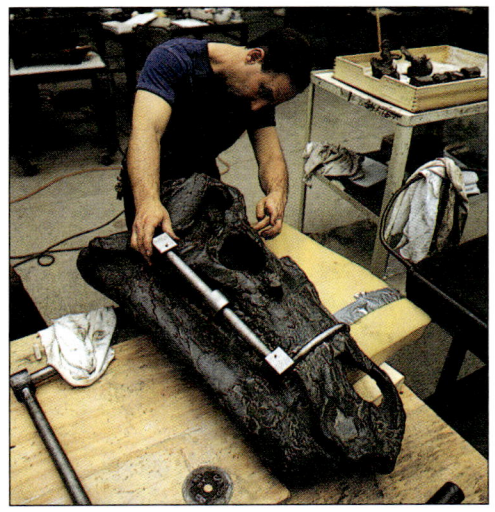

The skull (above) can be accurately attached to the skeleton with the help of a complicated system of chains and pulleys.

Members of the restoration crew contemplate the placement of a dinosaur foot (right) before repositioning it. The skeleton above needs only the addition of a skull to be complete.

PAST, PRESENT, AND FUTURE 187

received curatorial approval, the real tyrannosaur had to be remounted in that position. To disarticulate and reposition the dinosaur, each step had to be carefully planned in advance. Sometimes one small movement required weeks of engineering—calculating torques, stresses, and metal fatigue. Giant A-frames were constructed, complicated rigging was erected all around the mount, and individual sections were surrounded with specially built containers, or cradles.

The crew spent weeks rigging superstructures, until the pelvis, legs, and other elements were suspended from cables and could be moved around with one finger. (The tyrannosaur pelvis alone weighs 2,000 pounds [900 kilograms].) Once the bones' final positions were decided upon, the crew needed to construct an armature, or steel brace, that would follow each twist and turn in the new mount, support most of its weight, and lock everything in place.

Shaping an armature for small skeletons is a relatively simple, but exacting, task. Heating thin steel pipes with a torch, the preparator curves them so they will unobtrusively support the vertebrae, ribs, and legs. During the shaping process, the pipes must continually be placed alongside the bones to achieve a custom fit. To protect the original fossils from damage while working with hot metal, the artisan often substitutes plaster casts of the bones. With the full-scale dinosaur mounts, the process is similar but much more perilous. The legs and pelvis, for instance, must be supported by heavy steel pipes that are shaped using very high heat. Even plaster casts of the bones were prone to burn and crack as hot steel was bent around them. So the crew cast copies of the legs and pelvis in refractory cement, a substance used to line foundry furnaces. "It looks like concrete, but won't crack or crumble under very high heat," crew member Richard Webber explains. "With these special copies, we were able to form and weld our steel armatures without worrying about harming any real dino bones." Now the tyrannosaur, poised to hunt prey, faces the front of the hall, where visitors will enter.

A Relocation Project

Because species are now grouped by evolutionary relationships, the apatosaur and the tyrannosaur, which were not formerly exhibited near each other, have become

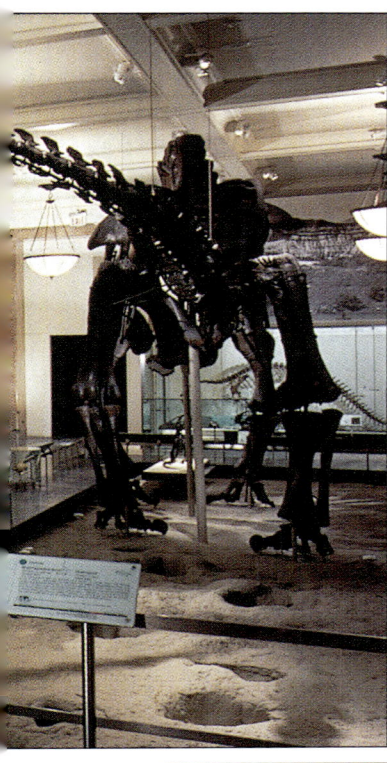

The renovated skeletons bear dramatic testimony to the attention to scientific detail and the meticulous reassembly work on the part of the restoration team. Among the skeletons sharing the Hall of Saurischian Dinosaurs (left) is the notorious Tyrannosaurus rex *(far left) and the gigantic but gentle* Apatosaurus, *formerly known as* Brontosaurus *(near left).*

In the late 1800s, a young artist named Charles Knight created paintings of dinosaurs and huge murals of prehistoric mammals for the museum—images so realistic that they ultimately became part of our popular culture. As part of the renovation program, Knight's murals were restored (below) to their original grandeur.

with walls, floors, joists, stairways, and doors—around the apatosaur. Before the skeleton was moved, the conservators carefully checked every bone for fatigue cracks and stabilized the surfaces with liquid plastic. Finally, the massive skeleton, "house" and all, was moved by electric winch to its new position on the opposite side of the hall. But now that the apatosaur was cleaned and preserved, it was in for some overdue changes.

Skeletons are rarely found complete, so museum mounts are usually composites of several individuals. Originally assembled in 1905, the apatosaur contained a few mistakes. For one thing, its tail had been reconstructed trailing on the ground, but subsequent discoveries of fossil trackways clearly show that the animal did not drag its tail. As more apatosaur skeletons were found, other inaccuracies in the old mount became

roommates. Although the two species lived at different times, as saurischian dinosaurs they are fairly close relatives.

With its head and tail removed, the apatosaur weighed about 7 tons, and Kelly's conservation team had the task of cleaning and preserving it in place. To give preparators easy access to the giant, Warsavage's installation crew built a two-story house—complete

apparent: the tail and neck were too short, and the dinosaur had been sporting the skull of a related species, *Camarasaurus*, for the better part of a century. Vertebrae were added to the now-suspended tail, increasing the skeleton's length to 86 feet (26 meters)—about 20 feet (6 meters) longer than the original version. The wrong skull was removed, and a proper one set in place.

PAST, PRESENT, AND FUTURE 189

A Giant Jigsaw Puzzle

Another major challenge was to move the trackway of fossil footprints that had long been exhibited behind the apatosaur. R.T. Bird had collected these fossilized footprints in 1938 at Texas' Paluxy River. His boss, the famed dinosaur collector Barnum Brown, had ordered him to focus on finding bigger and better skeletons, but Bird's passion was for trackways. Fossil footprints spoke to him about how the animals behaved: their habit of traveling in herds and their manner of locomotion. He eventually succeeded in quarrying for the museum an especially interesting sequence of footprints, in which a meat-eating dinosaur appears to be following or stalking a giant vegetarian.

But Bird's treasured trackways were improperly stored for 15 years, and their identification numbers had been lost. In 1953, when paleontology curator Edwin Colbert decided to place the remarkable sequence of footprints behind the apatosaur, he summoned R.T. Bird out of retirement to direct its assembly. The task was similar to solving a giant jigsaw puzzle in which each piece weighed a ton.

Now, more than four decades after the trackway was reconstituted, installation supervisor Steve Warsavage says, "we had to spin the whole thing around and put it at the other end of the hall. Well, how do you do that when it weighs approximately 22 tons and is falling apart?" Each block had been propped up with bits of plaster and blocks of wood. Although it looked like a unit because of a thin plaster coating, Warsavage knew it would not cohere. "First," he recalls, "we got thick steel plates, beveled the front of them, coated their bottoms with molybdenum grease, and drove them underneath the trackway blocks. You've seen how magicians pull a tablecloth out from underneath the silverware and plates? Well, this was like putting the tablecloth back underneath, without disturbing the plates and silverware." Finally, Warsavage's crew enclosed the trackway within a welded-steel frame, then covered the floor with dishwashing liquid. Despite the mumblings of some skeptics, they were able to slide the whole 22 tons as a unit and place it easily right behind the apatosaur.

Evolutionary-Tree Motif

The tyrannosaur and apatosaur are now part of a comprehensive sequence that, when completed, will form a consistent evolutionary scheme for the six fossil halls. Such exhibits can be organized either as a walk through time or as an evolutionary tree. Most older museum exhibits group animals of the same time period in each hall, regardless of their kin relationships. "We chose to do the evolutionary-tree motif," says paleontologist Dingus, "because it enables us to showcase current scientific research while demonstrating dramatically that we're all connected by this tree back to the original vertebrates."

In many museums, dinosaur halls contain mainly fiberglass casts, some of which are produced at the American Museum's reproduction studio and sent all over the world. "In many instances where they do have actual specimens," says Dingus, "they prefer to keep them off the exhibition floor—either for security reasons or because visiting scientists want the originals available for research. But the curators here want the real material on display. About 85 percent of what you see is the real thing, since some reconstruction of missing pieces is always necessary. Our tyrannosaur exhibit includes the first skull ever collected of the animal. Discovered by Barnum Brown and Peter Kaisen near Big Dry Creek, Montana, in 1908, it will be in a case below the mount, so that people can get up close to examine it."

Although many visitors cannot tell the difference between fossil bones and casts, Dingus insists that exhibiting original material matters. "If you went to your favorite art museum to see a Gauguin, and were told, 'Well, it's not the real thing, but it's a good reproduction,' you'd be disappointed. Putting the real specimens out there shows the public that there is hard, overwhelming evidence for evolution. Fossils, as well as modern genetic and anatomical studies, tell us that these animals really lived, that we share common ancestors with them, and that their descendants can still be seen visiting our bird feeders. People often ask, 'Are those dinosaurs real?' We're one of the few places that can answer, 'Yes, they are.'"

AN ELUSIVE MAP OF THE WORLD

by Simon Winchester

There came a point toward the end of the 19th century when humanity thought that, at long last, it had just about got the measure of the world. People had squared the circle, boxed the compass, and more or less knew where everything was. All the centuries of exploring and adventuring had borne fruit: the human race was in a proprietary, expansive mood. And just as any well-satisfied squires might do, they looked about and muttered: *Right, then—let's see just what we've got.* But 100 years ago, they were not talking about reviewing a mere scattering of acres; they were talking about the entire 58 million square miles (150 million square kilometers) of land and 140 million square miles (362 million square kilometers) of sea—an acreage of unimaginable vastness that had not been, up to that moment, comprehensively mapped.

REVOLUTIONARY PROPOSAL

Then, in Bern, Switzerland, came the decisive moment of change. It was in 1891, at the Fifth International Geographical Congress, that Albrecht Penck, a 32-year-old rising star from the University of Vienna in Austria, stepped up to the lectern and made what was at the time an utterly revolutionary proposal. The entire world, he declared, should and must be mapped, at a scale of 1 to 1 million—a fashionable scale ever since Napoleon's day. And it should be mapped in such a way that if all 2,642 sheets required by the gigantic undertaking were to be stuck together by their edges, they would form a paper spheroid that would be precisely one-millionth the size of the planet—roughly 42 feet (12.8 meters) in diameter.

The audience, 500 of the world's leading geographers, was stunned into euphoric silence. After all, the zeitgeist was one of international agreement. At another geographic conference around this time, the members had decided on time zones (24 of them, an hour apart, but with the island of South Georgia two hours seven minutes slower than London, England; and Bhutan

Although humans have surveyed the land and sea for centuries, serious efforts to produce a universal map of the world were not initiated until 100 years ago.

PAST, PRESENT, AND FUTURE 191

three minutes faster than Nepal). Another gathering had concurred on the establishment of a date line (though with one more zigzag than it has today, allowing it to swerve around two Pacific islands that have since been found not to exist). And diplomats were halfway to agreeing on where to put a zero meridian—Paris (France) or Greenwich (England)—having discounted Toledo, Spain; St. Petersburg, Russia; the Great Pyramid of Giza in Egypt; Cape Verde; and two dozen other places.

Now came this suggestion of the International Map of the World (IMW). The map Professor Penck proposed would have a uniformity to it, so that every sheet would have exactly the same typeface for the names of towns, the same colors for the heights of hills, the same conventional signs to depict roads or railways or swamps. The map would name the cities and rivers and mountains (in roman type) as the locals named them: Zhonghua, Suomi, and Hellas instead of China, Finland, and Greece. And the project would be titled, in the international language of diplomacy, *Carte Internationale du Monde au Millionième*.

Professor Penck sat down, breathless. The audience applauded wildly. And then began the ponderous dynamics of approval. The mills of geography grind exceedingly slow, however rich the grist: the vote taken that afternoon in Bern, while signaling the start of cartography's most nobly ambitious project, was just the first of many. Before any survey teams could fan out, before any engravers could begin to scratch away, before any intaglios could be carved and great gravure presses could begin to thunder out the sheets, there were nearly 20 years of tedious—although doubtless necessary—discussions and debate. The professor aged from youth to middle term before his work saw the light of day. His proposal had to be voted on again at congresses in 1893, 1895, 1899, and 1904. The map's formal standards were discussed and agreed upon only in 1908, and were officially accepted in 1909. In 1913, it was agreed that an office be set up in England, and the project was finally begun. Penck was suddenly recognized throughout the world as the father of the

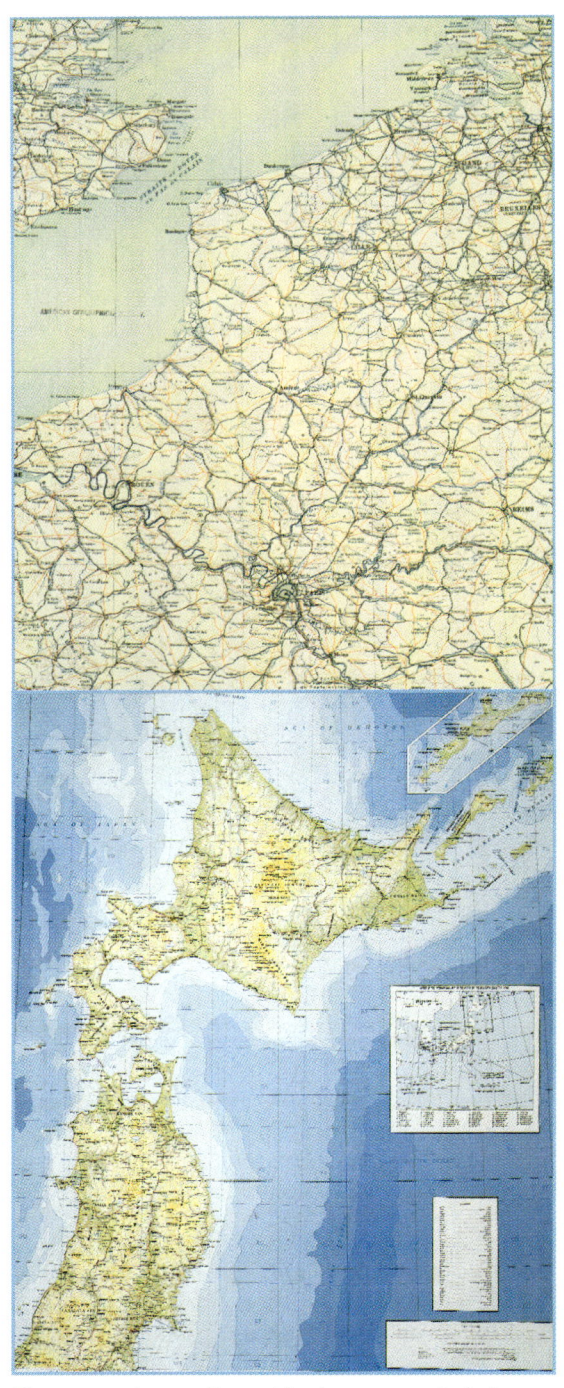

The program to map the world put an unprecedented emphasis on uniformity. A map of northern France (top), completed in 1911, used the identical scale, coloring, and typography as those made a half century later. Many nations participated in this sweeping effort. Japan, for example, contributed a map of its islands of Hokkaido and Honshu (above) in 1966.

A map of the Atacama plateau of Chile and Argentina (facing page), rendered by the American Geographic Society, was acclaimed for its beauty. By 1939, U.S. cartographers had mapped all of South America.

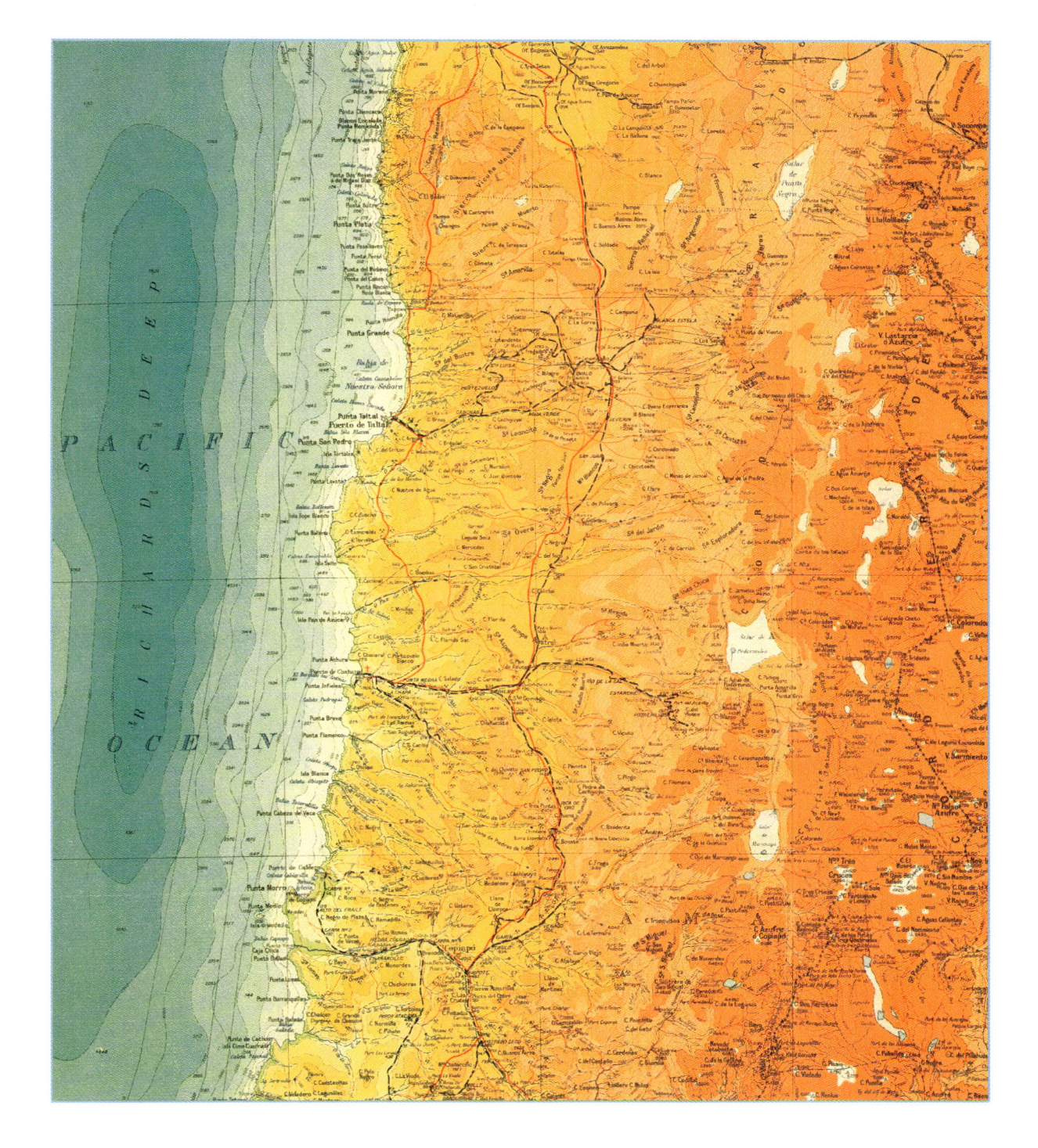

Millionth International Map. Governments and his peers showered him with honors.

A GLORIOUS UNDERTAKING
Now fast-forward to reality. The 1995 edition of the *Encyclopædia Britannica* has an entry, just one paragraph long, on the map. The entry talks of "little interest" by the world community in the aftermath of World War II, and it ends on a moderately sanguine note: "By the mid-1980s, the project was nearing completion." But this note, sadly, is quite wrong. For the truth is that, by the mid-1980s, the project was dying; and in

PAST, PRESENT, AND FUTURE 193

January 1987, it was formally pronounced dead. Albrecht Penck's grand idea survived for just 96 brief years.

Yet all those who were caught up in the excitement of it left a rare and remarkable legacy. During its abbreviated lifetime, the IMW project and those who ran it created more than 800 of the most elegant maps ever made. The results of their work fill 23 rarely opened drawers of the vast map collection at the Library of Congress in Washington, D.C., and are in the collections at the Royal Geographical Society in London, the American Geographical Society Collection at the University of Wisconsin-Milwaukee, and the United Nations' Dag Hammarskjöld Library in New York City. The sheets, all meticulously engraved, are stored away, ready to delight anyone astute enough to seek them out.

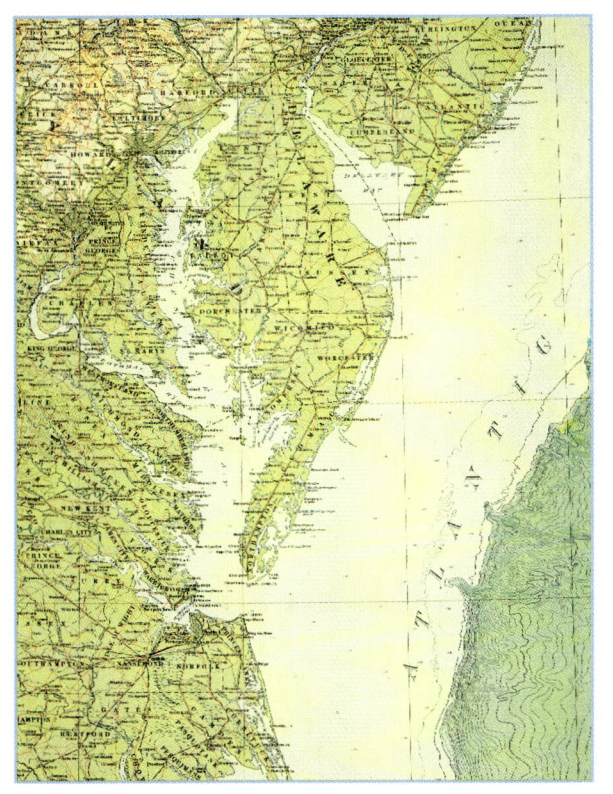

A map of the U.S. Mid-Atlantic region (above) extends from New Jersey to South Carolina and includes views of Chesapeake Bay and the Delmarva Peninsula.

In essence, these sheets are now no more than a memorial to what has become The Map That Almost Was. But in its salad days, the IMW was a glorious undertaking, and it is right to mourn the project's passing. Perhaps in the 20 years of labor that attended its trying birth, in all the fussing and bickering, we can spot some of the reasons for the IMW's premature death. The British and the French, foes eternal, left bleeding wounds: London wanted the scale to be in feet; the French demanded meters. The British required Greenwich to be the prime meridian; the French said Paris. Only after 1904, when the first trial sheets were seen to be creations of quality, did the squabbling begin to still.

The French produced sheets covering the Antilles and Persia; Germany did maps of China's Shandong Province; the British War Office showed 18 sheets it had made of sub-Saharan Africa, with 114 more promised. And these were maps to make the heart sing—precise, artful, properly sized, boldly tinted, adroitly projected sheets of the world's surface. The great American cartographer Earl Hanson, writing of some sister sheets produced 30 years later, remarked that they were "things of beauty and rugged integrity." He might as well have been writing of those first efforts that went on display.

WEARY OR WARY?

The mapmaking effort was led by Sir Charles Arden-Close, a soldier in the Royal Engineers whose life was devoted to surveying. Under his tutelage, real maps began to emerge. France did Marrakech. Britain did Basra and "Istambul." The French did Lyons. Austria-Hungary, in a final burst of Hapsburg zeal, did Budapest. Portugal, then still a major colonial power, offered up a Millionth map of Inhambane, the southern coastal province of Mozambique.

But the Great War (World War I) made nations hitherto interested or involved suddenly wary, weary, impoverished, secretive, or all four. It took until the early 1930s for the Millionth's vigor to return. During this next period, perhaps the greatest of all efforts took place; among other achievements, the privately run American Geographical Society produced the near-perfect collection of 107 sheets covering all of Hispanic America from the Rio Grande to Cape Horn, and which, if fully assembled, would stand 35 feet (10.6 meters) high and 28 feet (8.5 meters) wide. Sir Charles accepted this great gift and added it to the growing pile of sheets flooding into his office in Southampton from all over the world.

And then another world war—an infuriating interruption upon the apolitical civilities of mapmaking. So there were further, and now longer-lasting, frustrations: more secrets, more border disputes, a Cold War, and still more jittery countries ready to forbid all surveys and murder all surveyors. A number of war sheets were hastily produced, however—Germany's well-crafted maps of North Africa, for example, and American sheets of Siberia and parts of Japan. (Washington's mapping of the Siberian river city of Blagoveshchensk is thought to have been inspired, at least in part, by Stalin's decision to make it the capital of a new semiautonomous homeland for Soviet Jews.) But global conflict saps the will to do good works, and by the war's end, the project was in danger of foundering. In an effort to insulate the map from further political turmoil, the fledgling United Nations agreed to take it on.

Decades of Drifting

It would be agreeable to report that under the aegis of the UN, the project revived itself. But quite the reverse seems to have happened. Year by year, the pale-blue reports on the progress of the IMW, published under the hands of the various secretaries-general—Dag Hammarskjöld, U Thant, Kurt Waldheim, Javier Pérez de Cuéllar—listed the maps that had been produced during the previous 12 months; and year by year, the numbers would drop, the blank white spaces on the IMW's main index sheet remaining stubbornly unfilled.

There were many villains. America in particular was woefully derelict. There was a general belief that the United States was somehow unenthusiastic about the project; neither the Defense Department nor the Department of the Interior could agree on who should make the maps—with the result

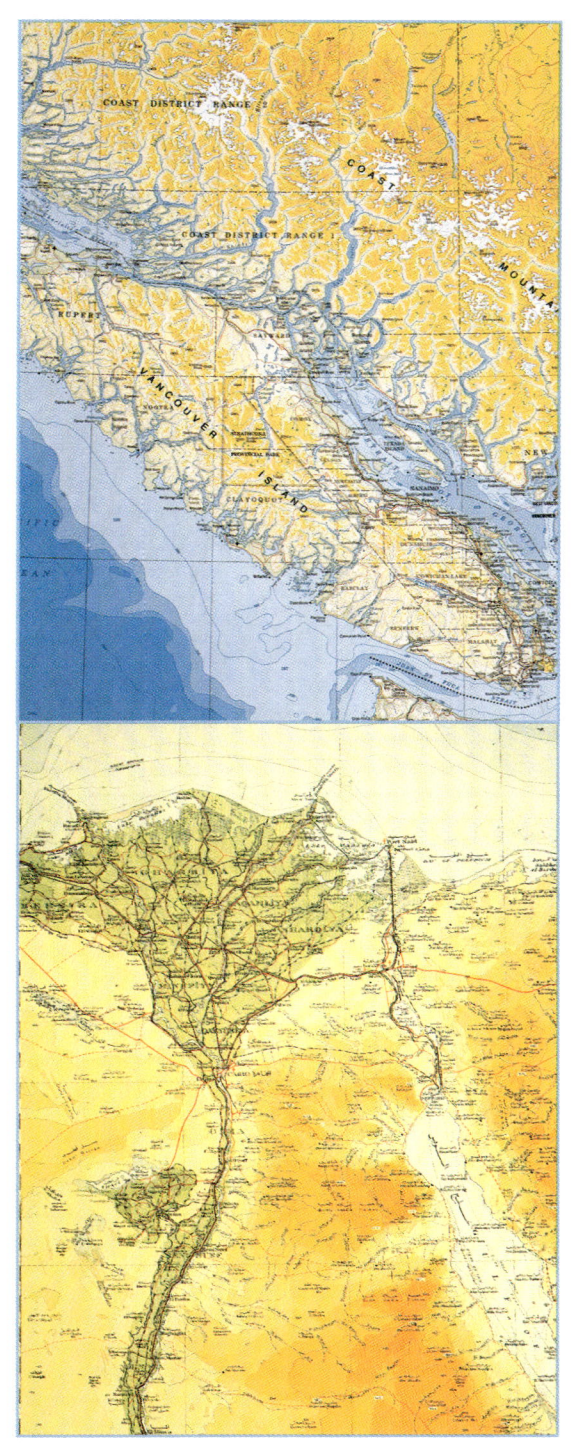

In the latter years of the program, some existing maps, such as Canada's rendering of Vancouver Island and coastal British Columbia (top), were revised using updated mapping techniques and coloring.

A rendering of Egypt's Nile Delta (above) and all of the other works produced for the Millionth Map program give the names of cities, rivers, mountains, and other geographical features in the local language.

that no one did. The Soviet Union did a far better job. Colonial powers mapped most of their colonies. But there were no sheets of northern Canada, none of the Arctic, not a single one in the South Seas. Who to make them? No one could agree.

Nor had many of those maps that did exist been brought up to date. In places like Portugal and Kenya and the Hadhramaut, details—heavy on railways and telegraph lines, but light on such newfangled things as airfields—recalled the now-historical days of King Edward and President Coolidge and maybe even of the czar. The maps looked exquisite, a joy to behold, but time had rendered them wrong and ultimately of little use.

After two more decades of drifting, in 1986 it was clear that the UN was eager—for financial and philosophical reasons—to rid itself of unnecessary responsibilities. The Millionth Map, the UN's Cartographic Section determined, was one such burden. At a conference the following January, and with almost indecent haste, a committee of wise men headed by an American, Rupert Southard, recommended killing and burying the whole exasperating white elephant of a project. The conferees, dull men to their very cores, raised their hands in muted agreement, a pale echo of the crowd at Bern a century before.

"A Terrible Irony"
A final report—not a printed blue book this time, but a single typewritten page—showed evidence of a last spurt of energy, a death rattle, in that terminal year. In 1987, Australia and New Zealand had each sent in two new sheets; Canada, a gallant 32; France, a dozen; and Norway and Malawi, one apiece. But these 50 sheets, though bringing the total assembled in library drawers to more than 850, were still shy of the 974 necessary to cover the entire land surface of the world.

Today, more than nine years after killing it with insouciance and bureaucratic spitefulness, those who were in power at the time of the decision seem almost shamed by what they did. The former head of the map library—now retired and living on Manhattan's Upper East Side—wishes the Millionth Map were still around, saying, "There's nothing half so beautiful, nothing half so useful." The former chief of the Cartographic Section, now living in Switzerland, says the same: "It was a matter of great regret." The present keeper of the map room at London's Royal Geographical Society—a body that was deeply committed to the IMW throughout its life—insists that what has unofficially replaced it, the Digital Chart of the World, is quite incorrect. "It is American—not international. It is based on wrong information. It is not uniform. It is opposed in every way to what the believers in the IMW first stood for, and for which we here still stand. A good, elegant, universal map of the world, produced by the world, for the world. We still don't have one. It is a terrible shame. A terrible irony."

Finish the Job!
And of the present? The United Nations is filled with circumspect men and women, officials who are reluctant to offer a definite opinion about anything that is remotely contentious. But a private document from a senior UN official suggests that the soul of the Millionth Map might well be revived. The map was, he writes, "one of the most monumental cartographic undertakings. . . . It also points to the political, social, economic and scientific benefits that such projects bring about. In my view, this is a role that the UN could and should play. The questions that may now be rhetorically asked are—should the Millionth Map be revitalized? Should it be replaced by a larger-scale series? And would the UN be a logical organization to coordinate such an effort?"

He awaits, he says, someone to formally suggest that the project be revived. What greater gesture could be made than to bring back from the grave this noble and astonishingly beautiful demonstration of the extent of humanity's estate? So let me first say it here: Let us begin a new cartographic revolution! To all the world's mapmakers let us proclaim: Take up your compasses and your chains again, unpack your theodolites and your plane tables, away with you to take the measure of our world—and this time, finish the job!

Watery Grave of the Azores

by William J. Broad

In the Age of Exploration, the Azores were an obligatory last stop for ships returning to Europe with the wealth of the Americas and the Orient, full of gold and silver, silks and spices, gems and diamonds, porcelains and fine steels. Over the centuries, thousands of galleons and other ships stopped at the lush volcanic isles for rest and refreshment, bracing for the final push home—often to no avail.

For centuries, an amazing trove of sunken ships has lain deep beneath the waters off the Azores, a volcanic island chain in the North Atlantic. New salvage technology has finally allowed archaeologists to gain access to these long-hidden treasures.

Attacked by pirates, destroyed in battle, and ravaged by storms, many hundreds of vessels sank to form a hidden museum off-limits to even the deepest divers.

Until now. The doors are opening for the first time, exciting salvors and archaeologists around the world as they jostle for position in line. By some estimates, the deep waters of the Azores hold not only the world's greatest concentration of treasure wrecks, but also countless warships and artifacts that offer an unusual window on Western history.

"It's a turning point," Margaret Rule, Ph.D., a prominent archaeologist in England, says of the opening, which she is aiding. "The Azores were the crossroads of the Atlantic."

HIGH STAKES

Robert F. Marx, a marine archaeologist who has investigated hundreds of shipwrecks around the world, calls it an

PAST, PRESENT, AND FUTURE   197

Private marine archaeologist Robert Marx (above) is among the salvage experts and academics vying to recover treasures from the shipwrecks of the Azores.

archaeological prize. "I've found so much gold and silver in my life that I'm sick of it," he says from Lisbon, Portugal, where he is vying for a license to explore in the Azorean waters. "This is different. It's my childhood dream come true. It's the only place I know where there are hundreds and hundreds of intact ships. It's history come alive."

Francisco J. S. Alves, Ph.D., director of the National Museum of Archeology in Lisbon, says the stakes are extraordinarily high, calling the Azores "a kind of world sanctuary of underwater culture."

Protected by unusually deep and icy waters, the shipwrecks of the Azores are now attracting attention partly because technical advances are opening the abyss. Lost ships have long been salvaged in shallow waters, where divers go down 100 feet (30 meters) or more. But deeper ones, like most of those in the Azores, have become accessible only with the wide availability of advanced gear that is beginning to illuminate the sea's inky depths for the first time.

The other factor is Portugal. Proud of its maritime past, happy to aid the Azorean economy, and eager to increase revenues from taxes and fees, Lisbon is opening the Azorean depths to commercial exploitation while striving to create a public showpiece that sheds as much light as possible on the nautical heritage of Portugal and the West.

A Delicate Balance

As such, the Azores are a case study in whether an ocean state can foster both private gain and public knowledge, a balancing act that has failed in some countries and invariably upsets partisans.

"The question is whether these riches go to the antiquities markets and private collections or to museums and scholars for scientific study," says the National Museum of Archeology's Alves, who has accused the government of wrongly favoring commercial interests over scholarly ones.

But Rui Gomes da Silva, the member of Portugal's parliament who wrote the shipwreck law governing the opening, says the issue was how to strike a middle ground rather than one extreme or another, marshaling as many forces as possible to expedite the opening and enhance the nation's reputation.

"The law is good," he says. "It can make Portugal the center of all the underwater archaeology in the world."

Experts say Lisbon's wreck debate is haunted by the threat of theft, just as in bygone days when pirates looted the Spanish and Portuguese treasure fleets.

"The government recognizes that there'll be piracy unless they get control," says Rule, the British archaeologist. "It's an administrative and practical question of trying to resolve a problem that already exists in Portuguese waters."

Kevin J. Crisman, Ph.D., an archaeologist at the Institute of Nautical Archeology at Texas A&M University in College Station, who has made inquiries about the Azores, says the potential rewards for scholarship are vast, perhaps including discovery of the sturdy little ships known as caravels, a class that included the *Niña*, *Pinta*, and *Santa Maria*, but about which little is known.

Scores of Spanish galleons sank off the isle of Terceira during a violent 1591 hurricane. Over the centuries, hundreds of ships have been sunk in Azorean waters during storms, by pirates, and in naval engagements.

The limits of deep-sea exploration and recovery have been greatly expanded by the use of submersibles. These craft—often equipped with cameras, lights, and sonar—can operate thousands of feet beneath the surface.

"It has to be fantastic," Crisman says of the Azorean wrecks. "For the 16th, 17th, and 18th centuries, it's a very rich place in terms of archaeological potential. You even have a good chance of finding a ship from a time before Columbus. We have all these questions about the ships of that period."

VITAL CROSSROADS

Lying about 1,200 miles (1,930 kilometers) west of Lisbon, the Azores were stumbled upon by Portuguese mariners early in the 15th century and quickly became a vital crossroads because of the North Atlantic winds that blow past in opposite directions.

Through trial and error, mariners, beginning with Columbus in 1492, discovered that their sails could catch easterly winds in the lower latitudes and ride them across the Atlantic. To return, they simply moved northward to catch the westerlies that blew them back toward Europe—and into the Azores, which are generously spread over the sea and hard to miss.

Treasure fleets returning from the New World were joined by ones coming back from India, China, and Japan by way of a long sea route that took them twice around the horn of Africa. These ships returning from the Orient also used the Atlantic's wind system, and the Azores, to speed their homeward journey. Overall, the key port was Angra do Heroísmo, on the isle of Terceira.

Perils included hurricanes, fog, pirates, and war. In 1591, the English tried to seize a Spanish treasure fleet. They failed, but a hurricane soon sank scores of Spanish ships and tons of treasures. Also lost in that storm was the *Revenge*, a famous English warship that had been captured by the Spanish after an all-night battle and whose brave resistance and loss were later put to verse by Tennyson:

> And the little Revenge herself
> went down by the island crags
> To be lost evermore in the main.

The waters of the Azores are unusually deep because the archipelago is a mountainous arm of the Mid-Atlantic Ridge. Rocky shorelines quickly give way to depths of 1 mile (1.6 kilometers) or more. By some estimates, wooden ships there might be well preserved because of the deep's cold temperatures and lack of oxygen.

TECHNOLOGY ADVANCES

For decades, modern archaeologists worked only on shallow wrecks, mainly using scuba gear. Supported by public and private funds, Rule and hundreds of volunteers used such techniques off Portsmouth, England, to salvage the *Mary Rose*, the flagship of Henry VIII that sank in 1545. No treasure was sought or found, but the archaeologists recovered tens of thousands of artifacts that illuminated Tudor life, including longbows, arrows, dice, clothing, guns, and medicines.

The impenetrability of the deep began to change a decade or so ago as new kinds of robots, sensors, and submersibles made their debut and gained wide use for undersea exploration. Eventually the wave included gear cast off by the world's military after the Cold War.

The first big breakthrough came in 1985 when the *Titanic* was discovered under 2.5 miles (4 kilometers) of water. More than 4,000

PAST, PRESENT, AND FUTURE 199

of its artifacts were subsequently gathered up during three expeditions, with more than 150 of them now on display at the National Maritime Museum in Greenwich, England.

Marx, the marine archaeologist—working from 1989 to 1991 with Seahawk Deep Ocean Technology Incorporated, a recovery company based in Tampa, Florida—salvaged a 17th-century Spanish merchant ship about 1,300 feet (400 meters) down off the Dry Tortugas of Florida. Using a deep-diving robot with lights and claws, the team picked up not only pearls, gold bars, and jewelry, but also wooden beams, olive jars, and ballast stones, recording each item's position for future archaeological analysis. Even some human teeth were gingerly recovered, including the preeruptive molar of a child.

Seeing the opportunity at hand, and urged on by Marx, Portugal passed a law in 1993 that sought to recover and preserve the "undersea cultural patrimony of Portugal."

Under the shipwreck law, companies are invited to propose search-and-salvage operations and to prove their expertise and chances of success. If artifacts are judged to be particularly attractive to Portugal, the government has the right to buy them. In general, the reward for the finder is to vary between 30 and 70 percent of the wreck's value, depending on the difficulty of the recovery job and the estimated value of the artifacts.

Exploration licenses cover a maximum area of 100 square miles (260 square kilometers), and recovery licenses cover a zone 2 miles (3.2 kilometers) in radius.

Today a dozen or so companies and groups are applying to work in Portuguese waters, with Marx, Rule, and Alves focusing on Terceira. Seahawk is focusing on São Miguel Island, the surrounding depths of which are also thought to harbor many lost ships.

Archaeological Integrity

New regulations have just been issued that make sure Portuguese museums get a share of the artifacts. The regulations substantially tighten the government's supervisory role, giving it the right to take over salvage sites if it desires and to study all recovered artifacts as long as it wants. The tightening has infuriated Marx, who says he applauds the museum rule, but considers the other ones too onerous and is tired of waiting to start work.

But Rule says the new regulations will aid archaeological integrity. "We're at a rather crucial stage in history to find most everything in the deep and destroy it," Rule says. "My involvement in the Azores is to find a wreck and do an acceptable job to archaeological standards."

Scholars worry that private companies seeking to turn a quick profit for investors will focus solely on treasures and fail to do detailed recoveries that illuminate a ship's cultural cargo.

But Marx and like-minded entrepreneurs charge that academic archaeologists are unknowledgeable dreamers and are unable to attract the kind of money needed to pull off deep recoveries. All deep-salvage operations to date, including those of the *Titanic* and the wreck off the Dry Tortugas, have been accomplished as commercial ventures.

"The academics have never tried deep water," Marx says. "They haven't found a thing."

Alves of the National Museum of Archaeology insists that private funds can be raised for such projects, and he is working with a private Lisbon group to do so.

The cold, deep Atlantic waters that surround the Azores have preserved sunken ships and many of their priceless contents remarkably well.

A Catalog of Riches: A Century of Wrecks off Terceira

In all, 867 ships are believed to have been lost in the Azores during the Age of Exploration. Of these, 274 were reported lost on or near Terceira Island. Here are some of them:

✠ **1504** Caravel loaded with elephant tusks and six chests of gold from Africa.

✠ **1517** *Nossa Senhora do Livramento,* with 44 bronze cannons, described then as the richest ship ever to return from the East Indies.

✠ **1533** *Nossa Senhora de Piedade,* with 66 bronze cannons and a valuable cargo from the East Indies.

✠ **1542** Reported total losses up to this date of more than 30 Spanish ships off Terceira alone; individual records were lost in a 1551 fire in Seville.

✠ **1542** Portuguese ship returning from the East Indies with a treasure of porcelain, gold, silver, precious stones, and "other commodities of the Orient."

✠ **1549** Spanish ship, the *Santa Barbara,* returning from Panama with treasures from Peru.

✠ **1552** Two Spanish ships, the *Madalena* and the *Santiago,* returning from Panama and Havana.

✠ **1554** Flagship of Spanish fleet, laden with treasure, sunk by French pirates.

✠ **1555** Portuguese ship, the *Asumpçao,* returning from the Far East, lost in a storm with "a great quantity of treasure aboard."

✠ **1560** Two Spanish ships, the *Concepción* and the *Trinidad,* carrying treasure from Peru.

Bronze cannons and other artifacts of interest to archaeologists have been recovered from the Azores.

✠ **1586** Two Spanish ships, the *Santiago* and *Nuestra Señora del Rosario,* with valuable cargoes.

✠ **1589** Two Spanish treasure ships returning from the New World, the *San Cristóbal* and *Nuestra Señora de Begonia.*

✠ **1590** Portuguese ship with a valuable cargo from the Far East, and a Spanish treasure galleon.

✠ **1591** At least 88 Spanish ships lost, most near Terceira in a hurricane; six Portuguese ships returning from the East Indies and English warships also lost.

✠ **1593** Six Spanish ships in a treasure convoy lost in a hurricane off Terceira; other ships of the same convoy lost off São Miguel Island.

✠ **1609** A Portuguese treasure ship, the *São Salvador.*

"Universities can work with private donors and people who for humanitarian reasons are interested in giving money for underwater archaeology," he says. "We think that is the way to go, and not the way of treasure hunters."

For Portugal, an added factor giving the wreck issue some urgency is the approach of the world's fair, which will open in Lisbon in 1998. The theme of Expo '98 is the world's oceans, and the exhibition will feature a large walk-through aquarium. Planners would love to have a Portuguese wreck on display there, perhaps raised from deep Azorean waters.

Da Silva, the author of the shipwreck law, says that government critics are overreacting. He says that only a dozen or so of the thousands of wrecks in Portuguese and Azorean waters might be selected for recovery in this century, with the work carefully supervised by a shipwreck commission and the Portuguese navy.

"With good application of the law, and good oversight by the navy, we'll discover a way to be rich in cultural terms," he explains. "We don't even know now how they made Portuguese caravels. It's very important for our history."

The Weather On Trial

by Jane E. Brody

Weather or not is the question that Mark D. Shulman, Ph.D., asks himself whenever a famous case comes to trial.

As a forensic meteorologist, he is less concerned with who done it than with how the weather might have affected critical pieces of evidence or testimony. Was it sunny, cloudy, raining, dry, hot, or cold? Was the Moon shining or the wind blowing, and, if so, in which direction? If a storm was involved, was it a freak occurrence or one that people should have been prepared for?

COUNTLESS FACTORS

In the O.J. Simpson trial, for example, air conditions played a role in such important issues as the melting of Ben & Jerry's ice cream, the shrinkage of the bloodied glove, and the deterioration of unprotected DNA. Watching the trial on television, Shulman was momentarily tempted to offer his services as a scientist who could interpret meteorologic data and apply it to civil and criminal cases.

Forensic meteorologists can assist in such matters as the evaluation of footprints in the snow, the degree of visibility at the time of an incident, whether an instance of severe weather might have been expected or was an "act of God" that could not be anticipated, whether slippery conditions had existed long enough before an accident to have permitted property owners to alleviate a hazard, how far away screams of distress could have been heard, or how long a body might have been floating in a bathtub, based on air conditions and evaporation rates.

Shulman, professor emeritus and former chairman of the meteorology department at Rutgers University in New Brunswick, New Jersey, says the task was often not as simple as it might have seemed. In most cases, it

Meteorological factors—such as those that could shrink gloves or melt ice cream—were debated during the murder trial of football legend O.J. Simpson.

Forensic meteorologists bring the science of weather into the courtroom. Rather than focus on guilt or innocence, these expert witnesses evaluate how the weather affects important pieces of evidence or testimony.

goes well beyond consulting weather reports in the newspaper or the *Farmer's Almanac*, as Abraham Lincoln did when he discredited the sole witness in a famous murder case. The witness claimed to have seen the slaying by the light of the Moon, but Lincoln, citing the 1857 *Farmer's Almanac*, showed that the Moon was in its first quarter and riding low on the horizon at the time of the crime, so that it would have shed almost no light on the crime scene.

Interpretations

The more typical tasks of forensic meteorology, Shulman says, require an ability to interpret recorded weather data, as well as an intimate knowledge of local conditions.

"The job of the forensic meteorologist is to determine what the likely conditions were—temperature, humidity, wind speed and direction, ice and snow, etc.—at an area other than where weather data were recorded, often an area some distance away," Shulman says. The expert must understand how the "microclimate" at the site in question differs from circumstances at nearby weather stations.

For example, he says, "the weather at the site may be affected by cold-air drainage or other unexpected thermal effects, which are often a function of elevation and local topography."

Heat Islands

To determine, say, if conditions might have favored the occurrence of drifting snow, a meteorologist "may first have to examine prior air and soil temperatures, cloudiness, wind speed and direction, and humidity," Shulman explains.

"The determination of precipitation may require the use of surface weather observations from several stations, radar reports, satellite imagery, and a knowledge of the climatology of storms that are common to the

In a famous 1858 trial, Abraham Lincoln relied upon meteorological data from the Farmer's Almanac *to successfully defend a man wrongly accused of murder.*

PAST, PRESENT, AND FUTURE 203

Forensic meteorologist Mark D. Shulman, Ph.D. (left), has testified in many civil and criminal cases. His involvement in such proceedings has ranged from interpreting simple weather data on the stand to conducting sophisticated microclimate analyses.

region," he says, adding that it might also be necessary to factor in moderating influences from nearby urban centers. These centers are called "heat islands" because of greater fuel use, waste heat from transportation systems, and solar heat absorbed and retained by concrete structures and road surfaces.

In many of the cases in which Shulman provided expert testimony, his findings were strengthened by the climatological studies he and his graduate students conducted through the years, including an evaluation of the urban heat island in the New York City–New Jersey area, the frequency and severity of area thunderstorms, and the force and frequency of strong winds in the area.

In one case, the insurance company for a New Jersey warehouse near New York Harbor had refused to pay for extensive flood damage to stored equipment after a thunderstorm. The insurer asserted that the warehouse should have anticipated the risk of flooding and taken measures to protect the equipment. But Shulman examined data from the National Climatic Data Center and research reports from the National Weather Service (NWS), and concluded that the storm had been of extraordinary intensity, producing winds of hurricane force and record-high tides in the harbor.

"Even if the National Weather Service had issued early warnings of severe weather," he concluded, "the warehouse could not have anticipated a storm of this magnitude and would not have been able to protect the equipment that got damaged as a result."

ACTS OF GOD

When a sign blew down in a western New Jersey shopping center one January and injured a woman, was it negligence (did the owner fail to secure and maintain the sign properly?), or was it, in insurance lingo, an "act of God," the result of a wind so unusually intense and unpredictable that the owner could not have anticipated it? If negligence was determined, the owner would have been held personally liable for the woman's injuries, but if the accident was deemed an act of God, the insurance company for the shopping center would pay.

By checking the hourly data recorded at the nearest National Weather Station at Philadelphia International Airport, Shulman determined that based on changes in temperature, wind speed, and wind direction, a strong cold front passed through the area at the time of the accident. A further analysis of many years of data that the National Oceanic and Atmospheric Administration (NOAA) had collected about climatological conditions for January at the airport revealed that "a wind of this intensity was a rare occurrence," Shulman said. "So the intense wind that knocked down the sign fell within the category of an act of God."

When a display tent from a roadside business was blown from its moorings during a summer thunderstorm and flung into a passing car, severely injuring the driver, Shulman reached a different conclusion.

He analyzed the relevant radar data, wind direction and speed, and intensity of the storm, and compared the findings with the results of a five-year study of area thunderstorms that he had conducted with Paul J. Croft, then a doctoral student. Shulman was then able to show that "storms of this magnitude occurred on average 10 times a season in that part of New Jersey," and therefore, the storm that precipitated the accident was no act of God. The driver, who had sued the roadside business, was awarded all the money she had sought.

A meteorologist's testimony about lunar and tidal activity may have played a role in the acquittal of William Kennedy Smith (above, at left) in his 1991 rape trial.

FORENSIC RECONSTRUCTIONS

Many other factors also come into play during the reconstruction of probable meteorologic conditions at a particular place, including local topography (valleys, for instance, are typically colder at night than hillsides) and how close the site is to major bodies of water. Shulman explains that a sea breeze could make a significant difference in the temperature and humidity of a site. But depending upon conditions, there may or may not be a sea breeze.

In 1991, one of Shulman's former students, Herbert Spiegel, was an expert witness in a case far more notable than the professor himself has ever been involved in: the William Kennedy Smith rape trial in Florida. Smith was accused of raping a woman after a moonlit stroll along the beach at the Kennedys' oceanfront mansion in Palm Beach.

Spiegel, then a meteorologist at Miami-Dade Community College in Florida, testified for the defense. In an interview, he said that he was asked to determine, on the basis of the tides, whether there actually was a beach at the time that the woman said that she was attacked. Spiegel said he was also asked to determine whether visibility was good enough for people on the beach to be seen from the house, and whether a woman's screams might have been heard at the house above the wind and waves.

Spiegel, now a professor emeritus and owner of Hilkar Consulting, his own forensic-meteorology company in Fort Lauderdale, Florida, said he had testified that there was indeed a beach that night, that it was bathed by an unobstructed full Moon directly overhead, that the wave action was moderate, and that the wind blowing toward the house would have made it easy for cries of distress to be heard there.

Spiegel said he believed that his testimony had helped Smith be acquitted.

Ask the Scientist

▶ *I have heard that some of the roads built by the Romans are still in use today. Can Roman roads handle today's vehicular traffic? How vast was the network of Roman roads?*

Roman roads exist throughout the territory conquered by the ancient Romans. By the time they finished their elaborate highway system, the Romans had built more than 50,000 miles (80,500 kilometers) of roads. These roads stretched across Western Europe, the Balkans, North Africa, Greece, and the Middle East.

A number of these Roman roads have been in continuous use for more than 2,000 years, says David Schulz, director of the Infrastructure Technology Institute at Northwestern University in Evanston, Illinois. The Appian Way, perhaps the most famous Roman road still being traveled upon, was named for Appius Claudius Caecus, who began the military highway's construction in 312 B.C. Along sections of the ancient thoroughfare, modern tourists can view the ruins of tombs of notable Romans from classical times.

"In many ways," Schulz says, "the Romans built roads better back then than we build them today. To an extent we don't grasp, they understood and appreciated the importance of transportation infrastructure."

▶ *Christmas Day in 1997 can also be expressed as 12/25/97. But how will it be expressed in the year 2000—by 12/25/00? Also, will high-school and college students who graduate in the year 2000 refer to themselves as belonging to the class of zero?*

Christmas at the turn of the century will most likely be written 12/25/00, predicts Ash DeLorenzo, the trend director for BrainReserve, a marketing-consulting firm in New York City. No one, he asserts, will confuse this date with Christmas 1900.

What students call the class of 2000 is a more speculative question. DeLorenzo thinks students will embrace one of two options. The first is simply sticking with "The Class of 2000." "It's a mouthful, but it's futuristic and it's exciting," he says. Or the students might refer to themselves as the Class of 00 (pronounced *Oh, Oh*). "I think they will call it the Class of 00, because saying 'zero, zero' sounds too negative." This generation will strive to avoid the double-zero designation, he suggests, to protect its self-esteem.

▶ *During a recent visit to Scotland, I could not help but notice the fuss made over the Loch Ness monster. How did that legend come to be? Do scientists think that the tales of "Nessie" have any truth to them?*

The tale of a Scottish sea monster may well trace its roots to Celtic folklore. Sightings of a monster have been recorded as far back as the 6th century. The modern "Nessie" caught the public's fancy back in 1934, when Colonel Robert Wilson allegedly spotted and photographed the beast in the murky waters of Scotland's Loch Ness. This photograph, like others after it, was grainy and, for a lot of skeptics, inconclusive. Nonetheless, interest in the beast eventually mushroomed into a multimillion-dollar industry complete with submarine rides and a multimedia tourist center.

In 1994, a story in London's *Sunday Telegraph* suggested that the famous photograph was faked. Apparently, back in 1934, London's *Daily Mail* hired a self-described "big-game hunter" named Marmaduke Wetherell to find the creature. Wetherell induced his son, Ian, and stepson, Christian Spurling, to stage the photo using a toy submarine. Ian Wetherell actually shot the photo, although credit was given to Colonel Wilson to lend the photo greater credibility. The photograph made such a huge splash that the pranksters decided their best course of action was to keep quiet. In 1993, however, on his deathbed, Christian Spurling confessed to his role in the hoax.

▶ *When I was a child, futuristic scenes often included depictions of monorails. Now it seems that Disneyland and other theme parks are the only places that have them. Why did monorails fail to catch on?*

In 1959, Walt Disney unveiled the monorail at Disneyland, which he envisioned as the modern means of transportation for the future, according to Dave Smith, Disney's archives director. But Walt Disney's prediction for the monorail, which uses one rail rather than the traditional two, never materialized. Instead, cities came to rely upon such modes of transportation as subways and light-rail systems.

Even though monorails have been used extensively in Japan, a perception exists in the United States that this type of transportation can work only in a theme park. Also, city dwellers have complained that an elevated monorail system would create an eyesore and be more intrusive than other forms of mass transit.

▶ *It seems as though some people would pay a great deal of money to ride the space shuttle. Do scientists expect that someday space tourism will become a viable industry? If so, when?*

Within the next two decades, it is "completely conceivable" that tourists will be soaring into space, predicts David Steitz, a National Aeronautics and Space Administration (NASA) spokesperson. Space enthusiasts may be able to travel to space stations and even spend time at a space hotel.

Tourism will be a natural consequence of the push to commercialize space exploration and to reduce its exorbitant cost. Right now, the price tag for commercial space endeavors is prohibitive, but NASA expects that as new technology pushes down the expense, corporations will become involved. The pharmaceutical industry, for instance, might build its own space laboratories because of its need to develop zero-gravity drugs.

The first private citizens to regularly visit outer space will probably be scientists. But maintaining space resorts for tourists would be an obvious way to help defray the cost of this research.

▶ *I have read that Quebec City is the only walled city in North America. Is that true? Why were walls built around the city in the first place?*

Quebec City is, in fact, the only walled city that exists in either the United States or Canada. From its very founding in the early 17th century, Quebec was encircled, says Ron Stagg, the chair of the history department at Ryerson Polytechnic University in Toronto, Ontario. Quebec traces its roots to the construction of a fortified building that bore a resemblance to a medieval castle. By the late 1690s, the French inhabitants had built an earth-and-wooden wall around their community; the French fortified it with masonry during the mid-18th century.

The French raised the structure to protect themselves against the British and the Iroquois. Nevertheless, the British ultimately captured Quebec. Unable to withstand the assault of high-powered cannon fire, the wall became obsolete; it owes its survival to shrewd politicians who saw its potential as a prime tourist attraction.

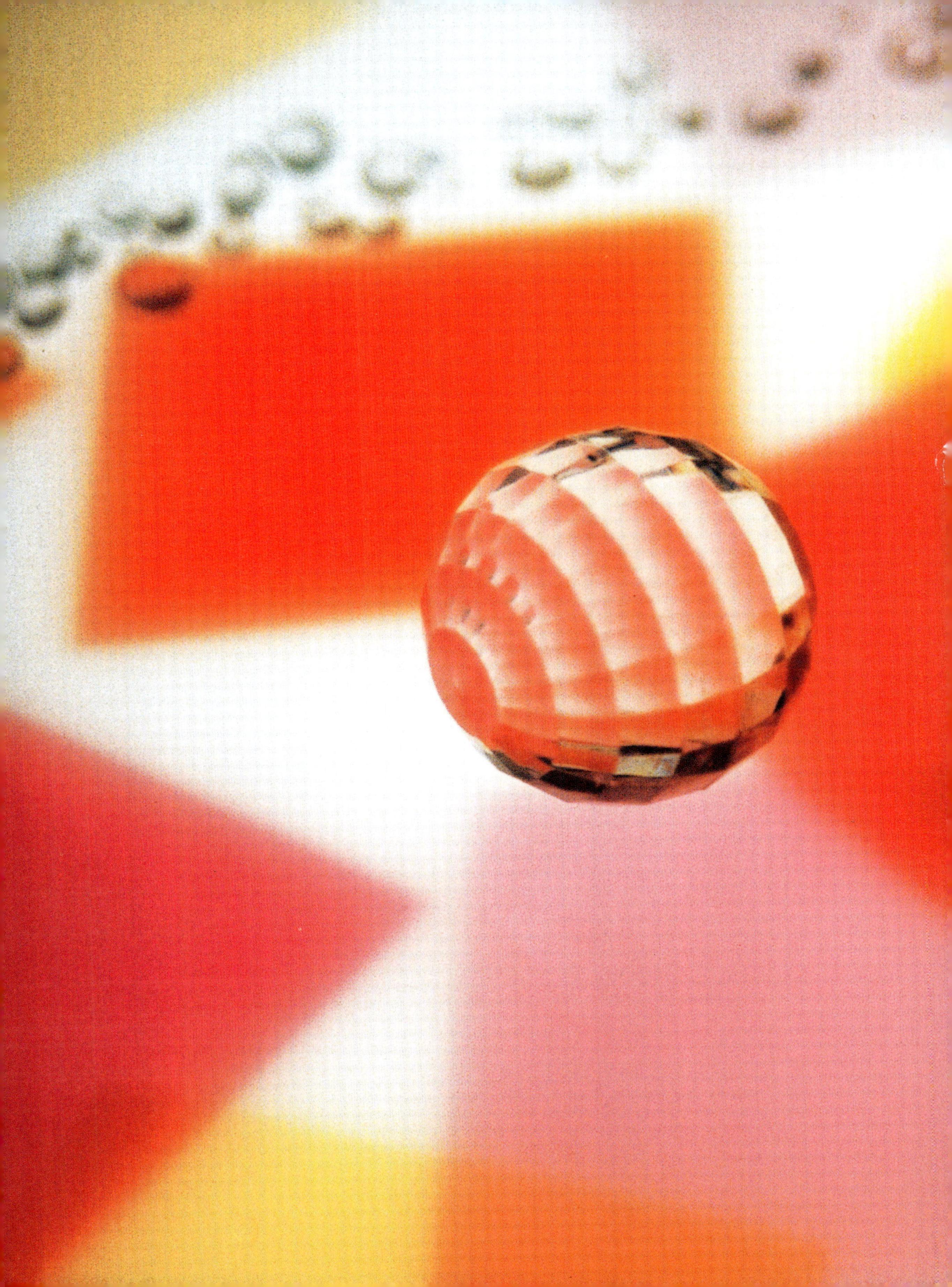

Physical Sciences

The term "plane" describes a mathematical idealization: a flat or level surface that has no thickness and an infinite length and width. In geometry, a plane figure is a figure that lies entirely within a plane. In theory, a sphere or any other three-dimensional shape is formed by the intersection of a virtually infinite number of planes.

Contents

Circus Science	210
The Noblest Metals	216
Dry-Cleaning Chemistry	222
Changing Times	226
Adding Up to the Presidency	233
Ask the Scientist	236

Circus SCIENCE

by Carl Zimmer

Number theory and neuroscience are probably not the first things that come to mind when you contemplate the circus. Yet along with the clowns and the fire-eaters, there is a place for science under the big top. Performers and their coaches have to know, either instinctively or consciously, what the laws of physics and biology will permit them to do. Moreover, a surprisingly large number of scientists are themselves circus fanatics, and a few are even performers—perhaps because the circus proves that science does not stop at the lab door.

THE TWISTED LAWS OF PHYSICS

The crowds are always awed when circus acrobats start twisting their bodies as they flip. But actually, such twisting is, in a purely physical sense, easy. Any object resists efforts to turn it—this is the quality that physicists call rotational inertia. Rotational inertia increases as the mass of an object is spread farther from its center of rotation: this seems obvious if you consider that a rotating object (or body) with long extremities has to sweep out a bigger circle than one that is more compact. Covering that extra ground requires more energy.

Performers can describe a circle by rotating around either a horizontal axis running through the hips (they can flip) or a vertical axis running from the head to the feet (they can twist). But an acrobat's body is centered around those two axes differently. From the axis through the hips, the performer's body might stretch about 3 feet (1 meter), giving him or her a high rotational inertia. From the vertical axis, however, the body may extend only a few inches, imparting a low rotational inertia. Thus, in a sense, an acrobat's body wants to twist more than it wants to flip. You can see this by taking a normally proportioned book and flipping it in the air—it is almost impossible to prevent the book from twisting. What makes twisting such a hard skill, in fact, is that it is too easy: if a performer cannot control it, his or her landing will be ugly.

Coaches know their acrobats cannot calculate Newton's laws of motion as they do acts like a back flip off a bar held by two other performers. Yet coaches have to help their performers take advantage of physics (rather than be taken advantage of). Some resort to white lies. For a back flip, for example, some coaches tell their performers to jump straight up and pull their knees to their chests. However, there is no way these performers could actually execute such a maneuver. For starters, if they jump up in a truly straight line, they have no angular velocity. Instead of going into a flip, they will simply come straight back down.

In fact, most performers learn unconsciously to lean back a little as they jump. That puts their center of gravity a little out of line with their feet, and the force of the jump can turn their body instead of simply pushing it straight up. When an acrobat then starts bringing his legs to his chest, he is actually just folding his body, bringing his chest to his legs at the same time. Explains Boris Verhovsky, head coach of Cirque du Soleil, a circus based in Montreal, Quebec: "Overall, the body has continued rotating. So it looks like the shoulders have stopped and the legs have accelerated, since they're going in the direction of the flip. That's why so many people get caught in the illusion and start believing it." Verhovsky himself trains his acrobats to lean comfortably back. But he spares

PHYSICAL SCIENCES 211

his performers the vector diagrams. "It's not something you directly tell; it's not an instruction. You're looking for a system of habits."

A Delicate Balance

In the early 1970s, Mikhail Tsaytin, a sports scientist at the Belorussian Institute of Physical Culture and Sport in Minsk, Belarus, performed an experiment. He had one performer stand on another's shoulders, then turned out the lights. They kept their position in the dark. Then he turned on the lights, brought in another acrobat, and had them form a three-person tower. Now when he turned out the lights, the tower collapsed.

To a neuroscientist, Tsaytin's results are not surprising, since a human tower pushes the sense of balance to its limits. The problem arises because balance (or, more generally, your sense of motion) is the product of at least three kinds of signals coming from distinct parts of the body.

The first sensor is your vestibular system, a labyrinth of tubes and sacs in your inner ear that is sensitive to quick, small changes in the position of

212 PHYSICAL SCIENCES

The relationship between juggling (upper left) and mathematics has been studied since at least the 10th century. Perhaps less scrutinized is the way the laws of physics play into many of the stunts performed at the circus. To execute a back flip, for example, a performer leans back slightly as he jumps, therefore putting his center of gravity slightly out of line with his feet. The force of the jump can then turn the body instead of just pushing it up.

your head. When your head tilts or rotates, tiny, sensitive hairs in the ear are nudged, triggering impulses to a part of the brain stem called the vestibular nuclei. This region converts the signals into a representation of your head's movement in space; it helps your body keep itself righted and lets your eyes track objects as your body moves.

Signals also come from nerves throughout your body that are sensitive to pressure and stretching. Such nerves register, for example, an ankle joint that suddenly sways or a back that gets pressed into a car seat. Finally, the retinas of the eyes also relay some information to the vestibular nuclei.

Under normal conditions, any one of these systems can give you enough information to sense movement. (That is why you can have the weird sensation of moving in a motionless train when the train on the next track is moving.) But each system works best at its own frequency. While the vestibular system can pick up the fast, little changes needed for eye control, the "touch" nerves work on slower adjustments, such as the shifts of weight you make while standing. "They complement each other—you want one system to be good where the other one is bad," says David Solomon, Ph.D., a neuroscientist at Johns Hopkins University in Baltimore, Maryland. (To get a sense of the different frequencies involved, move this book in front of your face fast enough so that it blurs. Now stop the movement and move your head back and forth at the same speed. This time you can read the words because when your head moves, your fast-working vestibular system is doing the tracking.)

Circus performers are always swaying slightly as they try to keep their tower aloft. If their swaying moves any of their centers of gravity too far from their base, they all fall.

PHYSICAL SCIENCES 213

Yet this swaying is too slow to be detected by the vestibular system. The stretch-sensitive nerves in the ankles work, but at a disadvantage: if someone sways underneath you, and you sway with them, your ankles do not bend, and you fail to perceive that you are in trouble. Tsaytin's experiment suggests that with two performers, there is just enough information available to stay aloft when the lights are out. But three performers introduce so much movement that they absolutely require sight, which can let them see the world shifting slowly around them—and before it starts shifting very fast.

Juggling Numbers

For centuries, mathematicians have had a special fondness for juggling (the 10th-century mathematician Abu Sahl, for example, liked to juggle glass bottles in a Baghdad marketplace), but not until the past decade did they invent juggling mathematics. One of the founders of this unique field is Ron Graham, a mathematician at AT&T Bell Laboratories in Murray Hill, New Jersey, and a former president of the International Jugglers Association.

Any juggling routine is made up of a repeating pattern of hand-to-hand tosses, each of which may take a different amount of time. Graham and his colleagues represent such a pattern with a string of numbers that indicate how long a ball is in the air: a quick pass from one hand to the other is a 1, a high arc may be as much as a 5. Thus, a three-ball shower—in which a juggler passes each ball from hand to hand (1) and then throws it high in the air (5) so that the balls seem to be tracing out a big circle—is 15.

But the mathematics of juggling is more than notation. Graham discovered that given some simple assumptions—such as, you cannot catch two balls with one hand at the same time—he could solve equations that revealed things about both juggling and mathematics. The average of the numbers in any sequence, for example, equals the number of balls needed to perform the corresponding routine. Not all sequences represent performable juggling routines, and with his equations, Graham can test whether they are legitimate or not. (By punching the numbers into a computer, he can even see what a routine will look like.)

Juggling math, Graham has found, is full of unexpectedly elegant laws. For instance, how many different sequences are there that consist of four throws with fewer than five balls? The answer is simply 5^4, or 625. "That's a really nice number," says Graham. "The question is, What is going on? Often, with a really nice answer, you suspect that something nice is going on more broadly." Graham is now trying to find general laws for such sequences. "It's all fitting into a general matrix, like a periodic table," he says. "These things are finding their place."

An acrobat maintains a stable center of gravity by swaying slightly. If the swaying becomes excessive, balance is lost and the acrobat falls.

SHOWING FLEXIBILITY
When we watch contortionists like Nomin Tsevendorj and Ulziibayar Chimed, two 12-year-old Mongolian performers with Cirque du Soleil, it is difficult to think of them as being different only in degree from the vast stiff majority of us. But physiologically, we are all sitting on the same spectrum.

Our skeletons need to be lashed together tightly enough so they do not collapse in a heap, yet not so tightly that they cannot take a step. Part of the solution lies in our muscles. Muscles consist of bundles of fibers that can contract and stretch as the filaments inside them slide over one another. But nerves monitor how stretched these fibers get, and if they reach a certain threshold, the brain commands the muscle to contract. Thus, muscles refuse to extend so far that they might damage themselves or a joint. It is possible, though, to retrain this reflex by extending muscles and holding them there; gradually, nerves get more accustomed to a longer muscle. This is how stretching makes you more flexible.

Still, a century's worth of stretching will not make most people contortionists, thanks to our body's other solution to its dilemma: collagen. This protein is the main ingredient of ligaments, which connect bone to bone, and of tendons, which connect bone to muscle. Cells scattered throughout this connective tissue produce collagen by twisting together three helices of protein and organizing them into long fibers. When these fibers are relaxed, they are loosely crimped up, accordion-style, and connected to each other by cross-links. When you pull on the fibers (say, by bending your knee), they unfold and stretch. The cross-links do not take kindly to stretching, however, and once they feel the force, they resist it. When you relax your knee, the collagen fibers crimp up again. There are many variations on the basic collagen recipe, and the body uses them all: some cells produce collagen fibers that are rather stiff, while others make collagen that can stretch quite far. By varying the proportions of each in different parts of the body, our ligaments and tendons can hold us together yet allow us elasticity where needed.

Each of us is born with genes that tell our collagen-making cells which recipe to use. Some of us, however, thanks to an abnormal gene sequence, have cells that produce particularly elastic forms of collagen. These people consequently have a wider range of motion than most of us. Among these folks, some produce an extremely elastic collagen, and they are the ones who can become contortionists.

For a few, though, flexibility can be too extreme. About 1 in 5,000 people is born with Ehlers-Danlos syndrome—a condition in which genes produce collagen that is so loose that joints can pop out spontaneously. In the syndrome's most serious forms, the tendons and ligaments are the least of a person's problems. Blood vessels, the bladder, skin, and the intestines all use collagen to stay elastic, and Ehlers-Danlos can make them dangerously pliable.

Fortunately, though, on the spectrum of flexibility, these troubles are at a far end. "We see only the patients who come with problems because of dislocating shoulders and things of that sort, but there are a lot of people who have it who are otherwise totally functional," says Leigh Ann Curl, M.D., an orthopedic surgeon at Johns Hopkins. "There are people who walk around with Ehlers-Danlos and don't even know it."

An extremely elastic form of collagen fibers gives contortionists their extraordinary elasticity and their remarkable range of motion.

The Noblest Metals

by David A. Pendlebury

Approximately three-quarters of the 110 or so known chemical elements are metals. Of all these dozens of metals, only four hold the distinction of being "noble"—a term used by chemists to describe an element characterized by an extreme lack of reactivity. Like the noble gases in the far-right column of the periodic table, the noble metals do not readily form compounds. Unlike the noble gases, however, the four noble metals are familiar substances. Three of them—gold, silver, and platinum—are precious metals, metals that we might associate with kings, princes, and other nobility. The fourth—mercury—is unique in being the only metal that exists as a liquid at room temperature. Some chemists also classify copper as a noble metal, although most do not.

Gold, silver, platinum, and mercury have exerted a powerful influence over many peoples for many years, and they have left a lasting impression on the world's historical record. These metals continue to be prized today for their beauty, their durability, and for their many specific and useful applications.

Noble Traits

Valuable insight into what makes these metals "noble" can be gained by looking at the periodic table, the chart that organizes chemical elements according to their atomic properties. The arrangement of elements in the same columns, or groups, often indicates a family relationship, with the elements exhibiting similar characteristics and similar behavior with respect to other elements.

Gold (chemical symbol: Au) is located in the periodic table in column IB, part of a large block of columns containing elements known as heavy metals or transition metals. Gold's atomic number is 79, which repre-

III A	IV A	V A	VI A	VII A	VIII A 2 He
5 B	6 C	7 N	8 O	9 F	10 Ne
13 Al	14 Si	15 P	16 S	17 Cl	18 Ar
31 Ga	32 Ge	33 As	34 Se	35 Br	36 Kr
49 In	50 Sn	51 Sb	52 Te	53 I	54 Xe
81 Tl	82 Pb	83 Bi	84 Po	85 At	86 Rn

‡Provisional Names

66 Dy	67 Ho	68 Er	69 Tm	70 Yb	71 Lu
98 Cf	99 Es	100 Fm	101 ‡Md	102 ‡No	103 ‡Lr

sents the number of electrons that orbit the nucleus of a gold atom. The electron configuration of an element is one major determinant of its tendency to form compounds with other elements; another is the element's atomic mass.

Another noble metal, silver (chemical symbol: Ag), appears just above gold in the periodic table, also in column IB. Its atomic number is 47. Above silver in the same column is the "near-noble" element copper. The Romans called copper *cuprum*, which explains the origin of its chemical symbol, Cu. The atomic number of copper is 29.

Each of these three elements—copper, silver, and gold—carries a single electron in its outermost ring, or shell. All three are relatively nonreactive. Copper, however, is the most reactive of the trio, silver somewhat less reactive than copper, and gold the least reactive. This hierarchy of reactivity is illustrated by what happens to each when exposed to air. Over time, copper changes from a reddish-brown color to green as it acquires a coating (called a patina) from its reaction with the atmosphere; this patina preserves the metal, which never rusts. Silver will tarnish over time from its reaction with sulfur in the atmosphere. Gold, however, is nonreactive with air and remains pristine in appearance.

Back in the periodic table, gold is straddled in the same horizontal row, or period, by platinum (Pt, atomic number 78) and mercury (Hg, atomic number 80), the two remaining noble metals. That the noble metals are arrayed in the periodic table around gold indicates that the properties that make these elements relatively nonreactive derive from a combination of atomic weight, which determines their sequential order in the periodic table, and their electron configuration, which determines their placement in the table's columns, or groups.

Quantum Forces

Recent research into the atomic structure of these metals explains precisely why gold and its near neighbors are so reluctant to form compounds—and why gold is the noblest metal of all.

In July 1995, Bjork Hammer and Jens K. Norskov, physicists at the Technical University of Denmark in Copenhagen, published a study in the journal *Nature* that described their use of sophisticated computer-modeling techniques to simulate the interactions of hydrogen gas on the surface of gold, copper, platinum, nickel, and other metals. The researchers found that gold happens to have an atomic structure and electron configuration that balances attractive quantum forces (hybridization) and repulsive quantum forces (orthogonalization) equally. The two cancel each other out, so to speak, to give gold its standoffish character with regard to other elements: it does not seek to bind with other elements, nor does it attract them to itself. Moreover, the attraction of gold atoms for one another is so strong that there is little opportunity for them to link up with atoms of other elements to form compounds.

As one moves upward from gold in column IB in the periodic table, the repulsive force of the electron configuration of each element decreases, which is why silver and copper form compounds more readily than does gold. For the first time, scientists now understand precisely why specific noble metals act the way they do.

When chemists discuss gold's nonreactivity, they are speaking of its reluctance to combine with common elements, such as hydrogen, nitrogen, and sulfur. Gold does, of course, form some compounds, called alloys, with other metals. Alloys of gold are common in jewelry. Only 24-karat gold jewelry is pure gold; 18-karat gold is three-quarters gold and one-quarter another metal, such as nickel, silver, zinc, or copper.

Since ancient times, gold has been fashioned into fine jewelry, sacred objects, and other decorative items that exploit the element's enduring luster and rich color. Undoubtedly the noblest of all metals, gold does not corrode, decay, or tarnish—even after millennia of exposure to the air.

Gold: The King of Metals

Gold's elite status derives in part from its attractive appearance in comparison to other metals, which are typically white, silver-colored, or reddish brown. Gold's warm, rich yellowish color is one of its most appealing characteristics. The Latin word for gold, *aurum*, means "glowing dawn" (this word is the origin of its chemical symbol, Au). Gold is also very durable and virtually indestructible. Unlike most other metals, gold does not decay, corrode, or tarnish, but remains bright and shiny, even after thousands of years of exposure to the air. Yet another reason for admiring gold is its ease of manipulation. Metalsmiths say that gold is malleable and ductile, since it can be pounded into sheets just 5 millionths of an inch thick, drawn into fine wire, and molded into specific shapes.

Gold's unique chemical structure makes it suitable for a variety of uses. Since ancient times, civilizations everywhere have fashioned their most precious and sacred objects from this lustrous yellow material. Gold has always been and remains today the metal of choice for fine jewelry. It is applied in delicate, tissue-thin sheets to highlight ornamental detail on important buildings, to illuminate inscriptions and illustrations in manuscripts, and to decorate the bindings of special books. In centuries past, a hunger for gold on the part of both monarchs and adventurers fueled exploration and colonization throughout the New World. In the 20th century, gold has found important applications in high-tech manufacturing, such as in the fabrication of computer components and electrical contacts in integrated circuits.

Gold, silver, and copper are excellent conductors of electricity. Copper is more commonly used for wire because it performs adequately for most purposes and is far less expensive than silver or gold. Since it does not corrode, gold makes for an excellent contact point in an electrical circuit. It is also heat-resistant. Many modern buildings use an ultrathin, transparent coating of gold on the surface of windows to reflect light outward in the summer and retain heat inside during the winter. The helmet visor of the Apollo astronauts had a transparent gold coating to deflect harmful ultraviolet radiation from the Sun.

Gold has also served as the most readily acceptable form of currency since ancient times. Today it remains a universal medium of exchange, whether in the form of coins or bars, called bullion.

Only the solution *aqua regia*, known since medieval times and used by alchemists in the vain attempts to turn lead into gold, is able to dissolve the king of metals. This liquid, whose name means "royal water," is a combination of hydrochloric and nitric acids. It is also used to dissolve platinum, which has many characteristics in common with gold.

Gold is extremely scarce relative to other substances, a characteristic that adds to its value. Although gold occurs in pure and semipure form as nuggets, it is more often encountered in—and recovered from—ores of silver, copper, lead, and nickel. Most gold obtained today comes from South Africa, the United States, Russia, Australia, Mexico, and Canada. Trace amounts of gold are present in seawater. Some researchers have contemplated practical ways for recovering gold from the oceans.

Approximately 75 percent of all gold recovered today is used for making jewelry. Another 15 percent is used in industrial settings. The balance becomes gold ingots and gold coins, which are exchanged or hoarded throughout the world.

SILVER

Like gold, silver has been known and prized by humans since ancient times. It, too, has been and continues to be fashioned into fine jewelry, sacred objects, and coins. Also like gold, silver is easily worked and is an excellent conductor of electricity. While silver does not corrode or rust, it does acquire a characteristic tarnish (silver sulfide) from exposure to pollutants in the air.

Because silver kills bacteria, it has many special uses relating to health and hygiene. It is used to make certain surgical instruments, for example. Silver nitrate, a compound of silver, is placed in the eyes of newborn children to prevent microbial infection. Silver is also employed in filtration systems to purify water. Dentists use silver in fillings to treat tooth decay; the filling is typically an alloy of silver, tin, and mercury, another noble metal.

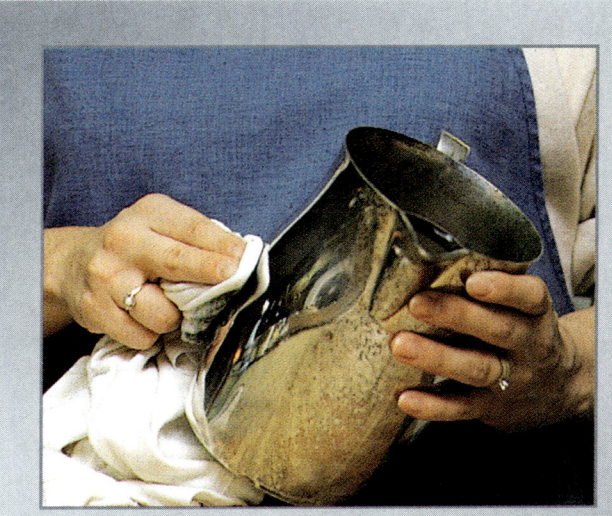

Objects made from silver must be polished to remove the tarnish that forms when the metal is exposed to hydrogen sulfide and other airborne pollutants. For household items, silver must be alloyed with copper, nickel, or other metals to give it the durability needed to withstand frequent handling.

Silver has many industrial uses, such as in the making of mirrors, in photography (where a compound of silver—silver bromide—adheres to a negative to record an image), and in storage batteries. It is sometimes used in electrical circuits that require efficient conduction and high reliability.

Sterling silver, as used for flatware and other tableware, is 92.5 percent pure silver and 7.5 percent copper. Utensils and trays require a hardening metal such as copper because pure silver is too soft to stand up well to day-to-day wear.

Mexico, Peru, Russia, and the United States are the world's main sources of silver. A sizable fraction (20 percent) is found in pure form; most silver is recovered from copper, zinc, and lead ores.

PLATINUM

Platinum, which is now more expensive per ounce than gold, was first recovered in South America in the 18th century. European explorers discovered the metal in alluvial deposits of the Pinto River. They called this silvery-white metal *platina del Pinto*, meaning "silver of the Pinto River," from which came the name platinum (chemical symbol Pt, atomic number 78).

Platinum shares many structural characteristics with gold, which explains its position adjacent to gold in the periodic table. Platinum maintains its structure and shiny appearance, even when exposed to chemical attack or heated to extremes. It is therefore used to fashion crucibles that are heated to extreme temperatures in industrial metal casting. Like gold, platinum is nonreactive and does not tarnish when exposed to air. But, unlike gold, platinum can prompt other chemical reactions. For this reason, platinum has found important uses as a catalyst. It is used in the catalytic converters of automobiles to reduce carbon monoxide output. It is also used as an additive in gasoline to increase the octane rating of fuel.

The Near-Noble Metals

Name	Color	Symbol	Atomic Number
Nickel	Silver	Ni	28
Copper	Reddish	Cu	29
Zinc	Bluish-white	Zn	30
Platinum Metals			
Ruthenium	Silvery-white	Ru	44
Rhodium	Silvery-white	Rh	45
Palladium		Pd	46
Iridium	White	Ir	77
Osmium	Black	Os	76

The other major uses for platinum include jewelry manufacture, especially for gemstone settings and wedding bands; electrical contacts and circuits; and some specialty photographic work (platinum-palladium printmaking). Alloys of platinum (generally platinum-iridium or platinum-ruthenium) are typically employed for gemstone settings and wedding bands, since the metal itself is extremely soft.

Platinum is mined mainly in the Ural Mountains of Russia, in South Africa, and in Canada. It is typically found in combination with other platinum metals (iridium, rhodium, ruthenium, palladium, and osmium) in ores of copper and nickel.

MERCURY

Mercury, also known as quicksilver, exists in liquid form at room temperature. Its chemical symbol, Hg, derives from its Latin name, *hydragyrm*, which translates as "silver water." The element has been known since ancient times. It is found in nature, generally in ores such as

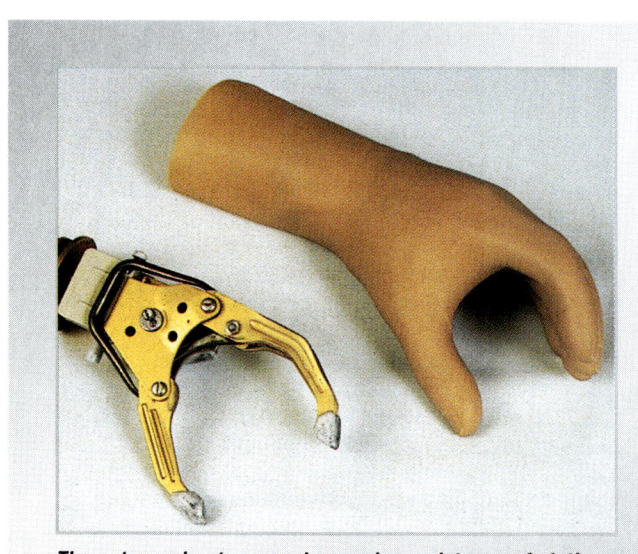

The extreme inertness and corrosion resistance of platinum make it an important ingredient in many industrial applications. Among the many specialized products that contain platinum are high-temperature crucibles, spark-plug electrodes, furnace thermometers, and mechanized prostheses (above).

HISTORY	PROPERTIES	USES/APPLICATIONS
Isolated in 1751 by A.F. Cronstedt.	Conductor of heat and electricity; resistant to corrosion; durable.	Used in many metal alloys, from jewelry to steel, to add strength and prevent corrosion. Also used as coating (nickel plating), and for coins.
Known since ancient times (ca. 8000 B.C.); mined in Sinai and on Cyprus.	Conductor of heat and electricity; on exposure to air, forms patina that protects metal.	Used in many metal alloys such as bronze (copper and tin) and brass (copper and zinc). Commonly used for electrical wires, piping, and for coins.
Known since medieval times; isolated in 18th century.	Highly reactive; combines readily with other elements; resistant to corrosion.	Most commonly used to make brass (zinc and copper). Zinc oxide used in suntan lotion to block ultraviolet radiation. Used as coating to protect pipes, wires, and nails. Also used in batteries and paint.
Isolated 1844. Isolated 1803. Isolated 1803. Isolated 1804. Isolated 1804.	All extremely hard, resistant to corrosion; palladium and rhodium have catalytic properties.	Used to fashion electrical contacts, in alloys to harden other metals (jewelry, tips for pens, certain engine components). Some used as catalysts in select chemical reactions.

cinnabar (mercuric sulfide). Mercury can be extremely damaging to humans if ingested, causing brain damage and deformities in the unborn. A severe outbreak of mercury poisoning occurred in Japan in 1956.

Mercury, also known as quicksilver, is the only metal that exists as a liquid at room temperature. Thermometers, barometers, and other weather instruments contain globules of this noble metal (above, suspended in oil) because they expand and contract evenly with changes in temperature and pressure.

Because mercury exhibits a high surface tension—clinging to itself more readily than to other liquids or surfaces—and because it expands evenly when heated and contracts evenly when cooled, it is used in thermometers, barometers, and other instruments that measure changes in temperature and pressure.

Another use for this unusual element—besides its aforementioned application in dental fillings—is in mercury-vapor lamps, used for sunlamps in tanning salons. Yet another is in specialty electrical relays (mercury switches).

Mercury is mined today in Spain, Italy, Russia, Mexico, Peru, Canada, and the United States.

THE NEAR-NOBLE METALS

In the periodic table near the noble metals can be found a number of other, lesser-known (and less esteemed) metals sometimes referred to as the near-noble metals. These metals nonetheless have a number of valuable uses; their characteristics and common applications are given in the table on pages 220 and 221.

A Brave AND WET New World for Laundry

by Daniel Pendick

Dry cleaners have long relied upon powerful solvents to clean items without shrinkage. Now, health and environmental concerns about these chemicals have prompted a search for less-toxic alternatives that will still keep customers happy.

Dry cleaning may be one of the more stubborn oxymorons in the English language. Although the process certainly involves cleaning, it is far from dry—and it's getting wetter all the time. Prompted by concerns over the health hazards of breathing fumes from the liquid solvents used in dry cleaning, the laundering industry is increasingly turning to an environmentally safe process based on soap and water.

Dry cleaning involves sloshing garments and other items in a solution of chemical solvents, detergents, and various additives. This process allows the cleaning of delicate fabrics that would shrink if they were washed in water. The solvent most commonly used in dry cleaning, perchloroethylene—popularly known as perc—emits fumes that are hazardous to human health.

Fortunately, new technologies are in the offing that will largely replace solvent-based dry cleaning. The reigning alternative to dry cleaning involves a fluid that people all over the world have relied on for thousands of years—water. In multi-process wet cleaning, as the new technique is called, items are cleaned by the controlled application of water, detergent, and mechanical agitation. Preliminary studies indicate that wet cleaning performs as well as dry cleaning and at a comparable cost.

THE FIRST DRY CLEANER

Dry cleaning traces its origins to the 18th century, when Europeans realized that petroleum-based solvents would remove greasy stains from fabrics, leather, and furs. According to one account, the "discovery" of dry cleaning as a commercially exploitable process was made in 1825. Allegedly, a domestic servant employed by French dyeworks owner Jean-Baptiste Jolly accidentally knocked over a kerosene lamp, spilling some of the fluid on a tablecloth. The kerosene seemed to dissolve dirt from the fabric. Armed with this observation, Jolly went into the stain-fighting business; he ultimately dubbed the process "dry cleaning" to distinguish it from washing with water.

For the next 100 years or so, various new solvents were introduced for use in dry cleaning. They all had one problem in common: like kerosene, they were flammable. The risks associated with such combustible chemicals limited the dry-cleaning process to the relatively controlled environment of the commercial cleaning plant. For dry cleaning to spread to Main Street, safer solvents were needed. In the 1930s, those solvents arrived: perchloroethylene and its close cousin, trichloroethylene. Today about 80 percent of the 34,000 dry cleaners in the United States use perc.

PERC AT WORK

Some of the toughest stains that garment cleaners face are those with a greasy base. These difficult spots are caused by perspiration, skin oil, machine grease, and the splatter from bacon, spaghetti, and other foods.

Fortunately, perc disassembles grease molecules, much like a demolition crew razes a skyscraper—floor by floor. This molecular disassembly is possible because greases and oils are made up of long chains of carbon atoms with hydrogen atoms attached. Cooking oil, for example, is based on carbon chains of a few hundred units; denser butter, by contrast, may have chains of thousands of units. Perc grabs onto these chains and breaks the chemical bonds between the carbon atoms. This reaction cuts grease chains into smaller units, sending the stain fragments into a liquefied suspension. This soiled liquid can be pumped out of the washing drum along with the perc.

Perc's ability to break chemical bonds is the secret of its success. Ordinary water-soluble detergents—the cleaning products that most consumers buy for use at home—work differently. The molecular structure of these detergents features a kind of clamping structure on each end. One of these structures bonds to a water molecule; the other grabs onto part of a grease chain. The mechanical action in a washing machine helps these detergents tear grease molecules into smaller chunks, but this process does not slice up long carbon chains as well as perc does.

THE CLEANING MACHINES

At a typical dry-cleaning establishment, any stains are "spotted," or pretreated with special chemicals, before the soiled items ever come into contact with perc. Each spotting agent is tailored to attack a particular type of stain. After spotting, the items are sorted by color and fiber type and loaded into a dry-cleaning machine.

The core of a dry-cleaning machine resembles the innards of a home washing machine. It has a drum assembly for agitating items, and piping and valves to route cleaning agents in and out of the drum. But at that point, the similarities end. Unlike home washers, dry-cleaning machines are "closed-loop" systems that filter and distill perc for reuse. They are also sealed to prevent the escape of toxic perc fumes into the environment.

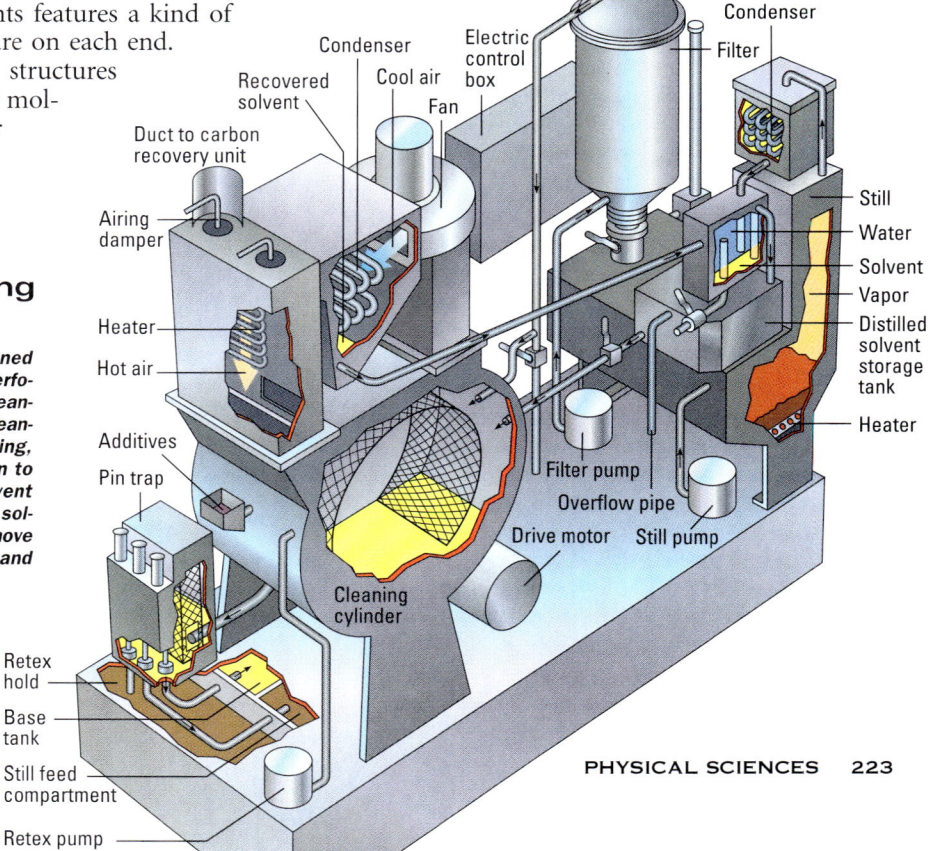

Dry-Cleaning Machine

Articles to be cleaned are placed in the perforated drum in the cleaning cylinder. After cleaning and solvent rinsing, the articles are spun to remove excess solvent and dried. The used solvent is filtered to remove solid dirt particles and distilled to purify it.

PHYSICAL SCIENCES 223

Dry cleaning is a complex process. Soiled garments are pretreated (left) with special chemicals formulated to attack a particular type of stain. Later, items are washed in sophisticated machines (below), some of which are run by microprocessor controls.

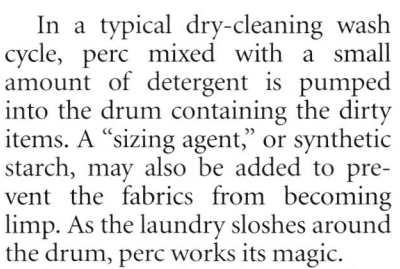

In a typical dry-cleaning wash cycle, perc mixed with a small amount of detergent is pumped into the drum containing the dirty items. A "sizing agent," or synthetic starch, may also be added to prevent the fabrics from becoming limp. As the laundry sloshes around the drum, perc works its magic.

After the wash cycle is complete, the load is spun at high speed to remove the soiled solvent, which is channeled away for recycling. Then the items are tumbled in warm air to remove the remaining dry-cleaning fluid. In the finishing stage of the process, the garments, draperies, and other items may be hung or pressed to remove wrinkles. However, if the shadow of a stubborn stain remains, the item may be taken back to the spotting board for custom cleaning.

THE VICES AND VIRTUES OF PERC

Anyone who has ever accidentally tossed a woolen garment into the wash at home knows the chief advantage of dry cleaning: if done properly, the process generally does not shrink, fade, or damage sensitive fabrics. Particularly with clothes made of natural fibers—such as wool, silk, and linen—the water, heat, and mechanical agitation of traditional laundering can spell doom. Under these conditions, natural fibers may absorb lots of water, swell, and stick together. The weave becomes clenched, which leads to shrinkage. Perc can prevent these problems before they ever occur.

For all of perc's cleaning virtues, however, it is also a health hazard. The solvent is toxic to the liver, kidneys, and central nervous system. Perc fumes have been classified as a hazardous air pollutant by the U.S. Environmental Protection Agency (EPA). There are even assertions, though controversial, that the chemical causes cancer.

Unfortunately, perc often escapes into the environment. The causes of such emissions are numerous: dry-cleaning machines may not be completely airtight; employees might spill solvents—or even deliberately flush used perc down the drain; and a small amount of perc always remains in clothing after the dry-cleaning process. The implications of this pollution are serious. One study funded by the EPA in the late 1980s found that perc consistently ranked near the top of the list of hazardous air pollutants in several cities in the Northeast. Then, in 1991, the perc really hit the fan when a New York State study found high concentrations of perc in apartments situated in the same buildings as dry-cleaning establishments in New York City. Perc pollution is now a serious environmental concern.

A Wet Alternative

Wet cleaning is now the reigning "green" alternative to dry cleaning. It relies on organic detergents that can be safely washed down the drain after use. A nontoxic chemical is used in the washing process to make sure that dyes do not run from fabrics as they are washing in water. A natural substance called collagen is also added to inhibit fibers from absorbing too much water and subsequently shrinking.

The critical challenge with wet cleaning is to achieve a proper balance of water temperature, mechanical action, and drying throughout the process in order to minimize shrinkage. Items are carefully sorted according to type and color; computerized wet-cleaning washers store dozens of combinations of water, heat, and mechanical action, each tailored to a particular fabric or garment.

Wet-cleaning machines contain sophisticated electronics to ensure that the items are dried carefully and without shrinkage. Most garments shrink when the moisture content of the fabric reaches 10 to 15 percent. Sensors located on the ribs of the dryer monitor the moisture content of the fabric. To get fast turnaround, the machine dries at 158° F (70° C). When the dryer's "brain" senses a moisture content of 15 percent, it shuts the machine off and sounds an alarm that signals workers to remove the items. Finally, in both dry cleaning and wet cleaning, the postwashing processes—pressing, bagging, and other finishing steps—are conducted in the same manner.

A Bright Green Future?

Wet cleaning has made dramatic inroads into the laundering industry in recent years. According to Ohad Jahassi, manager of the EPA's perc-reduction program, at least 90 commercial launderers nationwide rely on wet cleaning for part or all of their workload. Some clear advantages of the process are emerging. For one, it works better on many stains, particularly those that are caused by water-soluble foods and drinks. And many customers perceive that wet-cleaned garments have a cleaner, fresher smell than do their dry-cleaned counterparts.

The EPA has funded major field studies to find out how wet cleaning compares to dry cleaning in terms of cost and performance. According to Jahassi, the preliminary results are encouraging.

Overall, the wet process cleans as well as perc-based techniques, does not lead to excessive shrinkage if conducted properly, and can be done for a competitive price.

Despite all of the buzz over environmentally safe laundering, wet cleaning is unlikely to replace dry cleaning altogether, says Mary Scalco of the International Fabricare Institute, a major trade organization for professional dry cleaners. One important reason is economics: the wet-cleaning process usually requires more time and labor to obtain results comparable to dry cleaning. Clothing must be carefully sorted and pretreated to ensure thorough cleaning. And because of the colder, gentler wash cycle, some fabrics may have to wash longer. This time factor translates into a competitive edge for dry cleaning, which is quick and effective.

Another potential disadvantage of wet cleaning is the ever-present risk of shrinkage. Although new wet-cleaning equipment has lessened the problem, Scalco says, "wool will still shrink in water, just less." Right now, says Jahassi, all of the evidence is not in regarding the effect that wet cleaning will have over the life of a garment. Also, the EPA is working with the garment industry to change care labels to indicate which garments can tolerate wet cleaning.

The EPA is also evaluating some truly exotic dry-cleaning technologies. For example, the agency is funding research on a process involving liquefied carbon dioxide gas, a potent degreaser that has been used to clean computer chips and nuclear-weapon components. The EPA is also interested in clothes dryers that vaporize moisture with microwaves rather than hot air, which might reduce the shrinkage caused by heating garments.

Some ideas may not come to life for years. Perhaps the most notable futuristic laundry vision involves using high-frequency sound waves to blast away stains. Concrete designs for such a machine, however, have yet to emerge.

Changing Times

by Marcia Bartusiak

Time is an elusive notion. Poets often think of time as a river, a free-flowing stream that carries us from the radiant morning of birth to the golden twilight of old age. It is the intangible span that separates the delicate bud of spring from the lush flower of summer.

Physicists think of time in more-practical terms. For them, time is the means of measuring change—an endless series of instants that, strung together one after another like beads, turn an uncertain future into the present, and the present into a definite past. The very concept of time allows researchers to calculate when a comet will round the Sun or how a signal traverses a silicon chip.

In other words, time is a tool. Ancient astronomers meticulously tracked the Sun's march across the zodiac to mark off the seasons and determine when to plant and when to harvest. In this day and age, solar timepieces have been replaced by atomic clocks that, thanks to the steady pulsing of hydrogen or other atoms, do not gain or lose a second in millions of years. Time can now be sliced into slivers as thin as one 10-trillionth of a second.

But what is being sliced? Unlike mass and distance, time cannot be perceived by our physical senses. And yet we somehow measure it. Captivated by this conundrum, physicists are beginning to explore the very origins of time. In fact, some are wondering whether time is a fundamental property of the universe at all.

A Timely Problem

As a cadre of theorists attempt to extend and refine the general theory of relativity, Einstein's momentous law of gravitation, they have a problem with time. A big problem.

"It's a crisis," says mathematician John Baez of the University of California at Riverside, "and the solution may take physics in a new direction." Not the physics of the everyday world. Stopwatches, pendulums, and hydrogen-maser clocks will continue to keep track of nature nicely in low-energy earthly environs. The crisis arises when physicists try to merge the macrocosm—the universe on its grandest scale—with the microcosm of subatomic particles.

Gravity is the weakest of nature's forces: a toy magnet can easily pick up a paper clip against the gravitational pull of the entire Earth. But gravity gains collective strength as masses accumulate and exert their effect over larger distances. The force that causes one object to attract another eventually comes to control the motions of planets, stars, and galaxies. And the best description of how that happens is contained in Einstein's general theory of relativity. But the domain in which this theory works does not apply to problems at the subatomic scale. For decades, physicists have struggled to discern how gravity acts on the level of elementary particles, a realm governed by the rules of quantum mechanics. Arranging this rather curious marriage—an all-embracing theory of "quantum gravity"—is one of physics' last great tasks.

There is a vital reason for physicists' dogged pursuit of this problem. They believe that quantum gravity was the dominant force at the birth of the universe. It was an instant when all the matter and all the energy in the universe were squeezed into a space far smaller than a proton. The microcosm and the macrocosm, in effect, were crushed together in a "singularity," a state where density nears infinity and volume approaches zero.

A Clock at Creation?

By figuring out the physics of such a bizarre realm, theorists may at last find the key to the origins of the universe. Simultaneously, they would be learning what lies at the heart of a black hole, the gravitational abyss that is thought to result when the core of an exploding star is crushed inward until its size becomes atomic rather than celestial.

A solution to this mystery, it turns out, lies in understanding the meaning of time: how it acts—and whether it even exists—at the moment of creation or within a black hole. Telling time, after all, involves picking out something that is changing—the Sun rising and setting, pendulums swinging—and tracking those changes to establish a rhythmical chronology.

But how do you register time when the mass of a stellar core is squeezed into a subatomic speck? What kind of timepiece could physicists use to understand and deal with the crushing and featureless conditions that marked the universe's birth, when quantum gravity was in control?

The problem is really a mathematical one, but can be visualized in this crude way: imagine you could somehow shift a magical gear into reverse and travel back billions of years to the moment of creation. For most of the trip, a wristwatch would work just fine in keeping track of time. But upon reaching the very cauldron of creation, the watch would melt in a nanosecond. You could still keep track of time through the constant vibrations of individual atoms, the basis of atomic clocks. But go back far enough, and even atoms cease to exist. Soon there is no longer any means of measuring the progress of events. During that primordial moment when the force of quantum gravity was strongest and the cosmos was tinier than a nuclear particle, there was no room to place a clock, safe from interference, and gauge how the universe was evolving.

This dilemma summarizes the problem of time in physics. Either theorists come up with a "quantum clock," a means of understanding and dealing with the passage of time in that minuscule province where gravity and the quantum world mingle (at least on paper), or they do away with the concept of time altogether.

"The problem of time is one of the deepest issues in physics that must be addressed," says theoretical physicist Christopher Isham of Imperial College in London, England. There will be no Theory of Everything—no peek at "the mind of God," as the University of Cambridge cosmologist Stephen Hawking famously put it in *A Brief History of Time*—until this mystery is resolved.

Newton's Clock Takes a Licking

Time became a key word in the language of physics during the 17th century, notably when Isaac Newton wove the passage of time directly into his equations, as in *force = mass x acceleration*. Today it is difficult for any physicist to examine the universe without thinking of time in much the same way as Newton did more than 300 years ago. Most of the laws of physics continue to be designed to show how things change from one moment to the next.

Newton placed a clock upon the mantle of the universe. This timepiece chimed like

Many of the modern-day laws of physics still reflect Sir Isaac Newton's 300-year-old notions of time, designed to show how things change from one moment to another.

some cosmic Big Ben, in step with all celestial inhabitants, no matter what their speed or position. More important, the Newtonian clock was never affected by the grand events going on around it. Time was absolute, alike for all as galaxies collided, solar systems formed, and moons orbited planets.

This notion of time held until the beginning of this century, when Albert Einstein uncovered a glitch in Newton's clockwork. With his special theory of relativity, Einstein showed that a clock at rest and a clock in motion each register a different flow of time. Einstein transformed time into a true physical entity, one that was changed by what was going on around it.

Three years after this revelation appeared in print, Einstein's teacher Hermann Minkowski took Newton's clock and rolled it out like cookie dough to form the cosmic landscape called space-time. Minkowski, wanting to better explain some of special relativity's unusual properties, glued space and time together to form a seamless canvas, a new absolute framework in which time becomes physically connected to space. If you think of the space-time coordinates as the interwoven threads of a blanket, tweaking one set of threads will affect all the others: travel near the speed of light and space will shrink as time expands.

In 1915, with his general theory of relativity, Einstein shook up the classical, Newtonian view of time even further. He took Minkowski's image of space-time and warped it, and in so doing was able to explain the origin of gravity, long a mystery.

According to Newton, planets orbited the Sun because they were somehow held by invisible tendrils of force. Why should this be so? No one knew. But with Einstein's insight, the tendency of one object to attract another became a simple matter of geometry. It was the natural consequence whenever a mass distorted the space-time canvas. A massive body—the Sun, for example—indents the mat, and nearby objects must then circle it because they are caught in the deep space-time basin carved out by the Sun.

General relativity treats time very differently from the way it is handled in other areas of physics. Under Newton, time was tallied by a universal clock that stood apart from the phenomenon under study. But, in general relativity, no particular clock is special. Einstein declared that time is not absolute. His law of gravity looks the same no matter what timepiece is used as a gauge—whether a clock traveling near the speed of light or one sitting quietly at home on a shelf.

Albert Einstein's general theory of relativity upended the Newtonian view that time is absolute. Einstein declared that a clock in motion would measure time differently than a clock at rest.

Russian mathematician Hermann Minkowski, the first to postulate that space and time are not separate entities, proposed a new four-dimensional continuum called space-time.

The Subatomic Realm

The choice of clock is still crucial, however, in other areas of physics, particularly quantum mechanics. It plays a central role in Austrian physicist Erwin Schrödinger's celebrated wave equation. The equation shows how a subatomic particle can be thought of as a collection of waves, a wave packet that moves from point to point in space and from moment to moment in time.

According to the vision of quantum mechanics, energy and matter are cut up into discrete bits, or quanta, whose motions are jumpy and blurry. The behavior of these particles cannot be worked out exactly. Using Schrödinger's wave equation, one can calculate only the probability that a particle—a wave packet—will attain a certain position or velocity. This is a picture so different from the world of classical physics that even Einstein railed against its indeterminacy. He declared that he could never believe that God would play dice with the world.

Quantum mechanics introduced a fuzziness into physics: you can pinpoint the position of a particle, but there is a trade-off; its velocity cannot then be measured very well. Conversely, if you know how fast a particle is going, you will not be able to know exactly where it is. German physicist Werner Heisenberg best summarized this strange situation with his famous uncertainty principle. But all this action, uncertain as it is, occurs on a fixed stage of space and time. A reliable clock is always around to keep track of the goings-on, and thus enable physicists to describe how the system is changing.

At least, that is the way the equations of quantum mechanics are now set up. And that is the crux of the problem. How can physicists merge one law of physics—namely gravity—that requires no special clock with the subatomic rules of quantum mechanics, which continue to work within a universal, Newtonian time frame?

That's why things begin to go a little crazy when you attempt to blend these two areas of physics. Although the scale on which quantum gravity comes into play is so small that current technology cannot possibly measure these effects directly, physicists can imagine them. Place quantum particles on the pliable mat of space-time, and it will bend and fold like rubber. And that flexibility will greatly affect the operation of any clock keeping track of the particles. A timepiece caught in that submicroscopic realm would probably resemble a pendulum clock laboring amid the shudders of an earthquake. "Here the very arena is being subjected to

Time gained a key role in the realm of quantum mechanics thanks to the work of Erwin Schrödinger, whose wave equation describes how subatomic particles behave in space and time.

Werner Heisenberg's uncertainty principle states that a particle's velocity and position cannot be pinpointed simultaneously. Despite such ambiguity, all of the laws of quantum mechanics assume that time is fixed, unaffected by relativity.

quantum effects, and one is left with nothing to stand on," explains Isham. "You can end up in a situation where you have no notion of time whatsoever." But quantum calculations depend on an assured sense of time.

What to do? Different physicists answer the question differently.

Changes in Matter

For Karel Kuchar, a general relativist at the University of Utah, the key to measuring quantum time is to devise an appropriate clock. Kuchar has been seeking what might be called the submicroscopic version of a Newtonian clock, a quantum timekeeper that can help describe the physics going on in the realm ruled by quantum gravity. Unlike the clocks used in everyday physics, Kuchar's clock would not stand off in a corner, unaffected by what is going on around it. It would be set within the tiny, dense system where quantum gravity rules. This insider status has its pitfalls: the clock would change as the system changed—so, to keep track of time, you would have to monitor the variations.

The most common candidates for this special type of clock are so-called "matter clocks." Conventional timekeeping, after all, means choosing some material medium, such as a set of particles, and marking its changes. But with pen and paper, Kuchar mathematically takes matter clocks into the domain of quantum gravity, where the gravitational field is extremely strong and those probabilistic quantum-mechanical effects begin to arise. He takes time where no clock has gone before. Unfortunately, matter clocks tend to break down at some point.

More promising as a quantum clock is the geometry of space itself: monitoring space-time's changing curvature as the infant universe expands or a black hole forms. Imagine the tiny infant universe as an inflating balloon. Initially, its surface bends sharply. But as the balloon blows up, the curvature of its surface grows shallower. "The changing geometry," explains Kuchar, "allows you to see that you are at one instant of time rather than another."

Unfortunately, each type of "clock" that Kuchar has investigated so far leads to a different quantum description, different predictions of the system's behavior. "You can formulate your quantum mechanics with respect to one clock that you place in space-time and get one answer," explains Kuchar. "But if you choose another type of clock, perhaps one based on an electric field, you get a completely different result. It is difficult to say which of these descriptions, if any, is correct."

More than that, the clock that is chosen must not eventually crumble due to the extreme conditions. Quantum theory suggests there is a limit to how fine you can cut up space. The smallest quantum grain of space imaginable is 10^{-33} centimeter wide, the Planck length, named after Max Planck, inventor of the quantum. On that infinitesimal scale, the space-time canvas turns choppy and jumbled. Time and space, as we know them, are no longer easily defined, and theorists start walking on shaky ground.

Only a fully developed theory of quantum gravity will show what is really happening at this unimaginably small level of space-time. Kuchar conjectures that some property of general relativity (as yet unknown) will not undergo quantum fluctuations at this point. If that is true, such a property could serve as the reliable clock that Kuchar has been seeking.

"FORGET TIME"

While Kuchar has been trying to mold general relativity into the style of quantum mechanics, other physicists believe that quantum gravity should be made over in the likeness of general relativity, where time is in the background. Carlo Rovelli is a champion of this view.

"Forget time," Rovelli declares. "Time is simply an experimental fact." Rovelli, a physicist employed by both the University of Pittsburgh and the University of Trento in Italy, has been working on an approach to quantum gravity that is essentially timeless. To simplify calculations, he and his collaborators set up a theoretical space without a clock. In this way, they were able to rewrite Einstein's general theory of relativity, using a new set of variables so that the theory could more easily be adapted for use on the quantum level.

This was quite an accomplishment; relativists had sought to find a common vocabulary for these two diverse fields for decades. The new formulation may finally allow physicists to explore how gravity behaves on the subatomic scale. But is that really possible without any reference to time at all?

"First with special relativity and then with general relativity, our classical notion of time has only gotten weaker and weaker," answers Rovelli. "We think in terms of time. We need it. But the fact that we need time to carry out our thinking does not mean it is reality." Rovelli believes that if physicists ever find a unified law that links all the forces of nature under one banner, it will be written without any reference to time.

Getting rid of time in fundamental physical laws, says Rovelli, will probably require a grand conceptual leap, the kind of adjustment that 16th-century scientists had to make when Copernicus put the Sun, and not Earth, at the center of the universe. Divorcing time from physics seems equally incredible. No wonder, then, that Rovelli's ideas face resistance. Kuchar, for one, is not convinced that time can be dismissed. "We need rules to give the proper restraint to our imagination," he cautions.

But maybe, as Rovelli suggests, the true rules are timeless, including those applied to the subatomic world. Indeed, a movement has been under way to rewrite the laws of quantum mechanics without any direct reference to time.

A MATTER OF PERCEPTION

Of course, as Christopher Isham points out, "having gotten rid of time, we're then obliged to explain how we get back to the ordinary world, where time surrounds us." Quantum-gravity theorists have their hunches. Like Rovelli, many are coming to suspect that time is not fundamental at all. Time, they say, may more resemble a physical property such as temperature or pressure. Pressure has no meaning when you talk about one particle or one atom; the concept of pressure arises only when we consider trillions of atoms. The notion of time could very well share this statistical feature.

If so, reality would then resemble a pointillist painting. On the smallest of scales—the Planck length—time would have no meaning, just as a pointillist painting, built up from dabs of paint, cannot be fathomed close up. But as you move back, the dots begin to blend together, and a recognizable picture comes into focus. Likewise, space-time might take form and reveal

Physicist Carlo Rovelli believes that the fields of general relativity and quantum mechanics can be unified in a theory of quantum gravity by rewriting the laws of physics without reference to time.

itself only when we scrutinize larger and larger scales. Time could be simply a matter of perception, present on the large scale, but not on the smallest scale imaginable. Physicists talk of the universe "congealing" or "crystallizing" out of the chaotic quantum jumble that lies at the heart of the Big Bang. Time is not a physical entity, but rather a notion that emerges.

Stephen Hawking sees such an effect in his own work on quantum cosmology. To arrive at this conclusion, Hawking first had to circumvent the unique status of time in the space-time continuum. While time can be considered a fourth dimension, it is very different from length, width, and height. In space, an object can move freely in any direction—but in time, an object must always move forward into the future and away from the past. And this requirement makes the mathematics of quantum cosmology quite complicated. Hawking decided to get rid of this restriction by treating time as just another dimension of space—a mathematical procedure physicists often use to simplify what would otherwise be an intractable problem.

Cosmologist Stephen Hawking asserts that time did not exist in the crushing conditions of the Big Bang, emerging into the real world only after the universe expanded beyond the subatomic level.

Hawking's foray into the nebulous realm where general relativity meets quantum mechanics is suggesting that time, nonexistent at first, could have emerged into the real world from a domain of timelessness. Thus, there is no reason to inquire what came before the Big Bang. To Hawking, that is as senseless a question as asking what is north of the North Pole.

There is another way to interpret Hawking's result: time simply loses all meaning as you travel back, closer and closer to the Big Bang singularity, akin to the way a compass loses its ability to indicate a precise direction as you approach the north or south magnetic pole. A compass is accurate and useful only when it is far from a magnetic pole; likewise, time may be discernible only after you get far enough away from the Big Bang singularity.

Awaiting a Solution

Unfortunately, Hawking's mathematical procedure offers a glimpse, not a final solution. Physicists as yet only recognize the problem, and sense what must happen, but are far from postulating a mechanism. That awaits a full theory of quantum gravity.

It seems an impossible task, one that would appear to require re-creating the conditions of the Big Bang. But not necessarily. For instance, future generations of "gravity-wave telescopes," instruments now being built to detect ripples in the rubberlike mat of space-time, might sense the Big Bang's reverberations, relics from the instant of creation when the force of gravity emerged. Such waves could provide clues to the nature of space and time.

"We wouldn't have believed just 50 years ago that it would be possible to say what happened in the first 10 minutes of the Big Bang," points out Kuchar. "But we can now do that by looking at the abundances of the elements. Perhaps if we understand physics on the Planck scale well enough, we'll be able to search for certain consequences—remnants—that are observable today." If found, such evidence would possibly allow us to perceive at last how space and time came to well up out of nothingness some 15 billion years ago.

Adding Up to the Presidency

by Donald W. Cunningham

If, in the summer of 1996, you had asked U.S. President Bill Clinton for his opinion of the state of Illinois, he might have answered, "22 electoral votes." Given the same question, his Republican opponent, Bob Dole, might have offered the same reply. Both men would have been referring to the unusual system of voting that determines the winner of the presidential election.

Under the U.S. Constitution, the president is not elected by a direct popular vote of citizens, but rather by a majority vote of *electors* in an *electoral college*. In November of an election year, all of the electors within each state are committed to one candidate; this total is then submitted in an *electoral vote* for president.

So why does the public vote, if electors choose the president? Because within each state, the general vote determines the candidate to whom the state's electors are committed. In other words, the public elects the electors of a certain candidate; those electors then vote for president. The nation's founders placed this buffer between voting citizens and the choice of president to ensure representation and balance of power.

In creating the electoral college, the framers had other reasons that seem less compelling in modern times. For example, at the time of the Constitution's drafting, communication and transportation throughout the states were much less effective than they are today. As a consequence, a national campaign to inform the electorate and solicit popular votes was impractical. The founders thought that a small group of well-educated citizens in each state—the electors—would be better able to choose a satisfactory president.

Despite potential flaws, the electoral-college system has operated for more than 200 years with great success. It even has produced unintended benefits. This unique—some would even say odd—system for choosing the president has been changed only slightly since it was developed by the framers of the Constitution.

WHY AN ELECTORAL COLLEGE?

One of the geniuses of the U.S. Constitution is its attention to a balance among the powers of the states and the federal government. This attention extends to presidential elec-

Electoral Voting Strength of States, 1996

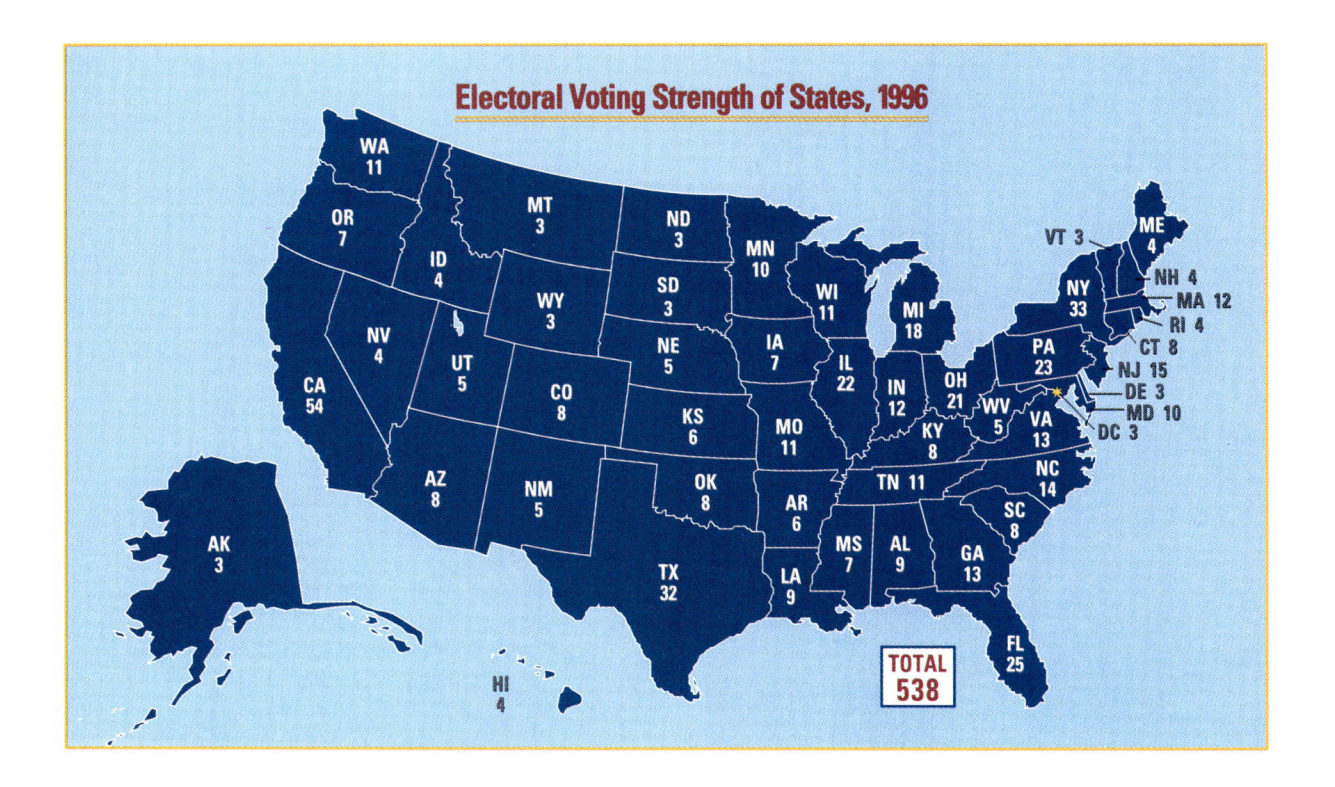

State	Votes	State	Votes	State	Votes
WA	11	OR	7	MT	3
ND	3	MN	10	VT	3
ME	4	ID	4	WY	3
SD	3	WI	11	NY	33
NH	4	MA	12	NV	4
UT	5	NE	5	IA	7
MI	18	PA	23	RI	4
CT	8	CA	54	CO	8
IL	22	IN	12	OH	21
NJ	15	DE	3	MD	10
KS	6	MO	11	KY	8
WV	5	VA	13	DC	3
AZ	8	NM	5	OK	8
AR	6	TN	11	NC	14
SC	8	AK	3	TX	32
MS	7	LA	9	AL	9
GA	13	HI	4	FL	25

TOTAL 538

tions. If the president were to be elected to office by a simple plurality (winning more votes than any other candidate) in a general election, then candidates representing the needs of a few large, densely populated regions or states would continually defeat candidates representing the interests of smaller, less populated regions or states. In such a voting system, smaller states would perennially be underrepresented.

In an effort to correct such imbalance, the electoral college dramatically reduces the disparity in voting power, or representation, between large and small states. For example, the numbers of electors of California and Vermont are 54 and 3, respectively, for a ratio of 18 to 1. In contrast, California has a population nearly 55 times greater than does Vermont. Note how Vermont's voting strength would diminish in going from an electoral vote to a direct popular vote.

The electoral college also reduces an individual's representation in presidential voting, particularly for voters in more-populous states. In a national, popular election, each individual vote carries an equal weight. Under the electoral-college system, however, voters in a state must share electors. In the larger states, more voters must share a single elector.

How Does It Work?

Each state has a number of electors in the electoral college equal to the sum of the state's senators and representatives in Congress. Illinois, for example, has 2 senators and 20 representatives, giving the state a total of 22 electors. New York has 2 senators and 31 representatives, for a total of 33 electors. In one notable exception, the District of Columbia—which has one delegate in the House of Representatives and no senators—has three electoral votes. (Each state's electoral representation is shown in the table above.)

Today a total of 538 electors comprises the nation's electoral college. The distribution of these votes can shift whenever states gain or lose seats in the House of Representatives. Such modifications are made if a national census indicates shifts in population distribution.

In each state, political parties appoint a slate of electors equal to the number of electoral votes of the state. In the general election, the public votes for the electors of a particular party (and candidate). In the 1996 election, a person voting for Bill Clinton would actually be voting for one of the Democratic Party's electors for president.

The electors meet to cast their ballots in December after a November election. The balloting is a formality, since the public's choices have been established in the general election. It is legally possible for an elector to change his or her vote from what was determined by the popular balloting. In 1988, for example, one West Virginian elector voted for Lloyd Bentsen, Jr., rather than for Michael Dukakis, the Democratic presidential candidate. Such defections, however, are rare and have never affected the outcome of an election.

The Constitution states that if no candidate receives a majority of electoral votes, then the House of Representatives must choose the president from the top three contenders. In balloting in the House, each state has one vote. If no candidate for vice president receives a majority of electoral votes, then the Senate must choose a vice president from the top two contenders.

These mechanisms have been used on several occasions. For example, the presidential winners of the elections of 1800 and 1824 were determined by the House of Representatives due to a lack of a majority. In 1876, with the electoral counts of a few states in dispute, the president was determined by a congressionally established commission.

UNDEMOCRATIC?

Many critics of the electoral system are quick to note the possibility that a candidate can win the electoral-college vote even though an opponent draws a majority of the popular vote. Such an improbable—some would say undemocratic—result occurred in the close contest between Benjamin Harrison and Grover Cleveland in 1888. Harrison won a majority of the electoral-college vote, while Cleveland won a slight majority of the popular vote. Cleveland won the popular vote by less than 100,000 votes (out of a total of more than 11 million votes); Harrison won the presidency by an electoral vote of 233 to 168.

The presidential election of 1876 was a far more controversial example of the electoral count overcoming the popular count. The loser, Democrat Samuel J. Tilden, received more popular votes than did the winner, Republican Rutherford B. Hayes, but Hayes won through a strange combination of electoral votes and special-committee votes. More than a century later, the integrity of this election remains the subject of great debate.

It should be noted that a winning electoral candidate commonly fails to garner a majority of the popular vote. All it takes is a close election and a strong third candidate. Recent examples include Richard Nixon's plurality over Hubert Humphrey in 1968 (with George Wallace running as a third-party candidate), and Bill Clinton's defeat of George Bush (and H. Ross Perot) in 1992.

ELECTORAL STRATEGY

Strategies for campaigning under the electoral system would seem to differ little from those used under a system of direct popular voting. In both cases, a candidate must try to appeal particularly to regions with many voters—namely, populous states. One difference is that, under the electoral system, there may be a proportionately greater benefit in appealing to midsized states as well.

In a popular election, an appeal to the most-populous states, subsequent strong support from them, and reasonable support from other states would likely put a candidate over the top. But winning under the electoral-college system requires that a candidate also win the popular vote in several midsized states to gain an electoral majority.

THE TWO-PARTY SYSTEM

An added consequence of the electoral college is that it bolsters the country's two-party system. The system makes it very difficult for a candidate who is independent or with a third party to win the presidency. Such difficulties are attributable primarily to the winner-take-all rule for state electoral votes. In the 1992 presidential election, for example, H. Ross Perot garnered about 19 percent of the popular vote nationwide, yet he failed to receive a single electoral vote.

One irony in the history of the electoral college is that the framers of the Constitution actually devised the system to reduce the influence of political parties. However, the founders might have found consolation in the fact that the two-party system and the electoral college have reduced the chance that a candidate with radical, if temporarily popular, views can win the presidency.

Ask the Scientist

▶ *A few nights ago, some friends of mine were having a discussion about Planck's constant. Could you define what it is and what it is used for?*

In 1900, the German physicist Max Planck announced a revolutionary idea: the quantum hypothesis. Before Planck, all physicists believed that radiant energy (such as light or heat, for example) emanated in continuous waves. Planck, however, argued that the smallest units of energy, which he called quanta, were not continuous waves, but rather were small packets that jumped from one energy level to another under certain conditions. The expression "quantum leap" refers to this type of sudden change.

Planck stated that the energy, E, of a quantum (also called a photon) could be found by multiplying the wavelength of the radiation, v, by a constant value h (6.626196×10^{-34} joules per second). The h came to be known as Planck's constant. The equation is $E = hv$. This formula and Planck's quantum hypothesis provided the basis for later work in physics by Albert Einstein and in physical chemistry by Niels Bohr. Planck received the 1918 Nobel Prize in Physics for his work in quantum physics and radiation.

Planck's constant has been used to construct atomic clocks, the most precise devices to measure time. Today, as scientists and engineers fashion smaller and smaller semiconductors, they are beginning to encounter quantum effects from the atoms passing through them. This represents both an engineering challenge and an opportunity to design new types of chips, whose switching mechanisms are activated by quantum-state transitions. The foundation of such a mechanism is Planck's constant.

▶ *When a gun is fired, does the air exert significant friction on the bullet? Is the bullet's course ever altered by the wind?*

Yes to both questions, which deal with ballistics, specifically exterior ballistics, the study of the behavior of projectiles such as bullets or missiles in flight.

Two strong forces act on a projectile and influence its course and speed: gravity and air friction. The force of gravity pulls the projectile toward Earth, creating a characteristic arc-shaped trajectory. Friction, caused by air molecules hitting the surface of the projectile, slows the projectile's speed. This force is significant. Imagine the force of air, even on a windless day, when you place your head or hand out of the window of a car traveling 60 miles (96 kilometers) per hour. So a bullet's course can certainly be altered by the wind.

Other factors that determine a projectile's behavior include the force used to launch the projectile, as well as its size, weight, and shape. Bullets, like airplanes and automobiles, are designed to create a minimum amount of air friction or resistance, which is also called drag. Air density, too, affects the amount of drag encountered by the projectile.

Classical physics tells us that in the absence of friction (as in a vacuum) and in the absence of gravity (as in a weightless environment), a launched projectile will continue on its course indefinitely. Artificial satellites that circle Earth maintain their speed thanks to the near-frictionless environment of space; they maintain their orbits because they have been placed on a precise trajectory that positions them in balance with Earth's gravitational pull.

▶ *Is there a formula that gives you the square root of a number without using a calculator or slide rule?*

The square root of a number is another number that produces the first when it is multiplied by itself. For example, the square root of 25 is 5, since 5 x 5 = 25. The numbers 1, 2, 3, 4, 5, 6, 7, 8, 9, and 10 are, respectively, the square roots of 1, 4, 9, 16, 25, 36, 49, 64, 81, and 100.

Few square roots are perfect, meaning they can be expressed as a whole number, such as 7, the square root of 49. But what is the square root of 52? It is a number that must be expressed as a decimal or as a fraction. Without a calculator, slide rule, or logarithm table, the square root must be calculated by hand. There is no single formula to do so. Rather, there is a series of steps to follow to approximate the square root.

A short method that quickly gives a good result works as follows:

To find the square root of 52, find the closest perfect root, which is 7, the square root of 49. Divide 52 by 7 to obtain 7.4. Then find the average of 7 and 7.4 (7 + 7.4 divided by 2 = 7.2). Since 7.2 x 7.2 - 51.84, 7.2 is the approximate square root of 52. For more precision, the process can be repeated by dividing 52 by 7.2 to obtain 7.222. The average of 7.2 and 7.222 is 7.211, which when squared is 51.998521. The process can be continued even further for greater accuracy, if required.

▶ *Where does gravity come from? Why is it that Earth has strong gravity, but other heavenly bodies, such as the Moon, have only weak gravity?*

No one knows where gravity comes from. It is, however, a measurable force in nature.

Gravitational force, put simply, is the attractive force that one object has for another object. The Sun exerts a gravitational force on Earth. Earth exerts its gravitational force on the Moon. The Moon, in turn, attracts the waters of Earth, creating tidal rhythms. Gravitational force holds objects in the universe together.

The amount of gravitational force of an object is proportionate to its mass. Generally, the larger the object, the greater its gravitational force, because larger objects typically have greater mass. The distance that one object is from another also determines the strength of the gravitational force between two bodies.

The effects of gravity on Earth are great. The Moon has a much smaller mass and weaker gravitational force than Earth. Of the bodies within our solar system, the Sun has the greatest gravitational force.

▶ *When was the element platinum discovered? Does platinum have any uses aside from jewelry?*

Platinum (symbol Pt, element number 78 in the periodic table) was first positively identified in Colombia, South America, in the 18th century. Spanish explorers called it *platina del Pinto* ("silver of the Pinto"), since the silvery-white–colored substance resembled silver.

Platinum is found worldwide, but not in pure form. It is usually found in combination with other platinum-group metals (such as iridium, rhodium, ruthenium, and osmium) in ores of copper and nickel at a concentration of approximately 1 part per million.

The form of platinum used in jewelry is actually an alloy, a mixture of platinum and other hardening metals. Platinum itself is fairly soft, so it requires another metal to give it strength. Platinum-iridium and platinum-ruthenium compounds are typically used.

Platinum exhibits several unusual characteristics, including a high melting point and a resistance to corrosion. Platinum is used to make containers, such as crucibles, that can withstand high temperatures. Because platinum is resistant to corrosion, it makes an excellent electrical contact. Platinum is also a catalyst for several chemical processes. As a gasoline additive, it increases the octane rating of fuel, for example. More on platinum can be found in the article "The Noblest Metals," which begins on page 216.

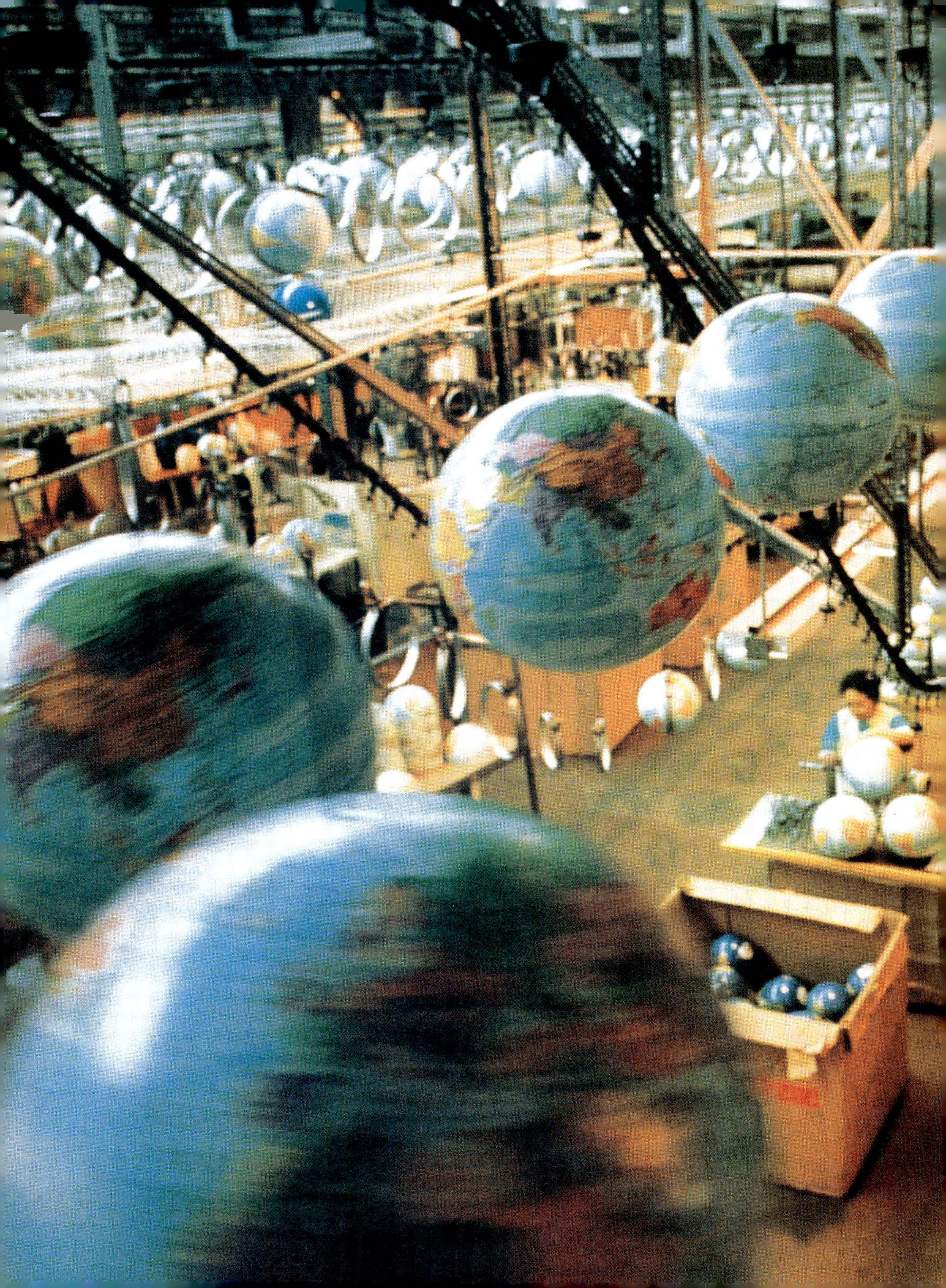

Technology

▪ All over the world, mass-production and assembly-line techniques continue to play vital roles in industry. Depending on the product, highly skilled hand labor might be required. Perhaps more often, computerization and automation help factories manufacture products that are identical in every way.

CONTENTS

3-D Computing	240
Why Airplanes Crash	247
The Internet	254
Steel Magnolias	262
After the Accident	270
Undersea Explorers	275
Ask the Scientist	280

3-D Computing
A WHOLE NEW DIMENSION

by Peter Coy and Robert D. Hof

Red troops poured across the Czech border in November 1994, trying to capture Germany. In the counterattack, U.S. Air Force Captain Reid D. Reasor flew an F-16 jet fighter out of Ramstein Air Base, dropping cluster bombs on bridges, bunkers, and Red regiments. "You got sweat beading up on your forehead," he recalls. "You could hear anxiety in your wingman's voice when you got separated."

Pretty exciting—for a battle that never happened. Reasor spent the whole time inside a three-dimensional (3-D) flight simulator in Mesa, Arizona. As he looked out the "canopy" of his F-16, computers worked furiously to generate the passing view of European rivers, hills, and towns. The U.S. Army–led exercise connected more than 300 sites, ranging from submarine simulators in Newport, Rhode Island, to infantry simulators in Grafenwöhr, Germany.

The key to the biggest military simulation ever was 3-D computing. Simulations in three dimensions are a cheap, safe supplement to conventional exercises in a time of budget

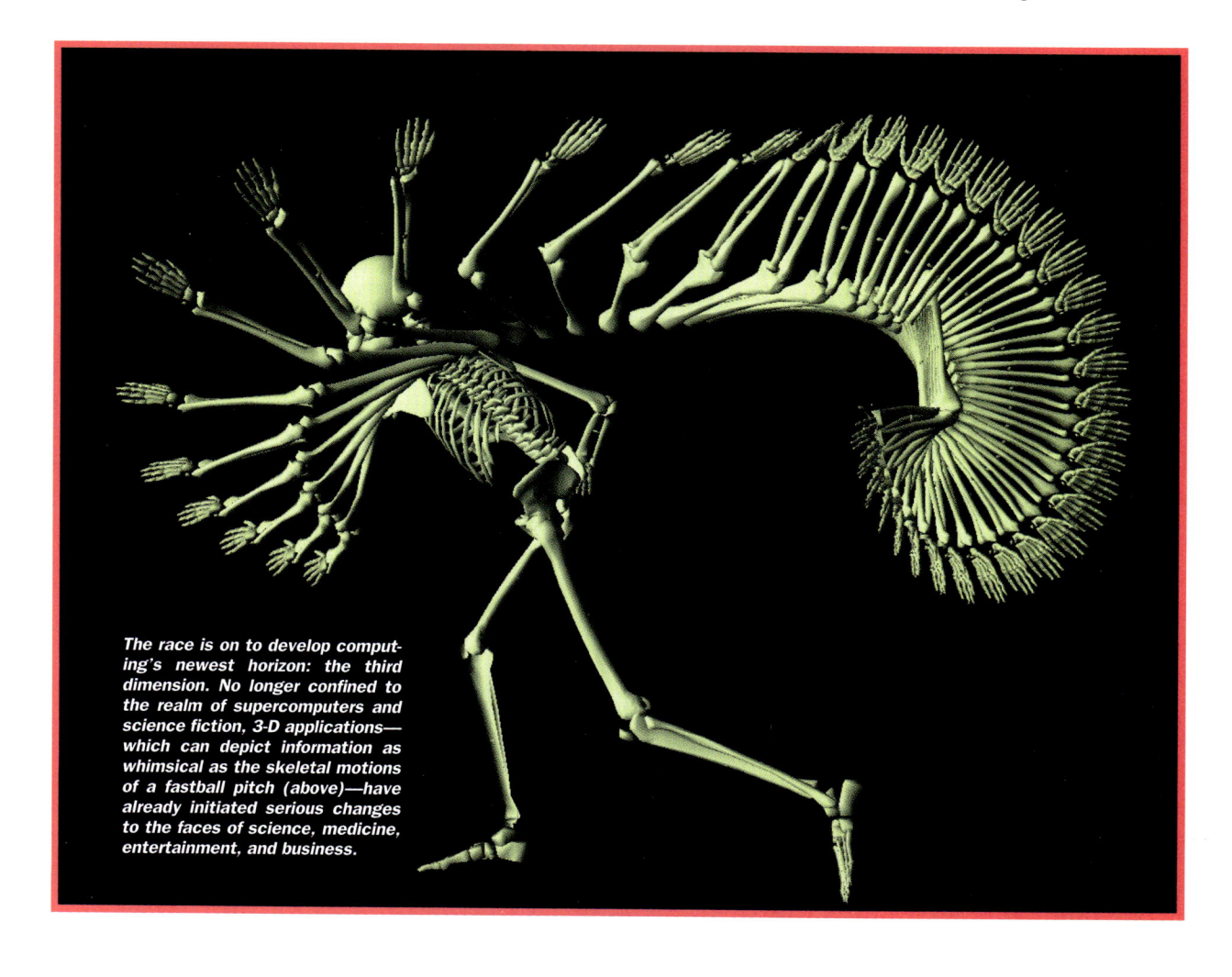

The race is on to develop computing's newest horizon: the third dimension. No longer confined to the realm of supercomputers and science fiction, 3-D applications—which can depict information as whimsical as the skeletal motions of a fastball pitch (above)—have already initiated serious changes to the faces of science, medicine, entertainment, and business.

austerity. The U.S. Department of Defense plans an even larger video war game for 1997.

EXPANDING HORIZONS

However, you do not have to work at the Pentagon to appreciate 3-D computing; it is pervading the civilian realm as well. Just a few of the fields expanding their horizons through 3-D vision: medicine, oil exploration, product design, and architecture—not to mention video games, advertising, and movies. While there is no good estimate of the overall 3-D market size, just look at Silicon Graphics Incorporated. SGI, the premier maker of 3-D computer workstations, posted revenue of $2.2 billion for the year ending June 30, 1995—triple the total earned only four years earlier.

People who want to try out 3-D for themselves need go no farther than the local toy store or computer dealer. Nintendo, Sega Enterprises, Sony, and others are relying on

Engineering software from Intergraph Corporation colorfully displays the elaborate routing of pipes beneath a ship's deck. The program automatically revises the spatial layout if the ship's design is modified.

dazzling 3-D effects to stimulate sales of a new generation of high-powered game machines. And 3-D games are coming to the home personal computer as well, thanks to 3-D graphics boards.

Today's 3-D is a new variation on a familiar theme: ever more powerful and inexpensive computers. The fanfold-paper printouts of yesteryear gave way to screens with monochrome text, followed by windowing software and simple graphics. Now the windows are being thrown open. "3-D was always seen as a very specialized, high-cost option, like an expensive spice from China. That day is passing fast," says Dan Mapes, president of SynergyLabs, based in San Francisco, which created special effects for the science-fiction movie *Virtuosity*. Mapes does all his effects on a network of Intel Pentium–based personal computers (PCs). Says Mapes: "We think of 3-D as simply the highest-quality way of transmitting information from one mind to another."

That is exactly why 3-D computing is so important. The brain absorbs three-dimensional information like a sponge: we have been doing it since birth. Italian painters realized that as long ago as the

The motion-picture industry has thrilled audiences with imaginative 3-D animation in such movies as Jurassic Park, Casper, *and* Toy Story *(above)—the first full-length feature film to be entirely computer-generated.*

TECHNOLOGY 241

15th century, when they added perspective to their paintings, aided by the geometric insights of Florentine architect Filippo Brunelleschi. He, in turn, had simply rediscovered the classical Romans' "vanishing point"—that faraway spot in a scene at which all parallel lines seem to converge. Today computers make it possible to step inside those perspective images and look around—say, wander around a virtual piazza and look back at the way you came in. Says Edward R. McCracken, Silicon Graphics chairman and CEO: "Someday all computers will be a window into three-dimensional space."

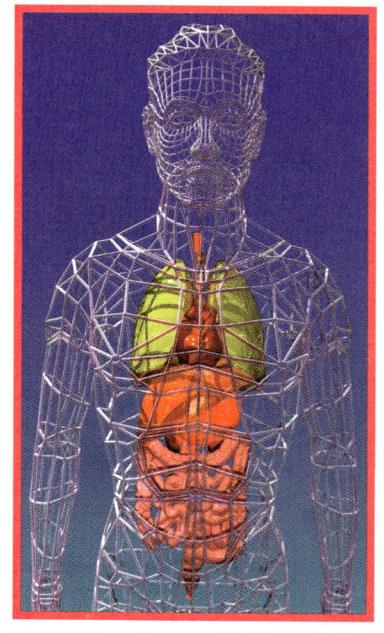

Using 3-D anatomy applications, doctors and medical students can review the precise location and interaction of body organs or practice a new procedure—all without textbooks or cadavers.

America's Cup with the help of a "digital wind tunnel"—a Silicon Graphics workstation next to the docks in San Diego, California, that allowed its team to analyze nearly 10,000 design variations in two months. In Japan, contractors mounted twin camera "eyes" on unmanned bulldozers in 1994, producing 3-D displays that let them remotely control the earthmovers as they cleared land around the Unzen Fugendake Mountain Volcano. In the ultimate confirmation of its hipness, 3-D made its debut in 1995 on the World Wide Web. Instead of pages, people are building "rooms" occupied by 3-D objects—finally adding spaciousness to cyberspace.

Smooth Sailing

3-D computing is not limited to virtual reality, whose devotees dress up in goggles and gloves to immerse themselves in an artificial environment. Most people who do 3-D computing are not interested in total immersion—just added insight.

These are early days for 3-D. Even the fastest workstations require hours to render a single complex, high-quality image using 3-D tools—say, one frame of an animated movie. But that limitation is a good sign: if people use 3-D now, despite the inconvenience, imagine how much they will use it when computers are vastly more powerful in 5 or 10 years.

Look how they are using it already. New Zealand sailed off with the 1995

The high sales volumes generated by video-game technology are driving down the price of 3-D equipment, and new 3-D products for home computers should accelerate the trend. These accelerator chips cost PC and board makers $30 to $40 apiece, one-fifth the price of their cheapest predecessors, says Jon Peddie Associates of Tiburon, California. There is still a market for the $150-and-up chips—mainly in

Cutting-edge 3-D graphics are now available at the toy store. Innovative video games such as Paradigm Simulations' "Egghead Shred" (right) feature dazzling effects adapted from scientific or military applications.

design and engineering, where geometric accuracy is paramount. 3-D chips are becoming widely available. While giants such as Digital Equipment Corporation (DEC) make them in-house, smaller designers hire any of several companies for fabrication.

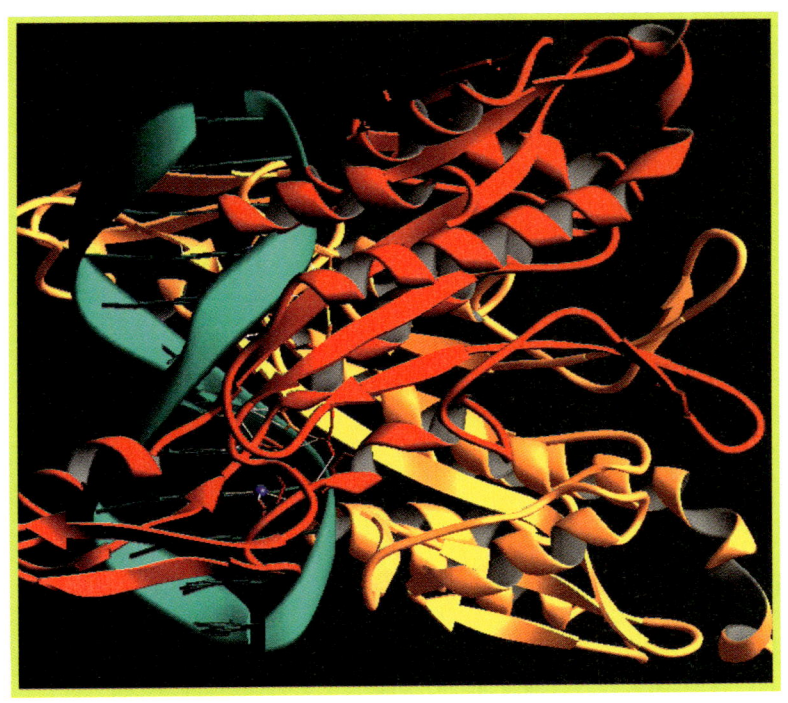

New 3-D visualization tools can display the structure of DNA (above) and other complex molecules. With virtual-reality software, scientists can even go "inside" a molecule to rearrange its atomic structure.

Plunging prices will, in turn, expand the ways businesses can speed analysis and design to lower costs. Voith Hydro Incorporated of York, Pennsylvania, uses 3-D computational fluid-dynamics software in place of scale models to test the efficiency of small turbines that the company sells to hydroelectric-power plants. That substitution is a huge savings, since physical models cost $500,000 apiece to build and test. The software is not accurate enough yet to supplant models for big turbines, but it helps engineers build models that require fewer costly tweaks, says Richard A. Fisher, Jr., vice president for technology. 3-D software also produces ideas for better designs—crucial,

Fisher says, since a 0.1 percent efficiency gain can be worth $1 million to a customer.

For scientists, the appeal of 3-D is the stimulation it provides for their imaginations. James Watson and Francis Crick tinkered with balls and sticks to figure out the helical structure of DNA. With computers, the toying is orders of magnitude more sophisticated. A good example is X-ray crystallography, a procedure for figuring out the placement of atoms in large molecules. Computers can make a rough guess based on the pattern made by the rays deflected off a crystallized molecule. But to get the positioning exactly right, scientists like to use a very special computer: the visual cortex of their own brains. Wearing virtual-reality goggles and gloves, they go "inside" a 3-D molecule model and rearrange the atoms by hand until the model fits the data.

Surgeons like 3-D for similar reasons. In the 1930s, radiologists created 3-D images by crossing their eyes as they looked at two offset X rays of the same object. Now computers do the job for them, using data not only from X rays but also from magnetic-resonance imaging (MRI) and positron-emission tomography (PET).

BRIEFCASE HIT

3-D is so new that many people who could benefit from it still are not using it. Most architects, for example, still prefer to deal with 2-D blueprints and cardboard mockups—arguing that computers spoil the design process. A few are catching on, though. Seattle, Washington–based Callison Architecture Incorporated uses 3-D not only for design but also for videotaped animations that give such clients as Boeing, Microsoft, and Eddie

Bauer a feel for what a new building will be like. Says James P. Rothwell, a Callison principal: "Clients love to show these videos around. They become a briefcase hit."

New applications are heating up the battle for supremacy in 3-D space. In 1994, Microsoft bought Montreal, Quebec–based SoftImage, one of the main suppliers of software for high-end special effects in movies and television commercials. While SoftImage software currently runs on Silicon Graphics machines, Microsoft is making it run on PCs that use its own Windows NT operating system as well. Microsoft has also snagged a British company called RenderMorphics, giving Microsoft what is likely to become the programmer's standard for creating PC-based video games.

Quietly, Microsoft Chairman William H. Gates III has assembled what University of Utah computer scientist Richard F. Riesenfeld calls "the largest collection of graphics talent under one roof in the world." James T. Kajiya, hired in 1994 from the California Institute of Technology (Caltech), is working on a new standard for fast, cheap 3-D software and hardware aimed at consumers. Another hire, Alvy Ray Smith, cofounder of computer animator Pixar Corporation, says he has wanted for years to pull together sound, pictures, and animation into one easy-to-use package. "Nobody was big enough to pull it off," he says. "I think Microsoft might be."

Besides Microsoft, giants taking aim at the high end of 3-D computing include Sun Microsystems, king of the Internet; Hewlett-Packard, with new graphics accelerators; DEC; and IBM, which is encouraging Hollywood animators to rent processing time on its giant SP2 supercomputer in Maui, Hawaii.

In these early days of the 3-D boom, demand is

Flight of the Synthetic Bumblebee

Here is how graphic artist Alan Henry built and flew a bee using a Commodore Amiga computer and Imagine software from Impulse Incorporated.

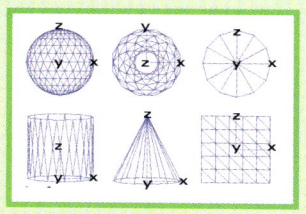

1 Henry picks some building blocks, known as wireframes, which are made from tiny triangles.

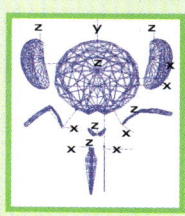

2 He reshapes the building blocks into bee parts.

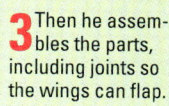

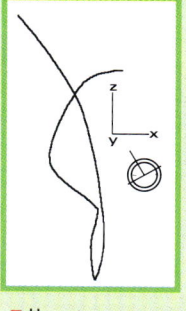

3 Then he assembles the parts, including joints so the wings can flap.

4 He sets an overhead-view "flight plan" for the bee and decides where the flight will be viewed from (camera, lower right).

5 He brings the bee alive by choosing colors, texture, reflectiveness, and transparency.

growing at such a rapid pace that plenty of suppliers should be able to get involved in the third dimension. Plenty already are. Autodesk Incorporated, the maker of computer-aided design (CAD) software for PCs, has become a major force by opening up its 3-D Studio entertainment software to makers of "plug-ins": software modules for special effects for movies and commercials—effects such as sepia tones, fog, or bubbles. MultiGen Incorporated specializes in extra-fast rendering of 3-D for video games and flight simulators—where there's no time to create each frame in lifelike detail. Its software draws the buildings that people "fly" over on a carpet in Walt Disney's knockout new Aladdin attraction at Epcot Center in Orlando, Florida.

Quantum Leap

3-D is generating new opportunities for smart companies such as Viewpoint DataLabs Incorporated in Orem, Utah. Viewpoint builds everything from helicopters to body parts out of "wireframes," which look like chicken-wire sculptures and are easy to manipulate by computer. The digital-prop house sends these 3-D objects to moviemakers, the military, and others. Viewpoint made the skeleton pictured in this article, and programmed it to throw a baseball by copying the arm motions of human volunteers. A moviemaker could simply dress the skeleton in cyberflesh and a baseball uniform.

The next step beyond viewing 3-D images on flat screens is stereoscopy—as in the 3-D glasses for 1950s horror movies such as *Creature from the Black Lagoon*. Stereoscopy mimics the real world, in which people gauge distance by the way their two eyes see the same object from different angles. Today's virtual-reality goggles are a quantum leap beyond the colored cellophane lenses of the 1950s. Says Linden Rhoads, president of Seattle-based goggles maker Virtual i-O: "Stereoscopy is just much more compelling—the same as we like talking films more than silent ones and color films more than black-and-white."

Goggles are a bit cumbersome, though, and certain poorly designed ones can cause

A combination of favorable technological and economic conditions—namely broadened 3-D software offerings and less-expensive and more-powerful personal computers—has enabled millions of consumers to journey into the third dimension without leaving home.

side effects ranging from headaches to disorientation. That is why inventors around the world are racing to develop displays that produce two different eye views without glasses. Australian inventor Donald Martin has one involving vertical bars that whiz by in front of the screen. In 1994, the U.S. Navy revealed that it can create 3-D images by firing 40,000 laser beams into something that looks like a round, transparent washing machine. Sharp Laboratories of Europe Limited has built a prototype 3-D display that remotely senses a viewer's head movements, and reacts by moving lenses to change the scene presented to each eye.

Holography is advancing as a 3-D medium, too, and not just as the pigeon on your credit card. Holograms can now be made in full, natural color, thanks to a 1992 breakthrough by Hans I. Bjelkhagen. In his new job as vice president for research and development at American Propylaea Corporation in Birmingham, Michigan, the Swedish researcher is working to make his full-color holograms bigger, brighter, and viewable from more angles.

Today's 3-D is miles beyond what was possible in the 1960s, when the most-sophisticated computer graphics in the world were being done in Salt Lake City. The University of Utah became a mecca thanks to faculty members David Evans and Ivan E. Sutherland. The two men later founded Evans & Sutherland Computer Corporation, which remains No. 1 in software for flight simulators. Utah alumni include Jim Clark, founder of Silicon Graphics; John E. Warnock, founder of Adobe Systems; Alan Kay, principal scientist of Apple Computer; Alan Ashton, founder of WordPerfect; and Ed Catmull, who cofounded Pixar with Alvy Ray Smith.

True Love

Computer memory in those early days was insufficient to hold animation, so programmers stored images directly onto film. To capture a frame, they would keep the camera shutter open as the computer generated the image, line by line. Since the screen was monochrome, they had to do multiple passes with filters to get color images. That took true love. Henry C. Christiansen, a Brigham Young University professor who did his research at Utah, says they borrowed the studio of the Mormon Church to make "movies" of such things as the vibration and flutter of experimental National Aeronautics and Space Administration (NASA) aircraft. Says Christiansen: "I used to work all night, until my programs crashed."

Creating good 3-D still is no snap, but lots more tools and building blocks exist. If photorealism is the object, artists can create reflections by letting the computer calculate the paths of light rays. They can also add "radiosity," so a hand next to a bright-yellow car will pick up a yellow glow. If speed is the object, forget computation-intensive tricks like radiosity. The processor in a shoot-'em-up video game is busy just trying to redraw tanks and missiles 30 to 60 times per second. Video-game programmers try to retain a respectable degree of realism while staying within the all-important "polygon count"—the maximum number of geometric shapes that the processor can draw per second.

Limitations

As people embrace 3-D, they must also understand its limitations. Computer simulations, no matter how seemingly accurate, are not replicas of reality. Case in point: when Boeing engineers probed the 1992 crash of an El Al Israel Airlines 747 cargo flight that slammed into low-income housing in the Netherlands, killing 43 people, they discovered that their finite-element-analysis stress simulation had not been detailed enough. It had missed several weak points in the design of a "fuse pin" that held the engines to the wings. One fuse pin broke, and the plane went into a fatal dive. A similar simulation flaw led to the sinking of a huge oil platform in a Norwegian fjord in 1991.

Sometimes 3-D pictures end up just being confusing. Pacific Northwest Laboratories in Richland, Washington, found that out when researchers wrote a data-visualization program called Galaxies. Each document was characterized by a point, or "star." You could tell how similar documents were by how closely the stars clustered on a 2-D chart. The system worked fine until the lab added another variable. When users put on virtual-reality goggles to explore the proximity of stars in three dimensions, they became disoriented—"lost in space," says staff scientist James A. Wise. The lab has had better luck with another program, Themescape, which portrays data as hills and valleys.

Like no other field of computing, 3-D hits people in the gut. Just ask the kids from Orville Wright Magnet School in Los Angeles, who got a preview of Nintendo's 3-D Virtual Boy at a 1995 trade show. Their one complaint was that the $180 machine displays only red on black. Wrote middle-schooler Aja Thrasher: "It would really be Da Bomb if you are able to make this in color." When it comes to 3-D, everybody's got something to say.

Why Airplanes Crash

by Jerome Greer Chandler

It is the end of a forgettable flight, an unremarkable day. Over the cabin public-address system comes a litany a frequent flier has heard 100 times before. It is delivered with the kind of casual authority only an airline pilot can muster. "Folks, from the flight deck. We should be on the ground in about 10 more minutes. Uh, sunny skies; a little hazy. Temperature, temperature's ah, 75°. Winds out of the west around 10 miles per hour. Certainly appreciate you choosing USAir for your travel needs this evening; hope you've enjoyed the flight. Hope you come back and travel with us again."

You yawn to equalize pressure in your ears. Muted clicks and snaps ripple through the cabin as fellow passengers stow laptop computers and return their seat backs to the upright position. The high priests of technology have brought you home safely again.

In reality, you and 131 other souls on board flight 427 have less than three minutes to live.

At 7:02:57 P.M., the cockpit voice recorder registers a sound similar to three fast electrical clicks. Captain Peter Germano exclaims, "Sheez." More mechanical sounds fill the cockpit. Germano inhales, and exhales quickly. Thump. "Whoa," he says. Clickety click. "Hang on." The Boeing 737-300 is rolling—inexorably—to the left. A strange sound shudders through the craft, presaging an aerodynamic stall. The forces that make flight possible evaporate. "What the hell is this?" asks Germano. "Oh God," he implores as the craft nose-dives. "Pull," exhorts Germano. "Pull." At 7:03:22.5—three-tenths of a second before the twin-engine jet buries itself in the ground—the voice recorder registers a final word from the cockpit: "*No.*"

Tragedy Strikes

In 1994, 264 people died in commercial-airline accidents in the United States. Half of them were on board USAir flight 427 when it crashed on approach to Pittsburgh, Pennsylvania, on September 8.

Following an airplane crash, investigators often must sift through enormous amounts of debris at the crash site in order to determine the exact cause of the catastrophe.

Parts That Fall Apart

Although they haven't yet been linked to any commercial air crashes in the United States, bogus aircraft parts threaten air safety. A. Mary Schiavo, the inspector general for the U.S. Department of Transportation (DOT), told a Senate subcommittee in May 1995 that "suspect unapproved parts" could have "potentially devastating consequences."

There are different species of suspect parts: counterfeit parts, substandard parts, and parts not manufactured in compliance with Federal Aviation Administration (FAA) regulations. Schiavo says the parts began to proliferate when high prices "made it extremely lucrative for unscrupulous parts manufacturers to prey on airlines experiencing huge financial losses."

In response to criticism from Schiavo and Congress, the FAA in August 1995 announced the creation of a special task force to "thoroughly review the issue of unapproved aircraft parts." Law-enforcement and aviation officials will be represented on the panel. The move follows another FAA initiative, one that creates an industry-operated authorization program for those who broker and distribute aircraft parts.

The deadliest crash attributed to unauthorized parts occurred in September 1989 when a Norwegian Convair, manufactured in the United States, crashed at sea, killing 52 people. The plane's tail fell off in flight. Bogus parts have also contributed to a number of general (private) aviation crashes.

More recently, tragedy struck two Boeing 757s. On February 6, 1996, a charter flight had just left the Dominican Republic when it crashed into the Atlantic some 13 miles (21 kilometers) off the coast. All 189 passengers and crew were presumed dead. Conclusive evidence of the cause awaited recovery of the flight-data recorders—sunken 7,000 feet (2,100 meters) beneath the ocean. Less than two months earlier, on December 20, 1995, American Airlines flight 965 crashed into a mountainside en route to Cali, Colombia. Only 4 of the 164 people aboard survived.

Heading down a narrow valley in the Andes, the pilots of flight 965 thought that a Spanish-speaking air-traffic controller had cleared them directly to the airport, and they punched in the autopilot code that would take them there. In reality, the controller intended the plane to head for the airport only after passing another checkpoint. By the time the pilots realized they were off course, they were closing in on a steep ridge that jutted into the valley.

"Terrain, terrain . . . pull up, pull up," announced the automatic-warning system. But even at full throttle, the plane failed to clear the ridge.

Safer Than Ever Before

Despite these recent disasters, commercial aviation overall has become safer. National Transportation Safety Board (NTSB) statistics show that in 1982, airliners with 30 seats or more (for the most part, that means jets) recorded 0.058 accidents per 100,000 departures. The number dipped to 0.013 in 1993. In 1994, the tally was 0.049.

Earl L. Wiener, an aviation researcher at the University of Miami, puts those numbers into perspective. In the United States, he says, "There are 15,000 to 20,000 takeoffs a day. Half a billion passengers a year, in a hostile environment, get to their destination quickly and safely."

More-Complicated Accidents

Impressive perhaps, but not good enough. For all the recent accomplishments—better aircraft, more-reliable engines, improved regulation of commuter airlines—airplanes still crash. Experts believe we have reached a critical plateau in aviation safety, a point we must push beyond if we are ever to attain the mythical goal of "zero accidents."

"We've solved the [causes of] easy accidents," contends Clint Oster of Indiana University in Bloomington, coauthor of the

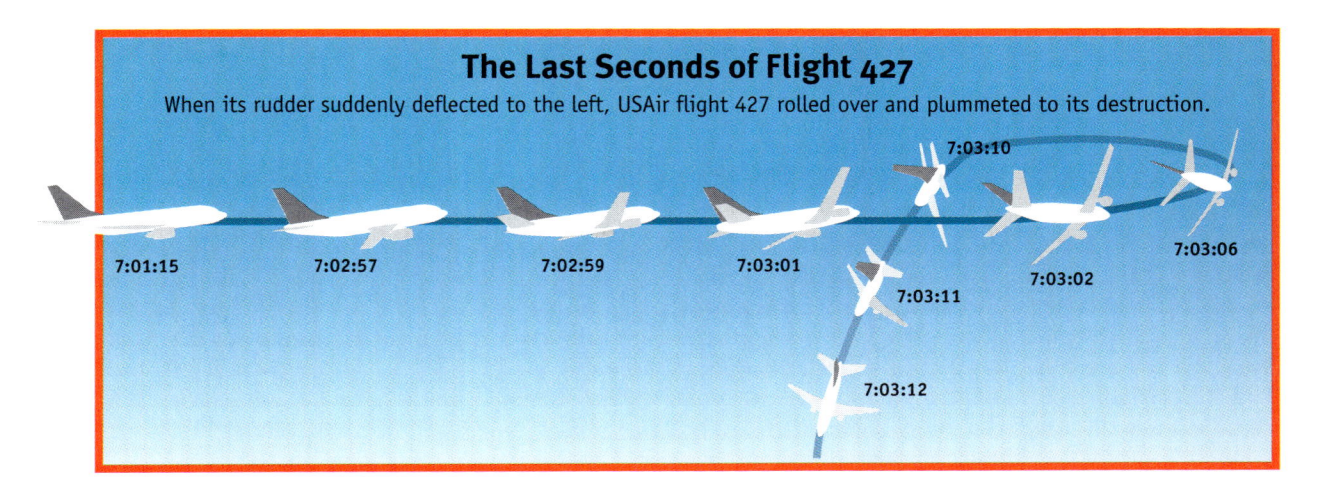

The Last Seconds of Flight 427
When its rudder suddenly deflected to the left, USAir flight 427 rolled over and plummeted to its destruction.

7:01:15   7:02:57   7:02:59   7:03:01   7:03:10   7:03:02   7:03:06   7:03:11   7:03:12

book *Why Airplanes Crash*. A former research director for the President's Aviation Safety Committee, Oster says that problems such as engines quitting, aircraft flying unknowingly into severe weather, and midair collisions have been all but eliminated. "So what happens," he contends, "is that we're getting into much more complicated kinds of accidents. It takes a much more unusual set of circumstances to lead to a plane actually crashing."

None more unusual than USAir 427. No one knows why the 737-300 rolled over onto its back and died. Somehow the rudder radically deflected to the left. Was it due to wake vortices wrought by the wingtips of a Delta 727-200 flying ahead of the 737? Such miniature horizontal "tornadoes" have been known to throw smaller craft out of control—but not a beefy 737-300.

Perhaps the rudder system was to blame. Boeing has tallied 187 possible incidents involving malfunctions of the jetliner's yaw damper, a mechanism attached to the rudder that prevents the airplane from fishtailing. But company spokesperson Liz Verdier says, "There's no way, even if it fails," that a yaw damper could flip a 737.

Last fall, the NTSB tried to duplicate the crash circumstances, flying a 737 in the wake of a larger 727. When the twin-engine jet hit the wake vortex produced by the larger 727, it rolled 60 degrees. Michael Carriker, Boeing's senior engineering 737 project pilot, told the safety board that the "seven-three" rolled at a rate of 36 degrees per second. That's some 12 times faster than the autopilot is programmed to roll the aircraft when banking for a turn. The tests have shed light on the fatal plunge of flight 427. But the specific link between vortex encounter and rudder deflection remains a mystery.

It is a mystery with far-reaching implications. Approximately 2,600 737s are in use worldwide. Different models of the popular twin jet have safely logged 55 million flights. While a number of 737s have crashed over the years, only two accidents have investiga-

The air-traffic-control system in the United States is beset with chronic staff shortages and woefully outdated equipment—problems that contribute to long delays and ultimately drive up airline costs.

tors really worried: the one at Pittsburgh, and the March 3, 1991, crash of United Airlines flight 585. It, too, was on approach, this time to Colorado Springs, Colorado. It, too, flipped over onto its back. Twenty-five people died. Five years later, investigators still don't know why. Since the crash of flight 427, Boeing engineers have spent almost 100,000 man-hours trying to make sense of it all.

Answers from the Black Box

One of the things 737 investigators are concerned about is not having enough information. Flights 427 and 585 were fitted with older digital flight-data recorders, so-called "black boxes." The data recorder in the tail of the USAir 737-300 measured only 11 flight parameters; the device on the United 737-200, just five. Neither kept track of the rudder position.

A clearly frustrated NTSB recommended that the Federal Aviation Administration (FAA) see to it that all 737s were wired with more-sophisticated data recorders by the end of 1995. The Air Transport Association (ATA), an industry trade group, labeled the timetable "unattainable." The FAA, while supporting new digital recorders, refused to go along with the safety board's "aggressive" target date.

It's 53 days after the September 8, 1994, crash in Pittsburgh. An American Eagle ATR-72 is heading from Indianapolis, Indiana, to Chicago's O'Hare Airport. Flight 4184 departs an hour and 18 minutes late on October 31, 1994. Weather in the Windy City is wretched. At 3:45:48 P.M., Pilot Orlando Aguiar turns on the cabin P.A.:

"Well folks, once again this is the captain. . . . You're, uh, do regret to inform that air-traffic control has put us in a holding pattern up here. We're holding for about 20 minutes out of Chicago, but, uh, I guess the congestion an' traffic's continued. . . . I do apologize for all these delays."

To steady the flight, the crew lowers the flaps to 15 degrees. On autopilot, the craft wallows along in the clouds at 10,000 feet (3,000 meters). At 3:56:27.8, air-traffic control clears flight 4184 to descend to 8,000 feet (2,500 meters). A minute later, the crew retracts the flaps. All hell breaks loose. "Oh," says Aguiar. "Oops," exclaims First Officer Jeffrey Gagliano as the prop jet rolls wildly to the right and the nose pitches up. The voice recorder registers three thumps followed by rattling. A trio of rapid chirps follows, indicating the autopilot has disconnected. The Eagle corkscrews, nose down. Captain Aguiar tries to coax some sani-

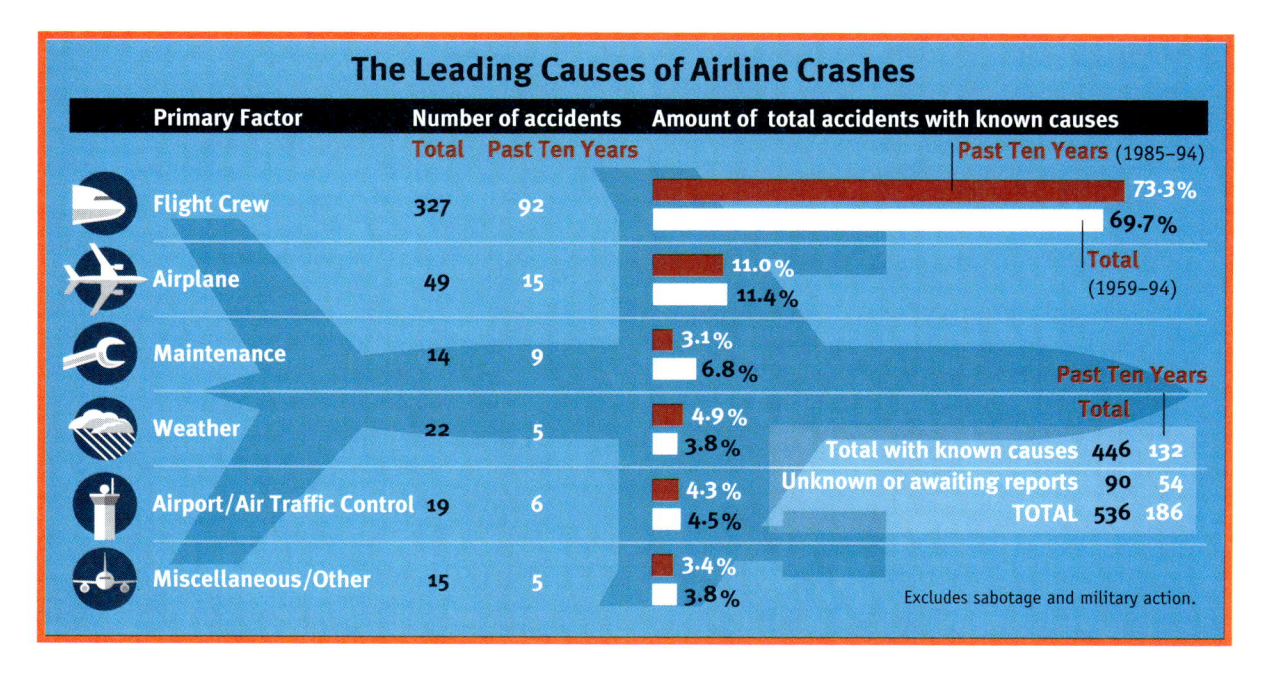

The Leading Causes of Airline Crashes

Primary Factor	Number of accidents Total	Past Ten Years	Amount of total accidents with known causes Past Ten Years (1985–94) / Total (1959–94)
Flight Crew	327	92	73.3% / 69.7%
Airplane	49	15	11.0% / 11.4%
Maintenance	14	9	3.1% / 6.8%
Weather	22	5	4.9% / 3.8%
Airport/Air Traffic Control	19	6	4.3% / 4.5%
Miscellaneous/Other	15	5	3.4% / 3.8%

Total with known causes 446 132
Unknown or awaiting reports 90 54
TOTAL 536 186

Excludes sabotage and military action.

On the ground, a plane can be readily deiced before takeoff. In the air, however, ice accumulation on the wings can compromise the craft's ability to stay aloft.

ty out of what's happening. "All right, man.... Mellow it out," he tells Gagliano. "Nice and easy." "Terrain, whoop, whoop!" warns an electronic voice. The terrain in question is a soybean field near Roselawn, Indiana. At 3:57:56.6, Gagliano concedes the short, terrifying fight: "Aw...." A tenth of a second later, his last word is truncated by a loud crunch.

Thanks to a data recorder that logged more than 100 parameters, investigators not only know what killed the 68 people on board, but they think they have a good idea *why* it happened. The NTSB, with almost unprecedented speed, found the most likely culprit—and the key accomplice. Ice, and the crew's reliance on the autopilot, most likely felled flight 4184.

The "upset" was apparently triggered when the flaps were retracted. This changed the airflow over the wing. Lift, which holds an aircraft aloft, is predicated on the differential flow of air over and under a wing. An examination of the cross section of an airfoil shows that it is rounded on top and somewhat flatter on the bottom. Air passing over the top takes longer to travel to the wing's trailing edge than does the air underneath. Because the air on top is "thinner," a vacuum is created. Higher pressure below it lifts the wing.

ICE: A COMPLICATING FACTOR

Ailerons are rectangular devices fitted on the trailing edge of wingtips. They control roll. In testimony before the NTSB, Andre Bord, chief engineer for the ATRs, said a buildup of ice could cause "aileron aspiration." When ice disrupts the airflow over the wing's upper surface, a vacuum can form beneath the airflow, which can snatch the ailerons out of position. When the Eagle's flaps were retracted, the theory goes, airflow aerodynamics were changed enough to cause aspiration.

After the Roselawn crash, the FAA temporarily prohibited ATRs from flying in known icing conditions. (After the installation of larger deicing boots on the remaining 175 ATR-42s and ATR-72s flying in the United States, the agency lifted the ban.) The FAA also required that pilots turn off the autopilots when encountering freezing rain or drizzle. That is critical. When a pilot "hand-flies" an airplane, he or she can literally feel aerodynamic changes. With an autopilot, the human pilot is out of the loop.

Case in point: April 29, 1993. Continental Express flight 2733 is bound from Little Rock, Arkansas, to Houston, Texas. The autopilot controls the EMB-120 prop jet as it climbs to 22,000 feet (6,800 meters). So relaxed is the captain that a passenger notices that he has propped a foot on the console.

At 3:32:28 P.M., the first officer remarks, "We're not climbin' very fast." "Heavy, really heavy," answers the pilot. They exchange a joke. Then: "Hang on. Somethin' ain't right," says the captain as the prop jet approaches 17,000 feet (5,200 meters). At 3:33:16.8, the autopilot-disconnect warning sounds. The pilot calls for "Airspeed." The sound of a stall alert pierces the flight deck. "Hang on," says the startled captain. "Hang on."

This time, they do. After plunging 12,000 feet (3,600 meters), the crew gains enough control of the crippled aircraft to manage an emergency landing at Pine Bluff, Arkansas. Thirty people are on board Continental Express 2733. All walk away. The NTSB report concludes that "the probable causes of this accident were the captain's failure to maintain professional cockpit discipline, his inattention to flight instruments and ice accretion, and his selection of an improper autoflight vertical mode."

THE COST OF HUMAN ERROR

Are pilots developing a "set it and forget it" attitude in which technology transcends basic airmanship? Some experts think so. "As airplanes become more and more sophisticated, as they make use of more and more computer capabilities," says Clint Oster, "how do we keep up?" In some respects, automation has made flying safer. "It can help pilots do things better than they do manually," says R. John Hansman, director of the Aeronautical Systems Laboratory at the Massachusetts Institute of Technology (MIT) in Cambridge. But "as the automation has become more sophisticated, it has become more complex. And now we're seeing people having some problems with the complexity."

- *February 14, 1990.* An Indian Airlines Airbus A320 crashes on landing at Bangalore. Ninety die. An Indian government report concludes: "The aircraft could not sustain the approach and also could not maintain the [required] speed because the engines were at idle thrust." Why? There was apparent confusion as to which "mode" the craft's high-tech autopilot was in.

- *January 20, 1992.* An Air Inter A320 on approach to Strasbourg, France, goes down. There are 87 fatalities. Evidence suggests that the crew inadvertently told the computerized twin jet to descend at 3,300 *feet (1,000 meters) per minute*, rather than on a far tamer glide slope of 3.3 *degrees*.

Most automation accidents occur when airplanes are landing or taking off. In the vertical channel, to descend, the autopilot pitches the nose down or reduces power. The opposite happens when climbing. "Sometimes it's appropriate to control with pitch," says Hansman. "Sometimes it's appropriate to control with power."

Instrumentation tells pilots what vertical "mode" the autopilot is in, but it doesn't communicate "the implication of that mode" terribly well, says Hansman. Will the airplane descend too fast and touch down short of the runway—or will the nose pitch up too steeply, bringing on a stall?

Today's advanced cockpit displays show the pilot what will happen when he chooses a

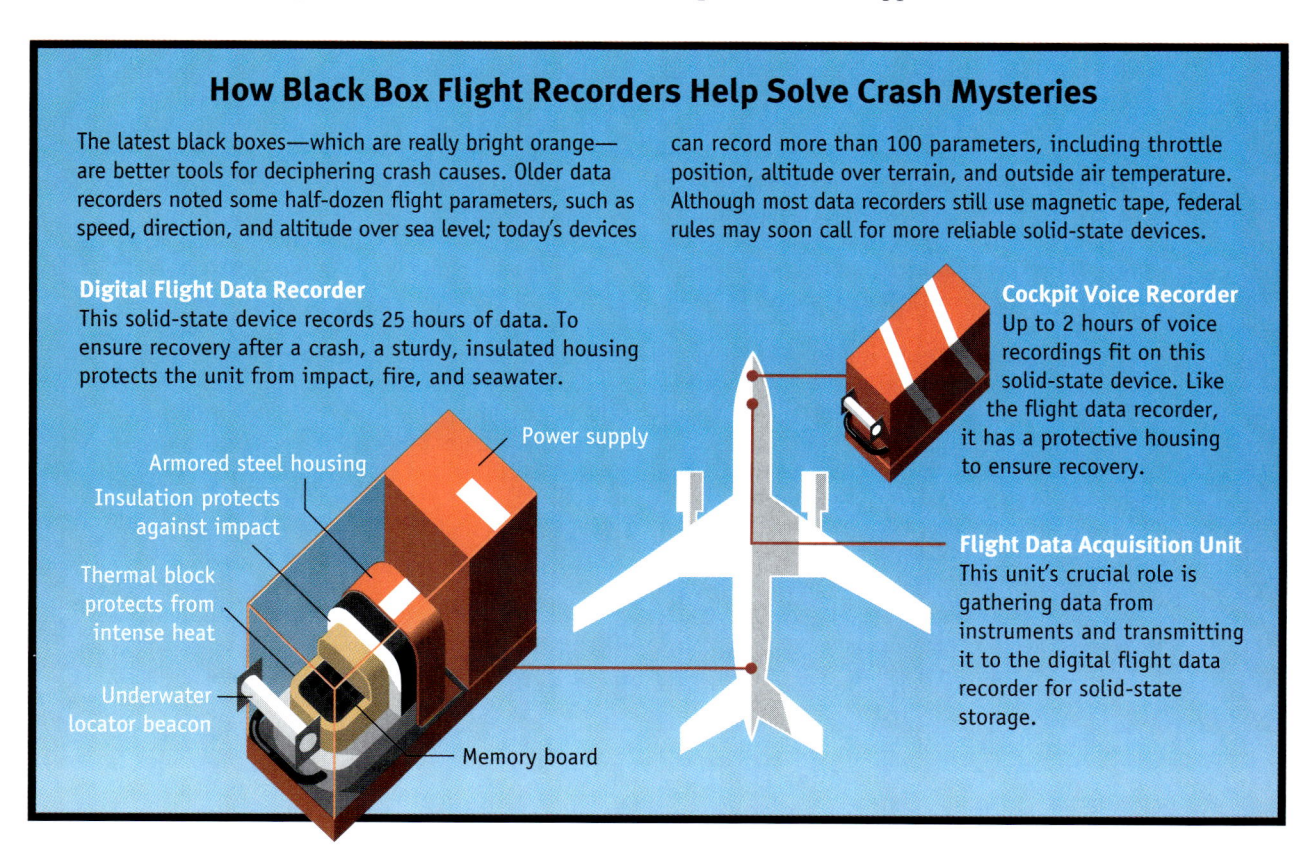

How Black Box Flight Recorders Help Solve Crash Mysteries

The latest black boxes—which are really bright orange—are better tools for deciphering crash causes. Older data recorders noted some half-dozen flight parameters, such as speed, direction, and altitude over sea level; today's devices can record more than 100 parameters, including throttle position, altitude over terrain, and outside air temperature. Although most data recorders still use magnetic tape, federal rules may soon call for more reliable solid-state devices.

Digital Flight Data Recorder
This solid-state device records 25 hours of data. To ensure recovery after a crash, a sturdy, insulated housing protects the unit from impact, fire, and seawater.

- Armored steel housing
- Insulation protects against impact
- Thermal block protects from intense heat
- Underwater locator beacon
- Power supply
- Memory board

Cockpit Voice Recorder
Up to 2 hours of voice recordings fit on this solid-state device. Like the flight data recorder, it has a protective housing to ensure recovery.

Flight Data Acquisition Unit
This unit's crucial role is gathering data from instruments and transmitting it to the digital flight data recorder for solid-state storage.

By undergoing training on advanced flight simulators, airline pilots gain the invaluable experience of reacting to a variety of potential emergency situations.

specific *horizontal* autopilot setting. These displays help crews steer around storms and plot the fastest routes. But commercial airliners are not fitted with *vertical* displays. Hansman and research assistant Sanjay Vakil are working on the software modifications to change that. But it won't happen anytime soon. The software is not mature. And, says Hansman, an airline's decision to outfit its fleet will be made on a cost-versus-benefit basis. To put this in context, U.S. airlines have lost $13 billion since 1990.

Refining the human/machine interface is important. But the root cause of most airline accidents continues to be human fallibility. From 1959 through 1994, 73.3 percent of commercial-jet accidents worldwide were caused by the flight crew, says Boeing, which maintains perhaps the most comprehensive registry worldwide of jet crashes. Consider recent history:

- *July 2, 1994.* A 70-knot (about 80-mile-per-hour) wind shear slams USAir flight 1016, a DC-9-30, into the ground at Charlotte, North Carolina. Thirty-seven people die. The NTSB cites as the prime probable cause "the flight crew's decision to continue an approach into severe convective activity that was conducive to a microburst." Microbursts, downdrafts of air up to 2.5 miles (4 kilometers) wide, may occur during thundershowers.

- *December 13, 1994.* American Eagle flight 3379 crashes short of the runway at Raleigh/Durham, North Carolina. Fifteen people perish. According to the safety board, pilot error was the cause. While the crew busied itself with a suspected engine problem, the Jetstream 32 prop jet went out of control. Ironically, both power plants were working normally. Captain Michael Patrick Hillis, who died in the crash, had been forced to resign from his previous piloting job because of poor flying skills.

SOMEONE TO BLAME

Much has been done in recent years to cut "human factor" accidents. Although no federal regulations require airlines to provide training for emergencies, some airlines are teaching pilots these skills.

Advanced flight simulators—costing tens of millions of dollars—duplicate emergency situations on the ground that can cost lives in the air. The effort has paid off. Over the past 10 years, the percentage of flight-crew-caused accidents is down somewhat. But at the heart of most airline crashes, you will still find someone to blame.

The General Accounting Office (GAO) has recently launched a study of the FAA's human-factors program. "As the technology has improved, and as planes have improved, one of the things that people have been looking at is the *source* of accidents," says Marnie Shaul, an assistant director at the government watchdog agency. The human factor, she continues, "has become increasingly more important."

Ultimately, it is human beings—our reactions and our interactions with the intricate machines we design—who determine the safety of commercial aviation. "The thing I worry about," says air-safety expert Oster, "is that [people] have a history of chasing headlines." Stories about crash A focus on ice; crash B, wind shear; crash C, the flame retardancy of the overhead bag-storage bins. The topics are journalistically sexy and rife with controversy. They are also, believe many men and women who make safety their profession, disastrously diversionary. "I think that as the causes of accidents get more complicated," says Oster, "it's going to get a lot more costly for us to chase the wrong stuff."

THE INTERNET:

by Christopher King

In March of 1876, the Scottish-American inventor Alexander Graham Bell was preparing a test of his new device for transmitting speech in the form of electrical impulses. Accidentally spilling battery acid on his clothing, Bell suddenly cried out, "Mr. Watson, come here. I want you!" In the next room, his assistant, Watson, clearly heard each word over his receiver and rushed in to help. Thus, the modern telephone was born, and with it a revolution that changed society.

ATTEMPTING THE UNPRECEDENTED
Less than 100 years later, in the fall of 1969, another revolution began. A group of computer scientists at the University of California, Los Angeles (UCLA) was attempting the unprecedented: logging on to a computer at Stanford University, some 400 miles (640 kilometers) to the north. The scientists began to type the words "log in" as they spoke by telephone to the other team, confirming that the letters were appearing, one by one, on the Stanford computer. Unfortunately, no sooner had the letter *g* appeared on the remote computer than the system crashed.

In terms of drama, the event may have paled beside Bell and his urgent plea for help. Ultimately, however, this first test of linked computers would also transform society. The experiment marked the birth of the Internet, the vast worldwide system of interconnected computer networks that has fundamentally altered the way millions of people communicate, spend their leisure hours, and acquire and share information.

The idea of linking computer resources on a worldwide scale has been around for decades. Years before the invention of the first modern computer, for example, the British science-fiction novelist H.G. Wells wrote in the late 1930s of an idea for a global information facility that he called the "World Brain." And during the 1940s, Vannevar Bush, a former vice president of the Massachusetts Institute of Technology (MIT) in Cambridge, and a leading scientific adviser to President Franklin D. Roosevelt, made a farsighted proposal for a desktop data center that he called "Memex." However, the real force behind the idea of joining computers in a far-flung network came from the U.S. military during the 1960s. Seeking to protect its computers from sabotage, or even from the devastation of a nuclear attack, the Department of Defense initiated the creation of a decentralized network of computers. The system was designed so that even if some of the computers in the network were shut down, the system could still function. The military funded the development of such a network through its Advanced Research Projects Agency (ARPA). Work commenced on both the systems and the software required for allowing computers to share information.

254 TECHNOLOGY

The expansive system of interconnected computer networks known as the Internet has transformed the ways in which millions of people around the world communicate, spend their leisure hours, and acquire and share information.

computer was linked to the Internet every 30 seconds. In 1983, there were approximately 560 host computers on the Internet; by 1994, the number had skyrocketed to nearly 4 million worldwide. Today the number of computers populating "cyberspace" is believed to exceed 10 million. Although estimates of users have varied widely and have been a source of disagreement, a 1996 survey by Nielsen Media Research placed the number of Americans with access to the Internet at more than 19 million—just a portion of the 30 million users in more than 60 countries throughout the world. According to some predictions, the Internet could have upwards of 400 million users by the year 2000.

In all the discussion of the system's incredible growth, and the breathless predictions about its projected effects on society, a simple question often gets overlooked: What exactly is the Internet?

A Network of Networks

Asked to define the Internet, one expert joked, "It's this vast amorphous blob—that's the technical explanation." In *The Internet Book*, Douglas E. Comer offers a slightly more helpful description: "Though it appears to a user to be a single, large network, the Internet consists of thousands of computer networks interconnected by dedicated computers called routers." Computers can be connected to the Internet in a number of ways: by fiber-optic cable, via satellite hookup, or through telephone lines using a modem (an abbreviation for *mo*dulator/*dem*odulator, a device that translates the analog data stream of electrical impulses carried by telephone into the digital language of 1s and 0s spoken

By the end of 1969, as the teams at UCLA and elsewhere worked the bugs out of the system, the fledgling network—dubbed ARPANET—consisted of four host computers, or "nodes," sharing data via telephone wires. By the middle of the next year, there were 10 such nodes. Soon, as more and more computers at universities and government installations were added to the system, the growth became explosive. Scientists and academics rapidly learned the value of being able to communicate with their counterparts around the nation and the world. The National Science Foundation (NSF) became a major force in the development of the Internet. During the 1980s, according to one estimate, the Internet grew at a rate of 10 percent per month, doubling in size every 10 months.

In the early 1990s, the growth of the so-called "information superhighway" accelerated. Throughout 1994, on average, a new

an on-line revolution

255

In 1969, computer scientists created ARPANET, the world's first decentralized computer network. Although the U.S. military funded the system's development, scientists and academics were the earliest users of the network.

by the computer). Special software known under the general names of *Transmission Control Protocol* and *Internet Protocol*—usually referred to as TCP/IP—assure uniform communication between the computers in the system.

The Internet's designers also pioneered a technology known as *packet switching*, a means of sending data rapidly from computer to computer. Instead of sending data in one complete unit along a specified route from one computer to another, packet switching breaks down data into small units. These units, sometimes referred to as datagrams, are each addressed to the destination (each computer linked to the Internet is assigned its own specific address). Routers pass the datagrams along, automatically bypassing electronic "traffic jams" and finding the quickest route to the destination. At the end of the trip, the datagrams are reassembled into what may be a text file, a graphic element, or a bit of audio. The whole process can take place in a matter of seconds.

The Internet is not a corporation: there is no "Internet, Inc." No one "owns" the system. Businesses, universities, and other institutions support the Internet by paying for their connections to regional networks. And, at least at this writing, no one controls or governs the Internet—although a volunteer group known as the Internet Architecture Board meets to discuss standards and the allocation of resources. Since its creation, the Internet has been free to grow and expand on its own.

One of the most popular applications of the Internet—from the very earliest days up to today—is electronic mail, or E-mail. It is now commonplace to see E-mail addresses on letterheads and business cards. A typical E-mail address consists of an abbreviation of the user's name—for example, "jsmith"—followed by the "at" sign (@) and the so-called "domain" name of the user's computer or local network. The domain name can also include an extension indicating whether the domain is, for example, a commercial business ("com"), an educational institution ("edu"), or an organization ("org"). Thus, John Smith of Widgetco, Incorporated, might have the following E-mail address: "jsmith@widgetco.com." Once a message is addressed and sent on its way, it will remain in the recipient's electronic "mailbox" until it is read.

Another popular feature of the Internet is the ability to link up to far-off computers in order to help users locate research materials and other information. An application known as Telnet, for example, allows a computer to remotely log on to other computers—such as the on-line catalog of a university library. Internet users can also employ an application called File Transfer Protocol (FTP) to seek out data files on other computers—such as text files or useful computer programs—and transfer, or "download," them to their own machines. Special systems known as Gophers (named for the Golden

256 TECHNOLOGY

Gophers of the University of Minnesota, where the system was developed) provide menus of available files and help users find the exact information they seek. Other useful programs, such as Archie and Veronica, allow detailed searching of FTP files and directories that are available at universities, government agencies, and other computers linked to the Internet.

The Internet also provides forums for people to share information more informally. Thousands of electronic bulletin boards and so-called "newsgroups" allow discussion of every conceivable topic—hobbies, sports, entertainment, the arts, current events, and more. Participants in newsgroups can post and read messages, and can have messages from other group members automatically posted to their E-mail addresses.

ON-LINE SERVICES

As the Internet was growing in the early 1980s, the personal-computer revolution was in full swing, with an increasing number of Americans buying their first home computers. Commercial on-line services such as CompuServe took aim at the home market to tap this new population of computer users. In recent years, the number of such commercial services has grown, and more and more users have gone on-line—without necessarily going onto the Internet. CompuServe, America Online, Prodigy, and other services offer access to many of the same kinds of utilities and entertainment as those available on the Internet. Subscribers to such commercial services can send and receive E-mail, download games and other files, participate in bulletin-board discussions, and even have

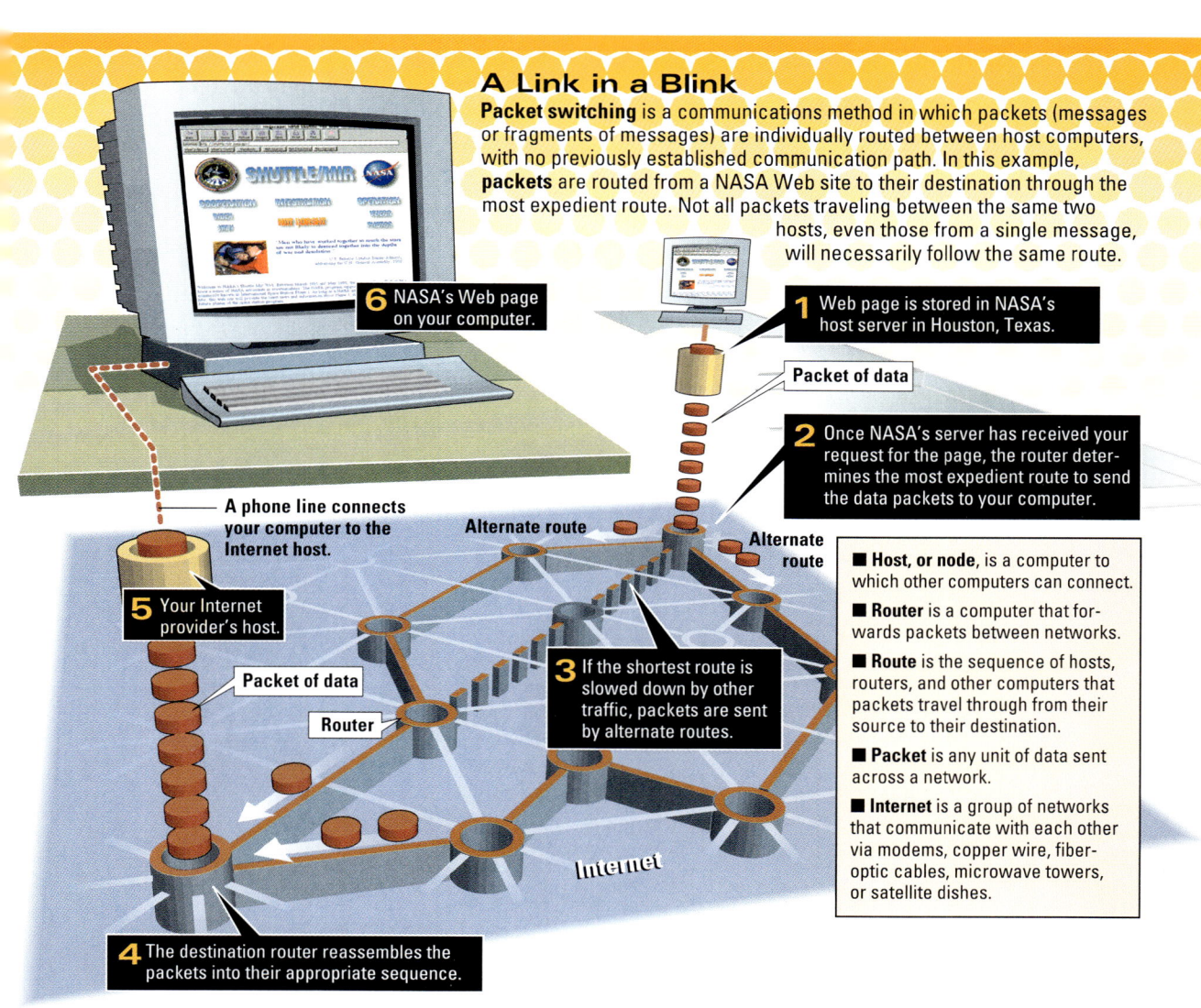

A Link in a Blink

Packet switching is a communications method in which packets (messages or fragments of messages) are individually routed between host computers, with no previously established communication path. In this example, **packets** are routed from a NASA Web site to their destination through the most expedient route. Not all packets traveling between the same two hosts, even those from a single message, will necessarily follow the same route.

6 NASA's Web page on your computer.

1 Web page is stored in NASA's host server in Houston, Texas.

Packet of data

2 Once NASA's server has received your request for the page, the router determines the most expedient route to send the data packets to your computer.

A phone line connects your computer to the Internet host.

Alternate route

Alternate route

5 Your Internet provider's host.

Packet of data

Router

3 If the shortest route is slowed down by other traffic, packets are sent by alternate routes.

Internet

4 The destination router reassembles the packets into their appropriate sequence.

- **Host, or node**, is a computer to which other computers can connect.
- **Router** is a computer that forwards packets between networks.
- **Route** is the sequence of hosts, routers, and other computers that packets travel through from their source to their destination.
- **Packet** is any unit of data sent across a network.
- **Internet** is a group of networks that communicate with each other via modems, copper wire, fiber-optic cables, microwave towers, or satellite dishes.

real-time "conversations" with other subscribers. America Online, currently the largest on-line service, also provides access to electronic versions of magazines, to stock-market information, and to coverage of the day's breaking news stories—among many other choices. This mix of entertainment, communication, and information has been very popular: in late 1995, America Online had nearly 4 million customers and was opening 14,000 new customer accounts per day.

In the early to mid-1990s, the question seemed to be up in the air as to whether services such as CompuServe and America Online should be counted as part of the Internet. As recently as 1995, a survey of on-line use by American adults drew a distinction between those individuals with full access to the Internet, and those who subscribed only to commercial services. In fact, until the middle of the decade, the major commercial on-line services did not offer full Internet access. That changed soon enough, however, due to the Internet's growing popularity. Subscribers to commercial services now have the option of accessing Gopher sites and using FTP protocols to search out files and mine the system's riches.

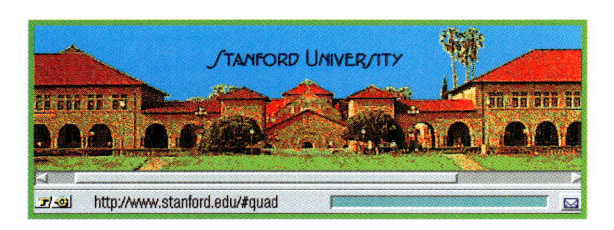

But even as the major commercial services were finding their way onto the Internet, a new development in on-line technology was exploding in popularity, achieving a dizzying growth (even by Internet standards), and changing all the rules of the on-line game. To many, this on-line domain has become the very definition of the Internet. It is called the World Wide Web.

No Missing Links

In the late 1980s, Tim Berners-Lee and his colleagues at the European Laboratory for Nuclear Research (CERN), located near Geneva, Switzerland, were looking for a better way to share information by computer. Since many of the researchers who participated in experiments at CERN were based at institutions throughout Europe, Berners-Lee and his colleagues wanted to develop a common store of information on CERN projects—one that would make it easy for colleagues to search, access information, and pool their knowledge. In developing a new network that would satisfy these needs, the team used a form of linking known as hypertext, in which documents contain automatic links to other documents. These links appear in the form of words that are underlined or highlighted in colored letters. Using a computer mouse to click on such text will automatically call up more information related to the highlighted term. In turn, the new information may contain more highlighted links that will lead to still other documents. A person can follow links at will, pursuing a trail of personal interests that can occasionally (and entertainingly) lead far from the originally intended path.

Hypertext documents are created with a language known as Hypertext Markup Language (HTML). HTML makes it possible to embed coded information within the text that is invisible on the hypertext "page." The codes contain the links that allow information contained on other pages to be transferred almost instantly from another computer, referred to as a "server." The location of each page is identified by a long electronic address known as a Uniform Resource Locator (URL). A URL typically consists of three parts, with the first part specifying the kind of network protocol used to transfer the information. The CERN designers made use of a protocol known as *h*y*per*text *t*ransfer *p*rotocol, or http; as a result, URLs typically begin with the letters "http." The second part of the address indicates the name of the computer on which the information is stored. The third part indicates the specific files holding the information. For example, the URL "http://www.widgetco.com/widgets" would transfer information about products available from Widgetco, Incorporated. Someone browsing a hypertext document,

The World Wide Web is the most popular part of the Internet. Using programs called browsers, one can "surf" among millions of Web sites—which provide graphics and information about sports, science, culture, and other subjects.

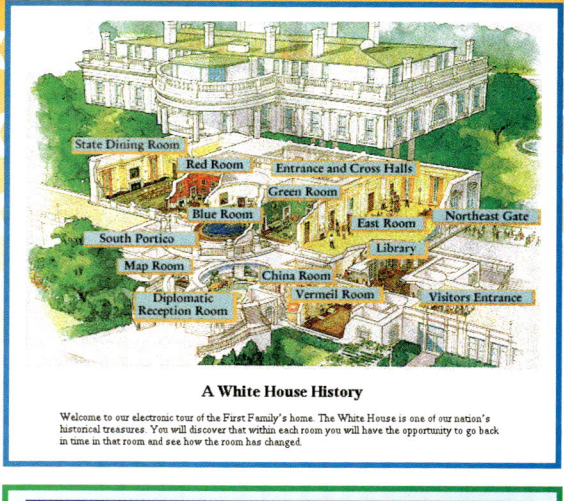

A White House History

Welcome to our electronic tour of the First Family's home. The White House is one of our nation's historical treasures. You will discover that within each room you will have the opportunity to go back in time in that room and see how the room has changed.

A Photo Gallery of the Universe

Hubble Space Telescope evokes a new sense of awe and wonder about the infinite richness of our universe in dramatic, unprecedented pictures of celestial objects. Like a traveler sharing their best snapshots, we present a selection of Hubble's most spectacular images.

however, would be unaware of the URL. The person would only have to click on the term "Widgetco Incorporated" to automatically link to the Widgetco computer. A URL can also be embedded in a graphic image.

The World Wide Web—widely known as "the Web"—was born in 1991. Originally used by a comparatively small number of scientists at CERN and elsewhere, the Web (or "W3" to some) quickly gained in popularity and soon became the most popular part of the Internet. One key to its rapid expansion was the development of special programs known as "browsers." The first such program, known as Mosaic, was developed in 1993 at the National Center for Supercomputing Applications at the University of Illinois at Urbana-Champaign. In contrast to the text-only browsers used by early Web surfers, Mosaic employed eye-catching graphics and brought out all the point-and-click ease of exploring linked documents on the World Wide Web. Some of Mosaic's key designers eventually formed their own company, releasing another browser known as Netscape Navigator (usually referred to simply as "Netscape"). By the mid-1990s, Netscape was the most popular Web browser, although many companies, including Microsoft, were offering browsers of their own.

SURFING AROUND THE WORLD

With a browser, a person exploring, or "surfing," the World Wide Web with a computer (in technical terms, a "client" computer) has access to Web servers all over the world. By clicking on links, surfers can jump to other Web pages instantly. (Well, not quite instantly: with each click, a new host computer must be contacted, and new files loaded onto the client computer. The delays and occasional broken connections that are an unavoidable part of Web surfing have prompted some users to complain about the

TECHNOLOGY 259

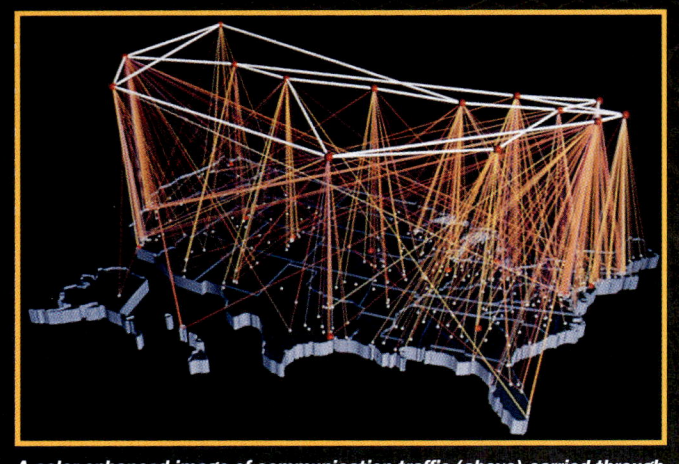

A color-enhanced image of communication traffic (above) carried throughout the United States by NSFnet, the U.S. backbone of the Internet, illustrates the scope and complexity of the so-called information superhighway.

"World Wide Wait.") Using hypertext links, Web surfers can call up not only text, but computer programs, photographs and other graphic elements, video, and sound.

At the end of 1991, there were approximately 500 "Web sites"—collections of HTML documents, or "pages," corresponding to a specific URL. By 1992, this number had grown to 50,000. In the next year, the number of Web sites increased by 300,000 percent. The number has since grown into the millions, as organizations of every conceivable kind—businesses large and small, government agencies, universities, religious institutions, newspapers, magazines, congressional offices, and even the White House—have set up Web sites. Entering a Web site, on-line visitors are greeted by a "home page" that features introductory information about the organization—usually with colorful graphics—and provides links by which visitors can reach selected pages on the site.

By the mid-1990s, a number of commercial applications had emerged. For example, many print and television advertisements feature a URL address for a Web site connected with the product. Record companies now offer CD-quality samples of new music releases. Automobile makers provide flashy images and specifications on new car models. Movie studios distribute promotional material, such as illustrated biographies of the stars, for new films.

Many entrepreneurs have flourished in this new business landscape. For example, the rush of corporations scrambling for a presence on the Web has established a career niche for freelance programmers and designers who create Web pages for a fee. Others have found success by providing guidance to Web users. For example, in 1994, two Stanford graduate students had the idea of creating a central Web directory with links to their favorite sites. They called their site Yahoo! (for "yet another hierarchic officious oracle"). It soon became one of the most popular search services on the Web. (Lycos, WebCrawler, and InfoSeek are others. Such services use sophisticated programs called "spiders" that automatically travel the Web in search of information.) By 1996, the Yahoo! site was attracting upwards of 750,000 visitors per day. Its two founders decided to turn it into a full-fledged business, making money by selling space to advertisers on their pages.

Yet much of the Web is not about business at all. Many sites are devoted to film, literature, sports—every imaginable pursuit. And countless individuals have used HTML and a good deal of imagination to create their own personal home pages, providing photographs of themselves, publishing artwork or poetry and other writings, and listing links to other Web sites that reflect their interests. In all, the Web is an astounding representation of human variety and eccentricity.

WIDE OPEN, AND HEADED . . . WHERE?
For now, all it takes to go on-line is a computer, a modem, and an Internet service provider. In addition to the large commercial services—all of which now offer access to the Web—there are 15,000 or so other firms that provide Internet access. In terms of hardware, experts recommend a computer processor with a clock speed of at least 75 megahertz, and at least 4 megabytes of random-access memory (RAM). Modems rated at 28,800 bits per second (28.8 Kbps)—indi-

cating how much information they can transfer—are also becoming the standard.

The electronic landscape, of course, is changing very rapidly. Soon conventional computers may be replaced by high-tech televisions that are equipped for Internet access. Someday built-in Internet connections in households might be as commonplace as telephone service is today. Many large and powerful companies are angling for a piece of the "on-ramp" to the so-called information superhighway. For example, in 1996, telecommunications giant AT&T began offering direct access to the Internet via its long-distance network. Cable-television companies also want to leverage their technology to get into the act: cable modems, which offer Internet access over television lines, began appearing in 1996.

Meanwhile, development of Web technology continues feverishly. Other programming languages, such as Java from Sun Microsystems, promise to enhance the multimedia capabilities of the Web. New software even offers the prospect of visiting three-dimensional, "virtual-reality" sites on the Web.

Decency Act on constitutional grounds attacked the law as too vague, charging that even legitimate discussions of sexuality could be outlawed. The dispute was headed for the Supreme Court of the United States. For now, the regulation of on-line free speech is still up in the air.

Other questions on the Internet's future are unresolved. Despite the stampede of advertisers to the Web, the Internet remains

In trendy "cybercafés," customers can surf the Internet while sipping cappuccino. Such establishments help expose networking neophytes to the technology and provide further evidence that an on-line revolution is afoot.

Unresolved Questions

Yet important questions remain. The Web, like the Internet as a whole, developed as a largely unregulated frontier. For better or worse, a considerable amount of material on the Internet is devoted to sexual topics. During the early 1990s, the lack of restriction on the transmission of this material over the Internet became a heated issue and a point of focus for politicians—resulting in the U.S. Communications and Decency Act that President Bill Clinton signed into law early in 1996. The law is intended to prevent the transmission of sexual material that might be seen by minors. However, opponents who challenged the Communications and

an unproven marketplace. And the incredible growth of the system has prompted some experts to wonder how much information the Internet can actually handle.

More than a quarter-century ago, when the Internet's designers were thrilled at the appearance of single letters of text on the first computer network, could they have envisioned a time when a volume of information comparable to the Library of Congress would be whizzing through cyberspace every few seconds? No one, it seems, knows exactly where the on-line revolution is heading. For now, however, most people seem happy to hold on to their computers and enjoy the ride.

TECHNOLOGY 261

Steel Magnolias
by Jay Stuller

It first hung only in the mind of Tung-Yen Lin, this roadway suspended high above a surging sea. The engineer imagined three soaring towers of high-strength concrete and steel, resting on massive concrete piers sitting in water 1,475 feet (450 meters) deep. Each tower sported a pair of huge cantilevered struts, slanting upward and outward like a bird's wings in mid-flap. Hanging on a webbing of wires from struts and huge cables between the towers were two steel-and-concrete spans. In his mind, each segment stretched over the waters for more than 16,000 feet (5,000 meters), or an incredible 3 miles (4.8 kilometers).

And then, when T.Y. Lin put this vision to paper and backed it with rigorous data, the engineering world gasped in wonder. A Chinese-born professor emeritus at the University of California at Berkeley, Lin had devised a means—a logical, practical, entirely feasible means—of bridging the Strait of Gibraltar.

Taking the shortest route from Spain to Morocco, the 8.6-mile (13.8-kilometer) span would be the first to link Europe and Africa. Funded by the United Nations and European governments, Lin's studies produced a design for a $10 billion structure. Unfortunately, there is not yet enough commerce between Spain and Morocco to justify the cost, so the Gibraltar suspension bridge remains a vision.

"We don't even have the technology to build such a span, at least today," concedes Lin. "But if we were to proceed with research and design work, I know that we would develop it." Although economic growth in North Africa will likely make the Gibraltar crossing an eventual necessity, Lin doubts the magnificent bridge will be built in his lifetime. At the age of 82, the fabled engineer is a pragmatic man.

A Sense of Wonder

Mortality doesn't dampen Lin's enthusiasm for an idea that received the 1993 Outstanding Paper Award from the International Association for Bridge and Structural Engineering. In his profession, folks often find comfort in the fact that large-scale projects—from the Great Wall of China to the "Chunnel" beneath the English Channel—are monuments that serve humanity long after the builders depart the corporeal world. But while sewers, freeways, and similar public works are useful, bridges are more than utilitarian objects.

"A bridge reveals itself instantly on first viewing," says Charles Seim, a senior principal with the civil- and structural-engineering firm of T.Y. Lin International

At their best, bridges reveal creative powers that transcend mere utility. Le Pont de Normandie (left and above) in France, for example, employs webbed wire cables to provide structural strength—and a monumental sense of grace.

and a longtime Lin colleague. "You see its form, materials, and function of carrying cars, trains, or pedestrians. But a truly great bridge also has proportions and textures that take it beyond its practical purpose. Every time I drive across or look at the Golden Gate Bridge, I still feel a sense of wonder at its aesthetic impact."

Because they become part of a landscape, the best of these steel magnolias bind their communities, literally and figuratively. "They also reveal," adds Seim, "important aspects of the society that built them."

If so, there is evidence to suggest that not every frontier of human progress resides in the etherworld of the Internet. While Gibraltar is

The striking color and symmetry of San Francisco's Golden Gate Bridge have long impressed engineers and tourists alike. A seismic retrofit is under way to bolster the bridge's chances of surviving a major temblor.

the most far-reaching design ever proposed, the art and science of bridge building is entering a new, daring, and very real era. In fact, today's leaps are quantum and even startling.

Consider that San Francisco's Golden Gate Bridge is still the world's third-longest suspension structure. Completed in 1937, its 4,200-foot (1,280-meter) main span is only a bit shorter than the 31-year-old Verrazano Narrows Bridge in New York City. The 4,626-foot (1,410-meter) Humber Bridge in England, finished in 1981, is now the longest. But in 1997, the 5,328-foot (1,625-meter) Store Baelt (Great Belt East) Bridge in Denmark is scheduled for completion. And only a year later, it will be relegated to second place by Japan's Akashi Kaikyo Bridge, with a main span that stretches 6,529 feet (1,991 meters).

Linking the Japanese islands of Honshu and Shikoku, this leviathan should stand as a triumph of suspension-bridge-engineering technology. Moreover, once thought to be limited to lengths of about 1,300 feet (400 meters), cable-stayed structures are also stretching. Norway's Skarnsundet Bridge, completed in 1992, has a 1,739-foot (530-meter) deck tied directly to tall, A-frame pylons with fans of cables. A cable-stayed span over the mouth of the Seine in Normandy has a deck that runs 2,808 feet (856 meters).

These are remarkable advances. However, should the forces of nature frown upon any of these artifices of humankind, London Bridge, my fair lady, might not be the only span in danger of falling down.

Inherent Uncertainties

Engineers push the boundaries of structural design despite haunting reminders that a bridge can indeed go too far. For them, the infamous 1940 collapse of the Tacoma Narrows suspension bridge in Washington State is not ancient history. Engineers, physicists, and mathematicians still discuss the way this slender, 2,802-foot (855-meter) bridge fluttered and twisted in high winds, then ultimately shook apart.

The fact is, any long span can behave in unpredictable ways. Joe McKenna, a University of Connecticut mathematics professor, has applied chaos-theory-type equations to bridge designs. He claims that a single large disturbance or sudden force that causes a large vibration can trigger wild oscillations in a span. Normally, a bridge design contains a host of compensating structures that dampen vibrations. But in a number of popular publications and scholarly journals, McKenna has claimed that civil engineers are not using sufficiently advanced models to study their new designs. He suggests that with some long suspension and cable-stayed structures, another Tacoma is highly possible.

This clearly rattles the bridge builders' cages; the mere mention of McKenna causes most to grind their teeth, Chuck Seim included. "A lot of my colleagues," he says diplomatically, "don't think much of his work."

Yet even with the most advanced computer models and tests, innovative bridge design carries inherent uncertainties. Seim is the first to admit that when a novel structure goes straight from the drawing board to the real deal, there is no way to know how it will fare until the completed bridge meets its ultimate inquisition—be it from freak winds, a ship colliding with a support pylon, or violent seismic activity.

Indeed, the 1989 Loma Prieta earthquake, a 7.1-magnitude temblor with an epicenter about 60 miles (100 kilometers) south of San Francisco, apparently gave the Golden Gate Bridge a frighteningly close shave. "Had the epicenter of a similarly sized earthquake been on the San Andreas fault just miles from the bridge, it would have suffered severe damage," says Seim. "Had it been the Big One, a reprise of San Francisco's 1906 quake . . . well, the bridge might have failed."

Seim is part of a scramble to get on with a $147 million earthquake-proofing retrofit of the Golden Gate—its towers, supporting piers, and, in particular, the approaches to the bridge. "Although we know we're in a race against time," Seim observes, "we're not going to shortchange the work. Just as important, we won't do anything that changes the character of the bridge."

High winds and poor design contributed to the 1940 collapse of the Tacoma Narrows bridge (above). Today, computerized models help engineers study how their designs will tolerate wind, earthquakes, and other disturbances.

WORKS OF ART

That attitude is well and good, for the world's truly great bridges are works of art meant to remain unchanged for the ages. They are the tangible product of rare individuals such as the legendary John Roebling, who (with his son Washington) built the Brooklyn Bridge; Joseph Strauss, the man behind the Golden Gate; and, to be sure, the likes of T.Y. Lin. Bridges designed by these men reflect the highly developed and precise left-brain mentality of an engineer; each individual also seems to feature a neural highway to an equally well-developed right brain, from which springs creative thought.

That combination is a hallmark that's been seen over a

Engineers have devised structures to span every imaginable obstacle. For example, plans have been drawn for a bridge linking Spain to Morocco over the Strait of Gibraltar (above) and for a "hanging arc" span over California's American River (right) that would tether into the surrounding hillsides. Neither plan, however, has left the drawing board.

couple of thousand years of bridge building, says Seim. "We don't know how far back simple rope-suspension bridges really go," he explains. But the seafaring Phoenicians were praised for their bridge-building skills. "Their cities were typically offshore, on fortified islands," he explains, "and their bridge designs and technology were extensions of ship rigging."

The son of an immigrant German laborer, Seim is a patron of the opera, ballet, and other fine arts. His wrists and fingers bear a flashing array of turquoise and silver jewelry.

The art of bridge building has been appreciated for millennia. The Romans, for instance, constructed many beautiful masonry arches (above) that still stand today.

He will talk at length about technical standards and engineering protocol, and then veer into poetic observations of how the Golden Gate Bridge "emerges from the hills on both sides of the strait and has near-symmetrical proportions and ratios which are most natural for human understanding."

That means it has sublime "architectonics," or a strong overall aesthetic impact. Great bridges, says Seim, have a symmetry akin to the compositions of Bach, Brahms, and Beethoven. But such world-class structures—with a new form, or new materials, new building techniques, and in a graceful setting—emerge infrequently.

Humble Beginnings

The road- and aqueduct-building Romans, for example, created long, arched structures, often in elegant tiers, most notably seen in the Pont du Gard Aqueduct, built over the Gard River in 19 B.C. The arched Pont d'Avignon, notes Seim, built in the late 12th century in France, was world-class. But until 1779, bridges were typically made of stone and timber and were constructed in place. Then Abraham Darby III, who ran a cast-iron foundry in Wales, designed and built a bridge over the Severn River, made of iron parts that were fabricated at the foundry and hauled to the site.

This forever changed the way bridges were made. In the 19th century, the use of steel enabled engineers to span greater distances with less material. Instead of using huge trusses and arches, lighter cables draped in parabolas between towers did the weight-bearing work. But the first lightweight suspension bridges were also rather flexible spans. Mark Twain complained about the "discomfiting" sway of John Roebling's railroad bridge over Niagara Falls. Roebling's Brooklyn Bridge, completed in 1883 and stiffened with many cables, is a masterpiece that is as famous now as it was then.

Cable and steel enabled engineers to create harplike visions in the sky. The architectonics of a well-designed and well-placed suspension bridge make it a nearly natural part of the environment. Folks may complain about the aesthetics of buildings, airports, and highway overpasses; you rarely hear objections to the looks of a suspension bridge.

Daring and Disastrous Spans

Beauty aside, suspension bridges do drive some folks queasy with vertigo. In the 19th century and the early part of the 20th, an alarming number of bridges of all kinds shook themselves apart in high winds or under the load of railway trains. Engineers began to make advances slowly.

But they grew daring during the Depression years. In 1931, at 3,501 feet (1,068 meters), the George Washington Bridge, which crosses the Hudson River and connects New York and New Jersey, nearly doubled the length of the previous world-record span. Joseph Strauss' Golden Gate Bridge was finished six years later, and for 27 years held the length record. But in 1940, the Tacoma Narrows Bridge put a chill on the profession.

Open for only four months, the structure was already known as "Galloping Gertie" for the way its span would rise and fall in waves, even in light winds. But on November 7 of that year, the bridge was hit broadside by wind gusts in excess of 40 miles (64 kilometers) per hour. The bridge normally oscillated up and down; on this day, its deck began to twist, corkscrew, and flutter. The classic film of the collapse—as cables snapped and a 600-foot (183-meter) section of the bridge plunged 190 feet (58 meters) into Puget Sound—has ever since been a staple of physics classes.

The collapse is often cited in textbooks as a case of simple resonance gone wild. Students are told that the winds amplified the bridge's natural period of vibration, to the point where the motion shook the structure apart. Engineers now claim that assessment is wrong. In fact, as parts of the deck turned sideways and were exposed to the wind, it set up an energy "feedback loop" much more complex than an amplified vibration period. Bridge designs allow a structure to move slightly when a force hits one point, but feature countertensions that usually dampen the oscillations.

At Tacoma, a twisting in the deck led to a state of "self-excitation." This "unfortunate wiggle," as one engineer has called it, set loose the feedback loop. As the wind pushed directly against the slanted deck, its own swinging forces made the back-and-forth

A wide variety of bridges—including examples of arch, beam, and suspension designs—traverse the three rivers of Pittsburgh, Pennsylvania (above).

twisting even more violent. The bridge self-excited itself to death.

Bridge builders learn from such mistakes. The side of the Narrows' deck had vertical sheets of supporting steel, which presented a flat surface against which the wind could push. The deck should have been stiffened and also made aerodynamic, so that wind would spill above and below it without causing sideways motion. Until recent years, fears of another Tacoma Narrows have prevented bridge builders from attempting radical design leaps.

Engineers also get in ruts, following tried-and-true forms. "These traditions stand up well," says Seim. "Following convention is also comfortable, takes less effort, and perhaps offers some legal protection in case of trouble." But he adds that structural engineers have learned more in the past half-decade than in the entire previous history of bridge design, thanks to computer programs, advances in prestressed concrete, better fabrication and welding techniques, and a clearer understanding of wind and seismic forces upon bridges.

Cable-stayed Bridges

Much of that innovation is going into cable-stayed bridges. In the mid-1980s, British Columbia's Peter Taylor came up with a 1,526-foot (465-meter), six-lane, cable-stayed span to cross the Fraser River in Vancouver. To build a suspended crossing at an economic cost, he designed a composite deck for what came to be called the Annacis Bridge. Until then, Taylor explains, suspended decks were typically made of either concrete or steel. Concrete is cheap but heavy; steel is light yet expensive. While engineers had been reluctant to mix the two on suspension spans, Taylor proved with the Annacis that the materials could provide a less-costly but still-strong deck.

Even so, such long, cable-stayed bridges carry risks. When Taylor developed his design, he admits that "there was a realistic chance of aerodynamic instability." Using the Tacoma Narrows as a cautionary benchmark, he gave the Annacis' deck a width-to-depth ratio that helped streamline its profile. But other cable-stayed bridges, such as the one in Normandy and a proposed span over the Strait of Messina between Sicily and the Italian mainland, seem highly risky.

Wind is but one danger; earthquakes are another. While working on the seismic retrofit of the Golden Gate Bridge, Chuck Seim has an unusually abiding faith in the strength of the venerable structure. In 1987, during the celebration of the 50th anniversary of the Golden Gate's opening, he was among the throng of pedestrians—estimated at more than 500,000—who packed the closed span. To some viewers, the party looked like it could suddenly turn into a monstrous catastrophe; under the weight of all that humanity, the Golden Gate's deck began to sag noticeably.

While Seim felt initial alarm at what engineers call "deflection," he began some mental calculations. "I figured out the average weight of the people in a given area, and came up with a load factor of 4,000 pounds per square foot [19,500 kilograms per square meter]," he recalls. "That was exactly its design load. At that point, I knew we weren't in danger."

But the Loma Prieta quake a couple of years later did bring down a small piece of the San Francisco–Oakland Bay Bridge. Motorists driving across the Golden Gate noticed that it was shaking quite hard, yet the bridge suffered no damage. Computer simulations of a 90-second-long quake as strong as the 1906 temblor suggest that while the Golden Gate would suffer failures in some areas, the upward force would still not be enough to cause deck cables to go slack and then fall.

Nevertheless, the retrofit is needed so that the Golden Gate Bridge remains usable—at least for emergency equipment—even after the Big One. Both anchorage housings are being reinforced. The arch over the historic Fort Point is being fitted with vibration-dampening additions and other reinforcements, while the north viaduct will get "isolators," devices that prevent the ground's shaking from vibrating through the structure. The concrete piers at the bottom of the towers will be confined with prestressed steel tendons.

None of this looks all that dramatic, nor will it change the Golden Gate's character. The $147 million retrofit investment is good insurance, given that one big oscillation can lead to feedback energy. It should protect the venerable bridge in the largest of quakes. But again, no one can be certain how a suspension span will behave under extreme duress.

The stark design of the Sunshine Skyway Bridge (above) in Tampa, Florida, reflects the trend of building structures with new forms, materials, and techniques.

Few human-made structures elicit such widespread praise as does a well-designed and well-placed bridge. The builders of these steel and concrete marvels continue to stretch the bounds of art and engineering in the quest to bring people and communities together.

An Engineering Dream

How far can suspension bridges go? Both Seim and Lin agree that the Akashi Kaikyo's 6,529 feet is not anywhere near the limit. (Incidentally, this bridge endured its first major inquisition when it survived the magnitude-6.9 Kobe quake in January 1995, which had an epicenter less than 0.6 mile [1 kilometer] from the span.) Build the Gibraltar Crossing and its 16,400-foot (5,000-meter) spans, and then we might talk limits. And while the Gibraltar design is certainly futuristic, it is based mostly on well-known technologies.

For instance, where the route between the continents is the shortest, the water's depth is 1,475 feet (450 meters)—initially considered too deep for conventional tower platforms. But Lin thought about "gravity-based" oil-production platforms in the North Sea, huge concrete cylinders that are built, floated, and sunk in depths nearly as great. And because the shortest route would require only three towers, they would present fewer navigational hazards than in other locations.

The Gibraltar design is also a hybrid suspension and cable-stayed span. While its deck would hang from parabolic draped cables, the towers' wing-like struts would also be connected to the deck with stays. The belt-and-suspenders approach, if you will, would allow the spans to stretch for such great distances.

"It's doable," says Seim, who helped Lin with the details. "I agree with T.Y. that we don't have the technology right now, but it would develop even as we built the bridge." And to Seim, who as a kid pored over accounts of the Golden Gate Bridge construction in the 1930s, the chance to work on such a structure would fulfill an engineering and artistic dream.

The economics of linking Spain and Morocco would drive the project. But to Lin, there are more-important considerations. As he often says, "Walls divide people; bridges unite them."

Leave it to a couple of old men with neural freeways between their right and left brains to imagine the unimaginable. But throughout the history of bridge building, it is these rare individuals who combine practical knowledge with creative leaps. The result is concrete and steel magnolias that, if only nature cooperates, will stand for the ages.

AFTER THE ACCIDENT

by Vincent Lytle

In the quest to make automobiles safer than ever before, the field of trauma-based research has brought renewed attention to the physics and to the medical consequences of real-life automobile accidents.

TRAUMA-BASED SAFETY STUDIES

Lindsey would eventually recover, but only after seven weeks in the hospital, many operations, and more than $300,000 in medical expenses. Even as she was being treated in the hospital, however, "crash reconstructionist" Mike Warner was at the scene of the accident, trying to determine why she sustained so many serious injuries. Surprisingly, he found only minor damage to the Jeep's front end. The Maxima, however, was a different matter.

"I was a little shocked to see the side impact as bad as it was," Warner recalls. He took out his tape measure and systematically went through the passenger compartment to get a precise record of the damage.

It was over in a matter of moments. Lindsey Gordon, an 8-year-old girl, was riding in the backseat of her mother's Nissan Maxima when a drunk driver drove her Jeep Grand Cherokee through a red light and into the car's side, crushing the rear door. Maryland police found Lindsey, pale and screaming, so tightly wedged in the wreckage that it took them half an hour to pry her loose. When a helicopter finally delivered her to Children's Hospital in Washington, D.C., doctors discovered that her stomach wall was ruptured, her spleen was lacerated, and her collarbone and several leg bones were broken.

Then he photographed the car and read the police report. Back at his office at Dynamic Science in Annapolis, Maryland, he plugged his data into a computer program that simulates accidents based on "crash profiles" of various makes of car. Ultimately Warner found that the girl's severe injuries resulted chiefly from the fact that the Gordons' car—like most passenger sedans—was not designed to withstand side impact from a vehicle like the Jeep, which, since it rides so relatively high off the road, missed the Maxima's steel frame and instead hit the much weaker door, crushing it inward.

Warner is engaged in the field of trauma-based safety studies, whose modest goal is to help lower the risk of serious injury in traffic accidents. Over the years, researchers have vastly reduced the fatality rate for automobile accidents by bringing us the seat belt, the air bag, and countless other safety devices. To make cars safer, however, they now need more-precise information than is found in traffic-accident reports, and more true-to-life data than they can get from crude crash-dummy tests. "Crash dummies don't really reproduce what happens to a passenger," says John H. Siegel, M.D., a surgeon and trauma researcher at the New Jersey Medical School in Newark. "Dummies have no physiology." Nor do they act much like humans. "Put crash dummies in a car, and they just sit there," says Catherine Gotschall, M.D., an epidemiologist at Children's Hospital. "Put children in a car, and they squirm, wiggle, and lie down."

CLUES TO THE PUZZLE
To home in on the specific causes of injury and death, trauma-based crash researchers like Warner begin their work as soon as a crash victim arrives in the emergency room. They go to the cars themselves to try to piece together precisely how the victim's injuries were sustained. Then they sit down with doctors and traffic-safety experts—and sometimes biomechanical engineers and rehabilitation specialists—to arrive at a total picture of the accident and its impact on the patient. "The easy answers have already been found," says Frances Bents, a research manager at Dynamic Science. "The kinds of changes required now are much more complex and require a multidisciplinary approach. That's the real beauty of trauma-center research. All parties sit around the table, and each provides a clue to try to solve the puzzle."

Trauma researchers have discovered that children often suffer injuries in accidents because car seats (left) and other safety devices are used improperly.

The main impetus for trauma-based crash research has come from the National Highway Traffic Safety Administration (NHTSA). Over two decades, the agency had put together an extensive research database of traffic accidents. When, in the late 1970s, some safety researchers began to clamor for more-detailed information from doctors about the kinds of injuries sustained in accidents, the NHTSA began funding trauma-based crash studies. One of the first was a 1988 study at the Maryland Shock Trauma Center at University Hospital in Baltimore that followed 144 patients with severe injuries from frontal and side collisions. At that time, government safety standards were based mainly on tests in which a car crashed head-on into a barrier. Researchers found,

By studying the aftermath of an automobile accident, investigators can construct a complete picture of the dynamics of the crash and determine the causes of any injuries that occurred.

Automakers rely on crash-test dummies, which mimic the effects of an impact on the human body, to ensure that their new cars comply with safety standards.

however, that most real-life crashes are not straightforward frontal crashes but offset, or corner, crashes. Because only a portion of the front end absorbs the energy of such a collision, the crash can be devastating: the car crushes easily, and the instrument panel and toe pan intrude into the passenger compartment. The good news was that air bags and seat belts were found to do an adequate job of preventing head and chest injuries in corner crashes; the bad news was that they did a poor job of protecting the legs.

Shifting Focus

These early findings underscored the need for safety researchers to shift their focus from preventing death, which in the past they had concentrated on almost exclusively, to preventing injury. "People who would once have died from head and chest injuries were surviving," says Bents, "but they were surviving with expensive, debilitating leg injuries." Since then, injury prevention has emerged as a major theme of trauma-based research.

Further studies at the Maryland Shock Trauma Center found that the people most susceptible to foot and ankle injuries were women—or, more accurately, short people. To learn why, biomechanical engineers at the University of Virginia in Charlottesville videotaped volunteers, both tall and short, in the act of braking. Shorter drivers, they say, lift their foot to step on the brake, whereas taller drivers tend to rest their heel on the floorboard. When the researchers

Computers can simulate the structural damage that a car will suffer in a collision with such objects as a utility pole (right) or another vehicle.

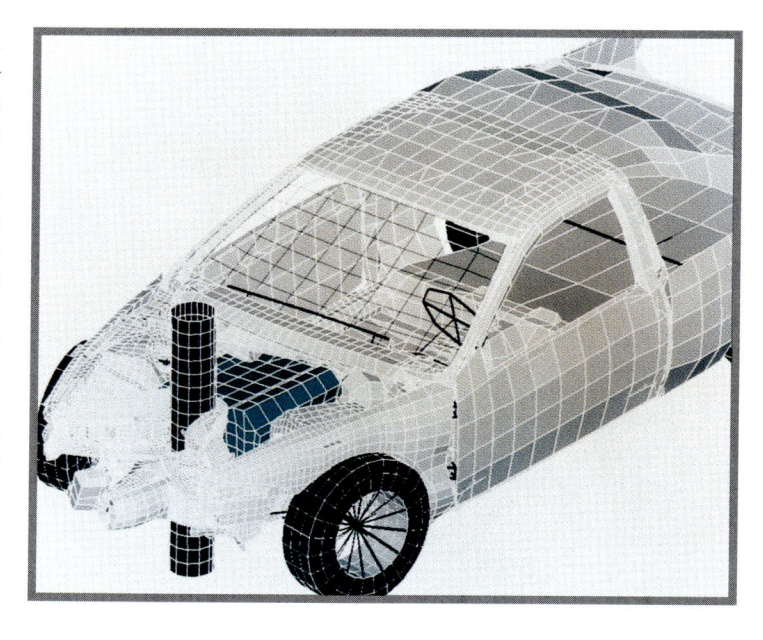

plugged these observations into computerized crash simulations, they found that as a crash pushes the floorboard inward, it slams into the foot of shorter drivers, resulting in injury. Taller drivers avoid this fate because they rest their heel on the floorboard and ride it up during the crash. The solution, it turned out, was simple: 1 inch (2.5 centimeters) of padding on the floor of the car below the brake pedal can cut the force on shorter drivers' ankles in half.

One of the trauma researchers' ultimate goals, of course, is to spur auto companies to design safer cars. Partly because of the work of the Maryland researchers and others, General Motors (GM) has begun to develop dummies with more-accurate legs, with more-lifelike joints, and with more sensors. Doctors have also begun documenting ankle injuries more precisely, giving biomedical engineers better data with which to work. And Mercedes-Benz is pioneering the use of pedals that bend under the stress of a crash, as well as new designs that better protect passengers' feet and that redirect the force of frontal collisions down the side rail and center of the car's frame.

CHILD SAFETY

Trauma investigators have also been taking a closer look at injuries that occur in spite of—and sometimes because of—safety devices. Work at Children's Hospital suggests that children between 40 and 60 pounds (18 and 27 kilograms) are at present particularly vulnerable in car crashes. The problem is that although these children are usually allowed by law to wear adult seat belts, the seat belts do not fit them very well. When a child squirms, the lap belt can ride up across the stomach, making the lower spine vulnerable. "Child safety would be promoted by the use of booster seats for children 40 to 60 pounds," says Catherine Gotschall, one of the principal investigators. Her groups also found some children to be at risk even when they have adequate car seats, because parents often use them incorrectly. As a result, the American Academy of Pediatrics, the NHTSA, and other groups are trying to educate parents about such dangers as putting a rear-facing infant seat

Crash-Test Dummies

Crash-test dummies are a lot smarter than they used to be. Today's dummy is a high-tech wonder, a far cry from the simple mannequins that began the crash-test movement in the 1950s.

A new generation of crash-test dummies incorporates far more sophisticated sensors and humanlike features than do older models.

As stunt doubles for motorists, state-of-the-art crash-test dummies bear an uncanny resemblance to their human counterparts. Dummies are now available with precise human proportions, breakable aluminum bones, and an anatomically correct steel and rubber spine. Most automakers run tests with dummies of both sexes and many sizes; one manufacturer has even subjected a pregnant dummy, complete with a vinyl fetus and a gel-filled amniotic sac, to a rigorous crash test.

These dummies also come wired with sophisticated sensors that can take thousands of measurements—assessing knee shear and sideways acceleration, for example—during a split-second crash.

Virtual dummies may be next. Engineers at General Motors are developing computer applications that simulate accidents with lifelike human cybermodels.

Peter A. Flax

in the front seat of a car with air bags (if the air bag inflates, the child can be struck in the back of the head or crushed against the car seat). The NHTSA recently decided to allow auto manufacturers to include a switch so that drivers can disable the passenger-side air bag when a child is sitting up front.

OCCULT INJURIES

The more-precise information available through trauma-based studies, researchers hope, should prove to be useful in the treatment of accident victims. Paradoxically, by cutting down on serious injuries, devices such as air bags have made it more difficult to diagnose less-apparent injuries. Researchers at Florida's University of Miami School of Medicine have found, for instance, that despite air bags, many injuries occur because riders are either not wearing their seat belts or are sitting too close to the steering wheel. "In the world of air bags, a patient who looks good can later turn out to have occult injuries," says Jeffrey Augenstein, M.D. As a result, paramedics, who usually quickly decide whether a crash victim needs urgent attention, may underestimate the extent of hidden internal injuries. To help them, Augenstein and his colleagues are trying to identify which accidents typically lead to hidden injuries. For instance, "if the driver is wearing an automatic shoulder belt without a lap belt, he or she is at increased risk of liver injury," Augenstein says.

Although air bags have helped many motorists escape accidents without serious harm, the devices have been linked to a rise in hard-to-detect internal injuries.

CONTROLLED STUDIES

Trauma-based studies are no substitute for epidemiological studies, in which researchers spot broad trends by gathering statistics on a great many accidents. Since trauma centers see only patients with bad injuries, their findings are not representative of all crashes. Trauma-based studies also do not replace studies in the laboratory, which are better suited for exploring certain specific issues in the biomechanics of accidents. "Field investigations of real accidents give you good injury data," says Lawrence Schneider, head of the biosciences division at the University of Michigan Transportation Research Institute in Ann Arbor. "In the controlled environment of the lab, you have less injury data, but you can control impact conditions. You have to work the two together."

Trauma research has been useful, however, in identifying which problems need laboratory work. Studies of real accidents have shown, for instance, that crash victims who sustain seemingly minor brain injuries can suffer long-lasting learning and behavior problems. In response, scientists are taking what they know about how much nerve cells can be stretched, and are using this knowledge to develop computer models of the brain during impact. NHTSA scientists hope the models can eventually be used to produce improved crash dummies, with dummy heads that yield more information about the effects of a crash on the human brain.

Ultimately, some trauma researchers want to automate themselves out of a job. Augenstein, for instance, sees no reason why automobiles cannot take their own trauma histories. Racing cars already carry a device, like the black box of an airplane, that records crash information. Why couldn't all cars provide information about the speed and direction of the crash and whether seat belts were being used at the time? That way, by the time the rescue crew arrives—after the car has automatically called 911 and relayed its location—paramedics would already have an idea of what injuries to expect. Although it makes perfect sense, Augenstein does not expect any of this to happen soon. Change comes slowly. After all, there are still people around who refuse to buckle their seat belts.

Undersea Explorers

by Peter Britton

Intensive ocean research is now being conducted using new generations of manned and unmanned submersibles. One such vehicle, Deep Flight, shown at right with its builder, scientist Graham Hawkes, is a single-occupant submersible capable of reaching a depth of 3,000 feet.

In August 1995, an autonomous submersible slipped into the waters over the Juan de Fuca Ridge off the coast of Oregon to monitor the changes in the magnetic properties of a fresh lava flow 6,500 feet (2,000 meters) below the surface.

A remarkable feat? The Woods Hole Oceanographic Institution's craft, the Autonomous Benthic Explorer (ABE), is but one of a remarkable fleet, thousands strong, of the world's ingeniously designed, constantly evolving submersibles, both manned and unmanned. Here is a quick guide to this swarm of subs:

Manned Submersibles

These free-"swimming" craft are controlled from inside by human pilots. Their performance is limited by the need for fuel, the need to maintain onboard systems that sustain human life, and the basic endurance of the pilots.

In 1960, the U.S. Navy's bathyscaphe *Trieste* took three men to a depth of 35,800 feet (10,900 meters) in the Challenger Deep of the Mariana Trench in the Pacific Ocean. That is the deepest spot in the ocean, with an ambient pressure of 16,000 pounds per square inch (1,125 kilograms per square centimeter). (The Japanese revisited this area 34 years later with an unmanned sub. See the sidebar on page 279.)

Now two scientists, Graham Hawkes and Sylvia Earle, want to visit the Challenger Deep themselves. Hawkes is building a pair of *Deep Flight* vehicles, single-occupant submersibles with wings. The pilot operates the unit from a prone position—on the stomach, hands extended toward joysticklike controls. For visibility, there is a clear acrylic bubble. Right now, the submersibles are able to dive only to 3,300 feet (1,000 meters). *Deep Flight-2*, a version of the craft Hawkes claims would be capable of reaching the sea's greatest depths, is on the drawing board.

Today the number "6,000" crops up frequently in the names of vehicles. That's because roughly 97 percent of the oceans have a depth of no more than 20,000 feet (or about 6,000 meters). So far, there are only five manned submersibles able to operate at this depth. They are the *Nautile* (France); the *Shinkai 6500* (Japan)—a second variation is planned for dives to 36,000 feet (11,000 meters); the *Mir I* and *Mir II* (Russia); and the *Sea Cliff* (United States).

Since the 1960 voyage of the *Trieste*, the most-dramatic uses of manned submersibles (excluding military submarines) have been the discovery and subsequent filming of some of the ocean's secret wonders. First, hydrothermal vents were discovered and filmed in the Pacific Ocean in 1977; then the sunken wreck of the *Titanic* was photographed for an Imax film in 1991.

These are the high-profile jobs. But recently one of the glamorous manned craft proved itself in a down-and-dirty industrial assignment. The *Deep Rover*, basically an acrylic bubble with strong manipulators, went tunnel inspecting. A Canadian firm fitted the vehicle with a special fiber-optic/power umbilical cord to inspect two 4.5-mile (7.25-kilometer)-long tunnels for the New York Power Authority's Niagara Power Project.

Remotely Operated Vehicles

ROVs are controlled from the surface by a tether that supplies unlimited power for propulsion and tool operation, as well as audio and video communications. These

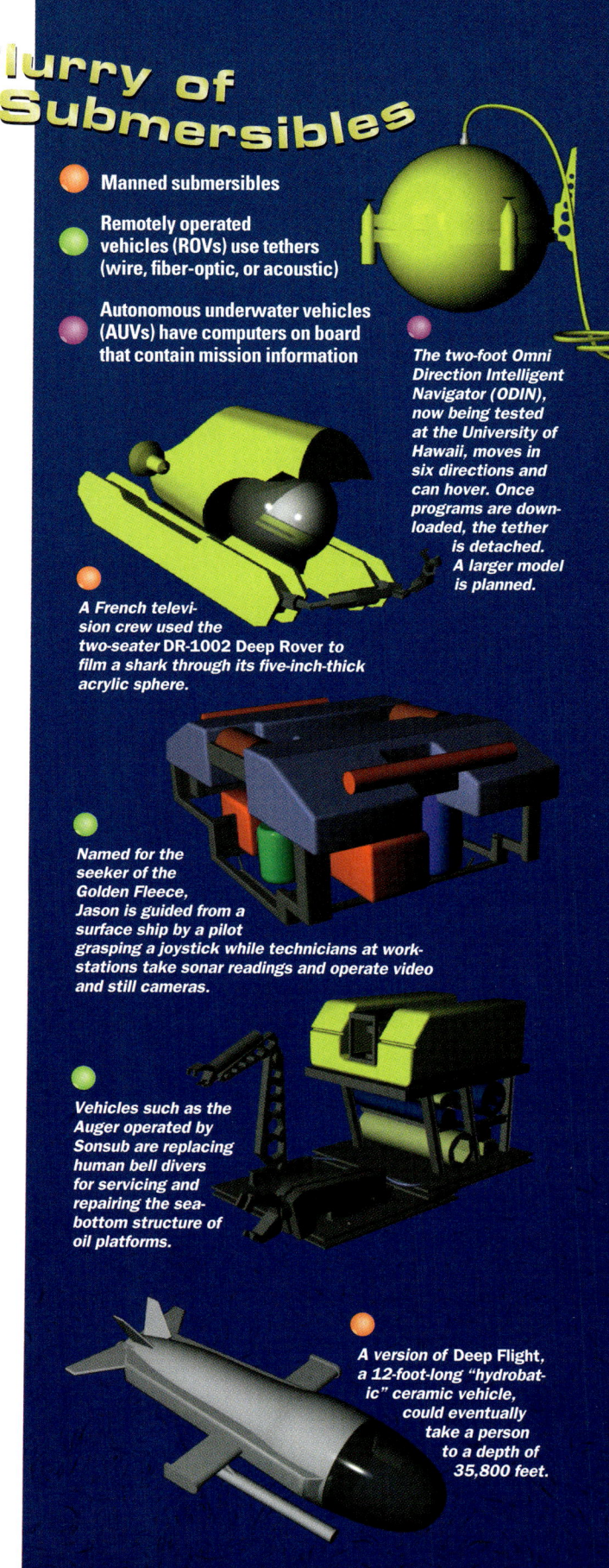

A Flurry of Submersibles

- 🟠 **Manned submersibles**
- 🟢 **Remotely operated vehicles (ROVs) use tethers (wire, fiber-optic, or acoustic)**
- 🟣 **Autonomous underwater vehicles (AUVs) have computers on board that contain mission information**

The two-foot Omni Direction Intelligent Navigator (ODIN), now being tested at the University of Hawaii, moves in six directions and can hover. Once programs are downloaded, the tether is detached. A larger model is planned.

A French television crew used the two-seater DR-1002 Deep Rover *to film a shark through its five-inch-thick acrylic sphere.*

Named for the seeker of the Golden Fleece, Jason *is guided from a surface ship by a pilot grasping a joystick while technicians at workstations take sonar readings and operate video and still cameras.*

Vehicles such as the Auger *operated by Sonsub are replacing human bell divers for servicing and repairing the sea-bottom structure of oil platforms.*

A version of Deep Flight, *a 12-foot-long "hydrobatic" ceramic vehicle, could eventually take a person to a depth of 35,800 feet.*

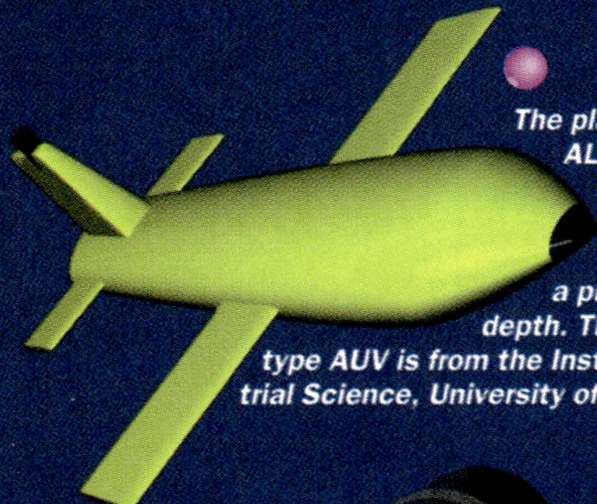

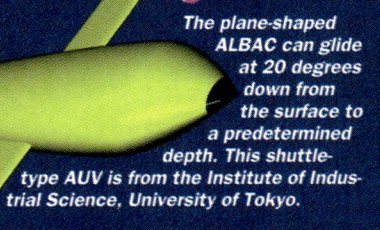

The plane-shaped ALBAC can glide at 20 degrees down from the surface to a predetermined depth. This shuttle-type AUV is from the Institute of Industrial Science, University of Tokyo.

The Probe for the Underwater Research Lab (PURL) is being tested by Simon Fraser University in Vancouver, Canada.

The autonomous EAVE III uses an acoustic telemetry program that enables it to communicate with kindred vehicles. Mission plans and obstacle avoidance are handled totally by onboard computers.

Only 61 pounds, the MiniRover MK II can operate in 1,000 feet and 1.5-knot currents. It carries a low-light, high-resolution color video camera.

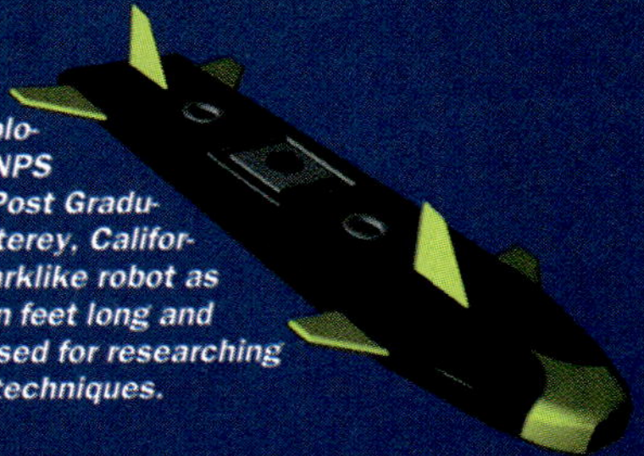

After a battery explosion wrecked the NPS AUV II, the Naval Post Graduate School in Monterey, California, rebuilt the sharklike robot as the Phoenix. Seven feet long and 434 pounds, it's used for researching advanced control techniques.

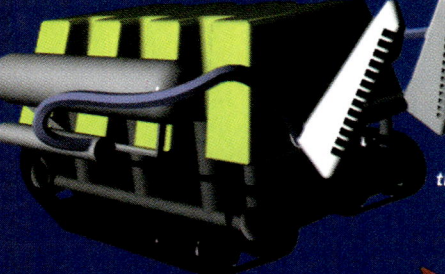

The remotely operated crawler, Trencher, from Perry Tri-Tech, could lay a pipeline along the ocean bottom.

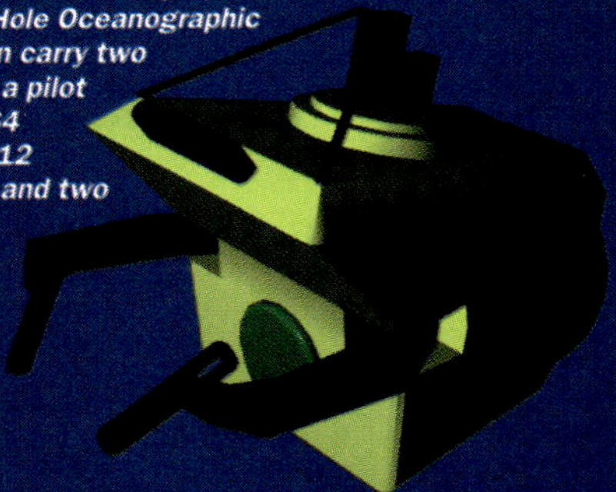

Alvin (named for its developer, Al Vine of the Woods Hole Oceanographic Institution) can carry two scientists and a pilot down to 14,764 feet. It sports 12 exterior lights and two robotic arms.

Japan's three-person Shinkai 6500 submersible (seen here) can achieve depths of four miles. The proposed Shinkai 11,000 would permit manned visits to the deepest ocean regions.

The one-ton MT88 was developed by the Russians for locating sunken wrecks (theirs and ours). It quickly descends, follows a search pattern, photographs the object, and releases its ballast to return to the surface.

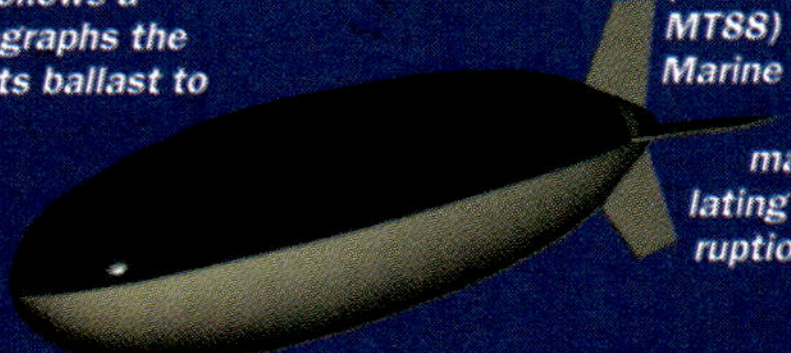

The torpedo-shaped Typhlonus (a streamlined version of the MT88) from the Institute for Marine Technology Problems in Vladivostok, Russia, maneuvers via an articulating tail. The name is a corruption of the word Teflon.

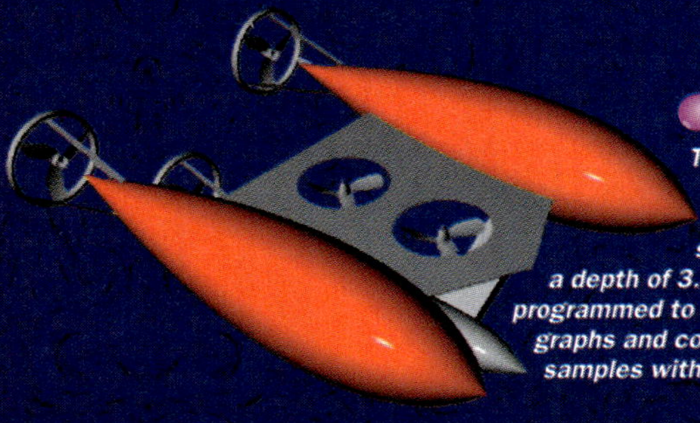

The Autonomous Benthic Explorer (ABE) can spend several months at a depth of 3.8 miles. It can be programmed to take daily photographs and collect data and samples within a specified area.

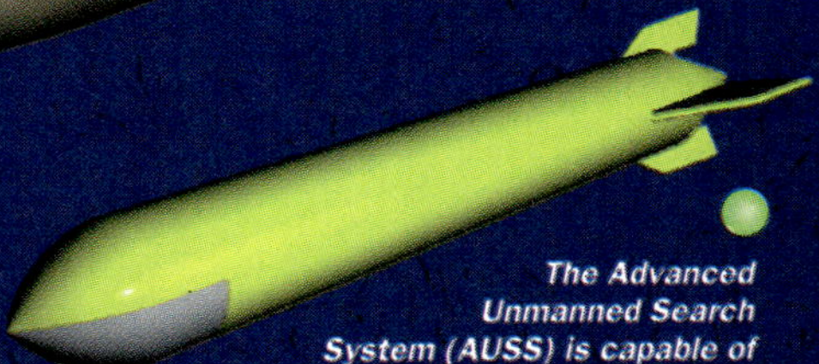

The Advanced Unmanned Search System (AUSS) is capable of inspecting the remains of an aircraft under 20,000 feet of water and sending sonar and optical images back to its mother ship.

Submersibles are carried out to sea on—and launched from (above)—specially equipped "mother ships." Most explorations conducted by manned submersibles are of relatively short duration.

ROVs can handle strenuous, continuous work with powerful manipulators, replacing and out-torquing divers on most tasks. But the tethers are unwieldy.

Through funding for research and development, the offshore oil and gas industry has figured prominently in the evolution of remotely operated vehicles—and now ROVs hold an important place in the oil and gas industries' plans. Fifteen years ago, divers did most of the underwater work around offshore rigs. But the oil companies' move into deeper waters and the costs of using divers gradually led the companies to develop machines to handle much of the work.

Shell, Amoco, and Exxon are increasing their use of ROVs. Shell uses them in the Gulf of Mexico for its Auger tension-leg platform; Amoco, for its Liuhua Floating Production System project off the coast of China. In December 1994, Shell, Amoco, and Exxon announced the Mars Powell project, involving a field of 15 billion barrels of crude located due south of Alabama under 3,220 feet (980 meters) of water.

AUTONOMOUS UNDERWATER VEHICLES

AUVs are "intelligent" vehicles that operate free of any tether constraints or commands. They use powerful onboard computers to run predetermined missions. Making this possible are artificial-intelligence (AI) programs for obstacle recognition and avoidance. Only a handful of AUVs exist. Limited power is a distinct problem.

There are also ROV/AUV hybrids. Acoustically and fiber-optically controlled vehicles are sometimes included as ROVs. And a subcategory of AUVs might be called the DGBs (drifters, gliders, and bounders).

A few years ago, a confused AUV named EAVE-East impaled itself on a metal rod in Lake Winnipesaukee, New Hampshire. (EAVE stands for experimental autonomous vehicle.) That early tendency has been programmed out of its repertoire. Now EAVE III has a clone with whom it "talks" in a language all its own. D. Richard Blidberg, the director of Northeastern University's Marine

Deep-diving minisubmersibles have helped researchers learn much about the geology of the ocean floor and the plants and animals that live there.

Japan Challenges the Deep

On March 1, 1994, Shinichi Takagawa nervously watched a row of video monitors aboard a support ship some 200 miles (320 kilometers) southwest of the Pacific island of Guam. The unmanned submersible Kaiko was relaying images to the video monitors as it approached the floor of the Challenger Deep, the deepest spot in the world's oceans. Suddenly, a heartbreaking 3 feet (1 meter) from the bottom, the monitors went blank.

"I could feel the muscles in my face and neck convulsing," recalls Takagawa, who had overseen development of the $50 million sub.

Luckily, the crew was able to retrieve Kaiko. It took months of testing to figure out what had gone wrong: a kink had worked its way into the optical fiber that carries video signals from the cameras to the ship.

With modifications to its cable system, Kaiko completed its demonstration dives, and will soon start its real work. As the world's only submersible capable of venturing below 21,000 feet (6,400 meters), it is expected to confirm whether or not marine life exists at the bottoms of the deepest ocean trenches, and it may help scientists better understand the movements of Earth's tectonic plates.

Developed by the Japan Marine Science and Technology Center, Kaiko (the name means "trench" in Japanese) is actually a two-part vessel. A 17-foot (5-meter)-long launcher connected to the support ship by a 7-mile (11-kilometer)-long primary cable descends to within 300 feet (92 meters) of the ocean bottom. There it releases a 10-foot (3-meter)-long rover tethered to the launcher by an 800-foot (240-meter)-long secondary cable. Both vehicles have titanium shells that can withstand the crushing water pressure in the Challenger Deep—about 16,000 pounds per square inch (1,125 kilograms per square centimeter).

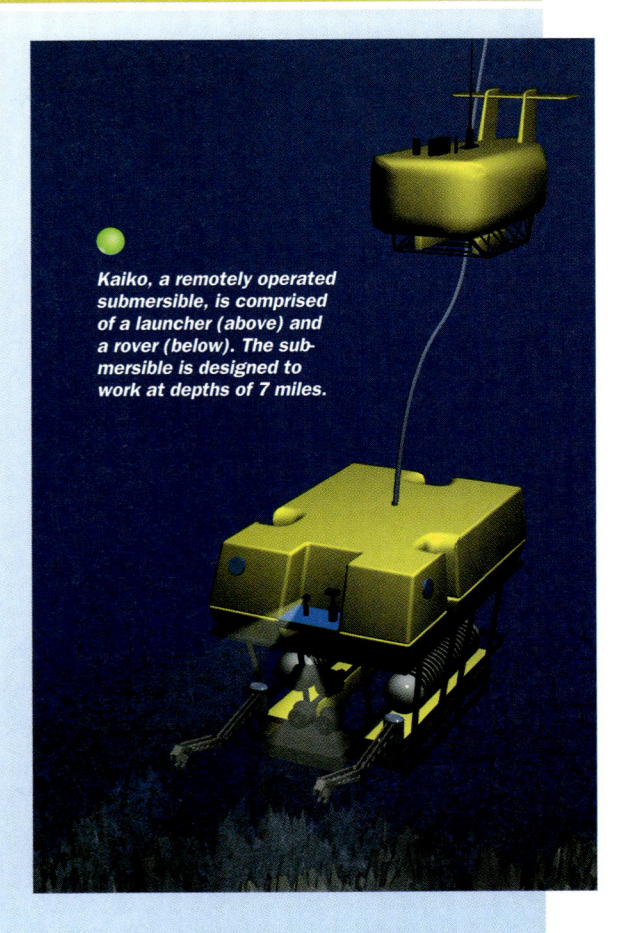

While the launcher's movements are restricted by the weight of the primary cable, the rover can swim freely through the water, propelled by three vertical and four horizontal thrusters. It carries video and still cameras, and it has two manipulator arms to retrieve samples.

Dennis Normile

Kaiko, a remotely operated submersible, is comprised of a launcher (above) and a rover (below). The submersible is designed to work at depths of 7 miles.

Systems Lab in Boston (and founder of the Autonomous Underwater Systems Institute), says communication is at the forefront of AUV development.

"We're deep into various kinds of computer architecture, neural nets, artificial intelligence—the things that will make these devices work. They must accept a mission, venture forth, avoid obstacles, do their work, keep from self-destructing, and come home with a report. Not unlike a kid going off to college."

Ask the Scientist

▶ *Is the silo on a farm simply used for storage? Is there a specific reason why most silos are cylindrical?*

Silos are the storage structures in which chopped green corn or forage crops are transformed into the livestock food called silage. This transformation is the result of a fermentation process that takes place when the cut and chopped green crops are stored in a relatively air-free, closed space. "Often molasses, uric acid, or other additives are mixed with the crops to speed up the fermentation," write Jerry Baker and Dan Kibbie in *Farm Fever*, a book about how to purchase and cultivate country land.

There are basically three types of silos: trench, bunker, and upright. Baker and Kibbie assert that the most common upright silos are cylindrical out of necessity. Because the green, prefermentation silage material usually has a high percentage of moisture in it, the building's shape must be able to support the tremendous pressure and the weight of the stored crops.

▶ *Were early cars equipped with such necessities as heaters, defrosters, and windshield wipers? Are such items required on all automobiles?*

According to Rick Lee, an antique-car restorer and salesman at West Shore Auto Sales & Service in Warwick, Rhode Island, "prior to 1928, these features were not standard in closed cars, let alone convertibles." In fact, the "open" Model T Fords, the most common cars of the Roaring Twenties, came with little more than a motor, transmission, rear end, chain drive, and brakes.

Model A Fords and other popular automobiles of the post-1928 period began carrying windshield wipers as standard equipment, but not heaters, and certainly not defrosters. "Gas heaters were available in the Model A's as early as 1922—but only as options," Lee says. "Most of these heaters did not defrost well at all, especially in winter conditions."

All modern road vehicles in the United States *must* have windshield wipers. Heaters and defrosters are standard but generally not required.

▶ *How does an elevator work? Why should an elevator be avoided during a fire? Why are there so few elevator operators left?*

According to Peter Wallack, director of operations at Beckwith Elevator Company in Boston, elevators are either cable or hydraulically driven—like an automobile lift or a "bottle" jack under a car. Elevators driven by systems of cables used to be analogous to a winding-drum machine (similar to a tow truck lifting a car), but these types of systems have become obsolete. The current "cabled" elevators are counterweighted—more like a grandfather clock, says Wallack, "where the counterweight balances the elevator's load."

Elevators, much as other complex machines, are comprised of a multitude of switches and relays. If soaked by a sprinkler system, any one of these switches or relays could bring the elevator into erratic operation, or to an undesired stop—trapping the occupants. In addition, an elevator shaft or hoistway is a particularly dangerous place in a fire. It becomes the natural draft chimney for the entire building—drawing smoke and poisonous gases into the shaft. (The elevator shaft is vented to the outside air at the top.)

Today there is little demand for elevator operators. In most modern elevators, merely pushing a button starts a sequence of events that ultimately brings the passengers to the desired floor.

▰ *My music teacher told me that the piano is considered a stringed instrument. How can that possibly be true? Also, do the pedals on a piano and the pedals on a harp have a similar function? Do any other instruments have pedals?*

The piano is a keyboard instrument in which depression of the keys causes the strings to be struck with felt hammers. In the sense that it *has* strings, it is also a stringed instrument. Generally speaking, foot pedals on pianos and harps both sustain notes and change pitch, but in the modern "double-action" harp, to alter their pitch, strings are shortened by means of seven pedals—instead of the two, three, or four found on various pianos. In addition, according to Bernie Houle, a master piano craftsman based in Warwick, Rhode Island, the left, or "soft," pedal on many pianos is purely ornamental.

Foot pedals are also found on pedal-steel guitars, organs and pipe organs (although organ pedals do not sustain notes and are more like shutters opening and closing), harpsichords, and the vibraphone played by world-famous jazz percussionist Lionel Hampton.

▰ *Following even the briefest power outage, I find that, if I have been working on my computer, the file is either missing or scrambled. Why does this happen?*

"The file has no way of saving itself unless you have battery backup configured with the appropriate software," says Darek Lorenz-Kruk, a technical-support supervisor at the Boston office of the Digital Equipment Corporation (DEC). What he is referring to is a UPS (that is, an *uninterruptible power supply*) system. When installed with a computer, UPS systems have the battery backup built in automatically for use during a power failure. With that optional software or monitoring kit in place, the UPS system will safely store your data and shut down your desktop or network operating system before the battery is fully discharged—whether you are there or not.

▰ *I have read that the price of paper is skyrocketing. This surprises me. I would have thought that recycling efforts would have produced a glut of paper, and caused paper prices to drop. Can you explain what is causing the paper problem?*

The price of paper has little to do with recycling, and everything to do with economics. In 1994 and 1995, paper prices "went through the roof." According to paper-industry officials, the sudden inflation in prices (in some cases, up 300 percent since 1993) was due to a boom in the global economy, combined with a slowdown in the construction of equipment and paper plants. The paper market had last boomed in 1989, and companies rushed to build plants, but the recession began just as many of the plants began operating, causing prices to plummet. "We invested a lot in capacity because the economy was very good, and then we went in the tank," says Barry Polsky, a spokesman for the American Forest and Paper Association, a trade group based in Washington, D.C. So the price jump was not caused by any kind of global paper shortage (as was often rumored), but by the desire of pulp manufacturers to make up for several years of losses from a depressed market and, according to the *New York Times* (February 20, 1995), "to recover some of the lost costs expended in upgrading mills to meet environmental clean-air and clean-water codes."

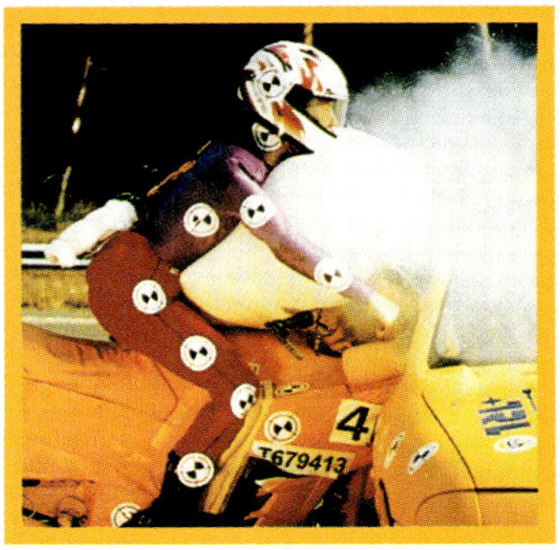
4
T679413

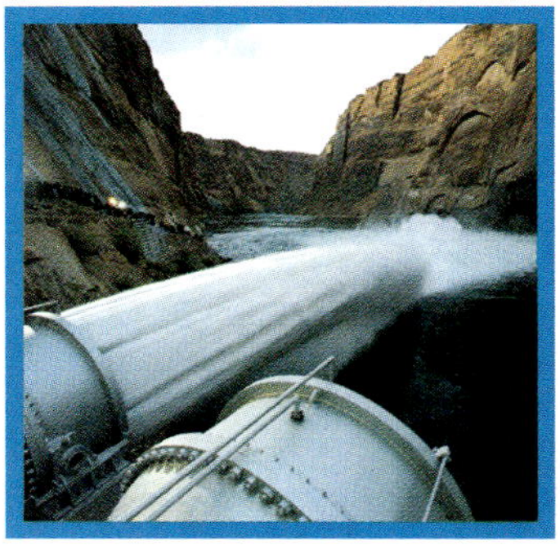

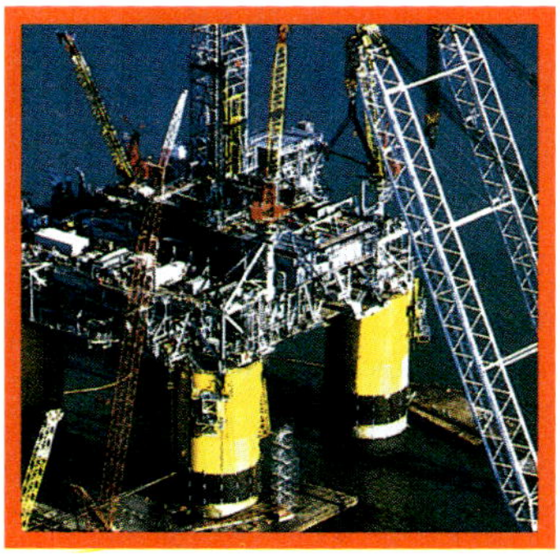

REVIEWS
1996

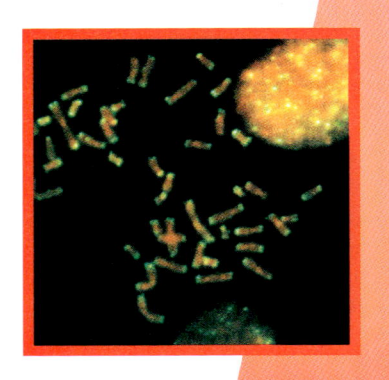

Agriculture

Production and Income

A difficult growing season in the United States ended with early frosts in much of the Midwest and South. Higher grain prices raised consumer costs for meat, milk, cheese, poultry, and grain products.

The 1995 North American wheat crop was reduced somewhat due to excessive rainfall in winter-wheat areas. A late-spring freeze and record-late plantings of spring wheat also contributed to the smaller crop. In California, severe flooding after a long drought reduced fresh-vegetable supplies, pushing retail prices sharply higher. Replanting later brought prices down to normal levels.

U.S. corn stocks were expected to equal only a three-week supply by the end of the 1995–96 marketing year. Feed grains grew in importance as a raw material for energy and other industrial products in 1995, with five new processing plants beginning operations in the western Corn Belt.

U.S. tobacco farmers looked for alternative crops in response to increased restrictions on smoking and tobacco-product advertising. As a rule, even though labor requirements for tobacco production are high, farmers in the U.S. Southeast and parts of Pennsylvania continued to plant tobacco, a major cash crop with a high value per acre.

Large factory-type hog-production units continued to grow in number in the United States and Canada. Some larger units were integrated vertically with meat processors. The economic impact of this trend raised major concerns about the declining number of farm families, the flow of profits to investors outside rural communities, increasingly limited competition, and the loss of traditional markets for pork produced on family farms. Environmental problems were also feared; concerns were heightened by several serious leaks and spills.

The dairy industry also began shifting toward larger farms and factory-type production. Some California dairy farms—strapped by high-priced land, high taxes, and growing environmental concerns—even moved operations to the Midwest.

Legislation

A seven-year omnibus farm bill, signed by President Bill Clinton in April 1996, reduced the costs of agricultural-support programs, with many subsidies being reduced or phased out. Farmers were also given the freedom to plant crops based on market incentives rather than on government regulations. Government storage of reserve grain supplies was de-emphasized, potentially leading to greater volatility of food prices. The new policies were expected to affect farmland values, rural incomes, the stability of food supplies, and the cost of food.

Programs and Trade Policies

For much of 1995, the United States restricted imports of Canadian wheat to avoid depressing the prices received by U.S. farmers. In early fall, the restrictions were lifted, but U.S. officials monitored the quantity of imports from Canada for possible adverse effects on U.S. farmers.

South of the border, Mexico increased its purchases of U.S. corn and soybeans in response to the North American Free Trade Agreement (NAFTA) and severe drought.

In the marketing year ending August 31, 1995, U.S. corn exports rose 68 percent from the previous year, while soybean exports increased by 40 percent. U.S. exports of high-value fruits, vegetables, and livestock products were also up sharply.

The net trade balance in U.S. agricultural products was strongly positive. Late in 1995, Japan placed tariffs on imports of U.S. pork under trade agreements that allowed Japan to protect its farmers from a rise in pork imports. Japanese imports of U.S. pork during the first three-quarters of the year were up 50 percent from a year earlier.

International Trends

In 1995, low stockpiles made global supplies of grains, oilseeds, and cotton quite dependent on favorable yields. Feed-grain stocks at the end of the 1995–96 marketing year appeared likely to fall to record-low levels. Tightening supplies pushed corn prices more than 80 percent above those of 1994.

Wheat production fell in 1995 due to severe drought in Australia, South Africa, southern Europe, the former Soviet Union, and China. Stockpiles were at their lowest levels relative to usage since the early 1970s.

Rice production was reduced in 1995 by erratic weather in India and China, the world's largest rice producers.

Great Britain. In the mid-1980s, British cattle were hit by an epidemic of a fatal brain disease called bovine spongiform encephalopathy (BSE), also known as mad-cow disease. Many scientists believe that the cattle contracted the condition by consuming feed containing tissue from sheep infected with scrapie, a similar disease.

In March 1996, officials announced the discovery of 10 unusual new cases of a rare, fatal human brain condition called Creutzfeldt-Jakob disease; these cases, admitted the officials, could have been caused by consuming BSE-infected beef. Panic set in, both in Britain and worldwide. Members of the European Union (EU) and other nations banned the importation of British beef. (The United States has prohibited the importation of British beef since 1989.) Sales of beef and its by-products plummeted in Britain and elsewhere. To assure a halt to possible infection and to restore confidence in British beef's safety, more than 4 million cattle were to be destroyed over five years. The EU agreed to shoulder 70 percent of the cost of compensating British farmers, but angered Britain by leaving the import ban in place indefinitely.

Former Soviet Union. Livestock production in the former Soviet Union continued to decline, but more slowly than in past years, as its farms struggled to become market-oriented. Russia sharply increased its imports of U.S. pork and broiler meat.

Grain crops in Russia were reduced by severe drought and shortages of key production inputs. Inadequate financing and lack of foreign exchange limited Russia's ability to import grain from Western nations.

Research and Development

Biotechnology research continued to improve productivity. New varieties of corn with resistance to corn-borer insects were developed; they were expected to reduce dependence on chemical insecticides. Insect-

British farmers were displeased when the country's agricultural officials ordered more than 4 million cattle to be destroyed in an effort to contain a possible epidemic of so-called mad-cow disease.

resistant varieties of potatoes were also developed. New varieties of corn with higher oil content became available, providing processors and livestock feeders with high energy content in rations. Soybeans containing more-healthful oil also were being developed.

In the livestock industry, pork producers took advantage of increased feed-conversion efficiency through early weaning and same-sex feeding. Early weaning helps control disease by moving pigs to nurseries and special rations while they still have disease resistance from their mothers' milk. Same-sex feeding allows livestock producers to match rations more specifically with nutritional needs of each sex.

Crop farmers began utilizing Global Positioning Satellite (GPS) systems to measure yield variability in individual fields and to adjust input-application rates to reflect varying soil conditions. The technology offers the potential to increase crop yields and reduce production costs.

Robert Wisner

ANTHROPOLOGY

WALKING UPRIGHT

The discovery of fossil bones near Lake Turkana (formerly Lake Rudolph) in Kenya provides the oldest evidence for upright locomotion among the ancestors of modern humans. The finds—21 objects—include limb bones, teeth, and jaw fragments; they date back about 4 million years.

The remains are of chimpanzee-sized creatures that exhibit traits similar to those found both in apes and in ancestors of modern humans. The shape of the jaw and the size and arrangement of the teeth resemble those of apes, while the thick tooth enamel is more like that of human ancestors. The shape of the limb bones indicates that the creatures were bipedal—they walked on two legs.

This discovery is important because it suggests that bipedal locomotion was a critical feature in the division that took place between the animals from which modern apes descend and those from which modern humans descend. This split in evolution is believed to have occurred between 5 million to 7 million years ago—before the brain size substantially increased and before the development of stone-tool technology.

Scientists debate why these creatures began to walk upright in the first place. Some view the change in terms of environmental variations, arguing that these primates left, or were forced from, forested landscapes into open savanna countryside. Upright posture may have aided them in seeing over the tall grasses, either to seek out food sources or to see potential predators. Others suggest that upright locomotion developed in response to a need to reach higher for foods that grew on the branches of trees. Still another argument suggests that upright posture was an adaptation that reduced the amount of the body exposed to the tropical sun. This factor would have been particularly important if these primates had left a forested environment for open grassland.

The shape of the limb bones and the pelvis had to change significantly in order for these creatures to walk upright. Changes in the form of the pelvis meant that the young had to be born at a reduced size in order to fit through the birth canal; therefore, they would need to be born at an earlier stage of development, and thus longer periods of intensive care by parents were required before the young could fend for themselves. Another consequence of upright posture was the freeing of the forelimbs, or arms, for purposes other than supporting the body. This change made possible the carrying of objects, including infants, in the arms, and it opened the way for making, using, and carrying tools more efficiently.

The discovery at Sterkfontein in South Africa of four fossil foot bones that provisionally date back 3.5 million years provides additional information on early bipedal locomotion. These specimens constitute the earliest known set of foot bones that join together from any hominid, or early human, and they show a combination of structures characteristic of humans and of apes. The shapes of the heel and the arch are similar to those of modern humans in that they are adapted for supporting the weight of the body as the individual walked upright. But the large toe of these ancient hominids was situated at a wide angle from the rest of the foot, and it was able to grasp, much like our opposable thumb. The creature represented

Anthropologists were amazed to learn that a mummy known as the Spirit Cave man (below, in an artist's sketch) is more than 9,400 years old, making it the oldest known mummy in North America.

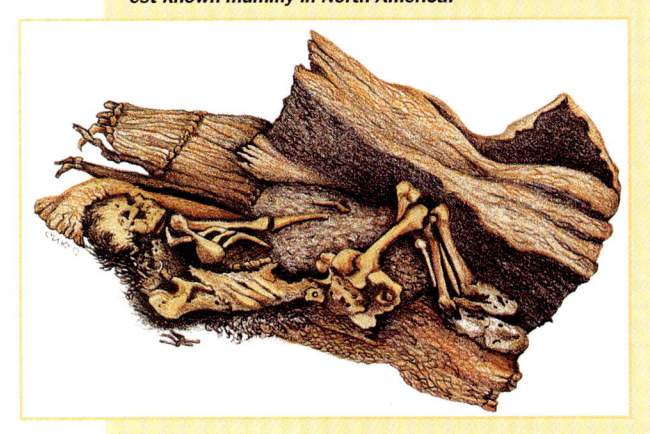

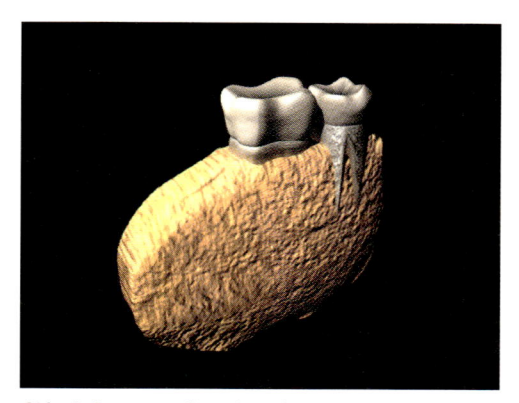

China's Longgupo Cave has yielded the oldest remains of early humans outside of Africa, including a lower jaw with some teeth intact (above).

by these foot bones was probably bipedal—it walked upright; but it also could climb trees in the manner that modern apes do, with a grasping big toe as well as an opposable thumb. From this single find, it is not clear to what extent this hominid actually walked upright, and to what extent it climbed trees. The find's importance is in showing that traits adapted to both forms of locomotion were present in a single hominid 3.5 million years ago.

FIRST HUMANS IN ASIA
Fossil bones found at Longgupo Cave in central China constitute the earliest evidence for the existence of humans outside of Africa. The remains—shown by new dating techniques to be about 1.9 million years old—include a lower jaw with premolar and molar intact, and a separate incisor. The structures are similar to those found in specimens of early forms of human ancestors that until now have been identified only in Africa, particularly East Africa. The Longgupo Cave layers that contained these fossils also yielded simple stone tools similar to some from Olduvai Gorge in Tanzania, as well as bones of extinct animals, including the mastodon. Until recently, the available evidence had suggested that the first form of early human that moved out of Africa was *Homo erectus*, believed to be the direct ancestor of modern *Homo sapiens*. This exodus from Africa is believed to have occurred some 1.8 million to 1.6 million years ago. The evidence from central China now makes it apparent that at least one form of hominid moved from Africa to eastern Asia at an earlier date. On the basis of this new find, some researchers have raised the possibility that *Homo erectus* evolved from earlier hominids in Asia, not in Africa. All evidence still suggests that humans first evolved in Africa, and that modern humans—*Homo sapiens*—also developed in Africa, perhaps around 100,000 years ago.

EARLY HUMANS IN EUROPE
At Gran Dolina Cave in Spain, a new discovery demonstrates the presence of humans in Europe earlier than any finds made to date. Fossil bones, including skull fragments, of four or more hominids were found together with stone tools in layers believed to date to sometime before 780,000 years ago. Preliminary study of the fossils suggests that they do not belong to the group known as *Homo erectus*, but perhaps represent an early form of Neanderthal. If the dating is supported by further tests, and ongoing analysis confirms that the fossils belong to the Neanderthal group, this find would be much earlier than any other known Neanderthal specimens. This discovery also contributes to the growing picture of a much more diverse spectrum of early humans than earlier studies suggested.

NEANDERTHAL/*HOMO SAPIENS* INTERACTION?
According to all available evidence, modern humans *(Homo sapiens)* did not arrive in Europe until long after they developed in Africa. Recent discoveries of fossil Neanderthal bones indicate that Neanderthals and *Homo sapiens* lived at the same time in parts of Europe. At Zafarraya Cave near Málaga in southern Spain, a Neanderthal lower jaw was discovered together with stone tools typical of Neanderthal implements; according to scientific dating techniques, the finds are about 30,000 years old. Similar Neanderthal fossils at Vindija, Croatia, date back 33,000 years. There is clear evidence of modern *Homo sapiens* living in these regions by 40,000 years ago, indicating that populations of Neanderthals and modern humans inhabited some of the same regions of Europe for at least 10,000 years.

These new discoveries raise important questions about interactions between the different types of humans, and bear directly on the long-standing problem of what happened to the Neanderthals. Did the two groups avoid each other, inhabiting and exploiting different microenvironments? Did they compete for the same resources? Did they interbreed, and through such interbreeding did the classic Neanderthal characteristics gradually disappear from the populations of ancient Europe?

Scythian Horseman Found

The frozen body of a man preserved in permafrost was found in a burial mound on the Ukok Plateau in the Altai Mountains in Siberia. The grave dates back to between 1000 and 500 B.C. The mound contained a burial chamber built of logs; inside, a man was buried next to a horse. The man was outfitted with leather boots, embroidered pants, and a coat of fur. His long hair was braided, and on his chest and back he bore a tattoo of an elk. A dagger, an ax, and a quiver accompanied him to his grave.

Frozen Women in the Andes

Archaeologists exploring 20,700-foot (6,300-meter)-high Mount Ampato in the Peruvian Andes discovered three frozen bodies believed to be about 500 years old. Highest on the peak was the body of a young woman who had been wrapped in woolen textiles and outfitted with a head covering made of feathers. With the body were pottery vessels and small figurines made of bronze, silver, gold, and shell, some of which had their own textile wrappings and head decorations. The Incas, the people who ruled this region of South America at the time, are known to have worshiped mountains, and also to have sacrificed young women to the gods. These bodies are thought to represent such sacrifices. Future analyses will provide information about the levels of health and nutrition of these individuals, and about the practice of religion on this mountain. The investigators also found an encampment apparently used by the people who performed the ritual sacrifice. Pottery, fragmentary ropes, and other settlement debris were recovered.

Peter S. Wells

Archaeology

Earliest Stone Tools

Stone tools believed to be about 2.6 million years old were found on a site near the Gona River in Ethiopia. These flaked cobbles are the earliest tools known anywhere in the world. No fossil remains of their makers were found on the site, but evidence elsewhere in East Africa from the same period suggests that the tools were created by australopithecines, an ancestral group to humans. Australopithecines lived before the first recognized members of the genus *Homo* appeared about 2 million years ago.

This discovery will be important in discussions about evolution. Some scientists believe that tools were first developed after brain size increased among members of the genus *Homo*. Others think that the bipedal gait evolved first (see page 286), thereby making the hands available for tasks other than walking. According to this argument, toolmaking developed once the hands were no longer needed to support the body. The new fossil evidence suggests that bipedal walking developed before the increase in brain size did; this archaeological evidence indicates that stone tools were being made well before the appearance of the larger-brained *Homo* genus.

Early Complex Tools in Africa

Finds from the valley of the Semliki River in Zaïre indicate that, about 80,000 years ago, people were making complex barbed points out of bone for spearing animals. The points are thought to have been made from rib bones of large mammals. The artifacts were found at a riverbank settlement where people harpooned and ate large catfish. The barbed bone points studied at Old Stone Age (Paleolithic period) sites in Europe are about 15,000 years old. Scientists do not know if the later European points represent a borrowing from earlier African forms, or a separate development.

Ancient Capital Discovered

Excavations at Tell Mozan in Syria have uncovered the remains of what is believed to be the city of Urkesh, the capital of the Hurrians, a people cited in an Egyptian inscription and in the Old Testament of the Bible. Research indicates that the city flourished around 2300–2000 B.C. as a commercial and political center. Among the excavated remains are a temple structure and a storeroom associated with royal authority. Seal impressions, inscribed clay tablets, and drawings provide rich information about the society. Many of the inscribed texts relate to the administrative affairs of the government. One tablet lists professions held by the citizens of the city. Festive banquets are portrayed in some of the drawings. Many of the seals belonged to an otherwise-unknown queen named Uqnitum, who appears to have wielded considerable power in the society.

ROMAN MARITIME TRADE

The Mediterranean Sea served as the principal transportation route for many ancient civilizations, including the Egyptian, Minoan, Mycenaean, Phoenician, Greek, Etruscan, and Roman. In recent years, a number of shipwrecks have been studied by underwater archaeologists, yielding discoveries that have provided important information about ancient ship technology, cargo capacity, and trade routes. Because of limitations in exploration technology, nearly all of the underwater archaeological research to date has been conducted in shallow waters along the coasts. Thus, nearly all of our knowledge of ancient Mediterranean shipping has been biased toward coastal carriers and has led some investigators to conclude that ancient mariners sailed close to the coasts for safety.

New research indicates that many ancient ships sailed the open seas. Until now, it has been difficult to find the shipwrecks in the deep regions of the Mediterranean. Recently Robert Ballard, the marine geologist who discovered the sunken *Titanic*, used a robot submarine to explore shipwrecks in the deep central part of the Mediterranean, over which numerous ships sailed between Rome and Carthage. Preliminary results indicate abundant materials from shipwrecks dating from the 4th century B.C. to the 12th century A.D. In the future, Ballard intends to employ special submarines to explore and recover materials from some of these deepwater shipwrecks. Organic materials, including the wood in the ships' hulls and items of cargo such as textiles, grains, and spices, are likely to be much better preserved in the deep-sea environment than in coastal areas; hence, this research may substantially change our understanding of ancient shipping. And, in any case, it is clear from these new discoveries that the ancient sailors often ventured across the central parts of the Mediterranean, and did not just hug the coasts.

Herod's Temple Base Found

Some 2,000 years ago, in a grand display of loyalty to the Roman Emperor Augustus, Herod the Great built an enormous temple in what is now Caesarea, Israel, on a prominent hill overlooking the city's harbor. Recently, archaeologists (right) uncovered the foundation stones of the magnificent structure. The remains suggest that the original building measured about 90 by 160 feet and soared to a height of 90 feet, making it one of the largest temples in that part of the empire.

ROYAL MAYAN TOMB
A tomb of a Mayan king has been found in a temple structure at the ancient urban center of Copán in Honduras. Inscriptions written in Mayan glyphs indicate that the individual buried there was Kinich Ah Pop, son of the king who established the dynasty at Copán in A.D. 426, and himself the second king in that dynasty. Inscriptions even tell the year—A.D. 437—that Kinich Ah Pop celebrated the completion of the temple building. Genetic tests are planned for the skeletal remains to compare the DNA of the individual buried in this grave with that from other royal skeletons to study the relationships between the different individuals. The body was accompanied by a rich assemblage of jade ornaments, pottery, and other goods. Beneath the grave was another burial, of an individual unidentified as yet.

COLLAPSE OF MAYAN CIVILIZATION
The Maya of Central America established one of the world's most spectacular civilizations between about A.D. 150 and 800. They built magnificent cities with huge temples and elaborate stone carvings, practiced specialized crafts in many different materials including jade and pottery, and developed a system of writing for recording political histories and keeping track of astronomical cycles. Around A.D. 800, Mayan civilization began a precipitous decline. The cities fell into ruin and were abandoned, and the fine crafts ceased; inscriptions recorded increasing warfare between cities. Finding an explanation for this collapse has been one of the great challenges of modern archaeology.

Two new studies bring important perspectives to bear on this complex problem. Examination of shells, plant remains, and other sediment from the bottom of Lake Chichancanab in Mexico indicates that a great drought began around A.D. 800 and continued until the year 1000.

The second study argues that the Maya brought on ecological crises by cutting down an excessive number of trees throughout their territories. The archaeological evidence before A.D. 800 indicates that the population grew rapidly in the Mayan lands, especially near the urban centers. In order to feed the growing populations, new fields had to be cleared, often by cutting away forests. Much of the stone architecture in the Mayan centers required stucco. The stucco was made by burning limestone, a process that required large quantities of fuel obtained by cutting vast tracts of woodlands for firewood. A result of this deforestation—both for agriculture and for fuel—was soil erosion. The erosion, most notably the loss of topsoil needed for agriculture, led to a general disturbance of the local ecology.

Most Mayan specialists argue that by A.D. 800, when the first signs of decline appeared in Mayan civilization, the society was subjected to political, social, economic, and ecological stresses. No single factor is likely to explain the complex process of decline and collapse that is reflected in the archaeological evidence. But the kinds of ecological changes suggested by the two new studies were probably important factors that contributed to the demise of this great civilization.

MEDIEVAL SRI LANKAN SMELTERIES
In medieval Arabic literature, reference is made to steel of exceptionally high quality from India and the island of Sinhala (later called Ceylon, which was renamed Sri Lanka in 1972). Field research in Sri Lanka has discovered the technological basis for this unusually fine product. Archaeologists have studied the remains of about 40 furnaces that were used for smelting iron to make fine steel between the 7th and 11th centuries A.D.

The furnaces, made of hardened clay, were placed on the tops of hills, where the strong and predictable monsoon winds provided the steady blast of air required to maintain the necessary temperatures. In most early smelteries, smiths used hand-powered bellows to pump oxygen, a method that required intensive labor by a team of workers. The newly discovered furnaces in Sri Lanka could operate without bellows, thanks to their exploitation of "monsoon power." Experiments carried out with replicas of the medieval furnaces, placed in the same situation on the slopes, show that adequate temperatures could be easily maintained.

Peter S. Wells

ASTRONOMY

OUR SOLAR SYSTEM

Images taken during a lunar eclipse revealed that the Moon's tenuous atmosphere is twice as large as previously thought. Astronomers recorded the glow of sodium atoms in the Moon's atmosphere beyond the shadow cast by Earth, and found that the glow extended to a height of about 8,700 miles (14,000 kilometers). They believe that sunlight somehow loosens the sodium atoms from the lunar surface to generate this thin atmosphere.

A rock was identified as the 12th known meteorite to have hit Earth from the planet Mars. The nearly 0.5-ounce (12-gram) chunk of rock was found in Antarctica's Queen Alexandra Range. Scientists drew their conclusions based in part on the similarity of the gases in the rock with material found by the Viking spacecraft on Mars in 1976.

Astronomers announced the detection of new volcanic eruptions on Jupiter's large moon Io. Photographs taken with the Hubble Space Telescope (HST) revealed a huge yellow-white spot some 200 miles (320 kilometers) across that may represent the freshest volcanic deposit ever seen on Io.

During a rare moment when Saturn's icy rings turned edge-on to our line of sight and nearly vanished from view, astronomers discovered two additional moons orbiting the planet. Ordinarily these bodies would be lost in the glare of the rings, and therefore would be invisible from Earth.

The Hubble Space Telescope found about 30 icy, comet-sized objects on the fringes of the solar system. Ranging from 4 to 8 miles (6 to 13 kilometers) in diameter, they appear to be the first evidence of the long-theorized Kuiper Belt, which may contain up to 10 billion comets in a disklike structure far beyond the orbits of Neptune and Pluto.

In July 1995, astronomers Alan Hale and Thomas Bopp discovered a comet that promises to capture the attention of sky watchers for several years. In measurements made by the HST, the comet's icy nucleus appears to be about 25 miles (40 kilometers) across, four times larger than that of Halley's comet. Comet Hale-Bopp has been compared to the Great Comet of 1811, and is expected to put on a dazzling light show when it swings past Earth in the spring of 1997.

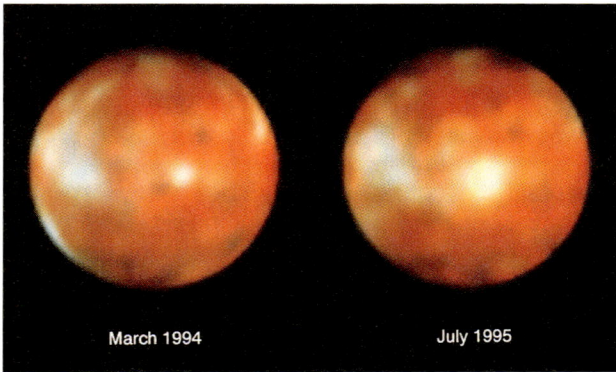

Hubble Space Telescope images taken 16 months apart clearly show the dramatic emergence of a volcano on the surface of Io, a moon of Jupiter.

IN SEARCH OF EXTRASOLAR PLANETS

Researchers reported finding a Jupiter-sized planet circling a typical star 57 light-years away. The planet is the first ever found around an ordinary star like the Sun. The planet has a mass between half and twice that of Jupiter, yet its orbit is only one-sixth the diameter of Mercury's around the Sun. It orbits the star, known to astronomers as 51 Pegasi, about every four days.

In subsequent months, astronomers found two new planets in orbit around two other stars. Near the star 70 Virginis, they found one about eight times heavier than Jupiter; and around the star 47 Ursae Majoris, one about three and one-half times more massive. Both are close enough to their parent stars to support the existence of liquid water.

The first photograph of a brown dwarf—a body significantly heavier than Jupiter but lighter than a star—has been taken. The faint object, orbiting the cool red star known to astronomers as GL229, was first revealed with the 60-inch (1.5-meter) telescope on Palomar Mountain in California and with the HST. Spectra indicate that the brown dwarf contains methane, which cannot survive the hot temperatures of even the coolest stars. The body therefore must have less than 8 percent of the mass of the Sun—the minimum needed to form a true star.

LIFE CYCLES OF THE STARS

One of the most breathtaking images taken by the Hubble Space Telescope shows gargantuan pillars of gas that mark the birthplace of stars in the Eagle nebula, only 7,000 light-years away. The spectacular photo shows the erosion caused by intense ultraviolet radiation emitted by hot, fully grown stars. The process strips away gas from the cloud, not only causing the newborn stars within to become visible, but also depriving them of the gas they need to become more massive.

Observations with the world's largest optical telescope suggested that the first generation of stars predated galaxies by 1 billion years. By using the 400-inch (10-meter) Keck Telescope on Mauna Kea, Hawaii, astronomers found low-mass clouds—long thought to be primordial—that contain carbon, an element that can be created only by the birth and death of stars. This discovery suggests that the universe may have been seeded with heavy elements from stars long before the formation of galaxies themselves.

THE MILKY WAY . . . AND BEYOND

The first "natural" laser in space was detected by scientists on board the National Aeronautics and Space Administration's (NASA's) Kuiper Airborne Observatory as they trained the aircraft's infrared telescope on a young, very hot, luminous star in the constellation Cygnus. The laser is created as intense ultraviolet light from the star "pumps," or excites, the densely packed hydrogen atoms in the gaseous, dusty disk surrounding the star. Then, when infrared light shines on the excited hydrogen atoms, it causes them to emit an intense beam of light at exactly the same wavelength.

A new galactic speed record was set when astronomers found a transient source of X rays and gamma rays ejecting gas at nearly the speed of light. Such powerful outbursts have been seen in distant galaxies and quasars, but never before in our own Milky Way. Because the source is relatively close—a mere 40,000 light-years—it may provide astronomers with a nearby example of the behavior of quasars and active galaxy nuclei.

Using the artificial intelligence of the new Palomar digital sky survey, astronomers found 16 new quasars—farther away than any ever seen before. These quasars appear as they did when the universe was only 1 billion years old, when the first structures were beginning to form.

Astronomers detected an ordinary galaxy—perhaps the most distant galaxy ever imaged—undergoing its first wave of starbirth. Using the 90-inch (2.2-meter) University of Hawaii telescope atop Mauna Kea, the researchers found the distant galaxy only a few hundred light-years away from a powerful quasar named BR2237-0607. The newly discovered galaxy appears as it was less than 1 billion years after the birth of the universe.

Researchers detected clouds of hydrogen gas in what were once thought to be giant voids between clusters of galaxies. Using the Hubble Space Telescope's high-resolution spectrograph to analyze the light of quasars that lie "beyond" the voids, astronomers found that their light was partially blocked by immense clouds of hydrogen. The clouds may be part of the outer halos of galaxies too faint to appear on photographs, or they may be pristine material, forged only a few minutes after the universe itself.

With the Hopkins Ultraviolet Telescope aboard the space shuttle, astronomers saw the spectral signature of singly ionized helium atoms lying along the line of sight to a distant quasar. The presence and density of this signature support astronomers' models for the nuclear reactions that occurred during the first few minutes after the Big Bang.

Astronomers discovered a new distant class of quadruple, or cross-shaped, gravitational lenses that might eventually provide them with a powerful new tool for probing a variety of characteristics of the universe. A gravitational lens is produced by the enormous gravitational field of a massive object, which bends light to magnify, brighten, and distort the image of a more distant object. This new discovery might contribute to our understanding of the distribution of dark matter and the number of supermassive black holes, and might even help scientists determine if the universe will expand forever or eventually collapse.

Dennis L. Mammana

Automotive Technology

Microcompacts

In an effort reminiscent of the 1960s, automobile companies are creating a new generation of small cars. These new compacts boast a difference—they are even smaller than the old ones. And they generally appear as variations on the shape of an egg.

"Smart" is now the official name of the tiny two-seater designed by Mercedes-Benz and SMH, the company that makes Swatch watches. (During development, this microcompact was referred to as the "Swatchmobile.") Smart will be produced in France beginning in 1997. It will join a host of new small cars by European and American automakers, including a second planned Mercedes mini, known as the "Vision A," which is only 136 inches (345 centimeters) long.

Renault, the French automaker, already sells the Twingo, which is 134 inches (340 centimeters) long. Fiat, the Italian automaker, offers the Cinquecento, which is even smaller: 126 inches (320 centimeters). Although tiny, the Twingo and Cinquecento seat, respectively, four and five passengers—by sacrificing almost all possible baggage space. Both models are selling extremely well. Volkswagen intends to market a smaller version of its Polo model (itself a subcompact) before the turn of the century. Ford and General Motors (GM) plan to produce microcompacts, although only for the European market. Ford's Ka will be smaller than its compact Fiesta, already a popular car in Europe. GM's Corsa is a current model that will be updated.

Europeans are particularly amenable to small and efficient cars for short trips in crowded cities. Unlike the small cars of the 1960s and 1970s, many of the new microcompacts seat four people comfortably and come with such luxury options as air-conditioning.

Electricity

GM's electric car, the Impact, has become a production vehicle. Renamed the EV1, it is available from a few Saturn dealerships in the U.S. Southwest and priced in the mid-$30,000 range. Like other electric vehicles, the EV1 suffers from a limited driving range between recharges. The fact that it is not being sold in the northern states points to a second drawback of electric vehicles: poor interior heating. Unlike internal-combustion-powered vehicles, which have plenty of heat, electric cars must rely on less effective technologies to keep their passengers warm.

In Vermont, scientists are studying a number of strategies for heating the interiors of electric vehicles. Funded by the U.S. military and the Northeast Alternative Vehicle Consortium, the Vermont Consortium has been testing such technologies as gasoline-powered heaters, heated seats, heat-absorbent materials for interiors, and solar glass.

One solution to both the range and the heating problems of electric cars is to develop a hybrid car, a vehicle that would contain a gasoline-powered engine that is boosted by an electric motor during acceleration. Ergenics Inc., a company in New Jersey, has developed a battery that could be used in such a hybrid vehicle. The battery uses a reversible nickel hydroxide and hydrogen reaction to deliver strong surges of power for use during the vehicle's acceleration. Acting in this supplementary role, the battery could increase a car's gas mileage to about 80 miles (130 kilometers) per gallon. A version of the battery has been used in space satellites for decades.

The Fiat Cinquecento (below) holds up to five passengers—an amazing feat for a microcompact vehicle that measures only 126 inches in length.

FEATURES

The days of car keys may be numbered. Two companies, Siemens-Automotive and TRW Inc., are developing keyless systems that unlock car doors as the driver approaches the vehicle. These systems use a card that contains a tiny battery which receives and sends signals. When the driver gets near, the device trips an electronic control inside the car. When control units in the car receive the proper signal from the card, they cause the doors to unlock. Once in the car, the driver presses a button or gives a voice command to start the engine. Siemens plans to install the device in some German luxury cars in 1998.

Automakers are improving traction-control systems for their midsize cars. Traditional systems respond to the driver's loss of traction by automatically applying rapid light braking to the affected wheel—in other words, by engaging the antilock-braking system. In new versions, traction-sensing systems connect to a computer that controls engine power. When traction slips, the engine slows, helping the driver to maintain control. Chevrolet Corvette, Ford Contour, and Cadillacs now have this system. More-sophisticated systems on some high-priced luxury cars have sensors and mechanisms that respond to understeering or oversteering during cornering.

Also now available on some high-priced cars are automatic transmissions that can revert to manual control. So-called auto-manual shifts are convenient for high-speed driving and negotiating winding roads. In Porsche's Tiptronic system, for example, when the transmission is in drive and the shift stick is moved over slightly, the operator can change gears using two switches on the steering wheel. Rather than switches, Acura's Sportshift system uses a second shift lever on the right side of the steering column. In Chrysler's Autostick (available on the Eagle), manual shifting is achieved by additional movements of the automatic shift stick.

SAFETY

Built-in child-safety seats increasingly are being offered in cars, especially in family minivan models. The belts of the built-in seats fasten more easily and fit the child more snugly than do those of cumbersome traditional seats, which are inserted and removed by an adult. The big three U.S. automakers all offer the seats in minivans as well as in a few other models. The newest versions allow the driver to check at a glance whether the child's belt is secured, based on an obvious color code in the buckle. The latest generation of seats features improved fits because of redesigned cushioning; some of the seats recline.

Ford announced that it will include side air bags in all of its U.S. and European car and light-truck models within a few years. Side air bags deploy when a vehicle sustains an impact on either the driver or the passenger side. Most bags will deploy from the inside of the door or from the outer front edge of the seat. However, BMW plans to introduce an additional feature within a couple of years. In its "side-impact system," an inflating tube bursts from around the edges of the window and contracts so as to block the window, preventing a passenger's head from hitting it.

Air bags will perform better in the coming years. Research companies such as Allied Signal Incorporated are developing technologies that improve safety by coordinating seat-belt pressure and air-bag deployment, by producing gentler and more effective deployment, and by deploying air bags outside the car.

Advances that will occur first include a system that takes up slack in the seat belt, then releases tension as the air bag deploys, and multiple detonators that cause an air bag to deploy in a series of quick steps, rather than in one large explosion. In combination, these two improvements will produce a more agreeable meeting between person and air bag.

An additional area of development is sensor technology. In the not-too-distant future, sensors will scan a car's interior, determine which seats are occupied, and measure each occupant's weight and shape. That information will be given to air-bag controllers, which will adjust air-bag deployment to accommodate each passenger. Air bags in front of empty seats will be instructed not to deploy. Sensor systems might use electric fields, radar, or infrared rays.

Donald W. Cunningham

AVIATION

FLIGHT RECORDERS

In February 1995, the National Transportation Safety Board (NTSB) urgently recommended to the Federal Aviation Administration (FAA) that all flight-data recorders on U.S. aircraft be upgraded to expand their capacity. The NTSB's action came after the crash of two Boeing 737s: the 1994 crash of a USAir plane near Pittsburgh, Pennsylvania, and the 1991 crash of a United Airlines plane at Colorado Springs, Colorado. In both of those instances, the safety board was not able to determine the cause of the crash, at least partially because the data recorders on the planes were not the most up-to-date available.

For instance, updated data recorders on the USAir plane might have been able to record whether movement of the plane's rudder had a role in the crash. Investigators suspect that the abnormal movement of the rudder may have helped lead to the crash. Subsequent NTSB studies of rudder movements on 737s have not led to any findings, again because their data recorders do not collect some crucial pieces of information. On the other hand, a French airplane that crashed in Indiana in 1994 had a flight-data recorder that provided investigators with information on 98 different parameters. Investigators were able to issue a report on that crash within days.

The NTSB recommended an upgrading of data recorders of all older 737s. In addition, the agency recommended upgrades for all other aircraft flown by U.S. airlines, to be completed by January 1, 1998. The recommendations would affect more than 1,000 737 aircraft and 4,000 other craft.

FAA PREDICTS INCREASE IN AIR TRAFFIC

Although many airports are already operating close to capacity, the FAA predicts that the number of passengers on U.S. airlines will double by the early part of the next century. The agency warns that the increase could lead to extensive delays and could cause carriers to lose revenues.

The FAA announced its estimates in March 1995, predicting that in 18 years, U.S. airlines will have to handle more than 1 billion passengers a year. As a result, by 2003, the FAA estimates that 32 major airports will experience more than 2,000 hours of delays each year. Only 23 airports fell into that category in 1994. The FAA also estimates that because airlines lose up to $1,600 for each hour that a flight is delayed, the increased air traffic would force carriers to either pass on their extra costs to customers or absorb them. In addition, to compensate for delays, airlines would be forced to increase the number of trips they make each day. An FAA spokesperson estimates that daily operations already increase 3 to 5 percent each year.

One solution to the probable increase in air traffic would be to build more airports. FAA Administrator David R. Hinson notes that only three new airports have been built in the past 30 years. Adding new runways to existing airports could also ease the problem. So far, 15 of the busiest airports nationwide are planning to build new runways. While most airports on average are equipped to handle 29 landings per hour, adding an additional runway would allow for 57 land-

In accordance with the 1991 Strategic Arms Reduction Treaty, the United States destroyed 217 B52 bombers—by slicing the planes into pieces using a guillotine-like device.

ings per hour. Using three runways for landing allows for 86 arrivals each hour.

New technology, such as improved runway monitors to allow planes to fly closer to each other, may also ease the traffic burden. However, according to the Airline Pilots Association, before these changes can be implemented, more research needs to be conducted on wake vortex, a phenomenon that occurs when planes fly too close together.

Air-Traffic-Control Safety

Several outages at major air-traffic-control facilities across the nation in 1995 raised concerns about the reliability and safety of the equipment used by such facilities. These concerns have prompted the FAA to promise that an update of air-traffic computer systems will be in place by 1998.

Many of the computers, radios, and radars used in air-traffic-control facilities are outdated, with some equipment dating back to the 1960s. In 1995, problems at several facilities highlighted the dangers of using such equipment. In February, for instance, a computer error caused an icon of a plane on the radarscope at San Juan International Airport in Puerto Rico to be 4 miles (6.4 kilometers) out of its actual position. When the icon suddenly moved to its correct position, the controller realized that the airplane was directly in the path of another plane. Fortunately, the two aircraft did not collide, but the incident pointed out the need for a system that could alert controllers to computer errors.

Another potentially disastrous incident occurred at the Oakland Air Route Traffic Control Center in Fremont, California, when both the radar and radio of a controller guiding 295 planes failed for 35 minutes. Again, no collisions occurred, although the incident caused hundreds of flights to be grounded.

The FAA had originally planned to update its computer systems by 1990, but that target date kept being postponed. This situation—combined with a nationwide increase in air traffic, plus a shortage of air controllers (there are still 1,500 fewer controllers than there were in the early 1980s, and they must handle 30 percent more flights)—leads some experts to believe that a disaster is almost certain. According to one FAA air-traffic manager, "the chances of a catastrophic failure in one or more centers should be considered likely." Other experts argue, however, that controllers will simply compensate for unreliable equipment by building in a greater margin of safety between planes. Such measures are likely to increase delays.

In the meantime, proposed congressional reductions to the FAA budget may force the agency to further cut back on both air controllers and the amount it spends on maintenance. The agency has estimated that such cuts would increase equipment-related flight delays to 5,000 per year from the current 4,000.

Miscellaneous Aviation News

Late in 1995, the FAA issued a security alert to airports nationwide. The "level-two" alert (based on a four-level system) was issued in October, and came after the verdicts against the coconspirators in the World Trade Center bombing trial and before the autumn visit of Pope John Paul II. The agency advised travelers to arrive at airports early, carry a photo identification, and be prepared to have baggage and carry-on items inspected. The security alert was lifted before the start of the Thanksgiving holiday travel season.

In an unrelated action, the FAA in October 1995 announced that it was beginning an aggressive program to detect the use of illegal aircraft parts used by the U.S. aviation industry. Although most U.S. airlines have programs that search for suspect or unapproved parts, the FAA's program is aimed at strengthening previous agency efforts to ferret out illegal parts.

Finally, United Airlines began in 1995 to offer commercial flights on Boeing's new 777 airplane. The 777, which is the first new plane produced by Boeing in 13 years, is a twin-jet aircraft. United's 777s will seat 292 passengers. A new stretch version of the 777, with room for 368 seats, is also being marketed by Boeing. In addition, the aircraft company is in the process of producing a 777 with a longer flight range than the current version.

Devera Pine

Behavioral Sciences

Creativity and Madness

Are creativity and mental illness related? A study from the University of Kentucky Medical Center in Lexington has poked some holes in this widely held assumption.

From Vincent van Gogh to Edgar Allan Poe, many famous artists, writers, and musicians have suffered from various forms of mental illness. In fact, some modern-day studies have found that painters, poets, musicians, and novelists have a higher incidence of mental illness, especially depression and manic depression. But in a study of 1,004 prominent men and women of the 20th century, researcher Arnold M. Ludwig found that many high achievers in various disciplines are actually emotionally stable. Moreover, he found that mental disorders aid creativity only in certain circumstances for certain types of personalities.

Ludwig drew his conclusions from his study of the lives of deceased people from the 20th century, all prominent in fields such as the arts, sciences, politics, military, business, or social activism. He rated the prominence of each person according to a scale that took into account the subject's reputation after his or her death, the international appeal of the person's work, whether he or she developed new ideas or trends in his or her field, the originality of the subject's creative efforts, and the total extent of all of the person's lifetime accomplishments. Next, Ludwig determined—from written biographical accounts—whether each person fit into broad categories of symptoms, including alcohol or drug problems, depression, mania, overwhelming anxiety, delusions and psychotic symptoms, preoccupation with the body, and suicide attempts. (Ludwig did not attempt to make a postmortem psychiatric diagnosis, however.)

The results were mixed, depending on the professions of the people studied: among poets, fiction and nonfiction writers, painters, and composers, 46 to 77 percent had suffered episodes of severe depression—twice the rate of people in other fields. From 11 to 17 percent of actors, poets, architects, and nonfiction writers experienced episodes of mania in their lifetimes. Yet only 3 percent of the scientists under study experienced the same patterns of turmoil, though the scientists received similar ratings of prominence.

Ludwig's explanation for the results is that some professions—such as painting, poetry, and fiction writing—help bring out emotional turmoil, while scientific fields tend to favor controlling emotions. In addition, Ludwig found that various elements play a part in how much prominence a person ultimately achieves. These elements include a capacity for solitude and self-reliance, physical trials such as a disability early in life, special talents that are displayed early on and encouraged by parents, and a restless, driven state of psychological unease that is relieved through creative problem solving.

"While mental disturbances may provide individuals with an underlying sense of unease that seems necessary for sustained creative activity, these disturbances are not the only source for inner tension," says Ludwig. "Mental illness is not the price people pay for their creative gifts."

Reexamining Delayed Recall

Delayed recall of trauma—first heralded as the great exposer of hidden childhood sexual abuse, then vilified as the questionable results

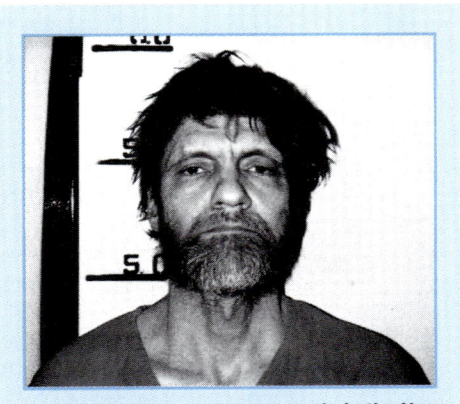

Long before any arrests were made in the Unabomber case, the FBI developed a psychological profile of the most-wanted serial bomber. This sketch matched with uncanny similarity the characteristics of alleged Unabomber Theodore Kaczynski (above)—a middle-aged, highly educated, antisocial white male.

of dubious psychotherapy—once again received a boost of sorts in 1995. A study of minority men and women who survived severe traumas found that many report a partial or total memory loss of their experiences.

Researchers at the University of California at Los Angeles (UCLA) School of Medicine drew their conclusions from a mailed questionnaire to which 280 women and 225 men responded. Twenty-seven percent of the respondents had survived a car accident; 23 percent had survived a natural disaster; 26 percent reported being assaulted as an adult; and 20 percent had lived through physical abuse as a child. Among all the respondents, 20 percent said that they temporarily lost all memory of a severe trauma, and 20 percent reported that they had at some time experienced a partial loss of the memory.

Blacks and Hispanics were more likely to report episodes of delayed recall than were whites. In addition, the study found that psychotherapy had little effect on the respondents' ability to recover lost memories. Instead, most people reported that an event—such as reading or seeing something in the media—helped their memory return.

Researcher Diana M. Elliot noted that the study shows that delayed recall of trauma is legitimate, despite the bad reputation it has gained from lawsuits that relied on dubious memory-recovery techniques. "The phenomenon of delayed recall of personal trauma occurs most often for violent and really distressing events encountered by both sexes."

PET Scans Reveal Schizophrenia

New medical technologies are helping researchers better understand the workings of the schizophrenic mind. Using positron-emission tomography (PET), a team of researchers in New York City and London, England, took the first-ever "snapshots" of the brains of schizophrenics who were having hallucinations. The PET scans allowed scientists to pinpoint exactly which parts of the brain were involved in schizophrenic hallucinations.

Schizophrenia, which affects approximately 2 million Americans, is a mental illness characterized by disordered thought, apathy and withdrawal, and hallucinations.

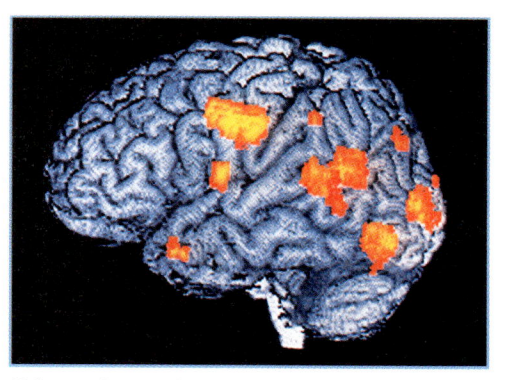

Using positron-emission-tomography (PET) scans, researchers are able to pinpoint which parts of the brain are activated during schizophrenic hallucinations.

In a study published in the journal *Nature* in late 1995, the researchers relied on a newly developed way to analyze PET scans. Standard PET scans capture pictures only of brain activity that lasts for several minutes. With the new technique, however, the scans can "see" even fleeting activity, such as that caused by hallucinations.

The PET study of six schizophrenics found that hallucinations generated activity in structures deep in the brain, including the hippocampus, thalamus, and striatum. Among other functions, these structures control brain circuits involved in emotion; meld experiences and emotions (both current and past); play a role in attention, motivation, and action; and are involved in thought and perception. The thalamus, for instance, uses sensory input to generate images of reality in consciousness and during sleep, making it likely that an abnormality in this structure might play a role in hallucinations. The researchers noted, however, that the prefrontal lobe, which oversees thoughts and actions, was not active during the hallucinations. "We've identified the areas that are responsible for the brain creating its own reality," says researcher David Silbersweig, one of the study's authors and a neurologist at New York Hospital–Cornell Medical Center.

Other research in 1995 also helped increase our understanding of schizophrenia. Scientists at Yale University in New Haven, Connecticut, for instance, found that the neurons in the brains of people with schizophrenia were denser than neurons in the brains of people without the disease. While the findings of both of these studies have not immediately led to new treatments, they do offer hope for the future.

BIOLOGY

DEPRESSION DIAGNOSIS

Contrary to a long-held belief, family physicians are effective at screening patients for depression, according to a study published in January 1995 in *General Hospital Psychiatry*. The study found that the physicians effectively distinguish between severe and mild cases of depression, but that they may not make a formal diagnosis in mild cases. This lack of a formal diagnosis may be what has led psychiatric researchers to conclude in the past that primary-care physicians do not treat depression enough.

Researchers at the University of Michigan Medical Center in Ann Arbor used interviews of 1,580 patients, from the practices of 50 family physicians, to arrive at their findings. The subjects completed a questionnaire that rated their emotional conditions; they were then rated by their own physicians. In a follow-up about two weeks later, approximately 25 percent of the subjects took part in more-extensive interviews with the researchers.

Of the 143 cases of depression that the researchers found, family physicians detected 40. However, the bulk of the undetected cases of depression were mild and did not interfere with the subjects' home or work life. The family physicians were able to diagnose three-quarters of the cases of severe major depression.

The official psychiatric definition of major depression now requires that the depression cause "significant distress or impairment" in work or social relations. Furthermore, studies have shown that supportive counseling and placebos are as effective at aiding mild forms of depression as are short-term psychotherapy or antidepressant drugs. Since family physicians have for many years used supportive counseling to help patients with mild depression, the doctors may have been unfairly accused of neglecting to treat depression, say the study's authors. "There's a huge reservoir of mildly depressed patients for whom appropriate treatment is not clear," the authors of the study state. Antidepressant drugs and psychotherapy may not be the best treatment choice for these patients.

Devera Pine

NEW ANIMAL PHYLUM

Most modern systems of biological classification divide organisms into categories based on basic differences in development and cellular structure. The highest ranking in these systems is kingdom; organisms are usually classified as Monera (single-celled microorganisms lacking organelles), Protista (protozoans, algae, slime molds), Fungi, Plantae, or Animalia. Within each of these kingdoms, scientists set up categories called phyla (singular: phylum), each of which contains all organisms that have certain fundamental characteristics in common. For example, based on arrangement of body structures and on embryological development, fish, amphibians, reptiles, birds, and mammals (humans included) are placed in the phylum Chordata.

Peter Funch, Ph.D., and Reinhardt M. Kristensen, Ph.D., of the University of Copenhagen in Denmark announced in December 1995 the discovery of an animal so different in its body structure and life cycle that they recommended it be classified in an entirely new phylum, which they call Cycliphora. The organism has been given the species name *Symbion pandora*.

Although microscopic in size, this newly discovered animal has a multicellular body

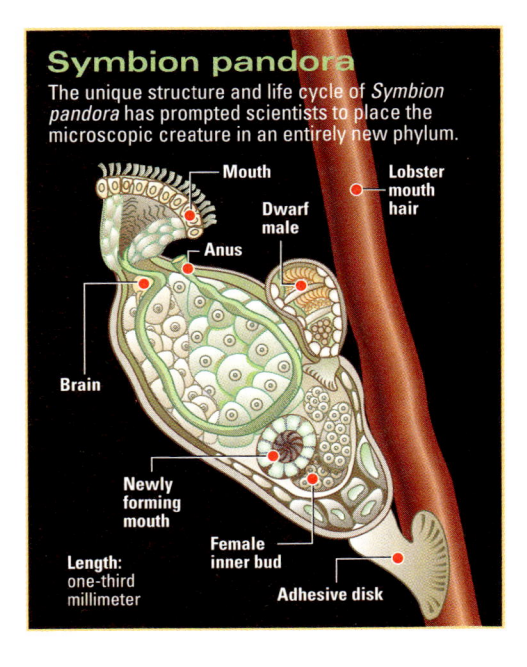

Symbion pandora
The unique structure and life cycle of *Symbion pandora* has prompted scientists to place the microscopic creature in an entirely new phylum.

Mouth · Lobster mouth hair · Dwarf male · Anus · Brain · Newly forming mouth · Female inner bud · Adhesive disk · Length: one-third millimeter that is quite complex. For most of its life, it lives attached to the mouthparts of the Norwegian lobster, where it sweeps up the lobster's tiny leftovers. The life cycle of *S. pandora* is very complex, including both asexual and sexual phases. Periodically, *S. pandora*'s feeding apparatus deteriorates completely; throughout this process, the animal remains adhered to its host, and soon grows a new feeding structure. The creature also undergoes a free-swimming dispersal phase, which is linked to the periodic molting stages of the lobster.

An area of the brain called the planum temporale is enlarged in people with perfect pitch—a rare ability to identify musical notes without any reference.

Perfect Musical Pitch

Scientists who specialize in the study of human talents are constantly searching for specific areas of the brain that may be responsible for the manifestation of various unique traits. One approach has been to look for areas of the brain that are enlarged in individuals with a particular mental ability. Such information can be deduced by the use of magnetic resonance imaging (MRI), which allows researchers to measure the volume of specific brain areas.

Musical talent is a multifaceted trait that has long attracted the attention of scientists. One characteristic that marks outstanding musical ability is perfect pitch—the ability to identify or sing any musical note without hearing a reference note. In a recent study by Gottfried Schlaug, Ph.D., and his colleagues at the Heinrich-Heine University in Düsseldorf, Germany, musicians (both with and without perfect pitch) and nonmusicians were examined using MRI. The scientists focused on a section of the cerebrum known as the planum temporale; research indicates that in all individuals, the planum temporale is larger on the left side than on the right side. However, Schlaug and his colleagues found that this asymmetry is twice as great in musicians with perfect pitch as in the others. By contrast, musicians who do not have perfect pitch show no more asymmetry than nonmusicians.

Primitive Living Organisms

Biologists have great interest in learning as much as possible about the first living material on Earth. This is not just a matter of intellectual curiosity, but rather, reflects scientists' need to understand how present-day organisms, including humans, came to develop as they did.

The most primitive organisms on Earth today are a group of single-celled forms that have been classified in the phylum Archaea within the kingdom Monera. Members of this phylum are able to live in environments characterized by high temperatures, high acidity, and the absence of oxygen—conditions similar to those that prevailed on Earth during the initial stages of organism development. Archaea species can be found in hot springs such as the Old Faithful geyser in Yellowstone National Park and in hydrothermal vents at the bottom of the oceans, where superheated water circulates through cracks in Earth's crust.

Early in 1996, Carol Bult, Ph.D., at the Institute for Genomic Research in Gaithersburg, Maryland, reported that her group had determined the entire nucleotide sequence—some 1.7 million base pairs—of the genetic material of the Archaea species *Methanococcus jannaschii*; by contrast, human beings have 3 billion base pairs. *M. jannaschii* lives near undersea hydrothermal vents. Researchers found that many of the genes are unique in their nucleotide sequences, implying that they are relatively primitive and may resemble an early form of living material.

Structures called telomeres (green tips, right) help determine the life span of human cells. Evidence suggests that cancer cells can divide rapidly by disrupting the normal shortening of telomeres.

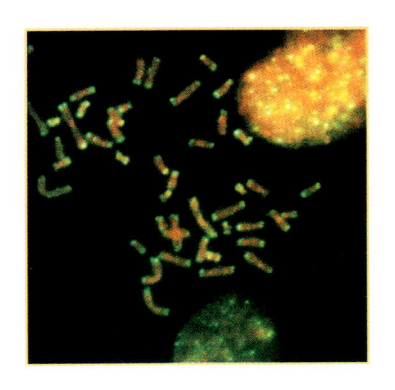

TOOLMAKING BIRDS

The ability to make and use tools for specific purposes is a well-recognized characteristic of human beings. Chimpanzees may use twigs or stems to "fish" for termites in holes. However, the chimps do not actually modify the twig or stem to make it more efficient in obtaining the termites. There are also instances in which animals use stones to break open nuts (chimpanzees) or the hard shells of crabs and oysters (sea otters), but these creatures also make no attempt to shape the piece of stone into a better hammerlike instrument.

Recently Gavin R. Hunt, Ph.D., at Massey University in New Zealand reported on his discovery of a bird species that actually modifies twigs and leaves into hooklike and sawlike tools for use in dragging insects out of holes. The bird, classified as *Corvus moneduloides*, is a member of the crow family. Hunt studied a population of the bird that lives on the New Caledonia island group, located about 900 miles (1,450 kilometers) east of Australia. The bird pulls a twig off one of a variety of trees, holds the twig with its claws, removes the leaves, and uses its beak to shape the twig so that it has a pointed hook on one end.

In fashioning a sawlike tool, the crow breaks off a piece of a stiff leaf from a particular plant and bites a series of sections out of the leaf, forming a serrated edge. Of particularly great interest is the fact that after using their tools, the birds leave them on secure perches in the trees, returning later to retrieve and reuse them. The similarity to human behavior is striking and will stimulate the search for other examples of toolmaking in the animal kingdom.

TELOMERASE AND CANCER

Some cells in the human body divide constantly, such as those of the blood, skin, and intestinal lining. Other cells, once formed during embryonic development, rarely if ever divide; nerve cells typify this type of development. Each of the remaining cell types—muscle, bone, and fat, for example—is characterized by some particular number of cell divisions. However, in all tissues of the body, it is always possible that some cells will begin to divide rapidly and in random fashion, forming cancers. Scientists have recently discovered that the number of divisions that a cell line will undergo is determined by an enzyme called telomerase, which indirectly controls cell division.

Capping each end of every chromosome are structures called telomeres. Human telomeres consist of multiple copies of a special sequence of six nucleotides that can be expressed by the abbreviation TTAGGG. Each telomere may contain as many as 2,000 repeats of this sequence. When cells are not producing telomerase, the number of TTAGGG sequences is reduced at each division. As the telomeric sections of the chromosomes become shorter, the cells divide more slowly, and finally not at all. However, if the enzyme telomerase is produced by the cells, it acts to replace the lost TTAGGG sequences, thereby permitting the cells to continue dividing.

One characteristic of cancer cells is their production of the enzyme telomerase. As a result, the telomeres of their chromosomes are maintained at the length needed for cell division to continue unabated. Research is being pursued by Calvin Harley, Ph.D., of Geron Corporation in Menlo Park, California, to discover a telomerase inhibitor that would specifically interfere with the enzyme's activities in cancer cells, but not in those normal cells that should continue to divide.

Louis Levine

BIOTECHNOLOGY

EDIBLE VACCINES

One way that a person can develop immunity to bacterial diseases such as typhoid fever or cholera or to viral conditions such as smallpox or hepatitis is by being vaccinated with a substance that resembles the particular infectious organism. This substance can be a relatively harmless but genetically similar organism, such as the cowpox virus used by the English physician Edward Jenner in 1796 to produce immunity against smallpox. Recent research has shown that the protein coat of the virus functions as the vaccine.

One of the more widespread disease-causing organisms among humans is the hepatitis-B virus. Although a vaccine for hepatitis B already exists, it is relatively costly and requires constant refrigeration. Recently, however, a group of scientists from Tulane Medical Center in New Orleans and from Texas A&M University in Houston inserted the protein-coat gene of the hepatitis-B virus into the cells of potato plants. These plants then produce the viral protein coat in their cells, despite the fact that they have no use for it. In experiments, researchers fed these bio-engineered potatoes to mice. The mice were later injected with the hepatitis-B virus and were found to be immune to the infection.

Vaccines You Can Eat

1. Virus DNA is put into potato leaf (Hepatitis B virus, Virus DNA)
2. Potato plant grown from leaf contains viral DNA in all of its cells
3. Potatoes produce viral protein
4. Mouse eats potato and develops immunity to hepatitis B virus

A CAUTIONARY WORD

An improvement in the nutritional value of our foods is one of the more important benefits that can result from the transfer of genes from one species to another. This is especially true of some plants that produce proteins known to be poor in certain amino acids. For example, the amino acid methionine, found in very small amounts in the proteins of beans and peas, is needed in large quantities for the formation of animal proteins, including those of humans.

The biotechnology company Pioneer Hi-Bred International in West Des Moines, Iowa, initiated one attempt to circumvent this situation by transferring a gene for the production of a protein rich in methionine from Brazil nuts to soybean plants. Soybeans, in various forms, are used in the diets of millions of people. However, a study at the University of Nebraska at Lincoln led by Steve L. Taylor, Ph.D., found that the protein produced by the transferred gene was the main cause of a common allergy to Brazil nuts. Should any of this protein be eaten by an allergic person, it could trigger a life-threatening reaction. Upon learning of the potential danger involved, the company canceled the gene-transfer project.

VEHICLES FOR GENE TRANSFER

Most procedures for gene transfer from one organism to another involve the use of harmless viruses as transfer agents. However, a number of mechanical procedures have also been developed. One such procedure has been devised by Wenn Sun, Ph.D., of Northwestern University in Evanston, Illinois, and Ning-Sun Yang, Ph.D., of the biotechnology company Agracetus, Incorporated, in Middleton, Wisconsin. The researchers use a gun, powered by pressurized helium, to shoot microscopic gold pellets coated with genes into skin-tumor cells of mice. The transferred genes produce proteins that stimulate a mouse's immune system to attack and destroy the tumor cells. If the procedure is approved by the U.S. Food and Drug Administration (FDA), the scientists plan to test it on human tumors.

Another mechanical procedure for gene transfer involves the use of liposomes—

spherical bodies of fatty material to which DNA can be attached. Liposomes are easily absorbed by cells and can serve as vehicles for the transfer of any attached genes. Natasha Caplen, M.D., and her colleagues at Royal Brompton Hospital in London, England, have experimented with this procedure in the treatment of patients suffering from cystic fibrosis (CF).

People with CF are missing a protein that keeps the mucous lining of the lungs moist and free-flowing. By the use of nasal sprays containing specially prepared liposomes, the normal form of the CF gene has been successfully transferred to the cells of the patients' nostrils, where the missing protein was subsequently produced. Researchers have not yet achieved similar results by delivering liposomes to the cells of the patients' lung passages, which would be necessary to actually treat cystic fibrosis.

Reproductive Biology

One reproductive option available to some infertile couples is a procedure called in-vitro fertilization (IVF). In this process, a mature egg (ovum) that has been shed from a woman's ovary is obtained from her fallopian tube and then fertilized in the laboratory using sperm from her partner.

In 1996, John J. Eppig, Ph.D., and Marilyn J. O'Brien, Ph.D., at the Jackson Laboratory in Bar Harbor, Maine, reported success in producing a mouse by IVF, using an egg that had earlier been obtained—while still an immature egg, or oocyte—directly from an ovary. Through the use of nutrients and hormones, the oocyte was stimulated to mature into an ovum ready for fertilization. After fertilization, the developing embryo was implanted into the oviduct of a different female mouse that served as a surrogate mother.

If researchers can accomplish a similar feat with other species, the procedure would help those involved in breeding programs designed to save endangered species from extinction. It would also be of tremendous benefit to human females who, for medical reasons, must have their ovaries removed, but want to be able to have children through IVF at some later time.

Louis Levine

Book Reviews

Animals and Plants

● Ackerman, Diane. *The Rarest of the Rare: Vanishing Animals, Timeless Worlds.* New York: Random House, 1995; 184 pp.—A collection of essays about short-tailed albatrosses, monarch butterflies, and other interesting species.

● Hoyt, Erich. *The Earth Dwellers: Adventures in the Land of Ants.* New York: Simon & Schuster, 1996; 319 pp., illus.—A fascinating ant's-eye view of life.

● Little, Douglas. *The Little Known Facts About Hippopotamuses.* New York: Ticknor & Fields Books for Young Readers, 1995; 48 pp., illus.—An amusing book that raises appropriate questions for younger readers, mostly about mammals.

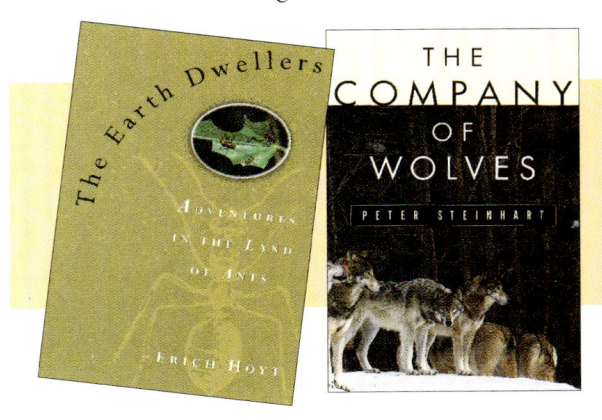

● Long, Matthew, and Thomas Long. *The Spectacled Bear and Other Curious Creatures.* San Francisco: Chronicle Books, 1995; illus.—A lively and informative look at flying foxes, frilled lizards, and other beasts. For younger students; includes a habitat map and glossary.

● Steinhart, Peter. *The Company of Wolves.* New York: Knopf, 1995; 240 pp.—A look at why wolves have taken center stage in the debate over the preservation and use of the wilderness.

● Tenenbaum, Frances, et al., eds. *Taylor's Master Guide to Gardening.* Boston: Houghton Mifflin, 1994; 624 pp., illus.—A profusely illustrated encyclopedia of the best plants to grow in every region, with details on cultivation.

Astronomy and Space Science

- Davies, Paul. *Are We Alone? Implications of the Discovery of Extraterrestrial Life.* New York: Basic Books, 1995; 160 pp., illus.—Cosmology for amateurs; a look at what might happen to philosophy, religion, and science if intelligent life were found elsewhere in the universe.
- Levy, David H. *Impact Jupiter: The Crash of Comet Shoemaker-Levy 9.* New York: Plenum, 1995; 300 pp., illus.—One of the world's foremost amateur astronomers describes the personal thrill of discovery and the efforts of scientists around the globe to record and understand the crash of the comet he and his colleagues discovered. 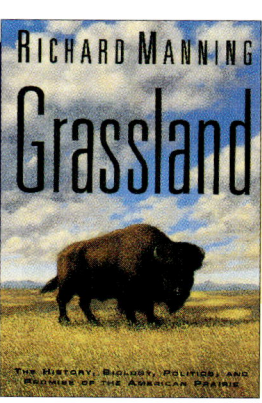
- Petersen, Carolyn C., and John C. Brandt. *Hubble Vision: Astronomy with the Hubble Space Telescope.* New York: Cambridge University Press, 1995; 256 pp., illus.—A superbly chronicled tour of the cosmos through the eyes of the Hubble Space Telescope.
- Vaughan, Diane. *The Challenger Launch Decision: Risky Technology, Culture, and Deviance at NASA.* Chicago: University of Chicago Press, 1996; 575 pp., illus.—An in-depth look at how the culture of the National Aeronautics and Space Administration (NASA) may have sent the space shuttle *Challenger* and its crew to their fate.

Earth and the Environment

- Chase, Alston. *In a Dark Wood: The Fight Over Forest and the New Tyranny of Ecology.* New York: Ticknor & Fields Books for Young Readers, 1995; 535 pp.—A gripping account of the battle over the forests of the Pacific Northwest.
- Cohen, Joel E. *How Many People Can the Earth Support?* New York: Norton, 1995; 532 pp., illus.—The author tackles the question of the carrying capacity of Earth.
- Earle, Sylvia A. *Sea Change: A Message of the Oceans.* New York: Putnam, 1995; 336 pp., illus.—An oceanographer talks about modern underwater exploration and her ideas for protecting the world's seas.
- Manning, Richard. *Grassland: The History, Biology, Politics, and Promise of the American Prairie.* New York: Viking, 1995; 320 pp.—A journey across the vast Midwest, tracing how America's prairie has been used and misused, and describing efforts to restore it.
- Raeburn, Paul. *The Last Harvest: The Genetic Gamble that Threatens to Destroy American Agriculture.* New York: Simon & Schuster, 1995; 269 pp.—A call to diversify crops to protect the world food supply.
- Salvadori, Mario, and Matthys Levy. *Why the Earth Quakes: The Story of Earthquakes and Volcanoes.* New York: Norton, 1995; 256 pp., illus.—Earthquakes, volcanoes, and other phenomena explained by the authors of *Why Buildings Fall Down* (1992).

Human Sciences

- Clark, William R. *In the Defense of Self: The Double-Edged Sword of Immunity.* New York: Oxford University Press, 1995; 240 pp., illus.—A summary of the rapidly developing field of immunology.
- Jamison, Kay R. *An Unquiet Mind.* New York: Random House, 1995; 223 pp.—A

world authority on manic depression describes her own battle with the disease.
- Maynard-Moody, Steven. *Dilemma of the Fetus: Fetal Research, Medical Progress, and Moral Politics.* New York: St. Martin's Press, 1995; 235 pp., illus.—A guide to the emotional debate over fetal-tissue research.
- Nathan, David G. *Genes, Blood, and Courage: A Boy Called Immortal Sword.* Cambridge, Massachusetts: Belknap Press, 1995; 288 pp.—A tale of efforts to treat and cure thalassemia, a rare blood disease.
- Scarf, Maggie. *Intimate Worlds: Life Inside the Family.* New York: Random House, 1995; 466 pp.—Eight years of research led to this explanation of what makes a good family.

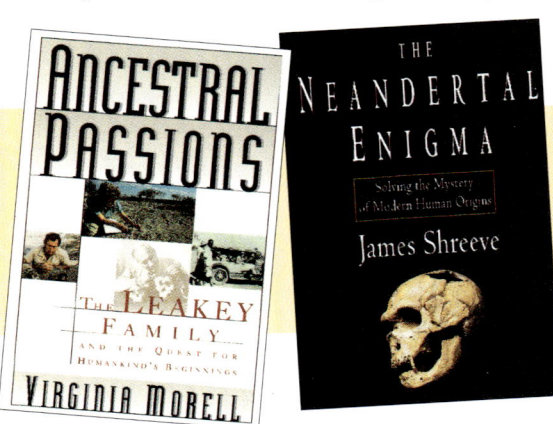

PAST, PRESENT, AND FUTURE
- Dawkins, Richard. *River Out of Eden: A Darwinian View of Life.* New York: Basic Books, 1995; 172 pp., illus.—An exploration of modern evolutionary theory, including the strategies various species have developed to survive.
- Eldredge, Niles. *Dominion.* New York: Henry Holt, 1995; 190 pp.—Explains how humankind became the first species to live outside the local ecosystem and what our evolutionary future may hold.
- Leakey, Richard E., and Roger Lewin. *The Sixth Extinction: Patterns of Life and the Future of Humankind.* New York: Doubleday, 1995; 271 pp., illus.—A primer on the new state of evolutionary and ecological science and an argument in favor of biodiversity.
- Morell, Virginia. *Ancestral Passions: The Leakey Family and the Quest for Humankind's Beginnings.* New York: Simon & Schuster, 1995; 638 pp., illus.—Traces the history of the family that has dominated the study of paleoanthropology in recent decades.
- Shreeve, James. *The Neanderthal Enigma: Solving the Mystery of Human Origin.* New York: Morrow, 1995; 369 pp., illus.—A clear presentation of the debate over the place of the Neanderthals in human history.

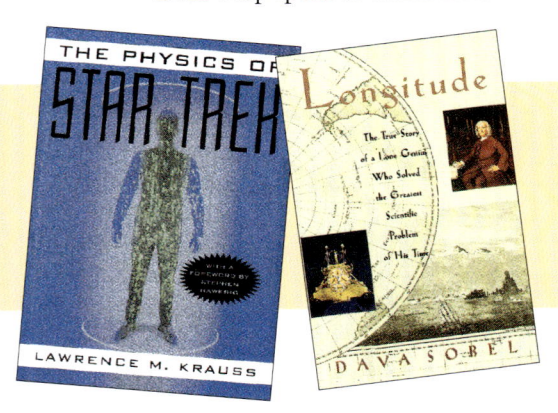

PHYSICAL SCIENCES
- Krauss, Lawrence M. *The Physics of Star Trek.* New York: Basic Books, 1995; 188 pp., illus.—A physicist explains his field by describing what is and is not possible in the television universe of the starship *Enterprise*, from warp speed to wormholes.
- Paulos, John A. *A Mathematician Reads the Newspaper.* New York: Basic Books, 1995; repr. 1996; 180 pp.—Features 52 vignettes deconstructing many mathematical "facts" commonly seen in the newspaper.
- Rhodes, Richard. *Dark Sun: The Making of the Hydrogen Bomb.* New York: Simon & Schuster, 1995; 731 pp., illus.—A powerful history of the development of the hydrogen bomb from the author of the award-winning *The Making of the Atomic Bomb* (1988).
- Sobel, Dava. *Longitude: The True Story of a Lone Genius Who Solved the Greatest Scientific Problem of His Time.* New York: Walker, 1995; 200 pp.—How James Harrison, an unschooled woodworker, solved a problem once considered synonymous with impossibility.
- Vogel, Shawna. *Naked Earth: The New Geophysics.* New York: Dutton, 1995; 240 pp.—The author gets under the hood of

"Spaceship Earth" and explains its inner workings, summarizing how modern scientific research has revealed the highly active history of our planet.

TECHNOLOGY

- Bourdon, David. *Designing the Earth: The Human Impulse to Shape Nature.* New York: Harry N. Abrams, 1995; 224 pp., illus.—Why and how Earth has been transformed by human efforts.
- Burke, James, and Robert Orenstein. *The Axemaker's Gift.* New York: Grosset/Putnam, 1995; 368 pp.—Explores the way people have used the gifts of the inventors, scientists, and innovators of world history, and concludes that only technology can rescue humans from the perils of technology.
- Meikle, Jeffrey L. *American Plastic: A Cultural History.* New Brunswick, New Jersey: Rutgers University Press, 1995; 403 pp., illus.—The technical history of a material that has transformed American life.
- Moody, Fred. *I Sing the Body Electronic.* New York: Viking, 1995; 311 pp.—A chronicle of a year in the life of a small team of software designers and developers creating a multimedia encyclopedia for children.
- Negroponte, Nicholas. *Being Digital.* New York: Knopf, 1995; repr. 1996; 243 pp.—The author argues that machines are taking over the world.
- Petroski, Henry. *Engineers of Dreams: Building Great Bridges.* New York: Knopf, 1995; 496 pp., illus.—Places the tradition of American bridge building in perspective, describing the achievements of civil engineers who were both dreamers and practical people.
- Turkle, Sherry. *Life On the Screen: Identity in the Age of the Internet.* New York: Simon & Schuster, 1995; 347 pp.—Case studies drawn from two decades of ethnographic fieldwork among regular and committed computer users.

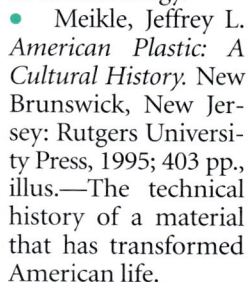

Jo Ann White

BOTANY

A GIANT LEAP FOR EVOLUTION

Charles Darwin put evolution theory on the map, but he may have been wrong when he said that evolution is made up of many small changes instead of a few big ones. H.D. Bradshaw and other botanists at the University of Washington in Seattle have found that just a few genetic changes can produce a new species.

In nature, two species of monkeyflower, *Mimulus lewisii* and *M. cardinalis*, cannot cross, because only bees pollinate *M. lewisii* and only hummingbirds pollinate *M. cardinalis*. The two have evolved flower forms that attract their specific pollinator only. For example, bee-pollinated *M. lewisii* has a lower lip on which bees can land, and yellow marks on its wide throat to guide bees to the nectar; hummer-pollinated *M. cardinalis* is red (hummingbirds love it; bees can't see it), and has a more tubular throat to accommodate the birds' beak and exclude bees.

Bradshaw and his colleagues crossed the two and, using genome mapping, figured out which genes influenced eight floral traits in the hybrid resulting from the cross, including the shape, size, and color of the flower; the amount and concentration of nectar produced; and the length of the flower's sexual parts.

For each trait, the scientists found that at least one small segment of genetic material accounted for 25 percent or more of the physical change in the offspring. One trait—the yellow marks on the throat of *M. lewisii*—mapped to just one chromosome region. In other words, the researchers found that major genetic changes mapped to just a few genes.

The research is important because it represents the first in-depth study into how much genetic change is required for the evolution of a new species. While there is still plenty of room for studying the impact of major genes, this research shakes Darwin's theory and scientists' long-standing belief that new species result from a large number of genetic mutations.

Boarding Ants Feed Epiphytes

Since they live in trees, away from moist, nutrient-rich soil, epiphytic plants have become adept at deriving the necessities of life from the air. But a new study shows that at least one species of epiphyte is also exploiting its ant guests for much of the nitrogen and carbon it needs.

Biologist Kathleen K. Tresder and coworkers from the University of Utah in Salt Lake City have found that one Malaysian epiphyte, *Dischidia major*, shelters *Philidris* ants in special sac-shaped leaves. In exchange for board, the ants leave feces, ant corpses, and other debris that supply 29 percent of the nitrogen the plant uses—the plant absorbs the nitrogen through special roots that grow through the leaf's opening and into the debris. In addition, 39 percent of the carbon the plant uses comes from the carbon dioxide exhaled by the ants. The carbon diffuses into the leaves through small pores in the leaf surface called stomata.

Although symbiotic, or mutually beneficial, relationships between plants and ants have been documented before, this is the first time the relationship has had hard numbers attached to it, giving a better idea of just how much plants benefit.

Sweetly Surviving Drought

Most plants store complex carbohydrates—starches—to provide energy during times of rapid growth or when unfavorable weather occurs. But botanists have been baffled as to why about 15 percent of plants store energy primarily in the form of water-soluble fructans, a simple sugar made up of several molecules of fructose, the sugar in fruit. A recent study by Elizabeth Pilon-Smits and colleagues at the University of Utrecht in the Netherlands suggests that fructans might help plants survive drought.

First, the researchers inserted a gene into tobacco (which normally does not make fructans) that made the tobacco produce the sugars. They then measured how well the altered tobacco grew under laboratory-controlled drought conditions, and compared it to normal tobacco. They found that the fructan-producing tobacco plants grew 55 percent faster, and weighed 33 percent more

Flower Fixation

The sudden appearance of flowers 150 million years ago has baffled scientists for decades. Now, geneticist Elliot Meyerowitz (below) and his colleagues at the California Institute of Technology in Pasadena are trying to find out how flowers evolved by studying the blossoms of a variety of mustard known as *Arabidopsis thaliana*, the workhorse plant of botanical genetics. Already, Meyerowitz has discovered the genes that tell certain cells to become petals, others to become stamens and carpels, and still others to become sepals. By inducing mutations in his blossoms, Meyerowitz has uncovered striking similarities in the way that all flowers develop, and may someday learn how flowers came to be in the first place.

when fresh and 59 percent more when dried. Most of the weight difference was in the roots. Under normal growing conditions, there was no difference in the rate of growth and weight of fructan and nonfructan tobacco plants.

Pilon-Smits and her coworkers still do not know just how fructans help plants thrive under water stress. They theorize that fructans might guard plant membranes from the adverse effects of drought, that the larger roots might increase water uptake, or that the fructans might enable plants to adjust osmotically. In any case, the implications for agriculture are enormous: since drought limits crop production worldwide, genetically altering plants to produce fructans could improve yields.

Putting Friendly Fungi to Work

As environmental problems and health concerns force an increasing number of synthetic pesticides off the market, plant researchers are looking to fungi to thwart pests. Separate research teams are working to use fungi to control weeds and fruit rot.

Kiwifruit, like many other types of fresh fruit, is prone to rot when stored. U.S. Department of Agriculture (USDA) researcher Horace Cutler, cooperating with food researchers in New Zealand, exposed thousands of kiwifruit to *Botrytis cinerea*, a mold-causing fungus. He then treated the kiwi with different concentrations of the fragrant extract of a fungus called *Trichoderma harzanium* to see if it would inhibit the mold. The extract—called 6-pentyl-alpha-pyrone, or 6-PAP for short—occurs naturally in peaches and also can be synthesized.

Cutler, of the Russell Agricultural Research Center in Athens, Georgia, found that nearly all the kiwifruit treated with 6-PAP was free of mold after one year in refrigeration. Only a small amount—less than 2 percent—of the fruit molded, and that fruit had been treated with the weakest concentration of 6-PAP. In the untreated control, about 55 percent of the fruit rotted. Not only did the treated kiwifruit stay mold-free, but it also retained its just-picked firmness, juiciness, and flavor. The researchers are now testing 6-PAP to see whether it can be used commercially for kiwi and other fruit; if so, it will provide a safer alternative to the chemical fungicides now used.

Plant microbiologists know that the fungus *Colletotrichum truncatum* attacks the weed hemp sesbania *(Sesbania exaltata)*, a southern weed that infests cotton and soybean fields. But for the microbe to have commercial value, it has to survive the dry conditions required to store, ship, and sell it, as well as the dry weather that may set in after the fungus is applied to farm fields. USDA microbiologist Mark A. Jackson and plant pathologist David A. Schisler are experimenting with techniques for mass-producing the sclerotium form of the fungus. Fungi change to hard, dry sclerotia under adverse conditions, such as intense heat and lack of moisture. But when water hits them, they jump back into action, growing and reproducing and, in the case of *C. truncatum*, infesting hemp sesbania. The researchers found that they can produce the most sclerotia by growing the fungus in a liquid medium, and that, when dried, most of it survives for up to eight months.

Insect Pests 1, Genetic Engineers 0

After a few days of being chewed on by insects, many plants fight back by naturally producing a protease inhibitor that interferes with the insects' digestive proteases—enzymes that enable the insects to digest protein. In hopes of preventing insect damage, plant geneticists have engineered experimental plants to *always* produce the protease inhibitor, even before an attack. But it looks like the insects are having the last laugh, according to Dutch researcher Maarten A. Jongsma and coworkers. The team inserted a gene into tobacco plants that enabled the plants to constantly produce the protease inhibitor, PI2; they then fed the leaves to beet armyworms *(Spodoptera exigua)*. Rather than becoming weak or dying off, the worms compensated by doubling the production of another digestive protease, which PI2 did not inhibit. The conclusions: genetically engineering insect resistance into plants may be more difficult than anticipated.

Erin Hynes

Chemistry

Superconductivity

Superconducting materials offer no resistance to the flow of electric current. Now research has yielded new superconducting materials capable of carrying higher currents than ever before. Scientists at IBM's Thomas J. Watson Research Center in Yorktown Heights, New York, reported developing a mercury superconducting film capable of carrying record-high current densities (about 10^5 amperes per square centimeter) at 110° K (-163° C or -261° F). This reported density value is at least an order of magnitude higher than what has been achieved with other (bismuth- or thallium-based) superconducting films. The discovery is expected to bring commercial applications of such films a step closer to reality.

In a second development, researchers at Los Alamos National Scientific Laboratory in New Mexico built a 1- to 2-micrometer (roughly 0.00005-inch)-thick superconducting tape capable of carrying more than 1.3 million amperes per square centimeter at liquid-nitrogen temperature (-196° C or -321° F). This current-density value is nearly 100 times more than superconducting tapes and wires developed elsewhere. Modified versions of such tapes could hasten the advent of superconducting motors, generators, and magnetic trains.

Novel High-Pressure Reaction

The explosive reaction of oxygen (O_2) and hydrogen (H_2) at ambient pressure to make water (H_2O) is well known to scientists. However, chemists at the University of Paris made the surprising discovery that the two gases reacted differently when pressurized to 76,000 atmospheres (about 1.1 million pounds per square inch) at room temperature. The gases clustered into a new 14-atom compound—$(O_2)_3(H_2)_4$—containing three molecules of oxygen and four molecules of hydrogen.

These findings illustrate how chemical reactions can produce radically different outcomes under high pressure. The discovery could lead to a novel energy-storage system (including better rocket fuels) or serve as a model for the interiors of the outer planets, such as Jupiter, whose cores contain much pressurized hydrogen, oxygen, and ice.

New Element

Following the discovery of elements 110 and 111 during 1994, scientists identified yet another element: a heavier relative of the metals zinc, cadmium, and mercury. The as-yet-unnamed element was created in early 1996 by a team of German, Russian, Slovak, and Finnish physicists, who made the element by bombarding lead (element 82) with zinc (element 30) until two atoms fused as a new substance. Element 112 contains as many protons as lead and zinc combined. It is unlikely to have any immediate practical application beyond the research laboratory, according to the scientists.

New Discoveries

- *Protein-based Optoelectronic Device.* Researchers at Syracuse University in New York report developing a prototype optoelectronic device made of cubes containing bacteriorhodopsin—a light-harvesting protein found in a bacterium that grows in salt marshes. The device may increase computer speed and memory by virtue of its ability to store up to 300 times more information than existing devices used in computers and other electronic equipment.

The device, which relies on laser beams to read and write information, uses the bacteriorhodopsin as an optical computational gate. Given their biological origin, protein cubes are considered by scientists to be more environmentally friendly than semiconductors.

- *Splitting Nitrogen Gas.* Despite the fact that nitrogen gas (N_2) is the most abundant element in Earth's atmosphere (occupying 78 percent of air by volume), it has been difficult for scientists to use it directly for industrial purposes. While nature utilizes gaseous nitrogen by "fixing" it with biological processes, industrial chemists have had few options so far.

Catalina E. Laplaza and Christopher C. Cummins at the Massachusetts Institute of Technology (MIT) in Cambridge discovered a novel way to split nitrogen gas. They used an intermediate molybdenum-containing molecule to cleave the triple bond that holds the two nitrogen atoms together. The discovery makes the element more accessible for industrial applications.

- **New Role for Proteins.** Scientists recently discovered that proteins play a role in the processing of cellular information in the human body. Dennis Bray, a chemist at the University of Cambridge in England, reported in *Nature* that proteins in cells of plants and animals carry information from the plasma to the genome, and are able to perform a variety of logical or computational operations.

"In unicellular organisms, protein-based circuits act in place of a nervous system to control behavior," reported Bray. "In the larger and more complicated cells of plants and animals, many thousands of proteins functionally connected to each other carry information from the plasma membrane to the genome," he added.

- **Self-cleaning Coating.** Chemical engineers Adam Heller and Yaron Paz at the University of Texas in Austin announced the development of a self-cleaning coating for glass surfaces. The scientists told the August 1995 national meeting of the American Chemical Society that the titanium oxide coating reacts with sunlight to break down and dissolve organic residues, dirt, and grime.

In March 1995, terrorists killed a dozen people and injured thousands more by releasing nerve-gas chemicals in the Tokyo, Japan, subway system.

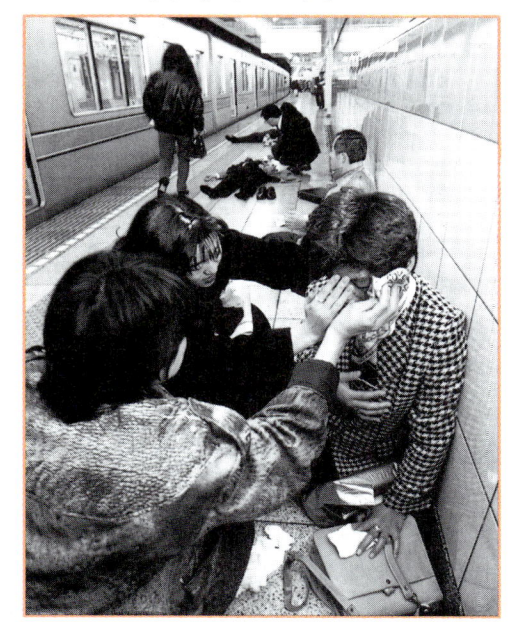

The self-cleaning coating also works on painted surfaces, and could lead to the development of new kinds of latex paints for household walls. The scientists envision wall coverings that can be cleaned of fingerprints and food stains by simply shining a light on the surface.

- **New Insight into Vitamins' Role.** Chemists at the University of Pittsburgh in Pennsylvania found that a vitamin E metabolite—called vitamin E quinone—is a potent anticoagulant that may be responsible for the vitamin's role in preventing heart attack and stroke.

Vitamin E quinone inhibits the chemical that controls blood clotting (vitamin K–dependent carboxylase), the scientists reported. Their recommendation: explore vitamin E quinone as an alternative or supplement to anticoagulants (such as warfarin) that are now given as a preventive measure to victims of heart attack and stroke.

The chemists also found new data about vitamin K's role in preventing hemorrhaging. Their research, published in the journal *Science*, details how an energy-transfer mechanism in vitamin K allows it to trigger coagulation. The research could extend the clinical use of vitamin K.

CHEMICALS LINKED TO TERRORISM

Two major terrorist attacks linked chemicals to human tragedies. The chemical sarin, a nerve-gas agent, was suspected to have been used in an attack in the Tokyo, Japan, subway system on March 20, 1995, that killed 12 people and injured more than 5,000. Sarin is an organic chemical that can be made using pesticide feedstocks. It kills by inhibiting cholinesterase, an enzyme the body produces to break down acetylcholine (a chemical that helps some neurons transmit signals to other neurons).

Ammonium nitrate, a common fertilizer, and fuel oil were identified as the chief ingredients of the bomb that devastated an Oklahoma City, Oklahoma, federal office building on April 19, 1995, killing 169 people. The event triggered calls to add inert materials to ammonium nitrate to make it less explosive, or to introduce nonexplosive fertilizers.

Vinod Jain

CIVIL ENGINEERING

AWARD-WINNING PLATFORM

An innovative deep-sea oil-drilling platform earned the 1995 Outstanding Civil Engineering Achievement Award, an annual honor given by the American Society of Civil Engineers. Shell Oil's Auger tension-leg platform, deployed in the Gulf of Mexico, became the world's tallest human-made structure. If it were located on land, it would stand about 3,280 feet (1,000 meters) tall—more than twice the height of Chicago's Sears Tower. It is also the world's deepest platform, reaching the seabed through 2,860 feet (870 meters) of water.

Engineers took six years to deploy the $1.2 billion platform, only the fourth of its kind. It is essentially a two-deck steel space frame sitting atop a hull. The hull is placed on four cylindrical columns that are connected to pontoons. This floating structure is held in place by lateral mooring lines and tendons latched into foundation templates.

WORLD'S TALLEST BUILDING

The Petronas Towers in Kuala Lumpur, Malaysia, appeared ready to take over the title of world's tallest building. The 88-story twin commercial towers, which are about 1,480 feet (451 meters) tall, were designed by Cesar Pelli. The framework of the $1.6 billion project was finished early in 1996; construction was expected to be completed by year's end.

There was some disagreement over whether the Petronas Towers, topped by a massive pinnacle, actually would displace the Sears Tower as the world's tallest skyscraper. Whatever the outcome, an even larger tower was under construction in China, due to be completed in 1997.

HIGHWAY MONITORING

Construction began in 1995 on an automated fog-detection and -warning system that will monitor a heavily traveled stretch of Interstate 75 in Georgia. Developed by the Georgia Department of Transportation (DOT) and the Georgia Tech Research Institute in Atlanta, this prototype could lead to

Kuala Lumpur, the capital city of Malaysia, is the site of the Petronas Towers—88-story twin skyscrapers whose soaring pinnacles make them the tallest buildings in the world.

automated visibility monitoring in any hazardous area throughout the country.

The system integrates 19 sensors, five sets of traffic-speed-monitoring loops, several weather-forecasting instruments, and an on-site computer. When the system detects a visibility problem, such as fog or heavy rain, it will notify authorities and post information on light-emitting-diode (LED) signs built over traffic lanes. The signs will warn motorists of potential hazards, suggesting reduced speeds and providing detour information when appropriate.

Boston's Central Artery Project

In December 1995, the first part of the most expensive U.S. public-works project to date was opened. The $1.95 billion, 1.6-mile (2.5-kilometer) Ted Williams Tunnel beneath Boston Harbor doubles the number of lanes connecting the city to Logan Airport and surrounding communities. The centerpiece of the Central Artery project, which is expected to cost more than $10 billion, is construction to take an elevated, eight-lane expressway that runs through the heart of Boston and rebuild it into an underground highway. The project, plagued with delays and cost overruns, is not expected to be completed until at least 2004.

New Waterway

More than 25 years after being authorized by the U.S. Congress, the Red River Waterway officially opened in 1995. The channel links Shreveport, Louisiana, to the Mississippi River. The $1.8 billion waterway, built by the U.S. Army Corps of Engineers, features a 200-foot (61-meter)-wide channel that straightens the river's course and makes Shreveport a viable destination for barge traffic.

New channels were cut across the river's curvy meanders by digging narrow pilot channels and letting erosional forces enlarge the channels to the width desired by engineers. Five sets of locks and dams help control the lower part of the river, lifting it 141 feet (43 meters). The locks are large enough to accommodate a string of six barges and a tugboat. The waterway shortened the 280-mile (450-kilometer) river by about 50 miles (80 kilometers).

Aqueduct Expansion

Construction was proceeding on a $300 million expansion of the California Aqueduct, which brings water from northern California to the metropolitan areas of Los Angeles and San Diego. The project, slated for completion in the summer of 1996, will bring treated water through 101 miles (163 kilometers) of pipeline, seven concrete water tanks, three pumping plants, and a water-treatment plant.

In 1995, a project milestone was reached with the completion of a $4.5 million upgrade of the Cuesta Tunnel. The 1-mile (1.6-kilometer)-long tunnel was originally excavated in 1941, and contained one open-flow channel 5 feet (1.5 meters) wide and 8.5 feet (2.6 meters) high. The upgrade more than doubles the tunnel's capacity.

Inventive Cleanup

Engineers at Lawrence Livermore National Laboratory in Livermore, California, successfully applied a new soil-cleanup treatment after learning that several thousand gallons of gasoline had leaked from an underground tank and pipe system at the lab. The gasoline had seeped down to the water table and spread along the top of the saturated zone.

Remediating the spill by conventional "pump-and-treat" methods would have taken many years to complete. Instead, engineers tried an extraction method called dynamic underground stripping to remove the gasoline quickly, causing less harm to the environment. The project, sponsored by the U.S. Department of Energy, removed virtually all of the gasoline in a matter of months—at a cost of $11 million.

Dynamic underground stripping combines on-site steam sweeping, electrical-resistance heating, and liquid and vapor extraction to recover contaminants from the subsurface. Steam-injection wells are configured around the subsurface plume to be extracted. Steam is then injected into permeable layers through wells constructed around the perimeter of the plume. The steam displaces contaminants and sweeps them toward extraction wells.

Teresa Austin

Communication Technology

Telecommunications Legislation

Undoubtedly the biggest news in communication technology was the passage in February 1996 of the Telecommunications Act, which overhauled the U.S. telecommunications industry. The legislation was debated throughout 1995, although discussions had been stalemated by budget talks and by two shutdowns of the federal government. Final action on the bill had been stalled until other budget issues were settled.

The legislation—the first major revision of U.S. telecommunications law in 62 years—had as a principal goal the promotion of competition among communications companies. It was hoped that increased competition would lead both to better prices and services for U.S. consumers and to the enhanced ability of businesses to compete abroad. Many traditional barriers between the various categories of communications companies fell, with cable-television operations allowed to provide phone service, long-distance phone companies permitted to compete in local markets, and the "Baby Bells" able to offer long-distance service. Network broadcasting companies would also be allowed to own more television and radio stations.

The main provisions of the legislation require local phone companies to share their networks with competitors, and to have competition before offering long-distance service; reduce cable-television price controls; remove many limits on television- and radio-station ownership; ease restrictions on foreign ownership of telecommunications carriers; set rules for local phone companies offering video services; restrict transmission of obscene material on computer networks; and, on all new television sets, require the installation of a so-called "V-chip" that allows viewers to block out violent or sex-oriented programs. The legislation mandates that the chip be installed in all televisions manufactured in or imported for use in the United States.

The most controversial point of the legislation is the provision restricting the transmission of indecent material over computer networks. The law was immediately challenged by civil-liberties groups, which filed a lawsuit arguing that the broad definition of indecency would lead to infringements on free speech. In June 1996, a panel of three federal judges unanimously ruled the law's attempt to regulate on-line content unconstitutional, stating that the Internet was a medium of "historic importance," and thus should be protected as much as possible from federal regulation.

Area-Code Creation

In the 1990s, there has been unprecedented growth in the demand for new telephone numbers, a need fueled by the increasing use of pagers, cellular phones, computer modems, and fax machines. As a result, the telecommunications industry introduced new three-digit area codes. Between September 1995 and December 1996, 16 new codes were scheduled to be introduced.

The new area codes have caused headaches for businesses and individual customers. Unlike traditional area codes, which have either "0" or "1" as their middle digits, the new codes can have *any* middle digit. Tens of thousands of computerized switchboards called private branch exchanges (PBXs) have long been programmed to recognize as area codes only three-digit numbers with "0" or "1" as the middle digit. The PBXs thus could not complete calls using the new codes, and needed to be updated or replaced. In states where the new codes were introduced, numerous businesses sued to block their introduction, but lost.

Technology

Cellular phones and pagers have burgeoned in popularity. The Cellular Telecommunications Industry Association reported during 1995 that 1 in 10 Americans owned a cell phone, with more than 25,000 signing up daily for service.

A new twist in wireless technology is called personal communications services (PCS). This technology eventually will make possible "personal" phone numbers: an individual could have one phone number that would access his or her home telephone, cell phone, or pager. A preliminary step in this direction was taken in September 1995 with the introduction by SkyTel Corporation of the first nationwide two-way paging system. The SkyTel pager allows the user to receive text messages. He or she can then respond with a preprogrammed reply or plug the pager into a computer and type a brief reply. The reply can be directed to another pager or sent over the Internet to another computer.

Market tests of modems that operate over cable lines are under way. Up to 1,000 times

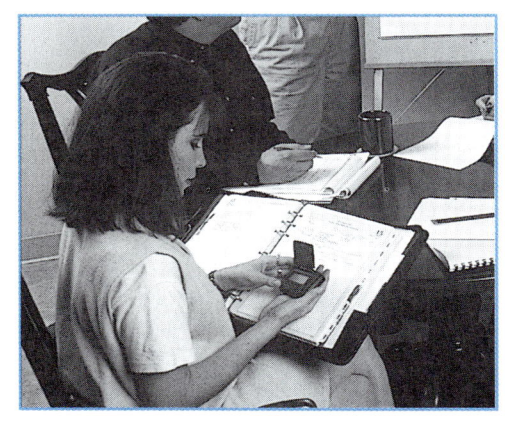

In 1995, SkyTel Corporation unveiled the first nationwide two-way pager. With this system, users can send a text message to a computer or another pager.

faster than most telephone modems, this communication technology soon may allow computer users to go on-line using existing cable-television systems.

A new product further blurring the lines between categories of telecommunications products is the screen phone. First marketed to help consumers take advantage of features like caller ID—for which a readout screen is needed—the phones have rapidly become more sophisticated. Next-generation models will include small keyboards and allow consumers to do electronic banking and bill paying and send and receive information or E-mail through the Internet. Others will feature credit-card swipers, touch screens, or voice recognition.

Robert C. Fiero, Jr.

COMPUTERS

THE INTERNET BOOM

The Internet global computer network continued to be a major driver in the growth of the personal computer (PC) market. The Software Publishers Association reports that the number of households with computers increased by more than 1 million, to an estimated 33.9 million, between 1994 and 1995. About 34 percent of U.S. homes now have personal computers. Seventy percent of PC owners report owning modems—a necessity for hooking up to the Internet. By one estimate, use of the Internet in 1995 grew at a rate of 20 percent every *month*. One part of the Internet, the graphics-oriented World Wide Web, was growing by leaps and bounds. Many companies (and a surprising number of individuals) have a Web home page, a sort of digital storefront where users can access whatever content—including text, sound, and graphics—that a provider creates.

A number of new software developments are helping to make the Internet more accessible and powerful. The development of HTML (Hypertext Markup Language) enables authors to link words or pictures in one document to other documents. HTML documents are created and interlinked using a standard called hypertext transfer protocol (http). These linked documents eventually formed what is now called the World Wide Web, the most user-friendly part of the Internet. "Browser" programs such as Netscape's Navigator and Microsoft's Explorer were developed to help users navigate the Web, and search programs like Yahoo, Excite, and WebCrawler help users find information on specific subjects.

A more recent innovation is Sun Microsystems' new software language, Java, which enables Web pages to include tiny application programs called applets. Java applets can help bring a page alive with animation and moving graphics.

In 1996, the popularity of the Web was reflected in the refocusing of commercial online services to serve as Internet gateways. Only a year or so earlier, it had seemed that such services were themselves the wave of the future, and would set the standard for electronic publishing and commerce. In March 1996, America Online (AOL) formed an alliance with Netscape that would lead to the distribution of Navigator software to AOL users. Microsoft announced that its on-line service—introduced with its Windows 95 operating system—would move entirely to the Internet by early 1997; CompuServe and Prodigy were said to be moving in the same direction.

On-line Banking and the Smart Card

On-line banking gained a foothold in 1995, and "E-cash" and "smart cards" entered the picture. Most major banks have established on-line banking options, which have become increasingly popular with consumers. By late in 1995, about 750,000 U.S. households were banking electronically, doing everything from verifying checking-account balances and paying bills to applying for mortgages on-line. Such services are expected to find a much larger audience in the future.

Taking the on-line services trend one step further, Security First Network Bank—based in Pineville, Kentucky—became the first "virtual bank" in late 1995. Customers can access the bank solely through the World Wide Web. Similar ventures soon followed.

Future plans go far beyond these modest incursions into cyberspace. Electronic money—also called digital cash, E-cash, or cybercash—is an important new buzzword. The cybercash concept requires the use of "smart cards"—refillable stored-value cards featuring embedded microchips. Ultimately, smart cards could replace cash, insurance cards, ATM cards, credit cards, and even driver's licenses. All information relevant to a specific individual could be encoded on one card. Other countries already use the technology. In France, for example, the nation's entire bank-payment system—involving more than 20 million people—has been converted to smart-card technology. And throughout Germany, medical-patient information is tracked and billing conducted through smart cards. In Swindon, England, a cybercash system called Mondex was tested in 1995, but with limited success.

In 1995, U.S. West introduced a refillable "cybercash" calling card for use in pay phones. Such so-called smart cards contain a microchip that can store encoded personal information.

In the United States, Visa will introduce a cash-card system at the 1996 Summer Olympic Games in Atlanta, Georgia; Chase Manhattan and Citibank also plan to try out smart cards this year. The Smart Card Forum, an industry group, is working to produce universal standards for the cards, and plans to test an experimental currency on the World Wide Web.

Critics of the smart-card concept are plentiful. Security is a prime concern. Critics point to DigiCash, an international electronic currency already in use. As untraceable cash, it holds vast potential for fraud, tax evasion, and money laundering. Opponents of digital cash assert that any on-line currency could be used in such schemes. In addition, the money supply could not be easily tracked or tightly controlled, potentially throwing a nation's monetary policy into disarray. Privacy issues have also gained prominence. Some worry that E-cash would be too traceable. With every transaction linked to an identifiable card holding valuable personal information, hackers could obtain huge amounts of privileged information about individuals. In response to such perceived vulnerability, encryption has become a hot area in smart-card development.

Faster, Smarter Hardware

On the cutting edge of the music and computer industries were "enhanced CDs," which can play music; display lyric sheets, video clips, and photos; and even link the user to a Web site. The downside to this new technology? Most CD-ROM drives bought before 1994 cannot play enhanced CDs.

Meanwhile, CD-ROM drives were being updated. In 1995, "quad speed" (4x) drives—which can read discs four times as fast as first-generation drives—emerged as standard equipment. Quad-speed drives came into widespread usage only in 1994, but already, newer and faster hardware has entered the market. In mid-1995, a 6x drive was introduced by NEC and Plextor, and 8x technology became available several months later. In early 1996, Pinnacle Micro introduced the 10Xtreme, a 10x drive with a claimed transfer speed of 1.5 megabytes per second. The value of these higher speeds was limited, however, since most CDs still were being produced to run on double- or quad-speed drives.

Modem technology has also advanced to meet the demands of the fast-growing Internet market. In 1995, 28,800-bits-per-second models emerged as a new standard, offering twice the speed of earlier models. Many users have found the higher price of a speedier modem to be worthwhile, since quicker transfer times result in lower phone bills and significant time savings.

Hard-drive storage capacity increased significantly. Some portable drives, like Iomega Corporation's Jaz drive, can hold up to 1 gigabyte (1 billion bytes, or characters) of data. Conventional hard drives, too, routinely began holding 1 gigabyte or more.

CORPORATE NEWS

The big marketing event of 1995 was the introduction of Windows 95, the long-awaited (and much-hyped) operating system from industry giant Microsoft. By year's end, sales were well below the first optimistic forecasts. Nevertheless, in the first quarter of 1996, Microsoft announced a 42 percent earnings gain over the same period in 1995. Fewer units than projected were sold as upgrades to existing systems; most sales came from PCs sold with the newer operating system already installed. The general consensus was that, despite some notable faults, Windows 95 was a great technical improvement over its predecessor, the Windows operating system.

Apple Computer, a pioneer of the PC revolution, fell on hard times. The nation's third-largest personal-computer maker faced a $69 million loss in 1995's fourth quarter.

THE CENSORSHIP ISSUE

The passage early in 1996 of the U.S. Communications and Decency Act ignited controversy. The legislation sharply restricts the dissemination of "indecent" and "patently offensive" material on the Internet. Anyone knowingly making such material available to minors would be subject to fines as high as $100,000 and imprisonment for up to two years. The law was challenged on constitutional grounds by the American Civil Liberties Union (ACLU) in a suit initiated by numerous free-speech groups and on-line services. The plaintiffs argued that the law's broad definition of "indecency" might lead to infringements on free speech. Meanwhile, proponents argued that the law would provide children with much-needed protection from pornography. In June 1996, a panel of three federal judges unanimously ruled the law unconstitutional.

In January 1996, an incident occurred that illustrated both the difficulty of controlling information sent over the Internet, and the problems introduced when such control is attempted. Prosecutors in Munich, Germany, asked CompuServe to suspend access in Germany to some 200 discussion groups that had violated German antipornography laws. CompuServe complied, but, in doing so, was forced to block access to the news groups worldwide. CompuServe later issued software enabling individuals to filter out unwanted material, and reported that it was developing software to block certain Internet sites in specific nations.

A consortium of on-line companies and software publishers was also working on a rating system, known as the PICS (Platform for Internet Content Selection) proposal. The system would allow individuals or groups to voluntarily establish ratings for Internet materials. The ratings tool would be incorporated into browser software; users could choose a browser based on the level of restrictions desired.

Meghan O'Reilly

Consumer Technology

Today's leading-edge consumer products are smarter, more compact, and safer than ever before. New technology, for example, allows us to instantaneously contact someone on the other side of the globe—for help, play, or just a simple chat. Indeed, the challenge of the future may be to isolate oneself from the reach of others and retain the skills that have been appropriated by the legions of new "smart" products. What will the consumer products of the future look like? It is a fair bet that they will have digital surround sound, use satellite data, and have computer chips to process incomprehensible amounts of data for sensing and correcting problems that have not even been noticed yet.

HOME ELECTRONICS

- *Hands-Free Displays.* Two new computer display systems can literally put information right before your eyes. The Hummingbird, by the Phoenix Group in Plainview, New York, is a full-function wireless computer monitor that can be coupled with voice-recognition software for totally hands-free personal-computer (PC) operation. The second system, i-Glasses, is a head-mounted display unit manufactured by a company called Virtual i.O. The display unit gives users the illusion of watching an 80-inch (203-centimeter) television or PC screen from a distance of only 11 feet (3.4 meters). Built-in headphones provide stereo sound, while stereoscopic technology gives the appearance of three-dimensional images. The PC upgrade adds head tracking, correcting images by factoring in roll, pitch, and yaw.

- *Digital Versatile Disc (DVD).* This new digital-videodisc technology provides for better picture quality than VHS tapes, and much greater storage capacity than existing compact discs. The 5-inch (12.5-centimeter) disc, designed largely by Toshiba, offers 300 percent better resolution than VHS—making it a likely replacement for both VCRs and laser-disc players in cutting-edge home theaters. The DVDs are designed to hold about 135 minutes of video with digital picture and sound quality. A proposed double-layering system would make even the longest of movies available in this new format. Also in the works are DVDs for computers that could hold as much as 9 gigabytes of data on a single disc side, allowing developers to create promising new and slicker games and multimedia applications.

- *Carbon Monoxide Eater.* An active system that combats deadly carbon monoxide gas may soon be available. A prototype catalytic system from the National Aeronautics and Space Administration's (NASA's) Langley Research Center in Hampton, Virginia, converts lethal carbon monoxide into harmless carbon dioxide. Originally developed for lasers that run only in a carbon dioxide environment, the device consists of a ceramic honeycomb coated with tin oxide and platinum.

- *Flat-Screen Plasma Monitors.* New technology has yielded screens only 3 inches (8 centimeters) thick—and even thinner. With such narrow profiles, televisions could literally hang on a wall, and monitors could be placed at the far edge of a desk. Several manufacturers are adopting this technology. Mitsubishi, for example, has introduced a 20-inch (51-centimeter) PC monitor, and has developed a prototype 40-inch (102-centimeter) television. Sony was expected to debut its Plasmatron monitor during 1996 in Japan, developing it for 20- to 50-inch screens.

- *"Smart" VCRs.* VCRs are bringing new flexibility to television viewers. Two new VCRs by Sharp, for example, can tape two shows simultaneously on the same tape. They can also play both shows back at the same time in smaller side-by-side pictures or with a picture-in-picture function. One of the Sharp models, the VC-BF70, allows sequential playback.

Two models from RCA, the mono VR542 and the stereo VR678HF, make watching commercials optional by analyzing a tape on playback, looking for such cues as black screens. Using these cues, the units map the location of commercials and then automatically scan past them on playback.

- *Wireless Video Playback.* Finally there is no longer a need for tape adapters or tangled wiring when playing back camcorder tapes on a television set. Hitachi's VM-H81A Hi8 camcorder transmits audio and video signals via an infrared receiver that attaches to the television. Viewing a self-recorded tape is then as simple as placing the camcorder within 10 feet (3 meters) of the receiver and hitting the "Play" button.

- **Clothes Processor EZ 1000.** Equator Corporation of Houston, Texas, has developed a single unit that can both wash and dry clothes. Simply place the laundry in the Clothes Processor EZ 1000, select the wash cycle, drying time, and temperature level, and press "On." The machine can handle up to 10 pounds (4.5 kilograms) of clothes per load.

AT WORK

- *Personal Satellite Communications.* Comsat Corporation of Bethesda, Maryland, has recently introduced Planet 1, a 6-pound (2.7-kilogram) satellite phone that looks much like a laptop computer. Once opened, a panel connects with an orbiting satellite, letting callers place and receive telephone calls from anywhere in the world. Fax, data, and paging services are expected to be available by 1997.

- *Canon NoteJet IIIxc.* This 9-pound (4-kilogram) laptop combines a computer with a 300-by-360-dot-per-inch (dpi) ink-jet printer and scanner. The scanner head can be swapped with a separate print cartridge, giving the printer four-color capability.

- *Instant Hot Lunch.* Simply add water to HeaterMeal's pad, and you have a hot meal at your desk in about 14 minutes. The water starts a heat-producing chemical reaction in magnesium, iron filings, and salt in the pad. The meals, which cost about $4.50, need no refrigeration and include a packet of water, plastic utensils, napkin, and salt and pepper. Originally developed for military use, the meals are now offered by HeaterMeal of Cincinnati, Ohio.

Apartment dwellers may now find space to do laundry at home, thanks to the Clothes Processor EZ 1000, which combines the functions of a washer and dryer.

ON THE ROAD

- *Motorcycle Air Bag.* Great Britain's Transportation Research Laboratory is developing an air bag for motorcycles. The bag is designed to inflate to stop riders from flipping over the handlebars, and then instantly deflate to prevent bouncing back. The folded bag would fit into a recess in the top of the gas tank.

- *Yaw Control.* Bosch Corporation has developed a device to counteract that disconcerting situation when a single wheel of an automobile slips. The system, which has more sensors than typical antilock-braking-system (ABS) technology, is based on a computer that uses both brakes and throttle to counteract slippery road conditions and bring a car back in line.

A British firm is developing an air bag to prevent motorcyclists from toppling over their handlebars—a leading cause of injuries in motorcycle accidents. In prototype models, the folded bag is stored atop the gas tank.

- ***Automotive Security and Tracking System.*** Cars can be watched 24 hours a day by installing a hidden ATX OnGuard tracker in combination with an integrated cellular phone. If a driver hits a "panic button" or fails to enter a code upon starting the car, ATX Research Incorporated's 24-hour response center in San Antonio, Texas, will start tracking the car using Global Positioning System (GPS) technology; meanwhile, an emergency communications specialist (ECS) uses the car's cellular phone to monitor what is occurring in the car. If the vehicle is involved in a criminal situation—such as a carjacking or a bank robbery—an ECS can contact local law-enforcement agencies and provide them with the car's exact location and description. This direct connection between a car and an ECS also helps drivers to seek assistance in more-mundane situations, such as getting directions in unfamiliar areas.
- ***Automotive Generator.*** Aura Systems Incorporated of El Segundo, California, has designed an internal-combustion engine that can also serve as an electrical generator. Aura combined the functions of the starter and alternator into a compact new motor that contains electromagnetic coils surrounding a gearless flywheel. The coils create an electrical force that turns the flywheel, which then can be used to start the car or generate up to 7,000 watts while idling.
- ***Clarion Car MultiMedia.*** Clarion has adapted the concept of the computer bus, a component that transports data inside computers, to automotive electronics with its Car MultiMedia product line. The data bus, called a C-bus, can integrate an in-dashboard cellular phone, voice-activated CD-ROM–based navigation system, and a video system that can accommodate a TV tuner, a rear-vision camera, and other video source units. All of these video inputs can be displayed on a 6-inch (15-centimeter) active-matrix liquid-crystal-display (LCD) unit. The C-bus system is also modular, allowing potential buyers to mix and match any or all of the available pieces.
- ***Jam Spotter.*** Drivers in Great Britain no longer have to rely on radio traffic reports to avoid congestion. Trafficmaster PLC has developed a system that monitors traffic with 2,400 sensors, relaying data to a control center. This information is then routed to individual Trafficmate units that vocally alert the driver of any impending problems. The dashboard-mounted units have a 200-word vocabulary.

AT PLAY

- ***Autobike.*** Autobike, from CSA of South Easton, Massachusetts, aims to help casual cyclists overcome the problem of knowing when to switch gears. Three weights slide along Autobike's rear-wheel spokes as pedaling quickens, using centrifugal force to shift among the bike's six speeds. This allows the rider to maintain a steady pedaling cadence of about 65 revolutions per minute (rpm) without having to worry about shifting.
- ***"Smart" Skis.*** Active Control Experts (ACX) Incorporated of Cambridge, Massachusetts, and K2 Corporation have developed a ski that vibrates less and makes better contact with the snow compared to conventional skis. Tiny control circuits embedded inside the skis can measure bending caused by terrain and other conditions; the circuits can then electronically modify the ski's stiffness to optimize performance.

Solar cells integrated into the lens cover of Canon's Sure Shot del Sol generate enough power to operate the camera and its flash.

- ***Canon Sure Shot del Sol.*** Dead camera batteries are a thing of the past with the new Canon Sure Shot del Sol. Solar cells built into the lens cover provide enough power to operate the camera's flash unit. When the cover is open, the camera activates and the cells generate power. A second battery, which never needs replacing, stores electricity for later use.
- ***Casio QV-10 Digital Camera.*** This pocket-sized camera has an LCD screen for viewing pictures already taken or about to be taken, and ports to connect it to a computer or a television. Users can resize stored pic-

tures and delete them. The camera can store 96 still images. The QV-10 also works as a presentation device, allowing downloaded images to be displayed on a television screen or copied onto videotape.

- **Skywalker.** Reebok has developed an exercise machine that takes all of the impact of exercising off knees and joints, making it a wise choice for older people and those recovering from injuries. Exercisers simply step into the arcing footpads, grasp the poles, and begin a total-body gliding motion that can range from easy to vigorous.

PORTABLE WONDERS

- **WOLVES Rescue Support.** Rescue operations are often hampered by not knowing exactly where victims are and by having no way of finding them. Cobra Support Services of Great Britain has developed a prototype rescue-technology system called WOLVES (Wireless Operationally Linked Electronic and Video Exploration System). The testing of this system involves strapping a tiny video camera atop the head of a specially trained police dog; a transmitter is also strapped to the dog's back. The dog, which can fit into smaller spaces than any adult human can, could then transmit images of people who have been trapped during an earthquake or other disaster.

- **Canon ES5000 Eye-Control Camcorder.** Photographers can program this high-tech camera to shift focus, quick-zoom, fade, and perform other functions by means of eye activation. A beam of light monitors the relative position of the photographer's iris to the image in the viewfinder. Functions are activated by simply looking at their name. Focus shifting is accomplished by using the eye to move an automatic focusing box in the field of view.

- **Accu-Master Golf Trainer.** ProActive Sports of Canby, Oregon, has developed Accu-Master, a liquid-crystal pad whose color lightens on impact, showing the golfer how close he or she came to the club's sweet spot. The spot of light on the pad fades in 15 to 20 seconds and is ready for the next shot. The Accu-Master is attached to the club with a reusable adhesive for easy removal.

- **Sportstrax Personal Sports Wire.** This handheld pager from Motorola gives more than just the up-to-the-minute scores of all the games in progress. The National Basketball Association (NBA) version, for example, gives time left, possession, and score. The baseball version keeps track of men on base, the team at bat, the inning, the number of outs, and the score. The data are updated every two minutes via satellite technology.

- **Kid-proof Remote.** Channel Control from TCI Incorporated directs kids toward child-friendly television programs. The units, which are available in two shapes—an orange dinosaur (Remote-A-Saurus) or a purple dog with sunglasses (Channel-Rover)—have only eight channel buttons. The buttons identify easily recognizable child-oriented networks, including Nickelodeon, PBS, and The Learning Channel.

David Scinto

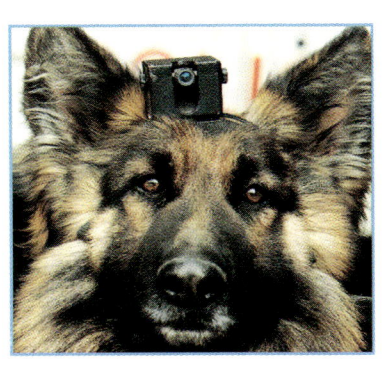

Trained police dogs equipped with the WOLVES wireless video system (above) can help rescue workers locate people trapped by earthquakes or other disasters.

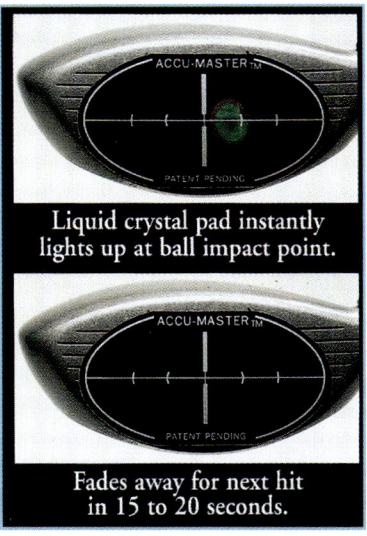

Golfers can hone in on a club's sweet spot with Accu-Master, a pad that illuminates the point of impact after each swing.

ELECTRONICS

GETTING SILICON TO GLOW

Silicon, the staple material of most computer chips and microelectronic circuits, stubbornly resists illumination by electrical stimulation. However, researchers have found that silicon, under certain conditions, can be coaxed into luminescence.

In a new method developed by scientists at the National Research Council of Canada in Ottawa, Ontario, extremely thin layers of silicon and silicon dioxide are stacked to form a superlattice that will emit light. The close packing of these layers causes electrons to become trapped in extremely small spaces, creating a condition called quantum confinement. As electrons zip around inside their nanometer-sized pens, they behave as trapped waves, a phenomenon that ultimately prompts the emission of a photon.

The glowing silicon may prove helpful in manufacturing new optoelectric devices for use in computing, telecommunications, compact discs, and holography.

MICROSCOPIC IMAGES IN 3-D

A computer-assisted microscope can now make sharp three-dimensional (3-D) images of living cells viewed in low levels of light. Scientists at the University of Massachusetts Medical School in Worcester have adapted optical methods used in astronomy and medical imaging to produce a system capable of imaging cell organelles with a resolution much higher than can be achieved with an ordinary light microscope. The new technique, which does not damage specimens, can produce an optical resolution four times greater than that achieved by an unaided light microscope.

The researchers joined a charged-coupled device (CCD) with an optical microscope, and then devised a computer program to compensate for the blurring caused by high magnification of light. The method resembles one used to correct the fuzziness of images taken by the flawed optics of the Hubble Space Telescope (HST) before its repair. Using a second computer procedure, the new imaging system scans many focal planes and constructs a 3-D model of the object—a process similar to the one used in computed tomography (CT) for medical imaging.

TERAHERTZ IMAGING

A new type of imaging system that uses terahertz waves, or T rays, has proved that it can effectively penetrate certain materials and produce pictures of an object's interior. The technique also gives clues about the imaged material's chemical composition.

Researchers at the AT&T Bell Laboratories in Holmdel, New Jersey, have demonstrated that the terahertz waves, which exist in between low-frequency infrared and high-frequency microwaves (and below the visible spectrum), can pass through most nonmetallic materials. This property makes T rays particularly useful for scanning the interiors of letters and packages (for security purposes, for example), as well as for performing some types of medical diagnoses and environmental monitoring.

A solid-state sapphire laser generates the terahertz waves by producing electromagnetic pulses at a frequency of 100 billion to 3 trillion cycles per second. Those pulses cause T rays to emanate from an antenna, where they are focused onto a target. As the terahertz waves pass through a material, they are changed in ways characteristic to the specific elements that constitute the material. An electronic receiver catches the emerging

T-ray Images

Water absorbs t-rays so the t-rays map out the water distribution inside the fresh leaf, (left). The water distribution has clearly changed in the same leaf after 48 hours, (center). Water has selectively evaporated from the leaf. The color bar, (right), indicates the relative water concentration inside the leaves.

FRESHLY CUT LEAF

TWO-DAY-OLD LEAF

COLOR BAR

waves, which are then decoded to determine the precise chemical composition of the material. The T rays pass readily through paper, for example, but are largely absorbed by water, making them more useful for analyzing thin, solid layers of material. In the presence of water, terahertz waves can penetrate only 0.04 to 0.08 inch (1 to 2 millimeters), meaning that biological samples must be extremely thin if they are to be imaged.

TINY TRANSISTORS

Using the ultrasmall tip of an atomic-force microscope (AFM), scientists at the Naval Research Laboratory in Washington, D.C., have adapted this extremely fine instrument for use as a lithographic tool to make microelectronic circuits. Designed to study a material's surface atom by atom, the AFM uses a minuscule probe to skate over a plane of atoms, sense each atom's electronic shell, and translate that information into an image of the material's atomic structure. Researchers have coated the tiny tip with metal and used it to draw circuit patterns on silicon surfaces. The silicon is then etched, yielding features less than 1 micrometer (0.00003 inch) thick and a few micrometers wide. As part of the process, the AFM can also produce images of the etched silicon surface.

As a demonstration of the technique's effectiveness, scientists have fabricated a component of a tiny field-effect transistor (FET), a microelectronic device, that measures only 30 nanometers (0.0012 inch) wide. Researchers at Stanford University in Palo Alto, California, are now fashioning a device that uses this lithographic process to draw many circuits in parallel at the same time.

PROBING A CELL WITH FIBER OPTICS

A fiber-optic sensor, one-thousandth the thickness of a human hair, now gives scientists a view into a live cell's biochemistry. With a fine tip drawn from a strand of aluminum-coated optical fiber, the probe can pierce a cell's membrane and venture into its liquid interior. A laser light shone through the tip enables scientists to analyze changes in the light's wavelength and intensity and to extract information about particular molecules and chemical reactions.

Researchers at the University of Michigan in Ann Arbor have used the probe to assay glucose, oxygen, sodium, calcium, and potassium ions inside a lung cell, as well as the cell's acidity level. Chemical changes occurring in as little as one-hundredth of a second can be monitored, permitting scientists to track a cell's chemical processes as they occur. Probing a live rat-embryo cell did not affect the cell's metabolism or rate of growth. In one test, scientists watched a cell die of poisoning while they recorded information about its biochemical changes. The probe's type is so small compared to the size of a cell—one hundred-thousandth the volume of a red blood cell—that a probe prick of a cell membrane has been likened to a mosquito bite. Researchers believe that the probe could prove useful for monitoring the effects of drugs on particular types of cells, identifying gene sequences, and checking embryos for birth defects.

CLEANING WATER WITH ELECTRONS

A system to cleanse wastewater using high-energy electron beams has proved effective in ridding tainted water of organisms and organic debris, making it safe to drink. Researchers have shown that even water polluted with sewage and sludge will yield to the power of the high-voltage blast.

When polluted water flows through the stream of high-energy electrons, which are propelled by an accelerator, the beam's charged particles facilitate chemical reactions that cause the water's organic pollutants to break apart. The process kills bacteria, viruses, and all other organisms in the water, while producing no hazardous by-products.

To test the process, engineers installed a mobile water-treatment unit in a 48-foot (15-meter) trailer. Pilot plant tests indicate that the technique also cleanses water of toxic organic materials, disinfection by-products, and some parasitic microbes such as *giardia* and *cryptosporidium*. The new disinfection method produces, overall, fewer chemical by-products than do existing chemical-treatment techniques, which use such agents as chlorine dioxide, chloramine, ozone, and titanium dioxide.

Richard Lipkin

Endangered Species

Falcons Return

In the 1960s, the American peregrine falcon was on the brink of extinction, its population decimated by eggshell thinning and other reproductive problems resulting from contamination with DDT and other chlorinated hydrocarbon insecticides. The use of DDT was banned in the United States in 1972, and the falcons were placed under federal protection in 1975, but by then only 50 nesting pairs could be found in the United States.

Since then, the U.S. Fish and Wildlife Service and other agencies have raised hundreds of falcons in captivity and released them into the wild with excellent results: there are now an estimated 1,233 nesting pairs in the contiguous United States, and another 300 pairs in Alaska. Peregrines are cliff nesters that feed primarily on other birds, often capturing them in spectacular midair attacks. They received national attention when some of the reintroduced birds successfully nested on ledges of skyscrapers in major cities, where they prey primarily on pigeons.

In June 1995, Secretary of the Interior Bruce Babbitt announced that the peregrine falcon had recovered sufficiently to warrant taking the first steps toward removing it from the Endangered Species List. Only 22 animals have been removed from the list since the Endangered Species Act was passed in 1973.

When another endangered hawk, the aplomado falcon, was found nesting last year near Brownsville, Texas, it marked the first time since the early 1950s that the bird had nested in the United States. Biologists were elated when they discovered that the falcon's rare nesting activity had resulted in the birth of a healthy chick.

The successful nest was the culmination of a 10-year project by the Peregrine Fund, a private organization devoted to protecting birds of prey. When a remnant population of aplomado falcons was found in southern Mexico in 1982, the Peregrine Fund was given permission by the Mexican government to collect 10 chicks from 10 nests and use them as the foundation for a captive-breeding program. The chicks were transported to the Peregrine Fund's World Center for Birds of Prey in Boise, Idaho, where they were reared and bred. Since 1985, more than 60 aplomado falcons have been released at Laguna Atascosa National Wildlife Reserve near Brownsville, but biologists did not expect the released birds to begin nesting until 1997 or 1998. They are cautiously optimistic that the 1995 nest will be the first of many, and that the once-abundant falcon is back to stay.

Flying Squirrel Rediscovered

The world's largest species of squirrel, the woolly flying squirrel of the Himalayas, had not been seen by Westerners since 1924, so it was natural to assume the squirrel was extinct—that is, until a specimen showed up alive and well 70 years later in a remote and inhospitable section of the Himalayas in northern Pakistan. The woolly flying squirrel,

The woolly flying squirrel was rediscovered in 1994 by two amateur naturalists. The creature, the world's largest species of squirrel, lives only in a remote part of the Himalayas in northern Pakistan.

Eupetaurus cinereus, is almost 2 feet (54 centimeters) in length, has a 2-foot-long tail, and, like the flying squirrels of North America, has a loose membrane of skin and fur that spreads into a gliding membrane when the creature leaps. But unlike the more familiar flying squirrels, the woolly flying squirrel does not live in trees. It inhabits rocky, mountainous valleys near the tree line, where it glides from rock to rock to avoid predators.

The squirrel was rediscovered by two amateur naturalists, freelance science writer and editor Peter Zahler and college math teacher Chantal Dietemann, both of Watertown, New York. Zahler and Dietemann had made two trips to northern Pakistan to search for the squirrel, which they were convinced might still live in the region's remote, roadless valleys. Near the end of their second trip in 1994, they were approached by a pair of local men who offered to catch them a living squirrel for $130. The animal the men produced was captured, they said, from a cave high on a rocky slope. It was later confirmed to be a woolly flying squirrel.

The squirrel has remained elusive partly because so little is known of its natural history. Although it was first identified in 1888, only a few technical papers have been written about the species, and almost nothing is known about its diet and other habits. Zahler and Dietemann conjectured that the squirrel is nocturnal, which would explain why it is so seldom seen, and that it nests in small, inaccessible caves like the one where their specimen was captured and later released. Zahler and Dietemann discovered body parts scattered beneath an eagle owl's roost in 1994, and a second specimen was captured by Zahler in 1995, suggesting a fairly healthy population of the squirrels in the region. Nonetheless, increasing human disturbance of its fragmented and fragile habitat puts the woolly flying squirrel at risk of extinction.

Manatee Deaths

Marine biologists have been baffled by a surge of unexplained manatee deaths in Florida waters during 1995 and early 1996.

The West Indian manatee, one of three manatee species in the world and the only one in North America, is among the most carefully monitored of all endangered mammals. Measuring an average 10 feet (3 meters) in length and weighing up to 3,500 pounds (1,600 kilograms), the manatee is an herbivore that feeds on kelp, sea grass, and other vegetation growing along inshore coastal regions. The manatee's size and habitat requirements and its need to surface frequently for air make it vulnerable to collisions with powerboats, a leading cause of death.

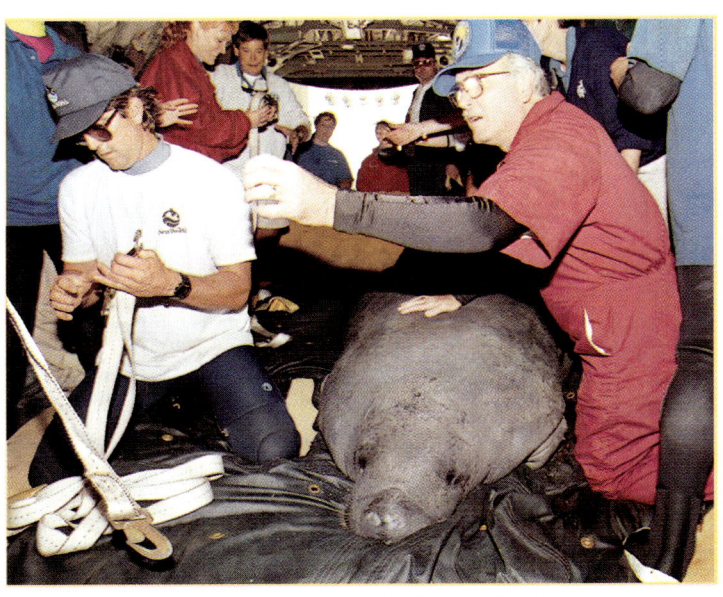

After being captured in Chesapeake Bay (above) and shipped back to Florida, Chessie the Manatee confounded scientists by swimming north again—this time all the way to New England!

Biologists expect one-quarter to one-third of the manatees that die around Florida every year to fall victim to powerboats and other human causes, but the recent spate of deaths seems to be unrelated to such incidents. A total of 201 manatees died last year, nearly 10 percent of the population and the second-highest total since records began being kept in 1974. Biologists are especially concerned about dozens of unexplained deaths, and are exploring the possibility of some as-yet-undiscovered environmental cause. During one two-week period in March 1996, more than 30 manatees died, most of them young, healthy adults. Manatees can live more than 60 years, but their rate of reproduction is extremely slow. A typical female gives birth to a single calf no more than once every four years. With a total population of only about 2,600, biologists worry that extinction is inevitable if the cause of the unexplained deaths is not discovered.

THE TIBETAN RED DEER

Another Asian mammal long feared extinct, the Tibetan red deer, was rediscovered in October 1995 by biologist George Schaller during a biodiversity survey in Tibet. Schaller and colleagues from Tibet and China stumbled upon a herd of about 200 of the subspecies of deer in a region of alpine meadows at an altitude of about 15,000 feet (4,600 meters).

Last seen by a Westerner in the 1940s, the deer had been hunted almost to extinction by local Tibetans. The herd of 200, almost certainly the last viable population of the subspecies, was put under the immediate protection of the Tibet Forest Bureau, which declared the area a preserve and appointed guards to protect the animals from poachers.

SERENGETI LIONS SURVIVE CRISIS

The outbreak of canine distemper that threatened the lions of Kenya's Serengeti in 1994 and 1995 has apparently run its course, but not before killing as many as one-third of the 3,000 lions in the region. The pathogen, introduced to lions, hyenas, leopards, and bat-eared foxes by domestic dogs living in Masai villages surrounding the Serengeti, was previously thought incapable of making a jump from canine to feline species. Researchers have since concluded that the organism that killed the lions is a new variant, mutated from the more familiar canine-distemper virus that infects dogs throughout the world.

Although the epidemic appears to be over, at least 1,000 lions died. The survivors probably developed immunity, but biologists predict that the virus will resurface every few years, much as influenza does in humans.

Jerry Dennis

ENERGY

FOSSIL FUELS

• ***Oil.*** Oil prices rode a roller coaster during 1995 and 1996. In 1995, the Organization of Petroleum Exporting Countries (OPEC) continued to lose its grip on the U.S. market. The cartel, which had forced a tripling of U.S. oil prices in the 1970s, tried to boost world oil prices by restricting production. The strategy failed because non-OPEC producers sold enough oil to meet world demand. Crude-oil prices hovered around $15 per barrel throughout 1995; as a result, U.S. consumers enjoyed the lowest prices for gasoline in decades. In many parts of the country, unleaded, self-serve gasoline cost less than $1 per gallon. The price break came despite the mandated introduction on January 1, 1995, of cleaner-burning, but costlier, "reformulated" gas in several highly polluted areas.

The good news did not last into 1996, however. Oil prices began rising early in the year, and soon reached their highest levels since 1991. By early May, regular self-serve gasoline was at a national average of $1.29 per gallon, up 14 percent in 13 weeks. Prices for full-serve premium gasoline exceeded $2 per gallon in parts of California, New York, and other states. Crude oil approached $25 per barrel.

The causes for the sudden price increase were varied. A rise in demand—partly due to harsh winter weather in the United States and Western Europe—followed the increasing popularity in the United States of gas-guzzling sport-utility vehicles and the repeal of the federal speed limit of 55 miles (88 kilometers) per hour. In addition, oil companies had been keeping their inventories low, anticipating the possible return of Iraq to the world oil market.

In response, on April 19, 1996, President Bill Clinton announced he would authorize the sale of 12 million barrels of oil from the U.S. Strategic Petroleum Reserve. In addition, the U.S. Justice Department began checking for evidence of collusion and price-fixing among oil companies, but economists said such a scenario was doubtful. In a related effort led by congressional Republicans, a

The Mars Oil Rig, shown above prior to being towed to its offshore site in the Gulf of Mexico about 130 miles southeast of New Orleans, is 31 stories high and has a deck nearly the size of two football fields.

repeal of the 4.3-cent-per-gallon federal gas tax passed in 1993 also seemed possible.

The future availability of petroleum products remained a cause of concern. The United States relies more on oil than on any other energy source to meet its energy needs; nevertheless, in 1995, the country completed its fourth consecutive year of falling domestic oil production. For the third straight year, the United States imported more than half of the oil it consumed.

With dwindling domestic reserves, U.S. oil companies expanded the search for new reserves overseas. Mobil Oil Corporation entered negotiations to join Chevron Corporation in a long-term investment to develop one of the world's last big proven oil reserves, in Kazakhstan.

The depletion of existing U.S. oil fields also spurred efforts to develop technologies to extract oil from less accessible locations. The world's deepest oil-producing rig—floating in almost 3,000 feet (914 meters) of water—was positioned in the Gulf of Mexico about 130 miles (210 kilometers) southeast of New Orleans in April 1996. Meanwhile, a consortium of four oil companies—Shell, Mobil, Texaco, and Amoco—announced plans to drill the deepest well yet attempted for oil exploration. The well, to extend to a depth of 7,625 feet (2,324 meters) into the seafloor, would be drilled in the Gulf of Mexico about 200 miles (322 kilometers) southwest of Corpus Christi, Texas. If any oil and gas are found, however, current technology has no way of recovering them. It is hoped that technology will catch up by the time any discovery occurs.

Technology based on a high-speed satellite system may soon allow those searching for oil sources to process offshore seismic data received from a ship in "real time" on land-based supercomputers. With such quick processing time, a ship still at sea could return to a likely area to gather further data.

• ***Natural Gas.*** Natural gas is the cleanest-burning fossil fuel, making it an attractive alternative for fueling trucks and automobiles. Utilities and taxi companies continued to add natural-gas-powered vehicles to their fleets. Because of the lack of service stations selling natural gas, however, conversion remained impractical for individual car owners.

As natural-gas prices continued climbing, U.S. producers found it profitable to extract gas from marginal wells, allowing domestic production to grow slightly. Although most natural gas consumed in the United States was from domestic reserves, imports also grew, accounting for more than 10 percent of consumption.

• ***Coal.*** By far the most abundant fossil fuel found in the United States, coal was the leading energy source produced in the country, accounting for almost 30 percent of all domestically produced energy. Coal remained the only energy source the United States exported in significant quantity. The main consumers of coal were electric utilities. Coal produced roughly 56 percent of the electrical power generated in the United States in 1995.

A major disadvantage of burning coal is acid precipitation. Scrubbers installed in plant smokestacks have diminished the impact of acid precipitation, but a technology introduced in 1994 may prove more effective. By greatly increasing the amount of electricity generated by burning coal, a new ceramic air heater produced by Hague International in Kennebunk, Maine, promised to reduce the cost and pollution associated with coal-generated electricity. Developed with support from the U.S. Department of Energy (DOE), the technology was expected to be installed in some 280 aging U.S. power plants.

NUCLEAR ENERGY

Construction of the Watts Bar I nuclear-power plant near Spring City, Tennessee, was completed in November 1995. The facility took 23 years and $7 billion to build. Watts Bar was expected to enter into service in 1996; it would be the 110th U.S. nuclear plant in operation. In 1995, nuclear power generated more than 20 percent of the nation's electricity.

Legislators failed to resolve an acrimonious debate over a permanent storage facility for the growing stockpile of spent nuclear fuel. Utilities called on the federal government to complete a permanent storage facility for nuclear waste as required by a 1982 law. But legislators from the leading candidate for such a repository, Nevada, resisted.

RENEWABLE ENERGY

Relatively low prices for fossil fuels during 1995 depressed production of renewable fuels such as solar, geothermal, and hydroelectric power. The production of geothermal energy was down almost by half from its peak level in 1987. Only hydroelectric power remained level with prior years. Hydroelectric dams accounted for about 9 percent of electric power in the United States.

Although they still are too expensive to compete with oil, gas, and coal, most renewable technologies hold promise as alternatives to fossil fuels in the future, when these finite energy sources become so depleted that their high cost drives them out of the market. One of the most promising alternatives, wind power, is already providing electricity in California at about the average national price for energy.

ENERGY POLICY

Energy policy was the focus of intense debate in Congress in 1995, as the Republican majority tried to reduce federal involvement in energy-related businesses and cut government spending on energy programs.

A 1995 push by House Republicans to dismantle the DOE failed. In the end, the department received $15.4 billion for fiscal 1996, only 1 percent less than the $6 billion funding level in the previous year.

Mary H. Cooper

ENVIRONMENT

RESTORING MISSISSIPPI WETLANDS

In the wake of catastrophic flooding of the Mississippi River in 1993, organizations such as the Upper Mississippi River Conservation Committee claimed that much of the damage could have been avoided had the river not been turned into "little more than a shipping channel." Over the years, large-scale engineering projects had straightened the river, enclosed it in dikes and levees, and drained adjacent wetlands. After the 1993 flooding, however, efforts were stepped up to restore wetlands, which soak up floodwaters, provide homes to a rich variety of plants and animals, and have fewer construction and maintenance costs than do dikes and levees.

A study of restored wetlands near the Des Plaines River north of Chicago found that a 5.7-acre (2.3-hectare) marsh can handle runoff from a watershed of 410 acres (166 hectares)—and reduce pollutants in the water by up to 99 percent. Extrapolating from data gathered there and at three other experimental marshes, hydrologist Donald L. Hey calculated that only 1.4 to 5.5 percent of a watershed needs to be in wetlands to accommodate runoff. On the upper Mississippi, for example, some 13 million acres (5.3 million hectares) of wetlands would be needed, less than 3 percent of the river's watershed.

GRAND CANYON RENEWAL

Major efforts are under way to restore U.S. ecosystems endangered by decades of human disruption. For example, in the spring of 1996, an artificial flood was created on the Colorado River to stir up sediment and redistribute it through Arizona's Grand Canyon. Ever since the Glen Canyon Dam was completed in 1964, sediment that would normally have been carried downstream by the Colorado River instead settled out of the water as it sat behind the dam. As a result, a warm, muddy river was replaced by a cold, clear river. This change made the river suitable for introduced species such as rainbow trout, but led to the disappearance of native species, including several endangered fish species.

In March 1996, an artificial flood on the Colorado River—created by releasing more than 117 million gallons of water from the Glen Canyon Dam—helped stir up sediment and redistribute it through the Grand Canyon.

"A new era has begun in the management of the Colorado River and the Grand Canyon," noted Bruce Babbitt, the U.S. secretary of the interior, as he reviewed the results of the government's experiment. Over the course of a week, more than 117 billion gallons (443 billion liters) of water were released from the dam. The water stirred up sediment in the river and redistributed it through the Grand Canyon. New sandy shores were created where plants could take root, providing food and shelter for fish, birds, and other wildlife.

Global Warming

Researchers announced that 1995 was the warmest year since reliable global measurements began in 1856. According to the British Meteorological Office and the University of East Anglia in England, the average global surface temperature in 1995 was 58.72° F (14.84° C), slightly higher than the previous record (set in 1990). The British researchers also reported that the period between 1991 and 1995 was warmer than any similar five-year span, despite the fact that airborne debris from the 1991 eruption of Mount Pinatubo in the Philippines caused significant cooling.

Many climatologists contend that pollutants from burning fossil fuels and other human activities are a primary cause of this alteration of Earth's climate. This theory received new support when the United Nations Intergovernmental Panel on Climate Change (IPCC) concluded for the first time that global warming is "unlikely to be entirely natural in origin," and that evidence "suggests a discernible human influence on climate."

One feature of the predicted global warming would be an increase in severe droughts, excessive rainfall, and other types of extreme weather. An analysis of U.S. weather data by researchers at the National Climatic Data Center found that the incidence of extreme weather in the period from 1980 through 1994 was significantly greater than during the previous 65 years. One recent example was a blizzard that struck the eastern United States early in 1996, dumping record snowfalls on many communities—even as Southern California was experiencing record-breaking heat.

Another anticipated impact of global warming would be a rise in sea level. The U.S. Environmental Protection Agency (EPA) released a report in 1995 that projected a likely sea-level rise along the U.S. Atlantic and Gulf of Mexico coasts of 10.2 inches (26 centimeters) by the year 2050, and 21.7 inches (55 centimeters) by the year 2100. The EPA said its projections were consistent with those of the IPCC, which projects that global sea level would rise in a range of 7.9 to 33.9 inches (20 to 86 centimeters) by the year 2100, with 19.3 inches (49 centimeters) the "best-guess" median.

A warmer climate is expected to impact living things in numerous ways. Already, increased water temperatures in the Pacific Ocean off Southern California have led to dramatic declines in wildlife populations. Scientists at the Scripps Institution of Oceanography in La Jolla, California, reported in 1995 that the water temperature had risen 2° to 3° F (1.1° to 1.7° C) since 1951. As a result, the population of zooplankton, the primary food for numerous species of fish and birds, had declined 80 percent. The area, which once teemed with living things, is now almost devoid of life.

A World Health Organization (WHO) study suggests that global warming will create major public-health problems. One concern is the spread of tropical diseases. For example, scientists note that malaria generally is found only in lands where annual average temperatures are at least 61° F (16° C). As temperatures rise, the disease could spread into regions that were previously malaria-free. Other diseases that have the potential to expand their ranges include schistosomiasis, yellow fever, onchocerciasis (river blindness), and dengue fever.

LAND USE

Much controversy has developed over the appropriate use of federally owned wilderness. Generally speaking, conservatives and business interests advocate development, while liberals and environmental groups support protection of habitats and wildlife.

A number of land-use issues arose in the U.S. Congress during 1995 and 1996. For example, Utah's congressional delegation led efforts to open about 20 million acres (8.1 million hectares) of federal lands in Utah to coal mining and other industrial development, while setting some 1.8 million acres (730,000 hectares) aside as wilderness. Environmentalists and their congressional allies opposed the legislation, proposing that up to 5.7 million acres (2.3 million hectares) be reserved as wilderness. A similar confrontation emerged after Alaska's congressional representatives began legislative efforts to open the Arctic National Wildlife Refuge to oil drilling and to increase logging of ancient trees in the Tongass National Forest.

Congress also debated broad land-use policy issues. The House of Representatives approved legislation to give hunting, fishing, and other recreational uses equal standing with conservation as central purposes of wildlife refuges. Western legislators also led efforts to preserve an 1872 law that requires the government to sell mining rights on federal lands for much less than the rights are worth.

Environmentalists fear that large tracts of public land in Utah (below) and elsewhere in the West would be critically endangered should proposals be enacted that open up this vast acreage to industrial development.

THE OZONE HOLE

Ozone levels plunged to record lows over parts of Northern Europe in early 1996. On some days, the levels were 45 percent below those recorded before 1978, when chlorofluorocarbons (CFCs) and other human-made chemicals began to destroy the ozone layer in the stratosphere. The problem mirrored that in the Southern Hemisphere, where the ozone hole that formed over Antarctica in the fall of 1995 covered a record 3.86 million square miles (10 million square kilometers).

In areas where ozone has been depleted, increased levels of ultraviolet radiation reach the ground; this radiation can cause skin cancers and cataracts, and disrupt environmental food chains. In late 1995, representatives of more than 100 nations met in Vienna, Austria, and agreed to phase out the use of methyl bromide, the only ozone-destroying industrial chemical not previously designated for elimination. Developed countries, which account for 80 percent of methyl bromide production, agreed to cut production 25 percent by the year 2001, 50 percent by 2005, and entirely by 2010.

Because chemicals often persist for many years in the atmosphere, ozone depletion is expected to continue for decades. However, Stephen Montzka of the National Oceanic and Atmospheric Administration (NOAA) reports that decreasing production of ozone-harming compounds is beginning to have a positive impact. Measurements at seven sites around the world show that levels of these chemicals in the lower atmosphere have peaked and are beginning to decline.

RECYCLING

A number of indicators suggest that recycling is gaining new momentum. A 1995 EPA report indicates that the United States recovered 24 percent of its municipal solid wastes through recycling and composting in 1994, up from 17 percent in 1990. And for the first time, composting of food scraps reached measurable proportions at the national level. Consumers were also having an easier time buying high-quality products with recycled content. For instance, in the early 1990s, paper made from 100 percent recycled postconsumer materials was of

Recycling has become a way of life for many Americans. In the past few years, paper made from recycled materials has improved greatly in quality.

poor quality and had a brown or gray tint. However, only a few years later, high-quality paper has become widely available at prices comparable to virgin paper. Similarly, paper made without chlorine bleach was becoming much easier to find.

Manufacturers have developed a variety of products and packaging that reduce waste and that incorporate recycled content. For example, Paramount Packaging Corporation of Chalfont, Pennsylvania, has designed laundry-detergent bags that contain 25 percent postconsumer plastic and use 80 percent less material than comparably sized detergent boxes. Fast-food giant McDonald's Corporation of Oak Brook, Illinois, has implemented nearly 100 initiatives to prevent waste, encourage recycling, and purchase recycled goods in its U.S. restaurants; these programs have eliminated 7,500 tons of packaging annually. By reducing the raised designs on napkins, for example, McDonald's found that 23 percent more napkins could fit into a shipping container, saving 294,000 pounds (133 kilograms) of corrugated cardboard and 150 truckload shipments.

Jenny Tesar

Food and Population

The United Nations (UN) estimates that in 1995, the production of cereal grains—the world's basic staple food—fell more than 3 percent from 1994's almost record harvest of 1.951 billion metric tons. The resulting reduction in world grain stocks brings them substantially below the 14 to 15 percent of expected utilization that the UN Food and Agriculture Organization (FAO) deems necessary for world food security. Adverse weather reduced output in most regions; countries affected by civil conflict—especially Afghanistan, Iraq, and several African countries—were especially hard-pressed. Meanwhile, the world's population again grew by about 95 million people, with nine-tenths of the growth taking place in the poorest and hungriest countries.

Food Security

At the same time that crops stagnated or declined almost everywhere, cereal prices rose significantly for all major grains, especially wheat—a fact that is reflected directly in higher prices for cereals, bread, and pasta, and indirectly for meat and dairy products. Although it is as yet unclear whether this escalation results from smaller supplies or greater demand, its impact will be felt most severely by those least able to cushion it—that is, those who are already "food-insecure" because of low incomes.

For the international community, food security means the ability of people either to grow or to buy enough food to maintain an adequate diet. The FAO says that about 800 million people—one-seventh of the world's population—lack that security and will continue to lack sufficient food supplies for at least the next 15 years unless there is a dramatic change in the international food system. These people are either too poor to meet the price, or their numbers overwhelm the portion of the food produced that is actually available to them through subsidies or donations. (Food aid has fallen by one-half in the past three years.) Poverty is the root cause of both their hunger and the excessive population growth that often accompanies it. What the FAO is describing is the hidden crisis of chronic hunger, which goes beyond the poignant images of refugees and famine victims that appear on our television screens from time to time.

This is not a new problem; more than two decades ago, in November 1974, it prompted the UN to convene a World Food Conference in Rome. More recently, in late 1993, the World Bank held a conference about world hunger in which the president of the World Bank, the secretary-general of the United Nations, and former U.S. President Jimmy Carter, among others, participated; the conference helped the hunger issue reemerge onto the international agenda. Several other international conferences have been held since then.

Now the recognition that the problem of food insecurity has not been solved has caused the UN to schedule a World Food Summit in November 1996, again in Rome. The summit will be preceded by five regional conferences that will focus on developing the agenda. Preparations for these conferences and for the summit are under way in most member countries.

Summit Goals

According to the draft plan of action, "The ultimate goal of the World Food Summit is to ensure the nutritional well-being of all people, today and in the future." This goal is set explicitly in the context of the situation described above; and efforts are needed not only by governments, but also "by all sectors of civil society, including social, cultural, and business institutions, NGOs [nongovernmental organizations], communities, and households."

The World Food Conference of 1974 concluded that the basic solution to the problem of world hunger was to grow more food in the food-deficient countries; but now, 22 years later, there is little evidence that this agreement led to significant improvement in the global food situation. Indeed, the reverse has been true: grain production has increased in the exporting countries of the industrialized world, while food production for local

In some food-deficient countries, the little rice (above) or produce that is grown locally must be exported in order to earn badly needed foreign exchange.

markets has largely stagnated or decreased in the food-deficient countries. One reason for this decline is that many of these countries must take what food they produce internally and export it to earn the foreign exchange needed to service their burgeoning foreign debt. This debt has more than quadrupled since the 1974 conference, and is owed mainly to creditors in the industrialized world.

The World Food Summit draft also notes that the political situation and conditions in the general economy have a powerful impact on the food situation: "A stable and predictable political environment free of war and civil strife is essential for fostering food security." And steady economic growth and "sustainable economic and social progress" are essential. The poverty and hunger of 800 million human beings indicate that these people are the victims not only of bad weather and technological disparity, but also of imprudent development policies, unfavorable global trade arrangements, and a growing debt (exceeding $2 trillion) that further impoverishes them.

The draft summit document also raises a broad set of issues that were not treated in any depth two decades ago: "the sustainable management of ecological processes, environmental services, and social goods." Drawing on the commitments made by the international community at the 1992 UN Conference on Environment and Development (UNCED, popularly known as the "Earth Summit"), it warns that continued unsustainable agricultural use of land, water, and energy, as well as inappropriate technologies, constitute a further threat to the world's food security.

Enormous Impacts

Two countries in particular promise to have the strongest impact on the global food situation for the foreseeable future.

The first is China, whose increasing industrialization and urbanization have already accelerated the deterioration of its land and water and rendered it less able to feed itself. In 1993-94, according to the FAO, China *exported* 13.4 million tons of grain; in 1995-96, it is expected to *import* 23.3 million tons—more than one-tenth of all grain available in the world market.

With its population—now at 1.2 billion—increasing in numbers (by 1 to 1.4 percent per year) and in average income, this one-fifth of the world's population is likely to need even more imports in future years. China will be able to afford them—which many other food-deficient countries will not, especially as prices are bid up in view of the increased demand.

Finally, there is the United States. It is still the world's economic giant, with an enormous impact on the international food system. Nearly half of the world's approximately 200 million tons of annual grain exports start in the United States. U.S. agricultural exports reached $54.1 billion in fiscal year 1995, and are expected to increase by 7 percent annually for the remainder of this century. World prices for grain are, in effect, set on the floor of the Chicago Board of Trade (CBOT). What the United States does about food policy profoundly affects the global food system and the lives of those millions of hungry people in food-deficient countries, as well as the roughly 35 million people in the United States who live below the poverty line and have difficulty making ends meet.

Martin M. McLaughlin

Genetics

Gene Therapy

Even though more than 100 gene-therapy projects have begun in the United States since 1990, researchers are divided about the effectiveness of gene therapy. Gene therapists cannot yet claim to have cured even one patient, and critics charge that researchers have implemented the technology prematurely because of its great financial potential. Proponents, by contrast, argue that they have had some limited successes, and that the only way to determine whether new techniques will work is to try them on humans. What has become clear, however, is that the degree of success of the technique is restricted by the nature of the disorder being treated. For those genetic problems in which tissues can be removed from the body, subjected to gene augmentation, and reinjected, the technique appears very promising. For those disorders in which affected tissues must be treated in the body, the outlook is more problematic.

The first patients to be treated with gene therapy were Ashanthi DeSilva and Cynthia Cutshall. Both suffered from severe combined immunodeficiency disease (SCID), a condition caused by the lack of an enzyme called adenosine deaminase, or ADA. The disorder leaves its victims highly susceptible to infections.

Scientists at the National Institutes of Health (NIH) removed white blood cells from the girls' bodies and exposed the cells to a virus that inserted a healthy ADA gene. When they were reinfused into the girls' bodies, the cells began producing ADA. Both girls are now healthy and attending school. But the treatment could not be considered a complete success because the girls are also being treated with a new drug called PEG-ADA, which artificially replaces ADA; physicians have refused to withdraw the drug for fear that the girls might develop a disastrous infection.

Because white blood cells have a short lifetime, the treatment must be repeated periodically. To overcome this problem, the researchers changed their approach. Using new techniques, they isolated and treated stem cells, the primitive bone-marrow cells that are the precursors of all other blood cells. The team reported in October 1995 that the two girls have not been treated for 30 months—although they still receive PEG-ADA—and that they remain healthy. Virtually all of Ashanthi's treated stem cells are producing ADA, but only about 1 percent of Cynthia's are doing so. Nonetheless, physicians have reduced the dose of PEG-ADA for both, their immune systems are functioning normally, and the researchers are calling the experiment a success.

In May 1995, researchers from the NIH and Childrens Hospital of Los Angeles reported similar results in the treatment of three newborns who suffered from SCID. In these cases, however, stem cells were isolated from the umbilical cord immediately after birth, given the ADA gene, and infused into the infants within four days. White cells in all three children are still making ADA two years later, and the children are healthy. All are receiving PEG-ADA, however, and researchers consider the treatment a limited success.

Two other studies reported in September 1995 showed less-satisfactory results. A team from the University of North Carolina at Chapel Hill used a genetically engineered adenovirus—a common-cold virus—in an attempt to treat 12 patients with cystic fibrosis, a disease that impairs breathing. Although it will be necessary to use the virus to insert a healthy CF gene in the lungs, the first tests attempted to insert it in cells in the nose, which are identical to those in the lungs. The researchers found that fewer than 1 percent of the cells accepted the healthy gene; 10 to 100 times as many cells would have to accept it to provide effective therapy.

More-discouraging results were obtained with an unusual form of therapy called myoblast transfer, used to treat Duchenne muscular dystrophy, a normally fatal muscle disease. Experiments in animals have shown that injection of millions of healthy, immature muscle cells called myoblasts into the muscle allows the myoblasts to fuse with defective cells, inserting a healthy gene. Anecdotal reports have suggested that the technique works in humans as well, and a number of physicians have begun using it clinically. But a controlled trial by a team at Ohio State University in Columbus shows that the technique provides no benefits whatsoever.

HUMAN GENOME
Researchers reported in September 1995 that they had reached a milestone in deciphering the human genome, the collection of genetic information that serves as the blueprint for a human. In a special "Genome Directory" compiled by the journal *Nature*, scientists said that they have identified about 75 percent of the estimated 100,000 genes that comprise the genome, and have identified the function of 10,000 of them. The genes account for only 3 to 5 percent of the genome's total DNA, which carries the genetic code. The remainder is composed of "junk DNA," whose function is currently unknown.

Researchers, especially a team at the Institute for Genomic Research in Rockville, Maryland, used a new technique called expressed sequence tagging, or EST, to examine 335,000 pieces of human DNA and find those that correspond to genes. To date, the team has found that the largest fraction of the identified genes are used in the brain. The team hopes to have a complete map of the genome by the year 2005.

OBESITY
A team at Rockefeller University in New York City reported in July 1995 that they had cloned the protein produced by the newly discovered obesity gene and tested its effects in mice. The gene, identified by the team in 1994, is the blueprint for a protein that tells the body to stop eating once enough body fat has been stored. When the gene is defective, the individual keeps eating and gains weight. The protein, called leptin, produced dramatic effects in mice. Ten obese mice injected with leptin lost an average of 30 percent of their body weight after two weeks. In addition to losing weight, the animals' body temperatures rose, their appetites fell, and their activity levels increased. The researchers have found and cloned a virtually identical protein in humans; they believe that it could be used to produce weight loss in people. The rights to leptin were sold to Amgen Corporation of Thousand Oaks, California.

Researchers looking for the genetic basis of red hair inadvertently discovered the gene that increases a person's susceptibility to sunburning and to skin cancer.

SUNBURN
British scientists searching for the gene that causes red hair reported in November 1995 that they discovered a gene that increases the risk of sunburning and skin cancer in light-skinned people. The gene, called MC1R, was found on chromosome 16, and plays an important role in regulating the production of a pigment called melanin.

Human skin produces two types of melanin: black eumelanin, which protects skin from the damaging effects of ultraviolet (UV) radiation in sunlight; and red phaeomelanin, which apparently increases the skin's susceptibility to UV damage. Redheaded people normally have a predominance of phaeomelanin or a reduced ability to produce eumelanin. That may explain their increased risk of sunburning. But the team from the University of Newcastle upon Tyne in England found that people with a defect in MC1R have a still-higher risk. Although defects in the gene are most common among redheads, they also occur in people with other hair colors.

ALZHEIMER'S DISEASE

Two teams of researchers independently reported in 1995 that they had discovered genes that cause an inherited form of Alzheimer's disease, a disabling neurological illness. Alzheimer's normally strikes people over the age of 55, but the inherited form—which affects as many as 400,000 Americans—affects victims in their 40s. The two new genes account for an estimated 95 percent of all cases of inherited Alzheimer's. Investigators now believe that they have identified all the crucial genes that cause the disorder.

In July 1995, a team from the University of Toronto in Canada announced that it had discovered an Alzheimer's gene, called S182, on chromosome 14. Six weeks later, a team from the University of Washington in Seattle said that it had found the second gene, called STM2, on a different chromosome. Surprisingly, the two genes have similar structures and the proteins for which they are blueprints have nearly identical shapes. The discovery of two nearly identical genes that cause the same genetic disorder is virtually unprecedented and is strong evidence that the genes play a key role in the onset of the disorder. Furthermore, scientists believe that the identification of two such similar genes indicates that there are others as well.

BED-WETTING

Researchers from the Danish Centre for Genome Research in Copenhagen reported in July 1995 that they are close to isolating a gene that causes bed-wetting. As many as 7 million American children over the age of 6 wet their beds once a week, and some researchers believe that more than half of those cases are genetic in origin.

Studying 11 families with a history of the disorder, the team localized the gene to a small area of chromosome 13 carrying about 10 genes. They are now attempting to identify the specific gene, and believe that the best candidate is a brain-chemical receptor called HTR-2. Meanwhile, however, the localization will provide the basis for a genetic test for susceptibility. That should be particularly useful because evidence suggests that children whose bed-wetting is genetic in origin are the most responsive to drugs used to halt the problem.

"WEREWOLF" GENE

Scientists are also close to identifying the gene that causes an extremely rare disease that is probably the source of ancient werewolf legends. The disease is called congenital generalized hypertrichosis, and causes victims to grow excessive quantities of hair on their face and upper body. When the hair is shaved off, however, the victims appear normal. There have been about 50 verified cases of the disease since the Middle Ages; many of those affected have worked in circuses as "ape men" or "human werewolves."

Julia Pastrana (above), a 19th-century woman whose extreme hairiness made her something of a celebrity in her time, suffered from a form of hypertrichosis, a very rare condition now known to have a genetic basis.

A team headquartered at Baylor College of Medicine in Houston, Texas, reported in May 1995 that, by studying one Mexican family with 18 affected members, they had localized the gene to a small region of the X chromosome. The researchers will probably not be able to precisely locate the gene, however, until they find at least one more affected family. Identification of the gene should shed light on the regulation of hair growth, about which very little is known.

The "werewolf" gene is an example of an "atavistic" genetic defect, a mutation that frees a gene that has been suppressed during evolution. Human ancestors were once covered with hair from head to foot. During evolution, they did not actually lose the gene that caused the hair growth. Instead, the gene's activity was suppressed, probably by another gene. It is a mutation in the latter that allows the hair to grow once more.

Thomas H. Maugh II

Geology

Speed Record for Magnetic Shifts

By studying the remains of lava outpourings from 16 million years ago, scientists discovered that Earth's geomagnetic field has the capacity to shift its orientation wildly within a few days. Researchers from the University of California, Santa Cruz, and from the University of Montpellier in France found that the direction of the field rotated 6 degrees a day over a period of eight days. If a similar event were to happen today, compass needles that normally point toward magnetic north would end up pointing toward the latitude of Mexico City in just over a week's time.

The rapid reorientation occurred during a time of geophysical unrest known as a magnetic reversal. Every few hundred thousand years, Earth's geomagnetic field flips over, exchanging north pole for south. The transition period lasts about 10,000 years.

The research team from California and France studied the rate of field changes by examining magnetic grains within ancient lava flows at Steens Mountain in Oregon. When the molten rock was erupted during volcanic outpourings 16 million years ago, the magnetic grains aligned themselves in the direction of Earth's field at the time. Once the lava hardened, the grains became locked into place, providing a snapshot of the magnetic field's orientation at a particular time in geologic history. Because Steens Mountain erupted 56 times during the reversal, the volcano could provide a detailed picture of how the field changed as it flipped over.

The researchers discovered that Earth's field shifted position dramatically even in the few days it took one lava flow to harden. Their finding shows that the geomagnetic field is capable of remarkably fast changes, and it raises tough questions for scientists who are trying to understand what prompts the magnetic field to reverse itself.

Two Plates Tearing Apart

Geologists overturned established ideas about Earth's surface when they determined that the Indian subcontinent and Australia are riding on two separate tectonic plates rather than together on one plate.

According to the theory of plate tectonics, Earth's outer skin is broken up into a dozen large pieces, or plates, which slowly migrate around the globe. As different plates collide, they raise mountain ranges, generate earthquakes, and spawn volcanoes. In the past, researchers had thought that the Indian subcontinent and the Australian continent were embedded within one plate that carried the two landmasses northward. This idea was challenged, however, by researchers

The Indian subcontinent and Australia are riding on two separate plates instead of together on one, as was long thought. The plate boundary region is the hatched zone on the map below.

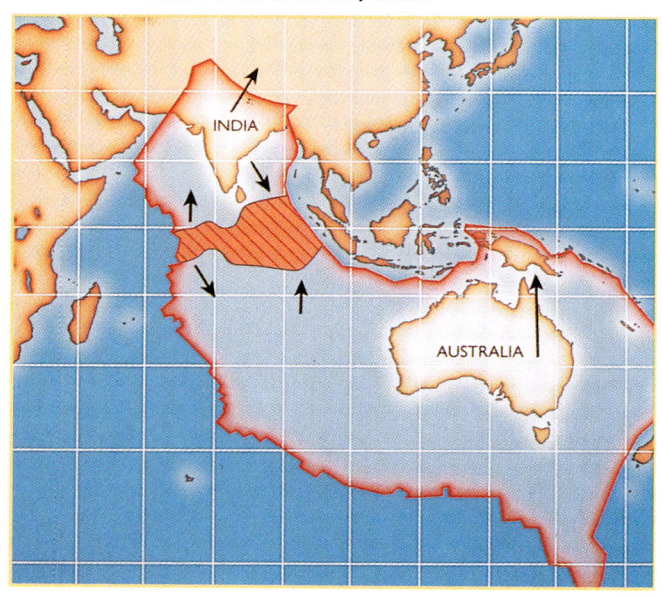

from the Massachusetts Institute of Technology (MIT) in Cambridge, the Lamont-Doherty Earth Observatory, and the École Normale Supérieure in Paris, France.

The international team analyzed seafloor measurements collected during scientific cruises in 1991 and 1986. Faults on the ocean bottom, they found, suggest that India and

Australia have been moving in different directions in the recent geologic past. The two continental regions must therefore sit on separate plates. The plates apparently began tearing away from each other around 8 million years ago.

The Earliest Continents

Australian scientists discovered evidence of the oldest known land, which basked in the sun of the early Earth more than 3.46 billion years ago. Their find pushed back in time the record of when continents first rose up out of the ocean.

The researchers hit pay dirt in northwestern Australia, where they uncovered a geologic feature known as an unconformity—a border formed when sediments are deposited on top of an eroded rock layer. From the erosion evidence, the scientists could tell that this rock existed above sea level and was exposed to the winds and rains. What made the find unusual was the age of the unconformity. At 3.46 billion years old, it is the oldest record of dry land.

The Australian find is not the oldest continental crust, because rocks from Canada have been dated to 3.96 billion years. The Canadian finds show that continents formed about one-half billion years after the planet's birth 4.6 billion years ago. But these Canadian rocks came from deep underground and offer little information about what was happening at Earth's surface. The Australian rocks have the potential to reveal what the atmosphere was like in the early days of the planet.

Renewable Oil Fields

Underground petroleum deposits may refill themselves even as oil is being pumped out of them, according to a controversial hypothesis proposed by a geochemist at the Woods Hole Oceanographic Institution in Massachusetts.

Oil and natural gas are created by the decay of plants and animals that get buried under layers of sediments. As the hydrocarbons form, they seep upward through the crust until they get trapped beneath an impermeable rock, thus creating a reservoir. According to conventional wisdom, the act of pumping petroleum from the ground depletes those reservoirs.

The Woods Hole scientist proposes that oil and gas can rise quickly from very deep deposits to refill reservoirs closer to the surface. The evidence for this came from a site in the Gulf of Mexico, where pumping has not depleted the reservoir as quickly as expected. Moreover, oil recently retrieved from the site does not have the same chemical characteristics as oil first pumped there, indicating that the source of the oil has changed over time.

If correct, this theory would suggest that oil fields may not be depleted as quickly as expected. Still, many scientists voiced skepticism about the idea.

Giant Iron Crystal in the Core

Recent studies using a supercomputer to model the planet's interior suggest that Earth's inner core may be one tremendous crystal of solid iron.

Geophysicists from the Carnegie Institution of Washington and the Georgia Institute of Technology in Atlanta derive their theory from experiments attempting to mimic conditions at the center of the planet, where temperatures can reach above 7,000° F (3,870° C). The scientists used a Cray supercomputer to simulate the crystalline structure of iron within the solid inner core. Part of their aim is to understand why earthquake waves travel faster when moving north-south through the inner core than when traveling east-west.

The computer calculations suggest that iron in the core is packed into a tight hexagonal pattern in which one atom has a dozen neighboring atoms arranged in the form of a hexagonal prism. To fit the earthquake-wave observations, the entire inner core would have to be one giant crystal. This theory has the potential to explain some mysterious aspects of the magnetic field issuing from the planet's core.

The Deepest Eruptions

Scientists studying rocks from Earth's largest volcanic eruptions found evidence that they tapped extremely deep layers of the planet. Researchers from the University of Rochester in New York, the Berkeley Geochronology

Center in California, and the Presidency College of Calcutta, India, discovered an unusual isotope of helium in volcanic rocks that erupted 65 million years ago in India and 250 million years ago in Siberia. These two eruptions—which produced rocks called flood basalts—are among the most massive outpourings that have ever occurred in Earth's history.

The isotope the researchers found, helium 3, is believed to be locked up within the deep mantle of the planet, some 1,800 miles (2,900 kilometers) beneath the surface. Its existence in flood basalts suggests that these lava flows were fed by rock rising from the lowermost mantle. Some scientists had previously argued that flood basalts come from relatively shallow sources of magma.

Reservoirs Steal Time

By storing vast quantities of water in reservoirs, humans have inadvertently changed Earth's rotation, according to a geophysicist at the National Aeronautics and Space Administration's (NASA's) Goddard Space Flight Center in Greenbelt, Maryland.

Since the 1950s, construction of new reservoirs has prevented roughly 10 trillion tons of water from reaching the oceans; this water is instead stored in inland locations. Because most reservoirs are constructed in the midlatitudes of the Northern Hemisphere, the practice of impounding water has shifted the globe's center of mass. Water that would have otherwise been spread around the tropics and other parts of the globe has been moved northward, thus altering Earth's rotation in slight but observable ways.

By transporting water away from the equator and toward the North Pole, humans have moved mass closer to the axis of rotation. This shift in mass tends to speed up the planet's rotation. The faster spin has shaved 8 millionths of a second off the day, calculates the NASA geophysicist.

The filling of reservoirs has also nudged the axis of rotation away from the North Pole. Reservoirs are not located symmetrically around the planet, so they have pushed the spin axis about 2 feet (60 centimeters) toward western Canada since 1960.

Richard Monastersky

Health and Disease

New Attacks on AIDS

In late 1995, the U.S. Food and Drug Administration (FDA) approved the drug saquinavir for use in the fight against human immunodeficiency virus (HIV), the virus that causes acquired immunodeficiency syndrome (AIDS). Saquinavir was the first of a new type of drugs called protease inhibitors; they act on a different part of the HIV life cycle than do older anti-HIV drugs—such as AZT, 3TC, and ddI—that are classified as nucleoside analogues. In March 1996, the FDA approved two additional protease inhibitors, ritonavir and indinavir. Other protease inhibitors were also being tested; at least some of them are expected to gain FDA approval.

Studies indicate that treatment using a combination of protease inhibitors and nucleoside analogues dramatically slows progression of the illness. For example, a combination of indinavir, AZT, and 3TC has been shown to reduce the amount of HIV detectable in a patient's blood by 90 to 98 percent. Such "triple-drug therapy" also boosts CD4-cell counts. These cells, which are the primary target of HIV, play a critical role in the body's immune system. (The commonly accepted definition of AIDS includes HIV-infected people with CD4-cell counts that are no more than one-fifth the level of those of a healthy, noninfected person.)

Cancer Research

A disfiguring and sometimes-fatal cancer found in many AIDS patients is Kaposi's sarcoma; its best-known symptom is blotchy, purplish skin lesions, which result from the uncontrolled growth of blood vessels. Scientists have suspected that Kaposi's sarcoma is caused by a type of herpesvirus, and, in 1996, virologists at the University of California at San Francisco reported that they had grown the virus in their laboratory and photographed it for the first time. The researchers hope to use the virus to develop a blood test that could be used to diagnose

Kaposi's sarcoma in AIDS patients and to conduct epidemiological studies to determine exactly how the virus is transmitted.

Various studies have suggested that the enzyme telomerase plays a central role in cancer by making cancerous cells "immortal." In a normal cell, the end segments of chromosomes, called telomeres, become shorter each time the cell divides. Eventually the telomeres are so short that the chromosomes can no longer divide, and the line of cells dies out. Telomerase is believed to renew telomeres, enabling the chromosomes to divide repeatedly. In 1995, researchers at the University of Texas Southwestern Medical Center in Dallas reported that they had detected telomerase in 80 percent of lung tumors tested. Work on developing techniques to neutralize telomerase was under way at a number of laboratories across the country.

Increasingly, flawed genes are being linked to cancers. In 1995, a team of researchers identified the RII gene, which is linked to about 20 percent of colon cancers, and BRCA-2, a second major breast-cancer gene. Such developments may make it possible to develop tests for detecting the cancers in their early stages, when treatment would be most successful.

UNDERSTANDING ALZHEIMER'S

The genetic underpinnings of Alzheimer's disease (AD), a neurodegenerative illness that afflicts mainly the elderly, are increasingly evident. People who inherit a mutated form of one of three recently isolated genes develop AD before age 65. Research laboratories also have reported that people with AD have heightened levels of a mutated form of the protein tau in their cerebrospinal fluid. The mutated tau is the building block of neurofibrillary tangles—abnormal cell formations that are a sign of AD. Researchers hope to develop methods to block the growth of tau.

There also is growing evidence that AD begins to do its damage long before signs of the illness are detectable. For instance, German researchers concluded that neurofibrillary tangles may be present by the time people reach the age of 20. And a provocative study of the writings of 93 young women about to become nuns predicted with startling accuracy which of them would have AD some 60 years later. Those women whose autobiographical writings were grammatically complex and full of ideas remained sharp-witted into their 80s; the women whose writings were simpler and comparatively void of ideas were more likely to develop AD.

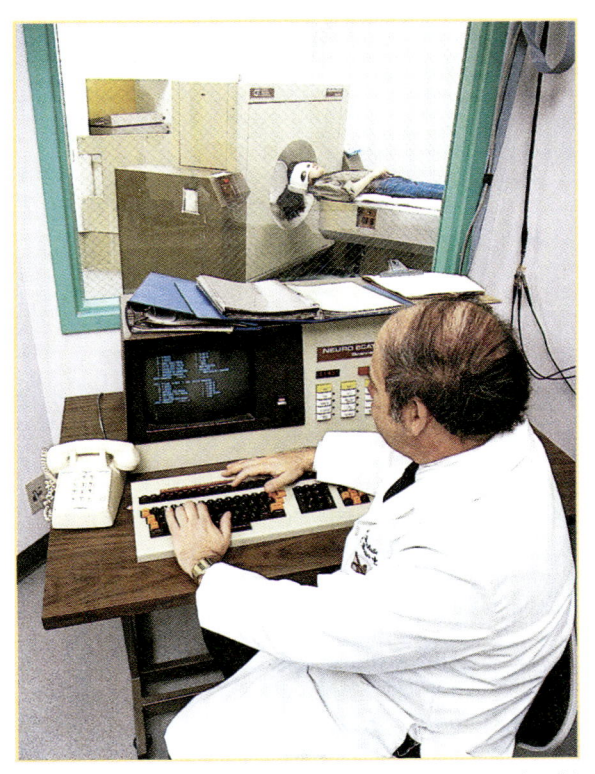

A PET scan (above) can show areas of the brain with low metabolic activity—one of several factors that may indicate a predisposition to developing Alzheimer's disease later in life.

A new technique that combines genetic screening with positron-emission-tomography (PET) scans enables doctors to see brain damage more than a decade before AD's effects become outwardly apparent. The PET scans are used to look for distinct areas of low metabolic activity within the brain, an indication that cells in those areas are dead or comparatively inactive.

FROM ONE SPECIES TO ANOTHER

The possibility that infectious agents can be spread from one mammalian species to another was raised anew in March 1996. British scientists reported a possible link between bovine spongiform encephalopathy (BSE)—a disease commonly called mad-cow disease—and a strain of Creutzfeldt-Jakob disease. Both are similar degenerative and fatal brain diseases. Creutzfeldt-Jakob, which is rare, normally afflicts middle-aged and elderly people. In Britain, however, at least 10 people under the age of 42, including some in their teens, have been infected by a new strain, and it was suspected that they may have been exposed to the bovine disease by eating beef from infected cattle. Creutzfeldt-Jakob disease has a long (10- to 40-year) incubation period. Thus, it was feared that many additional people may be infected, though they have not yet begun to show symptoms.

Mad-cow disease, discovered only in 1986, is most common in Britain; no signs of it have been noted in North America. It is theorized that the disease started in sheep, which are susceptible to a similar malady called scrapie. For decades, until the practice was banned in 1989, British farmers fed cattle a protein supplement made from ground-up parts of sheep. In the early 1980s, the process for preparing this feed changed, possibly allowing contamination through improperly treated parts from infected sheep.

The infectious agent that causes scrapie, mad-cow disease, and Creutzfeldt-Jakob disease breaks down brain tissue, giving it a spongelike consistency. The agent, which is extremely tiny and very hardy, is as yet unknown. Many scientists believe it is a type of protein called a prion. Prions, whose existence has not yet been proven, are thought to be abnormal variants of proteins that form part of the surface of nerve cells.

Trans-species concerns also were raised during 1995 when a man with AIDS received an experimental transplant of baboon bone-marrow cells. The objective of the procedure was to rebuild the man's immune system; scientists hoped that the baboon cells would take root in the man's bone marrow and produce cells that can fight AIDS. (Marrow cells produce 11 kinds of blood cells, including CD4 cells, and baboon marrow appears to be resistant to HIV.)

In an effort to rebuild his immune system, AIDS patient Jeff Getty (above, at right, leaving a San Francisco hospital) received a bone-marrow transplant from a baboon donor.

Other cross-species transplants are also being considered by scientists, in part because a shortage of human donors means that people who could benefit from transplanted hearts or other organs are instead dying. But such procedures present the risk of new infections among humans. Critics of the cross-species transplants point out that the AIDS virus is believed to have originally reached humans from monkeys.

NUTRITION AND HEART DISEASE

The link between diet and cardiovascular disease has become increasingly apparent. A six-year study of more than 43,000 men found

A recent study on the health benefits of fish suggests that the rate of heart disease is the same in people who eat fish six times weekly as it is in those who eat it just once a month.

that those who ate the most fiber had 35 percent fewer heart attacks than those with the lowest fiber intake. Several studies show that vitamin E supplements reduce the risk of heart attacks. And there is mounting evidence that folic acid, one of the B vitamins, limits the development of atherosclerosis, or hardening of the arteries, a leading cause of heart attacks and strokes.

Two recent studies look at the benefits of eating fish. A large study at the Harvard School of Public Health in Boston finds that the rate of heart disease is the same regardless of whether people eat fish six times a week or only once a month. A study at the University of Washington in Seattle does not contradict this, but finds that people who eat the equivalent of 3 ounces of salmon each week are 50 percent less likely to be stricken with cardiac arrest than those who eat no fish at all.

PEDIATRICS

The International Narcotics Control Board, an agency of the United Nations, reported in 1996 that use of methylphenidate, commonly sold under the brand name Ritalin, had soared to more than 8.5 tons in 1994, up from less than 3 tons in 1990. About 90 percent of the drug is consumed in the United States. Low doses of Ritalin are prescribed to combat impulsivity, temper and concentration problems, and other behavioral symptoms associated with attention-deficit disorder (ADD), a brain-chemistry syndrome. An estimated 1.5 million to 2.5 million U.S. children take the drug daily. However, in high doses, Ritalin can produce a sensation of euphoria, which has led to its abuse; some children sniff it like cocaine or even inject it like heroin.

A team of researchers at the University of Utah in Salt Lake City discovered two genes that cause long Q-T interval syndrome, a cardiac-rhythm disorder that can cause sudden death in otherwise-healthy teenagers and young adults. The researchers reported that the genes encode ion channels—proteins found on the surface of heart cells. Mutations in the genes result in cardiac arrhythmia, a disorder characterized by abnormal heartbeats that can disrupt blood flow to the brain and other vital organs.

The FDA ordered the food industry to fortify with folic acid bread, pasta, and other foods made from grains, in an effort to pro-

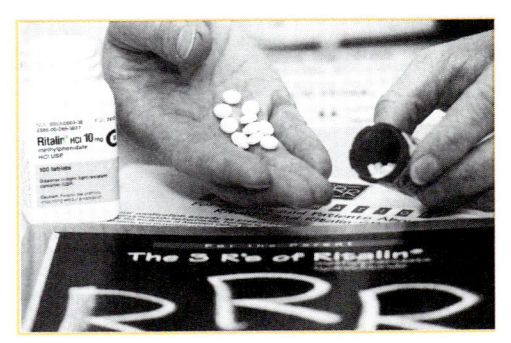

Some fear that the drug Ritalin, used to treat children with attention-deficit disorder (ADD) and other behavioral problems, is being overprescribed.

tect unborn babies from neural-tube defects such as spina bifida. The FDA had earlier recommended that women of childbearing age should consume 400 milligrams of folic acid daily. However, studies had indicated that, on average, women consume only about 200 milligrams daily.

Jenny Tesar

MATHEMATICS

MATHEMATICAL MODELING AND AIDS
Since the 1980s, the AIDS epidemic has been the subject of much scientific study. Scientists in wide-ranging fields have struggled to try to understand how the disease works and what treatments might cure it or stop it from spreading. Although many people might think that such research is restricted to the medical community, mathematicians have actually been working alongside biologists, chemists, and doctors, trying to mathematically model the behavior of HIV (the virus that causes AIDS) so that a successful treatment plan can be developed.

Mathematical modeling involves the belief that natural phenomena can be accurately described by numerical formulas. Put another way, mathematicians believe that numbers and their abstractions can represent the language of nature. This philosophy has been supported by evidence from physics, where forces such as gravity and electromagnetism have been accurately reduced to equations. Encouraged by the success of applying mathematics to the physical sciences, researchers have sought to employ similar techniques to solve problems in other areas.

At the heart of mathematical modeling lie assumptions about the relationships between two or more variables. Often these relationships cannot be reduced to a single cause and effect, but rather to an association that allows someone to predict values of one variable based on knowledge of other variables. For example, such relationships can be used to roughly predict the weight of an adult male based on his height. Clearly, a person's height does not solely determine his or her weight, but it is true that taller people tend to be heavier than shorter people. If research indicates that each additional 1 inch (2.5 centimeters) in height corresponds to an extra 5 pounds (2.3 kilograms) in weight on average, then mathematicians could determine an equation that roughly describes the relationship between height and weight. In the case of AIDS, mathematicians have gotten involved in charting the incidence and distribution of the disease in the general public—a field known as epidemiology. In their initial epidemiological models, mathematicians used assumptions that the rate of new cases of HIV would be proportional to the total number of cases. In other words, as more people become infected with HIV, a related number of new cases of the virus will occur. Similar assumptions—which imply an exponential rate of growth of a condition over time—are used in very simple population models.

Most realistic models, however, involve far-more-complicated assumptions. In particular, real-life models usually involve assumptions about the rate of change of one variable with respect to several others. As more was learned about the spread of AIDS, for example, mathematicians realized that the model for the number of cases required refinement. In population studies, an additional assumption is often included that states that there is a maximum sustainable population. Researchers are divided over whether the HIV virus satisfies this additional assumption. There also may be other variables that would provide a more accurate description of how many new AIDS cases there would be each year. One possible direction of study would be to try to measure the acceleration of the disease—that is, to understand how the rate of new cases changes from year to year.

Most of these epidemiological models require the study of differential equations, operations that require the use of calculus techniques to be solved. More-complicated models usually translate into more-difficult solutions. Many models are so complicated, however, that they require sophisticated computers—and even then, only an approximate solution is reached.

The mathematical techniques used to describe the epidemiology of AIDS can also be used to describe the internal nature, or pathology, of the disease within the immune system of an infected individual. Such epidemiological modeling could help doctors to determine more-appropriate treatment strategies for the condition. In order to monitor the pathology of a disease such as AIDS, mathematicians need medical researchers to identify which variables must be tracked. This is often a very difficult task, because different researchers will have their own lists of variables.

One such variable is a count of T cells, the scout cells of the immune system that recognize foreign agents and are attacked by the HIV virus. By modeling the number of uninfected T cells over time, researchers can see how the disease is progressing. Such modeling, however, is quite a complex procedure. Researchers must take into account factors unrelated to the disease that might affect an individual's T-cell count. For example, the natural production of T cells from the thymus causes the population of healthy T cells to increase. On the other hand, the cells have a finite life expectancy, a variable that demands a term to account for the natural deaths of T cells. Other variables are unique to HIV-positive people. The presence of the virus stimulates the immune system, causing T cells to proliferate and to attack the virus. The virus invades healthy T cells, causing a decrease in the number of uninfected T cells. While mathematicians are able to track changes in an individual's uninfected-T-cell count, medical researchers remain at odds over how to weigh the different variables that can affect the count. As a result, current mathematical models are essentially works in progress.

Modeling even sheds insights into AIDS treatments. One of the first drugs approved for treating AIDS was zidovudine, popularly known as AZT (azidothymidine). The drug is designed to slow the invasion of HIV into healthy T cells. This means that AZT therapy, if successful, would bring the rate of change of the number of uninfected T cells closer to zero than it would be without the drug. Thus, instead of having a sharp decrease in the ability of the immune system to fight diseases, there would be a more stable number of available T cells. Researchers still need more data to assess AZT's therapeutic value and to determine the appropriate timing and dosage of the treatment. Mathematical models should help doctors make such difficult decisions with far more insight than they would otherwise have.

James A. Davis

METEOROLOGY

The feverish pace with which technology is advancing around the globe is helping meteorologists to meet the weather-forecasting needs of a variety of interested groups. This technology has provided both individuals and large organizations with faster and more-accurate forecasting products than ever before. As we approach the year 2000, all of us have come to expect weather forecasts to be better than they were just a few years earlier. And it is not just the casual outdoor enthusiast who uses these products. From the 1996 Olympic Games in Atlanta to the last launching of the space shuttle, meteorologic technology has advanced to the point that weather is no longer the unpredictable variable it once was.

THE OLYMPIC WEATHER SYSTEM

As the 1996 Olympic Games in Atlanta approach, two support offices in Georgia have been set up whose sole responsibility will be to monitor the weather conditions during the events. This will be no small task: the Olympic Games will take place in 34 venues spread across the state of Georgia, where weather conditions are often volatile during the summer months.

To aid in this forecasting challenge, two support offices were established in Georgia by the National Weather Service (NWS): the Olympic Weather Support Office (OWSO) in Atlanta; and the Olympic Marine Weather Support Office (OMWSO), located at the Olympic Marina near Savannah. The heart of the Olympic Weather Support System (OWSS) is a state-of-the-art local area network (LAN) constructed exclusively for operational forecasting. Before the network's components were even set up, a detailed questionnaire was provided to the Atlanta Committee for the Olympic Games (ACOG) to help the NWS identify particular areas of concern for the Games. For example, a strong southeast wind is a concern to rowing officials since such a weather condition would tend to favor competitors in the wind-blocked lanes. And, considering the frequency of thunderstorms during July and

The National Weather Service established two special offices—one in Atlanta (above) and the other in Augusta, Georgia—to provide essential weather information for the 1996 Summer Olympics.

August in Atlanta, lightning is a major worry for any outdoor competition.

Once the concerns of the ACOG were known, the next task was to piece together the necessary forecasting tools and assemble a system to successfully monitor the weather. The data were drawn from a number of sources, including: the National Centers for Environmental Prediction; automated weather-observing systems (including the NWS' Automated Surface Observing Systems, or ASOS); meteorologic satellites operated by the National Oceanic and Atmospheric Administration's (NOAA's) National Environmental Satellite Data Information Service (NESDIS); weather-surveillance radar; a lightning-detection system; and upper-air observing networks.

The amount of information that forecasters will potentially have at their fingertips is overwhelming. In order to manage the flow of data, two software packages have been chosen: The N-A WIPS (National Centers Advanced Weather Interactive Processing System), and the ICWF (Interactive Computer Worded Forecast). This software quickly evaluates numerical data and renders the data in a high-resolution form never before achieved by the NWS. For example, the ICWF puts the forecasts in both text and tabular form. The text form is similar to what the public is currently accustomed to with the traditional "today, tonight, tomorrow" forecast format. The tabular form gives numerical values for variables such as thunderstorm probability, maximum and minimum temperature, wind speed and direction, lightning probability, and ultraviolet-ray index. The output from the N-A WIPS package is graphic in nature, and can be animated and contoured for forecasting purposes. Satellite images and other data can be stored and animated using the N-A WIPS package.

Another tool that will be used by the forecasters in Atlanta is the numerical model used for decades to forecast short-term weather. The Atlanta Olympics, however, will mark the first time that the latest and most-sophisticated models are run independently of the normal twice-daily run at the National Center for Environmental Predictions (NCEP). By running these models specifically for the Atlanta Games, forecasters will have four models per day—and therefore an unprecedented amount of data—from which to develop a forecast.

To successfully monitor the atmosphere for all those concerned, the OWSS will for the first time integrate surface data, satellite and radar data, and local and remote-run modeling data. The system that was finally devised should provide the most-advanced forecasts ever assembled for any Olympic Games, and give us a glimpse into the future of the operational forecasting environment of the NWS. Ultimately those advances—which should certainly help the athletes in Atlanta—will also help all of us in our daily lives.

FORECASTING FOR THE SHUTTLE

In order to launch the space shuttle, accurate observations and forecasts of the upper winds are critical. The sky may be blue and the ground breeze light, but the speed and direction of winds above the ground up to about 60,000 feet (18,300 meters) can produce turbulence and affect touchdown, rollout, and braking-energy margins on landing. With more-conventional aviation interests, such as commercial airlines, forecasters tend to bracket their upper-winds forecasts with worst-case scenarios. This technique is not prudent for shuttle forecasting, however, given that, in its simplest form, the shuttle may be considered "a payload in a controlled explosion" during launch, and an unpowered aircraft during landing. Erroneous wind measurements and forecasts could lead to excessive speeds, causing a runway overrun, or slow speeds, causing the shuttle to land short of the runway. Shuttle forecasters are asked to forecast wind speed to the nearest knot (nautical mile per hour), and wind direction to the nearest five degrees—demands that stretch the limits of our scientific capability.

Fortunately, these limits are being met—and even exceeded—by the Spaceflight Meteorology Group (SMG), which uses the Meteorological Interactive Data Display System (MIDDS) as the primary means for diagnosing and forecasting weather. Currently the MIDDS consists of an IBM mainframe computer running McIDAS-MVS, OS/2 personal-computer workstations, and an extensive list of supplemental software. This system provides forecasters with a variety of data from all potential shuttle-landing sites around the world, as well as with forecast products from a dedicated line to the National Meteorological Center (NMC).

The primary data sources for the upper-air forecast are numerical models similar to those planned for use at the Atlanta Olympics. To arrive at the best consensus forecast, the SMG uses several models, each with its own set of biases and observed trends. Using historical profiles of the data, the forecasters subjectively modify the data from these tools to produce an upper-wind forecast at 17 levels—from the ground up to 60,000 feet. While horizontal depictions of the forecast winds are helpful, the forecasters prefer vertical displays of the numerical data.

The forecasts are then sent to the Johnson Space Center (JSC) computer system, where they are used to calculate the shuttle's trajectory. Copies of the forecasts are disseminated to various personnel at Mission Control and ultimately to the orbiting shuttle crew itself. The crew manually enters the wind forecast into the flight computer and simulates the landing using the expected wind conditions.

Once the forecast is made, it must be periodically compared to any available wind observations, and then revised and disseminated as necessary. Eventually this final step of comparing the forecast to the current conditions will be automated.

WEATHER AND THE INTERNET

There is an inordinate amount of accessible meteorologic data available to large companies that depend on accurate forecasts. It may seem, however, that for the individual who has, say, 100 guests arriving for an outdoor gathering, there is nothing more he or she can do than listen to the local television or radio forecast in order to obtain the latest weather information. However, the Internet gives each of us the opportunity to access highly sophisticated forecasts right in our homes or offices. There are now over 250 Internet sites offering a variety of up-to-date observational and forecast weather maps, radar and satellite images, and the like. These sites have introduced computer-savvy weather-data consumers to information previously available only to professional meteorologists.

One of the Internet's original purposes was to exchange information between universities. Thus, some of the best sites to find meteorologic information are from schools that offer programs in meteorology. The fastest way to find out what type of meteorologic information is available is to perform a Net search using your favorite search engine from your local Internet provider. Although the interactive capabilities of these sites are still in their infancy, one can still download a wealth of knowledge. At Plymouth State College's Internet site, raw satellite and radar data are available every hour, as are surface observations from around the country. At the Internet site for Purdue University, one can access some of the same sets of model data that are used by the shuttle's forecasters and by other professionals around the globe. As more-advanced codes are written, quicker and more-interactive products will become available.

Even historical data is now available on the Internet. Various climate plots and related data sets are accessible from the Climate Diagnostics Center (CDC), a division of NOAA. By accessing the CDC's home page, the user can examine snowfall, rainfall, or other historical data for a particular city. This information is useful not only to researchers, but also to gardeners, students, travelers, or those planning to relocate.

David S. Epstein

The Year in Weather—1995

Overview

Spring 1995

A stormy spring was a fitting start to a year of active weather. In March, California was battered by heavy rains that caused widespread flooding, at least 15 deaths, and an estimated $2 billion in damages. By the end of the month, all of California's 58 counties had qualified for federal disaster assistance. Many communities in the Northwest also suffered severe damage from flooding. Farther east, tornadoes were wreaking havoc in the nation's midsection. In May, a total of 391 twisters were observed, a new U.S. record. On the busiest day of the month, the 18th, an amazing total of 86 tornadoes struck. Elsewhere, winter weather persisted in the Rocky Mountains until the end of May.

Summer 1995

The United States endured a summer of heat and hurricanes. From the Great Plains to New England, the nation sweltered through unusually hot temperatures in July and August; the Midwest region was particularly hard hit. In Chicago, more than 560 people died in July after temperatures exceeded 100° F (38° C) for five straight days. The heat also caused severe crop damage in the Corn Belt and Mid-Atlantic regions. Meanwhile, a record-setting Atlantic hurricane season was under way. Hurricane Allison reached Florida on June 5—the earliest named storm to reach U.S. soil in 27 years. By the end of August, 12 tropical storms had formed in the Atlantic—with more to come.

Fall 1995

The Atlantic hurricane season continued to make waves throughout September and October. In total, 10 full-fledged hurricanes and nine lesser named tropical storms raked the Caribbean Basin and parts of the southeastern United States and Mexico, making 1995 the second-most-active season on record since 1871. Hurricane Opal, the most destructive tropical storm of the year, left at least 59 people dead in the United States, Mexico, and Guatemala, and caused an estimated $3 billion in damage in the United States. Wintry conditions came early to the nation's eastern half. By the end of November, much of the Northeast was blanketed under record snowfall.

Winter 1995-96

Most of the United States experienced a winter that began early, ended late, and was harsh throughout. In mid-December, a massive storm system brought hurricane-force winds and widespread flooding to the Pacific coast, while the first of many strong winter snowstorms hit the Northeast. A particularly strong blizzard buried the region in early January, leaving up to 37 inches (94 centimeters) of accumulation. By season's end, most cities between Washington, D.C., and Boston had set new records for snowfall. Between the coasts, the High Plains and Midwest endured several stretches of arctic temperatures, while the Rocky Mountains and other inland ranges received record snowfall.

U.S. Highlights

- Mount Palomar, California, received more than 14.5 inches (37 centimeters) of rainfall during a three-day span in early March.
- The temperature rose to 75° F (24° C) in Juneau, Alaska, on April 27, the city's all-time high for that month.
- During a 24-hour span in mid-April, 30 inches (76 centimeters) of snow fell on the town of Ree Heights, South Dakota.
- An F4 tornado—with winds in excess of 200 miles (320 kilometers) per hour—flattened Ethridge, Tennessee, on May 18, killing three people and destroying more than 50 homes.
- In May, Washington, D.C., was hit by the first tornado to visit the nation's capital in 53 years.

- International Falls, Minnesota, famous for its low temperatures, set its all-time high, 99° F (37° C), on June 18.
- On July 13, La Crosse, Wisconsin, reached 108° F (42° C).
- Death Valley, California, recorded the year's highest U.S. temperature—127° F (53° C)—on July 29.
- The Southwest endured its driest July in more than a century.
- In early August, Hurricane Felix produced winds of 140 miles (225 kilometers) per hour and caused eight drownings in North Carolina and New Jersey.
- On August 27-28, a record of five tropical cyclones were simultaneously active in the Atlantic Basin.

- The mercury dropped to -3° F (-19° C) in Allenspark, Colorado, on October 24—the nation's lowest October temperature in 11 years.
- Vermont endured its wettest October in more than a century.
- On Hawaii's western island of Kauai, more than 8.9 inches (22 centimeters) of rainfall occurred on November 3.
- On November 25, Reno, Nevada, ended a record-setting spell of 129 straight days without any measurable precipitation.
- Astoria, Oregon, was soaked by more than 17 inches (43 centimeters) of rain in November.
- Syracuse, New York, received a record 34 inches (86 centimeters) of snowfall during November.

- A single December storm dropped 62 inches (157 centimeters) of snow in Sault Sainte Marie, Michigan. The same storm gave Buffalo, New York, a record of 38 inches (97 centimeters) of snowfall—in just 24 hours.
- A severe Pacific storm generated ocean swells as high as 28 feet (8.5 meters) from San Francisco to northern Oregon. Wind gusts of 119 miles (191 kilometers) per hour were measured.
- Hibbing, North Dakota, witnessed the nation's lowest temperature, -50° F (-46° C), on January 20.
- Record-setting seasonal snowfall totals in Hartford, Connecticut, eclipsed 9 feet (274 centimeters).

World Highlights

- During an unusual heat wave that baked Russia in May, Moscow saw temperatures rise as high as 91° F (33° C).

- Torrential rains began two months of flooding on China's Yangtze River in May. The Chinese government reported that 1,200 persons died and 5.6 million more were left homeless due to the flooding, which also caused an estimated $4.4 billion in damage.

- Northern Mexico suffered through a severe drought. Some reservoirs contained as little as 5 percent of total capacity by the end of spring.

- Heavy rains and rapid snowmelt in late May and early June led to severe flooding in Norway.

- Most of Europe suffered through an intense drought. Parts of England received only 1 inch (2.5 centimeters) of rain for the whole season—the lowest recorded total since the 18th century.

- More than 40 Spaniards died in July heat that reached 113° F (45° C). Similar temperatures were recorded in Greece and Italy.

- Temperatures hit 122° F (50° C) and hundreds died in northern India due to the late arrival of the monsoon. The abundant rains that came in July and August, however, caused flooding and left more than 1,000 dead in India, Pakistan, Bangladesh, and Nepal.

- In August, Seoul, South Korea, accumulated about 31 inches (78 centimeters) of rain—more than three times the normal amount.

- On September 11, the *Queen Elizabeth 2* encountered 98-foot (30-meter) waves near Newfoundland that had been spawned by Hurricane Luis.

- Hurricane Ismael, the year's deadliest tropical storm to hit the eastern Pacific, caused at least 57 fatalities in Mexico in mid-September. A total of seven hurricanes and three other named storms hit the region during the season.

- The strongest storm of the northern Pacific cyclone season, Angela, produced sustained winds of 160 miles (255 kilometers) per hour in early November. The powerhouse storm left more than 500 Filipinos dead and thousands more homeless due to widespread flooding and landslides.

- Several snowstorms pounded Eastern Europe in December. The Bosnian city of Sarajevo, for example, received nearly 2 feet (60 centimeters) of snow on December 13.

- A December snowstorm caused at least 99 deaths in Kazakhstan.

- A highly unusual wave of cold weather in Bangladesh left 45 people dead from exposure in December.

- A freak snowstorm occurred in western Mexico in early January. Millions of monarch butterflies died in the snow and cold.

- The season's coldest temperatures in Canada were reached on January 13. In the town of Watson Lake, Yukon Territory, the mercury dropped as low as -52° F (-47° C).

A record-setting heat wave scorched the Midwest in July 1995. In Chicago (above), for example, the mercury rose above 100° F for five consecutive days.

The unusually crowded streets of St. Thomas (above) testified to the power of Hurricane Marilyn, which hit the Virgin Islands and Puerto Rico in September 1995.

Washington, D.C. (above), and the rest of the Northeast were buried by a huge snowstorm in January 1996.

METEOROLOGY 347

Nobel Prize: Chemistry

Two American chemists and a Dutch meteorologist shared the 1995 Nobel Prize in Chemistry for their research on the formation and decomposition of ozone, a gas found in Earth's atmosphere. Paul Crutzen, Ph.D., received his share of the award for showing how atmospheric ozone is depleted by chemicals released from natural processes in soil. F. Sherwood Rowland, Ph.D., and Mario Molina, Ph.D., were recognized for demonstrating how the action of human-made chemicals known as chlorofluorocarbons (CFCs) has decreased the amount of ozone in the atmosphere. "By explaining the chemical mechanisms that affect the thickness of the ozone layer," noted the Nobel committee in its citation, "the three researchers have contributed to our salvation from a global environmental problem that could have catastrophic consequences."

The Ozone Layer

The workings of ozone in the atmosphere were first fully described by the English geophysicist Sidney Chapman in the 1930s. When oxygen (O_2) molecules absorb ultraviolet rays from the Sun, the molecules are split, releasing two oxygen atoms. The free atoms collide with complete oxygen molecules, forming the three-atom configuration that is ozone (O_3). Subsequently, the unstable ozone molecules are destroyed by the action of sunlight and by collisions with other free oxygen atoms. In one of nature's admirable balancing acts, the continuous cycle of destruction and creation leads to a relatively fixed quantity of ozone in the atmosphere.

In the region known as the stratosphere—roughly 10 to 20 miles (16 to 32 kilometers) above Earth's surface—ozone forms a layer. Even in Chapman's day, scientists understood that this ozone fulfills a critical protective function by acting as a buffer against deadly shortwave ultraviolet light from the Sun. Without the relatively thin but essential ozone layer, animal and plant life on Earth would soon perish.

By the 1950s, scientists had noticed that concentrations of ozone in the atmosphere were lower than they should have been. The reason for this deficit remained a mystery for many years. In 1970, Paul Crutzen offered a theory. Knowing that the decomposition of microorganisms in soil and other natural processes release nitrous oxide (N_2O) into the air, Dr. Crutzen proposed that this gas drifts up into the stratosphere. There sunlight breaks down the nitrous oxide into two nitrogen oxides. Dr. Crutzen theorized that these nitrogen oxides react with ozone, changing the three-atom molecules into O_2. Furthermore, the nitrogen oxides are not consumed in the reaction, but continue to break down ozone until they settle out of the atmosphere. Research in the mid-1970s confirmed Dr. Crutzen's scenario.

CFCs and Ozone

In 1973, Dr. Rowland decided to further explore his discovery that trichlorofluoromethane—one of the chemicals known collectively as CFCs—had been found in the troposphere, the region of the atmosphere 6 to 10 miles (10 to 16 kilometers) above Earth. Dr. Rowland set out to learn more about what happens to CFCs in the atmosphere. Joining him was Dr. Molina, who had recently joined Dr. Rowland's lab at the University of California in Irvine.

CFCs were invented in the 1920s by American scientist Thomas Midgley. Initially used as a nontoxic, nonflammable coolant in refrigerators and air conditioners, CFCs eventually found use as propellants in aerosol cans, as industrial cleaning solvents, and as a foaming agent in the production of Styrofoam and other packaging products. Known to be chemically inert, CFCs were believed to be harmless.

A very different picture emerged after Drs. Rowland and Molina determined that CFCs drift up into the stratosphere, where the Sun's ultraviolet light breaks them down, releasing chlorine atoms. These liberated chlorine atoms then react with ozone molecules, releasing O_2 and chlorine monoxide (ClO). In a self-sustaining chain reaction, a single chlorine atom can destroy as many as 100,000 ozone molecules.

Dutch meteorologist Paul Crutzen, Ph.D. (above, at left), and American chemists Mario Molina, Ph.D. (below left), and Sherwood Rowland, Ph.D. (below right), shared the 1995 Nobel Prize in Chemistry for their insights into the formation and decomposition of ozone in Earth's atmosphere.

CFCs are also highly stable, remaining in the atmosphere for upwards of 100 years. Drs. Rowland and Molina calculated that if the release of CFCs were to continue at its 1973 rate, as much as half of stratospheric ozone could eventually be destroyed, with total ozone destruction approaching 15 percent. With the protective ozone layer depleted, increasing quantities of the Sun's ultraviolet rays would reach Earth. In humans, this exposure would mean grave health risks, including higher rates of skin cancer. The damage to vegetation and other life-forms would be equally catastrophic.

Predictably, the manufacturers of CFCs criticized the theories as unfounded. Many scientists were equally skeptical. Gradually, however, the evidence mounted. The most-alarming findings came in 1985, when British scientists announced the discovery of drastic losses in the ozone layer above Antarctica. These results were confirmed by satellite readings.

The three Nobel laureates sounded an alarm on the fragility of the ozone layer. Fortunately, the alarm has not gone unheeded. In 1987, 23 industrialized nations signed a pact—now known as the Montreal Protocol—in which they agreed to phase out production of CFCs by the end of 1996.

Paul J. Crutzen was born on December 3, 1933, in Amsterdam, the Netherlands. He earned his bachelor's and master's degrees at the University of Stockholm, Sweden, where he also received his doctoral degree in meteorology in 1973. In addition to his postdoctoral work at the University of Oxford in England, Dr. Crutzen was a research scientist at the National Center for Atmospheric Research in Boulder, Colorado, from 1974 to 1980. In 1980, he joined the staff of the Max Planck Institute in Mainz, Germany.

Mario J. Molina was born in Mexico City, Mexico, on March 19, 1943. He earned his bachelor's degree in 1965 from the University of Mexico, and his Ph.D. in physical chemistry in 1972. After leaving the University of California in Irvine in 1982, Dr. Molina worked at the Jet Propulsion Laboratory (JPL) of the California Institute of Technology (Caltech) in Pasadena from 1983 to 1989, when he moved to the Massachusetts Institute of Technology (MIT) in Cambridge. He is now a U.S. citizen.

F. Sherwood Rowland, born in Delaware, Ohio, on June 28, 1927, received his bachelor's degree from Ohio Wesleyan University in 1948, and continued his studies in chemistry at the University of Chicago in Illinois, from which he earned his doctoral degree in 1952. Teaching positions at Princeton University and the University of Kansas followed; in 1964, Dr. Rowland joined the faculty of the newly established University of California in Irvine.

Christopher King

NOBEL PRIZE: PHYSICS

The 1995 Nobel Prize in Physics honored two Americans for their discoveries of key subatomic particles. Sharing the prize were Frederick Reines, Ph.D., of the University of California, Irvine, who provided the first experimental evidence of a neutrino, and Martin L. Perl, Ph.D., of Stanford University, California, who detected the tau lepton. The two scientists made their discoveries some 20 years apart and never collaborated, yet their work was instrumental in broadening knowledge of the families of fundamental particles that compose all matter. As the Nobel committee observed in its citation, "They have discovered two of nature's most remarkable subatomic particles."

TO TRAP A NEUTRINO

Forty years ago, few physicists believed that anyone would ever detect a neutrino. This particle had been proposed in 1930 by theorist Wolfgang Pauli in response to puzzling observations involving the decay of certain radioactive atoms—a process known as beta decay. Physicists observed that a tiny portion of energy seemed to disappear during this reaction. The finding appeared to violate the fundamental law of physics that governs conservation of energy. This law states that in any kind of activity or reaction, the sum total of energy must be the same before and after. What could explain the missing energy in beta decay? Would the conservation-of-energy law have to be scrapped in the case of subatomic reactions? In what he called a "desperate solution" to the problem, Pauli proposed that the energy was carried away by a particle that possessed no charge and no mass. This particle came to be called a neutrino, for "little neutron." Theorizing that the neutrino would barely interact with matter of any kind, Pauli himself expressed doubt that the ghostly particle would ever be detected.

Fortunately, the young technology of nuclear power provided hope for an answer. In the early 1950s, Frederick Reines, then working with the late Clyde L. Cowan, Jr., at the Los Alamos National Laboratory in New Mexico, had an idea. The two physicists knew that the reactors in nuclear-power plants should, in theory, be emitting trillions of neutrinos every second. They proposed that, by setting up a detector close to a reactor, it might be possible to "capture" a neutrino.

Beginning in 1953, near the reactor at Hanford in Washington State, and later at the Savannah River nuclear plant in South Carolina, Reines and Cowan set out to trap neutrinos. Their experiments paid off. Within the trap, the sequence of particle interactions unfolded just as the two physicists had planned: a neutrino would collide with the proton of a hydrogen nucleus in the water, creating two more particles—a positron and a neutron. The positron would be annihilated almost immediately with an atomic electron; the device's light detectors captured the subsequent release of light particles known as photons. The neutron, on the other hand, would be slowed by the water and captured a few microseconds later by a cadmium atom; this event would release high-energy gamma rays. The tiny delay between the release of the photons and the

Two Americans—Frederick Reines (above) and Martin L. Perl—shared the Nobel Prize in Physics for their independent discoveries of subatomic particles.

gamma rays signaled the presence of a neutrino. In their experiments, Drs. Reines and Cowan succeeded in detecting a few neutrinos every hour.

HEAVY LEPTONS

Some 10 years after the first detection of the neutrino, Martin L. Perl set his sights on a more massive variety of particle. By the early 1960s, when Dr. Perl joined the staff of researchers at the Stanford Linear Accelerator Center (SLAC), it was known that fundamental particles consisted of two main groups: quarks and leptons. Within each of these groups, two main families of particles were known: among the leptons, the electron neutrino and the electron constituted the first family; the second family consisted of the muon neutrino and the muon. Two corresponding families of quarks completed the picture. Theorists, however, had predicted heavier families of particles within these groups. Dr. Perl began a hunt for a new lepton—a relative of the electron, but considerably more massive.

Throughout the 1960s, Dr. Perl sifted through the records of hundreds of experiments at SLAC. In these so-called "collision events," concentrated beams of particles flash along the facility's 2-mile (3.2-kilometer) track. The beams are sent smashing into targets inside large, elaborate detectors; in the ensuing bursts of energy, new particles declare themselves to scientists who know how to interpret the evidence. Dr. Perl, however, was unsuccessful in his hunt.

Luckily for Dr. Perl, a new facility at SLAC became operational in 1972 that employed two beams rotating in opposite directions around a ring. Matter, in the form of electrons, constituted one beam, while antimatter, in the form of oppositely charged positrons, made up the other beam. The beams were created in the linear accelerator and then injected into a storage ring. A series of about 100 magnets deflected and focused the two counterrotating beams as they made more than 1 million complete trips around the ring every second. At an "interaction" zone along the ring, the beams were collided within a complex detector packed with sensing equipment. As the billions of particles passed one another, only a relative few collided head-on every second. The detectors recorded these events in detail.

Dr. Perl sifted through the records of some 10,000 collision events, searching for a particle that he knew would exist for less than a trillionth of a second before decaying. In 1974, he found evidence of the heavier class of lepton he was looking for: a lepton that was 3,500 times heavier than its electron relative. It was not until 1977 that other accelerator facilities managed to produce the new particle and confirm his discovery. Dr. Perl had designated the particle "U" for "unknown." Physicists soon named it with the Greek letter tau, the first letter of the word *triton*, for "third."

Dr. Perl's discovery demonstrated that there was a third family of fundamental particles. Among the leptons, the tau and the tau neutrino make up this third family, with two varieties of quarks—the so-called "top" and "bottom" quarks—rounding out the picture. Since Dr. Perl's discovery, physicists have continued their quest to bring these particles to light: not until 1994, for example, was evidence found for the top quark. And still other families of fundamental particles may yet be discovered.

Frederick Reines was born in Paterson, New Jersey, on March 16, 1918. He was educated at the Stevens Institute of Technology in Hoboken, New Jersey, earning a master's degree in 1941. He received his Ph.D. in theoretical physics from New York University in 1944. In addition to his research at the Los Alamos National Laboratory in the 1940s and 1950s, Dr. Reines was a physics professor at the Case Institute of Technology before joining the staff of the University of California, Irvine, in 1966.

Martin L. Perl was born in Brooklyn, New York, on June 24, 1927. He earned his bachelor's degree in chemical engineering from the Polytechnic Institute of Brooklyn in 1948, and his doctoral degree in physics from Columbia University in 1955. Dr. Perl taught physics at Columbia and at the University of Michigan before joining the staff at Stanford University and SLAC. He is now a professor of physics there.

Christopher King

Nobel Prize: Physiology or Medicine

The 1995 Nobel Prize in Physiology or Medicine honored two Americans and a German for their insights into the genetic processes that control how embryos develop into fully formed organisms. Sharing the prize were Edward B. Lewis, Ph.D., of the California Institute of Technology (Caltech) in Pasadena; Eric F. Wieschaus, Ph.D., of Princeton University in Princeton, New Jersey; and Christiane Nüsslein-Volhard, Ph.D., of the Max Planck Institute for Developmental Biology in Tübingen, Germany. Studying genetic mutations in the fruit fly *Drosophila melanogaster*, the three developmental biologists discovered how certain key genes play a critical role in embryonic development. Thanks to their groundbreaking work, scientists now know that similar kinds of genes also control development in higher animals, including humans. "Together," noted the Nobel committee in its announcement, "these three scientists have achieved a breakthrough that will help explain congenital malformations in man."

Genetic Workhorse

Once an egg is fertilized, it begins to divide rapidly, forming two cells, then four, then eight, and so on. Initially all these cells are equal, and their arrangement is symmetrical. Within a week or so, however, the cells begin to specialize, and it becomes clear which cells will form the head, which cells the torso, and which cells other regions of the body. The 1995 Nobel laureates made important discoveries about how this process is governed by genes. Their work provided an essential guide for current research in developmental biology and genetics.

For most of this century, the tiny fruit fly *Drosophila* has been the workhorse of genetics research. Proceeding in a matter of days from fertilized egg to embryo to new fly, *Drosophila* permits scientists to make relatively rapid studies of how genetic changes influence development through generations. Edward Lewis became fascinated with mutations in *Drosophila*. He chose to study a particularly bizarre set of mutations, in which cells intended for one part of the body seem to show up in the wrong place. Such mutations result in strange malformations, such as a fly developing two sets of wings. These kinds of mutations are known as *homeotic*, after a Greek word meaning "likeness." While some scientists dismissed homeotic mutations as simple errors in gene function, Dr. Lewis believed that they might be fundamentally important.

Beginning in the 1940s, Dr. Lewis spent decades investigating homeotic mutations. By breeding and crossbreeding successive generations of flies, he introduced selective mutations into the flies' genes. These mutations would cause changes in one system within the developing flies, leaving other systems to develop normally. By taking note of how specific mutations affected development, Dr. Lewis identified and classified a family of genes on the fly's third chromosome. These control genes, later named *homeotic selector genes*, governed the formation of various body segments.

In another key insight into these genes, Dr. Lewis demonstrated that the arrangement of these genes on the fly's chromosome matched the body segments that the genes control. The first genes control the development of the head, while the genes in the middle govern the formation of the body, and the last genes control the tail region. This phenomenon is known as the *colinearity principle*.

Classifying Mutations

Dr. Lewis published a major summary of his findings in 1978. His work intrigued two young scientists at the European Molecular Biology Laboratory in Heidelberg, Germany. Eric Wieschaus and Christiane Nüsslein-Volhard decided to concentrate on an even earlier stage of *Drosophila* development. They planned an ambitious project to identify the specific genes that help change a fertilized egg into a segmented embryo. However, sifting through the fruit fly's 20,000 genes would not be easy.

Using chemicals to alter *Drosophila* genes, the two scientists created nearly 40,000 families of mutant flies. For more than a year, Drs. Wieschaus and Nüsslein-Volhard painstakingly studied fly embryos. They shared a small workspace, using a dual microscope into which both could look at the same time. They found that most mutations had little effect on development. However, they soon began to close in on approximately 150 genes that produced extraordinary mutations in the embryos, such as missing organs, or skin consisting entirely of nerve cells.

Gradually, the two scientists recognized that the mutations fit into three distinct classes, corresponding to three different sets of genes. They named these three gene types "gap," "pair-rule," and "segment-polarity." Drs. Wieschaus and Nüsslein-Volhard published their findings in 1980. Subsequent investigation provided a detailed model of how the genes work. Once activated, the gap genes divide the embryo into broad regions. The pair-rule genes then cause these regions to be broken into segments. Finally the segment-polarity genes set up structures within each segment from front to back. At this point in the process, the homeotic selector genes, originally identified by Dr. Lewis, come into play. With their actions overseen by the pair-rule and segment-polarity genes, the homeotic selector genes direct the formation of different body parts within each segment.

The 1995 Nobel Prize in Physiology or Medicine was awarded to three scientists—Edward Lewis, Ph.D. (above), Eric Wieschaus, Ph.D. (below), and Christiane Nüsslein-Volhard, Ph.D. (bottom)—for their groundbreaking studies of the genetic processes that control embryonic development and produce mutations in all animals, including humans.

New Insights

In the wake of the discoveries by these three scientists, many other researchers began to investigate similar genetic functions in other organisms. The same types of genes have been observed in mice, chickens, zebra fish, and other higher animals—including humans. In fact, the work from Drs. Lewis, Wieschaus, and Nüsslein-Volhard is yielding insights into malformations that occur in human embryos. For example, it is now known that high doses of vitamin A during pregnancy will disturb the regulation of the same types of genes that the three Nobel laureates identified. This disturbance results in severe birth defects. Similarly, mutations in a human gene related to a *Drosophila* pair-rule gene can cause a condition known as Waardenburg's syndrome. This rare disease involves hearing loss, defects in the facial bones, and altered pigmentation in the iris of the eye.

The decision to award the prize to these three scientists was widely applauded in the scientific community for several reasons. For one, Dr. Nüsslein-Volhard is the first woman scientist in seven years to be honored with a Nobel Prize. She is one of only 10 women to achieve science's highest award since the prizes were first given in 1901.

The 1995 prize also marks the first Nobel in 60 years awarded in the field of developmental biology. Many observers believed that the field had been neglected by Nobel committees in the past. There was also universal agreement that the work of Drs. Lewis, Wieschaus, and Nüsslein-Volhard demonstrated the value of basic scientific research. Before making their discoveries, the three were able to pursue lengthy investigations into fundamental biological processes—work that carried no promise of immediate scientific payoff or commercial benefit. This kind of basic work has grown increasingly rare in today's climate of shrinking funds for scientific research, a troubling trend to many observers.

Edward B. Lewis was born in Wilkes-Barre, Pennsylvania, on May 20, 1918. He earned his undergraduate degree in biostatistics from the University of Minnesota in 1939. After taking a master's degree in meteorology at the California Institute of Technology (Caltech), Dr. Lewis received his doctoral degree in genetics from Caltech in 1943. He has been on the faculty there since 1946. At age 77, as a professor emeritus of biology, he is still active in research.

Eric F. Wieschaus was born on June 8, 1947. He received his bachelor's degree in biology from the University of Notre Dame in Indiana in 1969, and his Ph.D. in biology from Yale University, New Haven, Connecticut, in 1974. He did postdoctoral work at the University of Zurich, Switzerland, and at the Laboratory of Molecular Genetics in Gif-sur-Yvette, France. At the time of his prizewinning work, he was a group leader at the European Molecular Biology Laboratory in Heidelberg, Germany. Dr. Wieschaus joined the faculty of Princeton University in 1981 and is now a professor in the department of molecular biology.

Christiane Nüsslein-Volhard was born in Magdeburg, Germany, on October 20, 1942. She studied biochemistry at Eberhard-Karls University in Tübingen, Germany, and at the University of Tübingen. She earned her Ph.D. in genetics from the University of Tübingen in 1973. Her postdoctoral work included research at the Biozentrum Basel in Switzerland and at the University of Freiburg, Germany. She began her work at the European Molecular Biology Laboratory in Heidelberg in 1978. She is now director of the Max Planck Institute for Developmental Biology in Tübingen.

Christopher King

NUTRITION

OLESTRA APPROVED

On January 24, 1996, amid a flurry of controversy, the U.S. Food and Drug Administration (FDA) approved olestra, a fat substitute developed by the Procter & Gamble Company, for use in snack foods such as potato chips and crackers. Procter & Gamble's interest in and research on olestra began in the early 1970s; the company initially submitted a drug application to the FDA because it appeared that the compound lowered cholesterol. In 1987, Procter & Gamble submitted a food-additive petition to the FDA for the approval of the use of olestra as a fat substitute in shortening and cooking oil. Three years later, the company amended its petition to restrict the use of olestra as a total replacement for traditional fats in the manufacturing of snack foods.

Olestra is a human-made compound of sucrose (sugar) and vegetable oil. Most dietary fat comes in the form of triglycerides—one molecule of glycerol hooked to three molecules of fatty acids. In olestra, sucrose replaces glycerol and is attached to six, seven, or eight fatty acids. Fat-free olestra also does not provide any calories due to its unique chemical structure. Digestive enzymes need to reach the sucrose center to break down the compound for absorption into the body, but with the long chain of fatty acids, the enzymes are unable to reach the center in the time it takes the substance to pass through the digestive tract. Unlike other fat substitutes currently on the market, olestra is the first that provides no calories.

Although olestra may sound like an ideal fat substitute, the compound is not problem-free. First, olestra may give some people digestive problems, including bloating, cramping, flatulence, and diarrhea. According to the FDA, these gastrointestinal side effects will not cause other medical problems. The compound also inhibits the body's absorption of fat-soluble vitamins A, D, E, and K, and reduces the absorption of carotenoids, nutrients found in deep-orange and red fruits and vegetables and green leafy vegetables. Numerous studies indicate that diets rich in carotenoids are associated with a lowered risk of cancer and macular degeneration (a type of vision impairment). The FDA is requiring that Procter & Gamble fortify olestra with the fat-soluble vitamins, but not with carotenoids. The FDA believes that the connection between carotenoids and disease prevention remains unproven, although the agency will continue to review all new research on these phytochemicals.

All manufacturers using olestra are mandated to provide a label indicating its use as follows: "This product contains olestra. Olestra may cause abdominal cramping and loose stools. Olestra inhibits the absorption of some vitamins and other nutrients. Vitamins A, D, E, and K have been added." The FDA believes that the labeling provides consumers with necessary information so they can stop eating a food that contains olestra if it causes abdominal distress. Procter & Gamble is also required to monitor olestra's consumption and to look at its long-term effects. The FDA was slated to review these studies within 30 months.

According to the FDA, approval for olestra was based on evaluation of more than 150 studies conducted by Procter & Gamble. The FDA also asked for advice from outside experts through its Food Advisory Committee. A special working group of this committee met in November 1995 to review data from the FDA and Procter & Gamble. A clear majority of the working group concurred that the scientific evidence supplied reasonable certainty that the use of olestra would be safe.

Many consumer and health groups—including the American Public Health Association, the National Women's Health Network, and the Center for Science in the Public Interest—opposed the approval of olestra. Such opponents criticized the length and design of the Procter & Gamble studies and the fact that several members of the FDA working group had previously been consultants to the food industry. Critics argued that the FDA should do its own research on a new product. However, limited funding has forced the FDA to rely mainly on studies done by a company seeking approval for a new product.

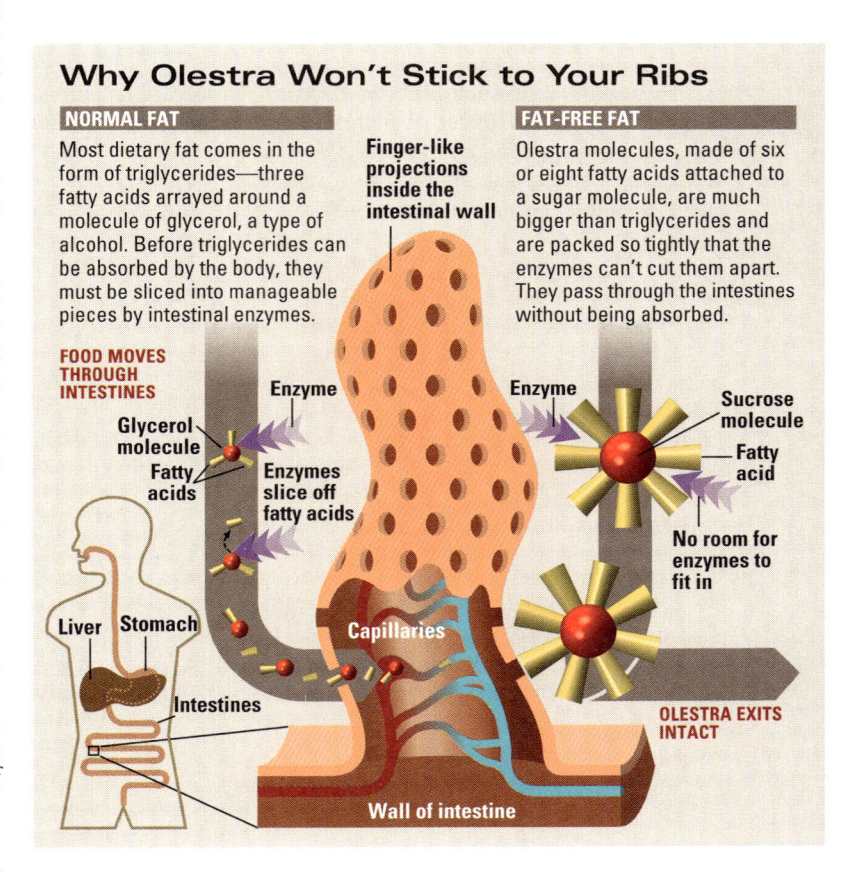

No one disputes the fact that high-fat diets are associated with increased risk for coronary-artery disease, certain cancers, and obesity. However, many health experts question the value of bringing to market a product that may cause its consumers unpleasant side effects. Moreover, other experts doubt that a fat-free fat will actually help Ameri-

Why Olestra Won't Stick to Your Ribs

NORMAL FAT
Most dietary fat comes in the form of triglycerides—three fatty acids arrayed around a molecule of glycerol, a type of alcohol. Before triglycerides can be absorbed by the body, they must be sliced into manageable pieces by intestinal enzymes.

FAT-FREE FAT
Olestra molecules, made of six or eight fatty acids attached to a sugar molecule, are much bigger than triglycerides and are packed so tightly that the enzymes can't cut them apart. They pass through the intestines without being absorbed.

cans lose weight. Studies have shown that people who eat fat-free cookies or foods containing artificial sweeteners such as NutraSweet have a tendency to overeat them. U.S. rates for obesity have increased since 1981, despite the introduction of NutraSweet and many fat-free products. Products made with olestra were targeted to reach the market by the middle of 1996, leaving the ultimate fate of olestra up to U.S. consumers.

New U.S. Dietary Guidelines

The federal government released revised dietary guidelines on January 2, 1996. "Nutrition and Your Health: Dietary Guidelines for Americans," a joint project of the U.S. Department of Agriculture (USDA) and the U.S. Department of Health and Human Services, was first published in 1980. The guidelines have since been updated every five years to include the most-recent knowledge derived from medical and scientific research. The latest guidelines are considered more user-friendly, and provide consumers with clear advice about choosing healthy diets and lifestyles. The dietary guidelines also form the cornerstone of the federal-nutrition policy and initiatives such as the nation's school-lunch program.

The revised standards include instructions on implementing the "Food Guide Pyramid" and the "Nutrition Facts Label." The guidelines continue to underscore the importance of balance, moderation, and variety in food choices, with an emphasis on grains, fruits, and vegetables. The text offers examples of specific American foods that provide given nutrients, as well as foods from other cultures. Although the guidelines have changed minimally in their wording from previous editions, the text contains significant changes that echo information gained by the scientific community in the previous five years.

Regarding weight, the dietary guidelines now recommend that adults maintain their weight in a single healthy range, replacing standards that allowed for increasing weight as an individual ages. Although the guidelines state the importance of weight control, the text adds that some overweight people may not be able to lose weight, and those individuals should set a goal of maintaining their weight. Consumers are advised to work toward a slow, steady weight loss of about 0.5 to 1 pound (0.2 to 0.4 kilogram) per week by moderately reducing calories and increasing physical activity. The guidelines specifically suggest 30 minutes or more of moderate exercise—such as walking, gardening, or housework—on most days of the week. The report also contains a warning about the dangers of crash-diet programs.

For the first time, the guidelines state that moderate consumption of alcohol may lower the risk of heart attacks, and that alcoholic beverages have been used to enhance the enjoyment of meals by many cultures throughout history. The text clearly states that "moderation is defined as no more than one drink a day for women and no more than two drinks a day for men." The 1995 guidelines warn that excess alcohol can be harmful, increasing risk for violence, suicide, accidents, and certain diseases.

The revised dietary guidelines also include a positive statement regarding the healthfulness of vegetarian diets. The panel that revised the standards agreed that a vegetarian diet is compatible with the advice in the guidelines. This bulletin is available to the public for 50 cents per copy from: Consumer Information Center, Department 378-C, Pueblo, Colorado 81009.

Soy Protein Lowers Cholesterol

Researchers from the University of Kentucky in Lexington, under the direction of James Anderson, M.D., report that soy protein substituted for animal protein in human diets significantly lowers blood-cholesterol levels. Although numerous investigations since the early 1900s have shown that soy protein lowers cholesterol in animals, the results in humans had previously been inconsistent.

The results of a meta-analysis of 38 earlier studies on soy protein were published in August 1995 in the *New England Journal of Medicine*. (A meta-analysis combines the results of multiple smaller studies, thus increasing the statistical power of the research topic.) The studies chosen for analysis all had a control group, and isolated soy protein or textured soy protein was substituted for animal protein.

The study reports that the total caloric intake, saturated fat, total fat, and cholesterol were comparable in the control diet and the soy-based diets in most of the investigations. Soy-protein intake averaged 47 grams per day in the test subjects. The meta-analysis showed that consumption of soy protein in place of animal protein was associated with a significant decrease in total cholesterol, low-density lipoprotein (LDL, or "bad" cholesterol), and triglycerides. A significant change in high-density lipoprotein (HDL, or "good" cholesterol) was not seen.

Studies indicate that forgoing animal proteins and instead eating soy protein—a nutrient found in soy milk, tofu, bean sprouts, and other foods—lowers cholesterol levels in humans.

Soy protein lowered cholesterol most effectively in people with high initial levels of cholesterol—greater than 250 milligrams per deciliter. Since more than one-third of the studies used 31 grams of soy protein per day, the researchers suggest that 31 to 47 grams of soy protein consumed daily can significantly lower total cholesterol and LDL cholesterol. Individuals can obtain 30 grams of soy protein by consuming two cups of soy milk and one 3.2-ounce soy-protein patty. Tofu, tempeh, soy milk, tofu franks, soy-protein patties, and soy-based convenience meals are now readily available in health-food stores and most supermarkets.

Maria Guglielmino, M.S., R.D.

OCEANOGRAPHY

OCEAN MAP

The creation of the first "good" map of the ocean bottoms was reported by the *New York Times* on October 24, 1995. Satellite measurements were used to develop a detailed picture of the global seafloor.

"It's like being able to drain the oceans and look at Earth from space," says David T. Sandwell, Ph.D., a geophysicist at the Scripps Institution of Oceanography in La Jolla, California, who helped make the new map. "We're having a data feast. It really is a time of celebration."

At a Washington, D.C., news conference, Sandwell and Walter H.F. Smith, Ph.D., of the National Oceanic and Atmospheric Administration (NOAA) revealed a striking multicolored map 12 feet (3.6 meters) wide and 8 feet (2.4 meters) high of the 71 percent of Earth that underlies the oceans.

Previously, seabed maps were generally made with the aid of surface ships that bounced sound waves off the bottom to get a glimpse of the wilderness below. The sound-wave method produced little more than a scattered patchwork of readings that were transformed into a map only with a great deal of guesswork.

The new map is far more accurate. It is based in part on secret U.S. Navy data that had been declassified as part of a post–Cold War peace dividend. From an orbit 500 miles (800 kilometers) up, a navy satellite in the 1980s made gravity measurements over the world's oceans as part of a quiet effort to increase the accuracy of long-range missiles fired from submarines.

Almost immediately, the existence of the map revamped how scientists viewed plate tectonics—a widely accepted theory holding that Earth's surface is made up of a dozen or so plates that float on a sea of molten rock and grind past one another in earthquake spasms. Dr. Jian Lin, Ph.D., a geophysicist at the Woods Hole Oceanographic Institution on Cape Cod, Massachusetts, says the new data would "almost certainly change our thinking about the active geological processes in the world's deep ocean basins."

Southwest Indian Ridge

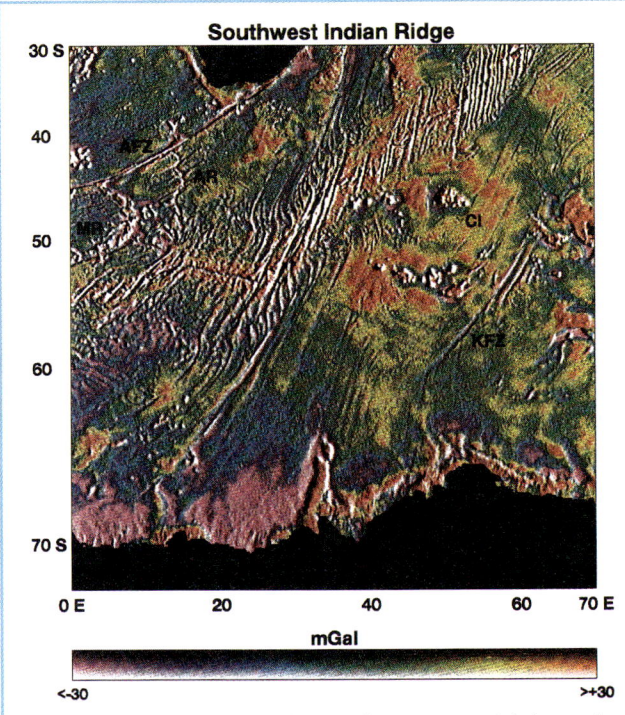

In 1995, scientists unveiled the first accurate global map of the seafloor. The map, based on satellite data, has called into question theories on plate tectonics and volcanology.

Other stated benefits were more specific; for example, the number of known volcanoes under the sea—many of which are strangely elongated—instantly doubled.

NAVY SECRETS

Without the release of certain declassified navy documents dealing with marine gravity, the new global seafloor map would have remained impossible. But those retrieved documents were only the "tip of the iceberg." The *New York Times* (November 28, 1995) announced the existence of a treasure trove of physical data about the sea that was gathered in secrecy during the long decades of the Cold War. This new data, finally released publicly by navy sources, deal with such intriguing topics as geomagnetics, seafloor sediments, ice shape and depth, marine bathymetry, temperature and salinity fields, and even ocean optics and bioluminescence.

OZONE ALERT

Steady depletion of the atmospheric ozone layer—causing some areas of the world to be exposed to above-normal levels of ultraviolet (UV) radiation—has become a proven fact, no longer just an arguable theory, according to a 1995 study conducted by the Scripps Institution of Oceanography.

The depletion of ozone in the lower stratosphere had been a global environmental concern for two decades, but until the Scripps study, little data had been collected on whether it was causing an increase in the amount of harmful ultraviolet-B radiation reaching Earth's surface.

While scientists had attempted to estimate the impact of declining ozone, the variability of clouds—which also screen out ultraviolet radiation—had been a major stumbling block.

Dan Lubin, a research physicist at the California Space Institute at Scripps, and Elsa Jensen of SeaSpace Corporation in San Diego, California, overcame that obstacle by developing a satellite-mapping method that takes cloud variability into account. Because it includes satellite measurements of global ozone levels and solar radiation, the new method is able to predict when the erosion of the ozone layer supersedes variations in cloud opacity in different parts of the world—thus causing an increase in UV radiation to reach Earth's surface.

"The scientific community has kind of assumed that any decrease in the ozone layer over temperate regions will automatically bring about an increase in UV radiation at Earth's surface that has some biological significance," Lubin says. "What our study shows is that whether or not there is a biological significance to the UV increase depends very much on where you are."

The results indicate that, as of 1995, large parts of North America, Central Europe, the Mediterranean, Canada, New Zealand, South Africa, and the southern half of Australia, Argentina, and Chile were subjected to significant increases in UV radiation.

ZOOPLANKTON DYING

Another Scripps study reported in the March 3, 1995, edition of the journal *Science* that the population of zooplankton in the waters off the coast of Southern California declined by 80 percent in the past 42 years. The decline was attributed to a warming of the ocean surface in that region over the same time period.

Zooplankton includes a wide variety of organisms that drift with ocean currents and feed upon planktonic plant life called phytoplankton. Because the tiny creatures form a vital link near the base of the food chain, a decline in their numbers is viewed as potentially ominous for other marine life.

"Zooplankton is the main diet of many species of fish, including sardines, anchovy, hake, jack mackerel, and Pacific mackerel," says John McGowan, a biologist in the Scripps Marine Life Research Group who coauthored the study with Dean Roemmich, a physical oceanographer at Scripps.

Sea-surface temperatures increased an average of 2° to 3° F (3.6° to 5.4° C) throughout the area between 1951 and 1993. Such a change could affect zooplankton growth by preventing nutrients such as nitrates and phosphates, needed for phytoplankton growth, from being brought to the ocean surface, where zooplankton tend to congregate.

ALVIN SETS ANOTHER DIVE RECORD

The first U.S. research submarine, the deep-diving submersible *Alvin*, made its 3,000th dive to the ocean floor on September 20, 1995—a record no other such submersible has achieved.

The 23-foot (7-meter), three-person submersible has been operated by the Woods Hole Oceanographic Institution since 1964 for the U.S. ocean-research community.

Dive 3,000 was made on the Juan de Fuca Ridge off the coast of Washington as part of a series of 12 dives to study geologic and geochemical processes of hydrothermal vents along the Endeavor Segment of the ridge.

In July 1986, the submersible gained worldwide attention when it explored the wreck of the ocean liner *Titanic* during the Woods Hole Oceanographic Institution's second expedition to the wreck, which had been discovered in 1985 using unmanned submersibles.

Gode Davis

PALEONTOLOGY

MONGOLIA'S NESTING DINOSAUR

During an expedition to Asia's remote Gobi Desert, a team of U.S. paleontologists discovered the bones of a dinosaur perched on a nest of eggs. Researchers from the American Museum of Natural History in New York City, who made this find, called it the first hard evidence that dinosaurs actually sat on their nests in a manner similar to modern birds.

The 80-million-year-old *Oviraptor* was unearthed atop a clutch of at least 15 eggs carefully arranged in a circle. The 8-foot (2.4-meter)-long carnivorous dinosaur had its forelimbs surrounding the eggs, and its hind legs tucked up underneath its body. The brooding animal may have died when it became entombed by a massive dust storm.

According to scientists, the *Oviraptor* specimen adds further support to the theory that birds evolved from theropod ancestors similar to *Oviraptor*.

THE FIRST SHARK

British paleontologists from the University of Birmingham uncovered scales from the earliest known sharks, pushing the records of these creatures back to the Ordovician period, some 455 million years ago.

The scales, measuring about 0.04 inch (1 millimeter) long, were discovered in Colorado. Although the scales clearly came from a shark, the exact shape of the animal will remain unknown until scientists discover fossils that show the outline of the entire body. It is even possible that this early shark lacked jaws; indeed, no known fish from that time had teeth or jaws.

Prior to the Colorado discovery, the oldest shark scales came from rocks 25 million years younger, dating to the Silurian period.

TEETH OF A CONODONT

Studies of mysterious animals called conodonts reveal that they may be the earliest known vertebrates, which would make them some of our oldest relatives.

Conodonts were eel-shaped creatures that lived from 520 million to 205 million years ago. In their mouths, they had tiny, hard plates resembling teeth.

When scientists first discovered these isolated plates in the 1800s, they could not identify to what kind of animal the plates belonged. More than 100 years later, in the early 1980s, paleontologists found the first complete conodont fossil, thus showing the creature's true shape.

In 1995, researchers filled in more details about conodonts. By studying the animal's eyes, a group from Leicester University in England found that conodonts had a distinctive type of muscle that allowed the animal to look up and down. Because this muscle appears only in vertebrates, its presence in conodonts suggests that they were the earliest vertebrates.

Other researchers from the University of Leicester investigated the hard, toothlike plates in the mouths of conodonts. Some scientists had wondered whether these features functioned as teeth or whether they were filters for capturing small plants and animals from the water. The Leicester group discovered microscopic scratches on the conodont pieces, indicating that they served for crushing and shearing food.

The two discoveries showed that the earliest vertebrates were active predators. Conodonts could track potential prey by using their movable eyes, and then could catch victims with their teeth.

Largest Carnivorous Dinosaur

Bones from the largest known carnivorous dinosaur were discovered in Argentina. The fearsome animal, named *Giganotosaurus carolinii*, stretched nearly 42 feet (12.8 meters) in length and weighed an estimated 9 tons.

Scientists have found the remains of the largest known carnivorous dinosaur. Giganotosaurus **weighed about 9 tons and had 10-inch-long serrated teeth.**

The Argentine giant usurps *Tyrannosaurus rex* as the king carnivore from the Cretaceous period. *T. rex* weighed 6 tons and reached 39 feet (11.8 meters) in length.

Giganotosaurus lived 100 million years ago, when South America was completely separate from North America, with no land-bridge connection. The separation allowed dinosaurs to evolve entirely different species on the two continents.

For unknown reasons, dinosaurs in South America grew to gargantuan sizes. Six years ago, Argentine paleontologists discovered the remains of what may be the largest known dinosaur, a plant-eating behemoth called *Argentinosaurus*. This sauropod dinosaur may have weighed as much as 100 tons, beating out the next-largest dinosaur by 20 tons. *Argentinosaurus* remains a mystery, however, be-

Fossil remains suggest that the herbivorous Argentinosaurus **was the largest animal to ever walk on Earth. An Argentine researcher estimates that the massive creature weighed 100 tons.**

cause researchers have not yet completed the herculean task of pulling the fossils out of the ground.

ANCIENT BACTERIA REVIVED

A pair of microbiologists claim to have resuscitated bacteria from the time of the dinosaurs. Their report, however, was greeted with skepticism by scientists who believe that bacteria cannot remain alive after millions of years.

The two scientists from California Polytechnic State University in Pomona extracted bacterial spores from insects encased in amber. The insects date back to as long as 135 million years ago. The researchers revived the spores by placing them in a nutrient bath; the scientists then grew colonies of the purportedly ancient microbes.

According to the California scientists, the bacteria survived because they went into a dormant, dehydrated state and were protected by a tough protein shell. The amber provided further protection by keeping the cells dry.

Other microbiologists, however, doubt that organisms could have survived so long. The "revived" microbes may have been modern species that contaminated the experiments, say skeptical scientists.

JURASSIC BIRDS

Chinese and U.S. paleontologists announced the discovery of 140-million-year-old bird fossils belonging to a new species they named *Confuciusornis*. This feathered flyer lived during the Jurassic period and ranks as the second-oldest known bird.

The pigeon-sized creature had a number of primitive features not seen on modern birds. Its fingers, for instance, were long and ended in large, curved claws; modern birds have stunted, clawless fingers that are hidden by the wing feathers.

The Chinese bird is only slightly younger than the oldest known bird, the 150-million-year-old Jurassic creature called *Archaeopteryx*. Despite the proximity in age, considerable evolution occurred between the two species. Although *Archaeopteryx* had teeth, *Confuciusornis* had a toothless beak similar to those of modern birds.

Richard Monastersky

PHYSICS

A NEW STATE OF MATTER

In 1924, Satyendra Nath Bose set forth a new statistical framework for the behavior of photons. By applying Bose's statistics to gases, Albert Einstein soon after predicted a mysterious state of matter, appropriately dubbed a Bose-Einstein condensate (BEC). More than 70 years later, in the frigid confines of the Joint Institute for Laboratory Astrophysics (JILA) in Boulder, Colorado, physicists glimpsed a Bose-Einstein condensate for the first time and rejoiced.

What Einstein imagined but never lived to see was the behavior of atoms chilled to near-absolute-zero temperatures—0° K (-460° F or -273° C). Under such conditions, the atoms lose their momentum and grow sluggish. In a quantum-mechanical trade-off known as the Heisenberg uncertainty principle, if the momentum of an atom is well defined—in this case, nearly zero—uncertainty in the atom's position grows large. The wave function, which describes the atom's position in terms of probabilities, spreads out.

If enough atoms have been cooled, Einstein reasoned, their wave functions would merge and overlap, so that individual atoms would no longer be distinguishable. A coherent state of matter would form that scientists recognize today as being similar to light from laser beams, which consists of photons in the same phase.

During the past 15 years, physicists seeking the BEC holy grail have devised several ways to freeze atoms. Laser cooling involves bombarding the atoms on all sides with photons. Atoms absorbing a photon head-on are slowed down, and thus cooled. Evaporative cooling is more passive—chilled atoms imprisoned by magnetic fields grow colder as higher-energy atoms in the brew are allowed to fly out.

Eric Cornell, Carl Wieman, and colleagues at JILA used these techniques in tandem to cool 2,000 rubidium 87 atoms to 170 nanokelvins (170-billionths of a degree above absolute zero) and create a 15-second BEC burst, as reported in the July 14, 1995, issue of *Science*. What separated the evapo-

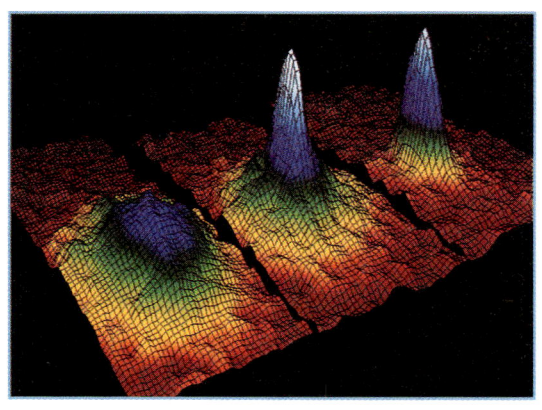

Physicists have finally glimpsed a Bose-Einstein condensate, a new state of matter that was predicted in the 1920s. Scientists condensed atoms into a single entity (dense white area above) by chilling them to temperatures approaching absolute zero (-460°F).

rative-cooling technique of the JILA scientists from that of competing research teams was a plug for the bottom of the magnetic trap, where a magnetic field of zero allowed some colder atoms to escape. Cornell used an additional magnetic field to rotate this zero point and gather enough atoms for the condensate.

On the heels of the Colorado experiment, two other groups published BEC sightings of larger condensates from different atoms. In August 1995, Randall Hulet and colleagues at Rice University in Houston, Texas, formed a BEC out of 200,000 lithium atoms. Their work provoked controversy because lithium atoms were thought to be unlikely candidates for condensation, according to theorists, and the team had no direct way to measure the condensate. Three months later, Wolfgang Ketterle and coworkers at the Massachusetts Institute of Technology (MIT) in Cambridge announced condensates of as many as 500,000 sodium atoms.

As BECs get bigger, physicists anticipate using the trapped, cooled atoms to study such related low-temperature phenomena as superconductivity (when a metal loses all electrical resistance) and superfluidity (when cooled helium creeps up and rises out of a beaker). Ultimately scientists hope to create an atomic-scale laser for precise measurements and procedures.

G Whiz

When is a physical constant not constant? When three independent groups of physicists measure the gravitational constant G and produce different answers above or below the current benchmark (6.6726×10^{-11} cubic meter per kilogram-square seconds). The ultraprecise measurements were presented at a 1995 joint meeting of the American Physical Society (APS) and the American Association of Physics Teachers in Washington, D.C.

Introduced in Isaac Newton's 1687 masterpiece *Principia*, the gravitational constant G is part of a simple mathematical relationship that describes the attractive force between two objects. A value for G remained unknown until 1798, when Henry Cavendish observed a torsion balance—a device with one small sphere attached to each end of a horizontal rod suspended by a wire. The gravitational attraction between the small spheres and a pair of large spheres causes the suspended rod to twist. In his experiment, Cavendish determined a value for G by measuring the degree of twisting.

In the tradition of Cavendish, two teams presenting at the APS meeting assembled sophisticated torsion balances that harnessed electric fields to compensate for the gravitational force. In the same units as above, Mark Fitzgerald and colleagues at the Measurement Standards Laboratory in Lower Hurt, New Zealand, arrived at a value of 6.6656 for G, while a team led by Winfried Michaelis of the Federal Physical-Technical Institute in Brunswick, Germany, measured G at 6.6685.

Hinrich Meyer and his collaborators at the University of Wuppertal in Germany took another approach. They measured the deflection of two pendulums attracted to large masses to determine a value of 6.71540.

Unable to explain the differences, the teams could agree only that measuring G is difficult. The gravitational forces involved are weak—roughly 10,000 times smaller than the weight of a cat's whisker. And shielding the sensitive laboratory equipment from gravitational forces proves practically impossible.

Additional research suggests that the torsion balance may not be the most appropriate tool for measuring G. In the October 9, 1995, edition of *Physical Review Letters*, Kazuaki Kuroda of the University of Tokyo pointed out that the elastic properties of wires used in torsion balances could inflate the measured value of G.

As physicists continue to ponder the discrepancies in G, Gabriel Luther of Los Alamos National Scientific Laboratory in New Mexico, who established the currently accepted value in 1986 with a torsion experiment, reports plans to assess G again, this time on a remote desert plateau.

FLIP-FLOP NEUTRINOS

If only the stereotypical mother-in-law could be more like a neutrino—neutral, massless, and having about one chance in 2 million of visiting. That is how physicists describe these particles created by the Sun's fusion.

But recent experiments with the liquid scintillator neutrino detector (LSND) at Los Alamos National Scientific Laboratory by a team of 39 scientists from a dozen institutions may prove that the neutrino indeed has a tiny mass, which would have sweeping implications for the Standard Model of particle physics and the field of cosmology.

According to the Standard Model, neutrinos come in three varieties—electron, muon, and tau—along with their antimatter counterparts. All are massless and travel at the speed of light.

Physicists believe that nearly all of the Sun's energy comes from the fusion of protons into helium, a reaction that produces electron neutrinos. But solar-neutrino detectors consistently find fewer neutrinos than expected. One explanation for the lack of evidence is that electron neutrinos change type, or "oscillate," into tau or muon neutrinos before reaching Earth. Such oscillations could not occur unless neutrinos have mass.

Cosmologists are keen to know a neutrino's bulk in order to account for mass in the universe. Contributions from celestial objects and interstellar matter represent only 10 percent of the mass needed to support an event like the Big Bang, which is believed to have formed the universe. Scientists call the missing 90 percent "cold dark matter." Neutrinos are so abundant that even with a tiny mass, they could far outweigh all the stars, planets, and galaxies in the universe and account for much of the dark matter.

The Los Alamos experiment aimed an accelerator beam of protons through successive targets of water, copper, and a steel wall to create an assortment of pions, muons, muon neutrinos, and muon antineutrinos. However, only the neutrinos and antineutrinos could penetrate the steel wall into a 184-ton vat of mineral oil lined with photomultiplier tubes. Physicists know whether or not a muon antineutrino has oscillated into an electron antineutrino because of a signature flash of light that occurs when the new particle interacts with a proton.

What the team found in five months' worth of data were nine events instead of an expected two. Controversy surrounded the results, which first appeared in the *New York Times* rather than in a scientific journal. When collaborators disagreed on how to interpret the data, one author withdrew his name from the paper—which was eventually published in *Physical Review Letters*.

The Los Alamos group has stood by its findings, but plans to continue taking measurements to get a better handle on neutrino masses. Meanwhile, to further study neutrino oscillations, other research groups in the United States and Japan have proposed cross-country neutrino beams that run hundreds of miles from accelerator to detector.

Therese A. Lloyd

The subterranean Sudbury Neutrino Observatory, scheduled to open in 1997, may help scientists confirm that neutrinos have mass, a finding that would have profound cosmological implications.

Public Health

THE EBOLA-VIRUS OUTBREAK
After remaining fairly quiet for almost two decades, the Ebola virus has reappeared, reminding us of its potential for spreading quickly and for causing horrible deaths. The virus is named for the Ebola River in Zaïre, near where the first outbreak was identified in 1976. The spring 1995 outbreak of Ebola virus, in the form of Ebola viral hemorrhagic fever, was centered in the city of Kikwit and the surrounding Bandundu region of Zaïre. This virus is the same strain as in the original outbreak, even though the Ebola River is hundreds of miles from Kikwit.

The first identified case in the 1995 outbreak was in a laboratory technician who entered Kikwit Hospital with fever and bloody diarrhea. He underwent surgery for a possible perforated intestine, but died. Some hospital staff who assisted with the surgery were infected with the virus and became ill. Most of them died of the disease.

Zaïre soon issued a plea for help. Quick responses came from the U.S. Centers for Disease Control and Prevention (CDC), from the Pasteur Institute in Paris, and from the World Health Organization (WHO). Specialists in epidemic control were flown in, along with medical equipment and supplies, especially protective gloves, gowns, and masks for health-care workers. Cars, bicycles, and motorcycles also were brought in to help the epidemiologists go from house to house looking for new cases that needed isolation and treatment.

A field laboratory was established to test large numbers of blood samples for antigens from, or antibodies to, the Ebola virus. Antibodies to the disease begin to be detected in the blood about 8 to 10 days after infection. Other tests were used to help confirm each diagnosis, and to test for the routes by which the epidemic was spreading.

The first people to have blood tests were hospital workers in Kikwit. Those with recently developed antibodies were isolated so they could not spread the disease to others. When the outbreak was finally controlled, there had been more than 300 known cases

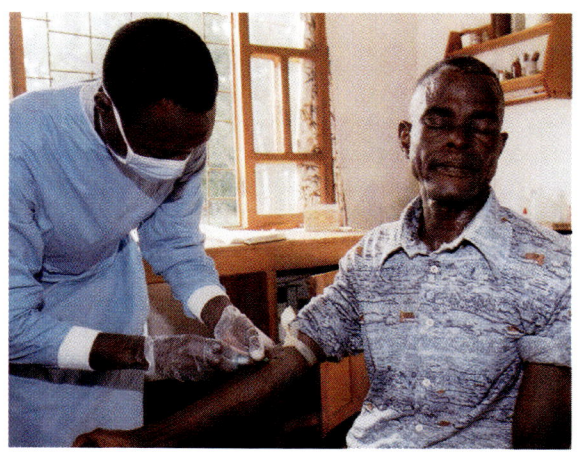

The fatality rate for people infected with the Ebola virus is extremely high. Specialists from around the world helped victims of the 1995 Zaïre epidemic.

of Ebola virus hemorrhagic fever. A large majority of these cases (79 percent) proved fatal. (Of the 283 cases for whom the person's occupation was known, 32 percent were health-care workers.) In later outbreak locations, the death rate of those who became ill varied from 50 to 90 percent.

Ebola viral hemorrhagic fever is known to spread by close personal contact, especially with blood or bodily fluids. Because of this, 45 percent of the spouses of infected individuals also acquired the disease. Only 14 percent of others living in the household became infected. By contrast, HIV-1, the virus that causes AIDS, is also spread by blood and bodily fluids, but it is seldom, if ever, spread by ordinary nonsexual contact.

Ebola hemorrhagic fever can cause considerable suffering. It starts with flulike symptoms (headache, chills, fever, muscle aches), and proceeds to cause abdominal pain, vomiting, and diarrhea. Hemorrhages then develop throughout the body.

The incubation period for Ebola virus infection may be as short as two days or as long as three weeks. Although transmission of the virus can occur early in the disease, it appears to be most infectious during its later stages, when vomiting, diarrhea, and bleeding from multiple sites often occur.

It is not clear how the virus survives between outbreaks. Some suspect that wild animals, perhaps monkeys, may harbor the virus and spread it to human beings when the animals are killed or otherwise handled. One Swiss graduate student—who was completely protected by a gown, gloves, and a mask—assisted with the autopsy of a monkey and, not long afterward, became infected with a milder variant of the Ebola virus. Despite this incident, an intensive search for the virus in a variety of vertebrate and invertebrate animals has proved futile. It is possible that the laboratory technician who died at the start of the Kikwit outbreak became infected while handling human blood from an infected person.

Aggressive intervention helped to control this epidemic. Medical workers searched extensively for new cases of the illness and then isolated each infected patient. Medical personnel were trained to protect themselves with gloves, masks, and gowns, and to protect others by sterilizing needles and instruments and by screening blood. Educational efforts in the affected communities helped some people avoid being infected. In addition, attempts were made to limit the migration of people into and out of infected areas, further limiting the virus' spread.

SPREAD OF TWO TICK-BORNE DISEASES

Since 1986, researchers have known that ticks in the United States spread *Ehrlichia* bacteria, which attack white blood cells. Two forms of these bacteria cause two illnesses in humans: human granulocytic ehrlichiosis (HGE) and human monocytic ehrlichiosis (HME). HGE received attention in 1995 when at least 29 cases were identified in New York State, and other cases were suspected but could not be proved. About one-third of the infected people (62 percent) lived in Westchester County, just north of New York City. Their ages ranged from 21 to 90 years, and about half were men and half were women.

Both forms of ehrlichiosis cause fever and a drop in the cells that initiate blood clotting. Infection-fighting (white) blood cells (either monocytes or granulocytes, depending on the invading bacterium) are also diminished. The patient usually has a history of known or possible exposure to ticks.

If HGE or HME is a possible cause for illness, the patient should be started immedi-

Tick-borne Diseases: Not Just Lyme

A newly identified disease, human granulocytic ehrlichiosis, or HGE, is transmitted by the same tick that carries Lyme disease. Many symptoms of the diseases overlap, but in HGE, the symptoms tend to reach a peak very quickly, while Lyme disease often develops slowly.

Lyme Disease

In over 60 percent of cases, a circular blotchy rash appears within days or weeks after the tick bite. There may be flulike symptoms, such as fever, headaches, sore throat, sleeping difficulties, swollen glands, dizziness, muscle aches, joint pain, and fatigue. *Treatment:* The antibiotics most often used are amoxicillin or doxycycline.

Ehrlichiosis or HGE

Flulike symptoms, including fever, headache, chills, malaise, muscle and joint pain, nausea, vomiting, and weight gain or loss. A blood test may indicate a low white-blood-cell count or low platelet count or even both. *Treatment:* Tetracycline or doxycycline.

A deer tick can more than triple its size when engorged with blood.

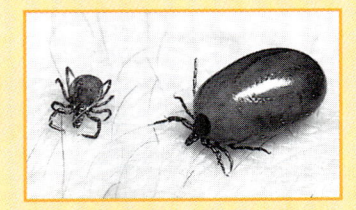

How to prevent tick-borne diseases

• Know how to spot a deer tick.
• Wear light-colored clothes (to help spot ticks) and long pants tucked into socks.
• DEET, a pesticide used in repellents, is effective against ticks but can be hazardous to children. Use according to label instructions.
• Check for ticks on clothing, skin, and hair. Check pets and remove ticks before allowing them indoors.
• Remove a tick with a single, gentle motion using finely pointed tweezers held as close to the skin as possible. Don't squeeze, twist, burn, press ice against, coat with petroleum jelly, or otherwise disturb a biting tick. Save the tick.
• If you've been bitten by a tick, see a doctor. If you have the tick, take it with you.

ately on the antibiotic doxycycline. Delay due to waiting for laboratory results can increase the harm caused by either disease. It is especially important that physicians consider the possibility of ehrlichiosis in any feverish patient who has been exposed to ticks.

The tick responsible for the spread of ehrlichiosis is the small deer tick *Ixodes dammini*. This is the same tick responsible for the spread of Lyme disease in the eastern United States. As a precaution, persons in tick-infested areas should wear protective clothing, use tick repellents, and inspect their bodies for ticks after exposure.

HUMAN RABIES

A 13-year-old girl from Connecticut died of rabies in October 1995, increasing a public-health concern about bats. (The strain of rabies that caused her death is a strain associated with bats.) Two years before, a young girl in upstate New York also died of rabies and was found to have been infected with a bat-related rabies strain.

What is most alarming is that neither girl was aware of any contact with bats. It is not known why most cases of human rabies without known exposure are due to bats. Perhaps contact with the bat is not felt at the time it occurs, especially if the person is sleeping. It is also probable that some kind of airborne transmission can occur under certain circumstances, such as when people explore caves in which bats live. Fortunately, there has never been any evidence to suggest airborne transmission from bats to people other than in bat caves. It is important to realize that even remote contact with a bat may involve the transmission of rabies, and that medical advice should be sought immediately.

A rabies scare in New Hampshire involved a kitten purchased at a pet shop. The kitten died of rabies, but only after many people had been exposed to it. Public-health authorities tried to find those who might have had contact with that animal or with the other cats in the pet store. Many of the exposed people received the rabies-vaccine series, because rabies is almost always fatal once symptoms develop. The vaccine is the only method of prevention.

When wild animals can be given an oral rabies vaccine, they become immunized and serve as a barrier in the wild to further spread of the disease (another example of herd immunity). Recent trials of an oral vaccine in New Jersey, Texas, and on Cape Cod, Massachusetts, have provided evidence of its effectiveness. This approach to rabies control also has been successful in Europe.

SICKNESS FROM SEAWEED

Seaweed is sometimes eaten as a health food in the United States; overseas, it is a common side dish among people of the Pacific Rim and the islands of the Pacific Ocean. Unfortunately, some seaweed from the Pacific Ocean occasionally causes a toxic illness. In 1995, an outbreak of gastrointestinal illness and a burning sensation in the mouth or throat were investigated by the Hawaii Department of Health. The cause was seaweed, identified as *Gracilaria coronopifolia*, from which a previously unknown toxin was extracted. The toxin is heat-resistant, indicating that it can survive cooking or other processing. During testing, it produced a similar illness in mice.

CIGARETTE SMOKING AND CANCER

Actor David McLean, the "Marlboro Man" of television-commercial fame, died of lung cancer at the age of 73. Another man who claimed to be a former Marlboro Man in print advertisements, Wayne McLaren, died of lung cancer in 1992 at the age of 51. After his diagnosis, McLaren worked against cigarette smoking until the time of his death.

In another cancer-related matter, the tobacco industry was astonished when a jury gave a $1.3 million award to a man who sued, claiming he had developed mesothelioma—a kind of cancer that comes almost exclusively from asbestos exposure—from the asbestos in cigarette filters. This was one of only two anti-cigarette-company lawsuits that were won by the plaintiff, and it represents a new approach to lawsuits against cigarette manufacturers. In this case, the plaintiff successfully proved that he had no other known exposure to asbestos, and that he did in fact smoke asbestos-containing filter cigarettes.

James F. Jekel, M.D., M.P.H.

SCIENCE EDUCATION

SCIENCE STANDARDS

In 1995, the National Academy of Sciences (NAS) published the National Science Education (NSE) Standards, which outline what scientific concepts students should know by the time they graduate from high school and how they should learn them. These guidelines propose to bring science to all students, not just high achievers.

The standards took four years to develop, as educators, parents, and scientists argued over the content requirements. For example, an early draft stated that students should learn the details of relativity theory and biochemistry. The finished NSE standards do not contain such requirements; instead, they suggest that pupils learn fewer facts and more concepts. Less contentious was the development of new methods to teach the revised curriculum. The standards propose two main changes to current practice:

- Science topics should be integrated in the curriculum, rather than taught individually in separate grades. Physics, chemistry, biology, and earth/space sciences will be taught together for at least four years to allow students to learn how the different areas of science interact and overlap.
- Students should perform science experiments first and then formulate what they have learned from the process. Only then will the laws that govern the results be taught. This method will teach young investigators how to learn—that is, how to ask the questions that provide the information needed to solve problems. Science as inquiry, as this method is called, lets pupils know that the process is more important than coming up with the right answer.

While adopting the standards is voluntary, the NAS believes that schools will comply. The National Science Teachers Association (NSTA) has received a grant from the National Science Foundation (NSF) to introduce the standards in 15 schools from Washington, D.C., to Houston, Texas. Some of the money will be used to train teachers in the new methodology and to develop teacher instruction manuals and student work materials. No textbooks currently exist that integrate the different disciplines and key concepts in one course.

NONCOMPUTER CLASSROOM ACTIVITIES

Creativity in developing hands-on activities for learning continues to proliferate in schools, on television, and in government agencies across the United States.

School Programs. A national program for teaching mathematics, known as the Interactive Math Program (IMP), uses problems from literature, history, science, and daily life to teach math to high-school students. Traditional topics such as algebra, geometry, and trigonometry are taught together as needed to solve a problem, rather than as separate subjects. For example, by planning a 19th-century pioneer family's journey on the Overland Trail, students learn algebra and linear regression. IMP, funded by an NSF grant, is now in its seventh year. It is part of a four-year curriculum used in 100 schools in 12 states. The program was developed by California educators in response to a report from the National Council of Teachers of Mathematics that called for new methods in mathematics teaching.

Less-structured—but not less-effective—methods are continually being developed in individual schools throughout the country. For example, every fall and winter in Haddonfield, New Jersey, fourth graders now take part in Project FeederWatch, a national bird-observation study run by the Cornell Laboratory of Ornithology, Ithaca, New York. The students build bird feeders, fill the feeders with seeds provided by a local merchant, and keep records of which species of birds appear. The results are than tabulated and sent back to Cornell. By participating in Project FeederWatch, the children learn not only about individual bird species and their feeding and migratory habits, but also about how to collect scientific data.

Plant physiology is the focus of a hands-on science project now in its second year at a middle school in Bala Cynwyd, Pennsylvania. By planning an organic garden and then growing the selected plants from seed, stu-

New Jersey fourth graders build bird feeders, fill the feeders with seed, and keep records of what bird species come to the feeders as part of a Cornell-sponsored project that helps students learn about avian behavior and gain experience on how to collect scientific data.

dents learn how to work together as a team; they also collect data on weather and soil to determine what conditions are best for individual plants. The program has been so successful that when the school year ends, many of the children continue to care for the garden they planted.

Television. In 1990, Congress passed the Children's Television Act, which stipulated that television stations increase the amount of educational programming available to children. Stations that do not comply risk not having their licenses renewed.

A number of television shows now in production aim to teach math and science to young children. One of the most successful of these debuted nationwide in September 1994. Titled *Bill Nye, the Science Guy*, the program has an audience of over 4.4 million a week and is shown weekdays on 250 public-television stations; on weekends, it is syndicated on commercial channels. Nye, a former engineer, tailors the subject matter to a fourth-grade audience. The format—fast-paced and funny—shows children how to perform simple experiments. According to Nye, viewers should remember at least one or two key points of the experiments when the show is over. For example, a program on buoyancy teaches two essential facts: that an object put in water displaces water, and that if that object weighs less than the water it displaces, the object will float. The NSF, Walt Disney, and the Public Broadcasting Service (PBS) all fund *Science Guy*. As part of this funding, program-related science kits and teaching guides are sent to every fourth-grade teacher in the United States.

NASA. The National Aeronautics and Space Administration (NASA) is sponsoring several middle- and high-school educational endeavors in conjunction with the space-shuttle program. One took place in November 1995, when Al Sacco, an astronaut and engineer at Worcester Polytechnic Institute in Massachusetts, taught science to 40,000 classrooms while in orbit aboard the space shuttle *Columbia*. Students with a live hookup to the shuttle could participate in experiments conducted by the shuttle crew, and they could talk with the astronauts.

In 1996, another shuttle program will commence: KidSat, short for "kids' satellite." Cameras will be installed in strategic positions on the space shuttle to allow students in participating schools to study various aspects of Earth. For example, a seventh-grade class in San Diego, California, will be able to focus the camera on Earth's bodies of water to study water-related phenomena such as hurricanes and erosion. KidSat was created by Sally Ride, the first U.S. woman in space. She envisions eventually putting a KidSat camera on *Mir*, Russia's space station, so that students can study the Earth from space whenever they want, not just when the shuttle is aloft.

Abigail W. Polek

Seismology

Among the thousands of earthquakes that shake the globe each year, only a few dozen cause any damage. During 1995 and early in 1996, large and destructive earthquakes rocked Russia, Indonesia, Mexico, Peru, Egypt, and Greece, among other places.

Sakhalin Island
A particularly deadly earthquake struck the Russian Far East on May 28, 1995, destroying the oil town of Neftegorsk on Sakhalin Island. The magnitude-7.1 tremor killed nearly two-thirds of the 3,000 people living in the town. Occurring at 1:03 A.M., the quake leveled 17 five-story apartment buildings in the center of Neftegorsk, instantly entombing most of the sleeping residents.

The apartment complexes were built in a notoriously flimsy construction style during the post-World War II oil boom in the former Soviet Union. At that time, Soviet scientists had not expected such a strong earthquake to occur in the region. Rescue efforts were hampered by nighttime temperatures that fell below freezing.

Greece
Three strong earthquakes rattled different parts of Greece during the spring of 1995, tremors that an Athens scientist claims to have predicted in advance.

The earthquakes occurred on May 4, May 13, and June 15. The third quake, killing 26 people in the Aiyion area, measured magnitude 6.3 and caused an estimated $660 million in damages.

Panayiotis Varatsos, a solid-state physicist at the University of Athens, reported that he had forecast the quakes using a technique of his own called the VAN method. Because Varatsos regularly faxes his predictions to seismologists around the world, his apparent success captured the attention of U.S. seismologists, who invited him to a conference at the University of California, Berkeley, in October 1995.

The VAN method relies on electromagnetic sensors placed at select sites in regions of frequent earthquakes. Varatsos began experimenting with the method in 1983, following laboratory experiments that showed that rocks emit electric signals prior to breaking. The Athens scientist claims to have predicted 10 of the 14 strong Greek tremors during the past nine years.

Detractors, however, say that Varatsos makes overly vague predictions of the time and location of future earthquakes. Because earthquakes occur in Greece throughout the year, Varatsos can always claim success with his unspecific predictions, critics charge.

Nonetheless, American scientists were intrigued enough to begin studying the VAN method more closely to gauge whether the technique has potential for predicting earthquakes in the United States.

China
The southwestern province of Yunnan, China, was battered by a strong earthquake on February 3, 1996. The magnitude-6.5 earthquake killed at least 251 people and injured 4,000 in the mountainous Lijiang-Zhongdian area. The tremor destroyed more than 300,000 houses and left roughly 1 million people homeless.

Yunnan suffered a smaller, but still-damaging, quake on October 24, 1995. The magnitude-6.1 jolt claimed more than 36 lives and injured 200 in the Wuding area.

At least 24 people lost their lives during a magnitude-6.0 earthquake on March 19, 1996, that struck along the ancient Silk Road in far western China.

On July 12, 1995, a magnitude-6.8 tremor struck along the border between China and Myanmar (Burma). This event killed six people, injured 99, and destroyed more than 100,000 homes.

The Chinese province of Gansu was hit by a magnitude-5.6 earthquake on July 22, 1995. Fourteen people lost their lives.

Indonesia
A strong tremor rocked the Indonesian island of Sumatra on October 7, 1995, killing 84 people. The magnitude-6.8 earthquake injured more than 2,000 persons and left 65,000 homeless.

A much larger quake measuring magnitude 7.9 hit the Irian Jaya region of Indone-

The comparatively minor (magnitude-5.4) earthquake that shook the Seattle, Washington, area on May 2, 1996, caused no injuries or deaths and did little more than nuisance damage.

sia on February 17, 1996. The quake killed at least 96 people. Centered underwater at a shallow depth, the shock caused a large tsunami, which reached heights of 23 feet (7 meters) in some locations. The waves dragged hundreds of homes into the ocean.

A magnitude-7.8 earthquake killed eight people on the island of Sulawesi (Celebes) on January 1, 1996.

U.S. Quakes

Seismologically, the year was a quiet one in the United States. The largest earthquake within U.S. territory occurred off the coast of northern California on February 19, 1995, and measured 6.8 in magnitude. A magnitude-6.5 shock hit the Aleutian Islands on April 23, 1995.

Brewster County, Texas, was the site of a magnitude-5.7 quake on April 14, 1995. The vibrations injured two people.

On May 2, 1996, a magnitude-5.4 quake shook the Seattle, Washington, area. No injuries were reported.

Other Notable Jolts

More than 100 people died in a magnitude-6.0 earthquake that struck western Turkey on October 1, 1995. The strong vibrations destroyed buildings in the Dinar area, leaving 50,000 residents homeless.

More than four dozen people lost their lives during a shock in Mexico on October 8, 1995. The magnitude-7.6 event hit near the coast of Jalisco.

An earthquake on November 22, 1995, claimed at least eight lives in Egypt's Sinai Peninsula. The magnitude-7.0 tremor caused damage as far away as Cairo, and killed several people in Israel and Saudi Arabia.

A very large shock occurred near the northern coast of Chile on July 29, 1995. Measuring magnitude 7.5, the tremor killed three people.

Quake Precursors

Several groups of researchers studying the devastating Kobe earthquake, which struck Japan on January 17, 1995, found evidence that the ground gave hints of the disaster long before it occurred.

One team of scientists from the University of Tokyo discovered that the chemistry of the groundwater in the Kobe area changed prior to the earthquake. Bottled water from mineral springs in the region showed that the concentrations of dissolved chloride and sulfate ions began to rise five months before the quake. Another researcher from Hiroshima University found that levels of dissolved radon gas in the groundwater started rising three months before the quake and jumped markedly only nine days before the event. The scientists speculated that rock in the region began developing microscopic cracks long before the catastrophic faulting that produced the quake. These microcracks may have allowed radon and other chemicals to leach into groundwater.

An astronomer from Hyogo College also picked up precursory signs. Forty minutes before the Kobe earthquake, sensitive receivers 48 miles (77 kilometers) away detected unusual radio signals. The astronomer suggested that building underground stress squeezed the rocks and caused electromagnetic emissions.

The water and radio clues were detected long after the earthquake, so they could not be used for prediction. The findings nonetheless intrigued seismologists, who have long tried to find reliable precursors that would herald an upcoming quake.

The Key to Subduction Shocks

Two researchers developed a theory to explain why certain ocean trenches host frequent tremors while others remain silent. The theory focuses on so-called subduction zones, where one of Earth's outer plates slides beneath another and dives into the planet's interior. The act of subduction creates the great trenches that rim many of the ocean basins.

Scientists with the U.S. Geological Survey (USGS) and with the University of Chile in Santiago discovered that subduction zones spawn earthquakes when the two colliding plates are moving in opposite directions—like cars in a head-on crash. Subduction zones do not generate jolts, however, when the two plates are moving in the same direction—as might happen when a fast car runs into a slower one on a highway.

According to the subduction-zone theory, the head-on crash generates greater friction and thus more chances for earthquakes. The Chilean coast has this type of geology, as does the Pacific Northwest region of the United States.

Richard Monastersky

Space Science

International cooperation between spacefaring nations helped to make 1995 a year of major progress. Around the world, 74 launches were successfully executed from the United States, Russia, Canada, India, Israel, Ukraine, Japan, China, South Korea, and the European partners. Nine crewed missions carried 48 humans into space.

United States Space Activities

In 1995, the National Aeronautics and Space Administration (NASA) celebrated 100 missions in space and flew seven successful shuttle missions. The shuttle fleet carried 42 astronauts into space, with foreign crew members from Russia and Canada. An *Endeavour* mission, launched on March 2, became the longest shuttle flight to date, with a flight time of 16 days, 15 hours.

Twenty years after the first in-space meeting of American astronauts and Russian cosmonauts, the space shuttle *Atlantis* docked with the Russian space station *Mir*. The two crews remained in each other's company for five days, during which time they ran medical experiments, shot photographs, and practiced flying the huge station. It was a first step in a series of dockings that will help both countries prepare for the construction of an Earth-orbiting international space station.

Astronaut Norman Thagard set a U.S. record when he spent 112 days in orbit aboard *Mir*. He became one of only five people ever to fly in space five times.

Russian Space Activities

Russia again led the world in the number of spacecraft launches (74), but, of those, only 32 succeeded. Among them were about two dozen military, scientific, and telecommunications satellites.

The new partnership with the United States began to play a major role in shaping the Russian space program. Three cosmonauts flew on U.S. space shuttles during 1995, and one U.S. astronaut flew aboard the Russian *Soyuz* craft.

By the end of 1995, the *Mir* space station had been in continuous operation for 3,602

After the historic docking of the U.S. space shuttle Atlantis with Russia's Mir space station in June 1995, astronauts and cosmonauts gathered together in the space station for an impromptu sing-along.

days, having circled Earth approximately 56,380 times, and having been inhabited continuously for 2,308 days.

Cosmonaut Valeriy Polyakov, a 52-year-old physician who had been on the Russian space station since January 1994, returned to Earth on March 22, 1995, setting a new space-endurance record of 438 days. Elena Kondakova also returned that day, having set a new women's space-endurance record of 169 days.

EUROPEAN SPACE ACTIVITIES

In a major decision by the research/technology ministers of its membership countries, the European Space Agency (ESA) affirmed its prior commitments to supplying the International Space Station with the Columbus Orbital Facility (COF) and the Ariane 5–launched Automated Transfer Vehicle (ATV).

In cooperation with Russia's space agency Rossiyskoe Kosmicheskoe Agentstvo (RKA), the ESA undertook the EUROMIR 95 mission when German cosmonaut Thomas Reiter was launched as a member of the *Mir*-20 crew. He was still on board at year's end, and became the first guest cosmonaut given the rank of "flight engineer."

Despite a serious economic slump throughout Europe, the ESA and Arianespace conducted brisk business at Kourou, French Guiana, with the launch of 11 Ariane rockets carrying a variety of payloads—including TV satellites; Europe's first military-reconnaissance satellite, Helios 1A; the giant infrared space telescope ISO; and the remote sensing/environmental satellite ERS-2.

ASIAN SPACE ACTIVITIES

On January 15, 1995, Japan's space agency ISAS launched a four-stage Mu-3S-II from Kagoshima, carrying the German Experiment Reentry Space System (EXPRESS) and a Russian-built reentry vehicle. Due to land in Australia, its thrust system failed during ascent and the mission failed. Given up for lost in the Pacific Ocean, EXPRESS landed safely on its orange parachute two and one-half orbits later near a partially inhabited area of Ghana, West Africa.

The People's Republic of China suffered a serious setback on January 25 when an LM-2E satellite blew up nearly a minute after launch. The explosion killed six people near the Xichang Satellite Launch Center; 27 others were injured by falling debris. A second and third LM-2E successfully carried communications satellites into orbit later in the year.

South Korea employed its first communications satellite with the partially successful launch of Mugunghwa (Hibiscus) on a Delta 2 rocket. The country also entered negotiations with the RKA for future participation in the Russian cosmonaut program.

INTERNATIONAL SPACE STATION (ISS)

During 1995, a $5.63 billion design-and-development contract with the Boeing Company, Seattle, Washington, was finalized and signed after 16 months of negotiations. Overall, U.S. contractors delivered nearly 80,000 pounds (36,300 kilograms) of hardware—including solar-array panels, truss segments, rack structures, hatch assemblies, and various mock-ups. Canada, Japan, and the ESA all remained on schedule with their contributions to the station.

COMMERCIAL SPACE ACTIVITIES

In a first small step toward "privatization" of the shuttle, NASA selected the United Space Alliance, a joint venture formed by the two firms Rockwell and Lockheed Martin, to operate the space-shuttle fleet, thereby replacing 85 separate contracts.

Lockheed has developed a smaller rocket, the Lockheed Launch Vehicle (LLV), for use with lightweight commercial payloads for low-Earth orbit. The LLV's first launch, however, failed when the vehicle pitched out of control shortly after liftoff. Additional newcomers to the launch-services market are two small companies, EER Systems and Orbital Sciences Corporation (OSC).

The International Launch Services (ILS), a joint venture between Russian and U.S. launch companies, firmly booked about $1 billion worth of Proton rocket launches by midyear.

PLANETARY EXPLORERS

After taking a circuitous route of six years, which involved one gravity-assist swing-by of Venus and two of Earth, the Jupiter probe Galileo finally reached its target. On July 13, 1995, the entry probe left the main ship and spun toward the Jovian atmosphere. On December 7, the probe entered Jupiter and, until vaporizing 58 minutes later, measured intensely turbulent winds of up to 330 miles (530 kilometers) per hour. It found less helium, neon, carbon, oxygen, water, and sulfur than scientists expected. And the probe discovered that lightning occurs on Jupiter only about one-tenth as often as on Earth. The orbiter continues circling Jupiter on a two-year mission to monitor the giant planet and its satellites.

After a journey of more than two decades and about 4.1 billion miles (6.6 billion kilometers), the Pioneer 11 craft began to die a slow death. Now in the farthest reaches of our solar system, far beyond the orbit of Pluto, the craft can no longer power its detectors. Listening to its ultimate failure may help scientists prolong by a few months the life of Pioneer 10, its sister craft.

The NASA-ESA solar-polar explorer Ulysses, having overflown the Sun's southern pole in November 1994, passed over the solar north pole and concluded its main mission successfully in September 1995. The spacecraft used radio waves to take the first "aerial photo" of the Sun's immense and pinwheel-shaped magnetic field.

EXPLORING THE UNIVERSE FROM AFAR

During 1995, the Hubble Space Telescope (HST) shook the foundations of knowledge about the birth, evolution, and fate of the universe, and provided the best evidence yet for the existence of black holes. It also peered deep within the star-forming regions of our galaxy to photograph the birth of stars and planetary systems.

The three ultraviolet telescopes of Astro-2 flew in the cargo bay of the space shuttle *Endeavour* for 16 days, longer than any other ultraviolet-imaging telescope had been aloft. Astro-2 collected ultraviolet spectra of about 300 celestial objects and measured intergalactic gas abundances.

The X-Ray Timing Explorer satellite (XTE), the largest X-ray telescope ever placed in orbit, was launched by NASA on December 30, 1995. It is designed to study exotic objects such as white-dwarf and neutron stars, X-ray novas, black holes, and quasars throughout the universe.

Dennis L. Mammana

United States Manned Spaceflights—1995

Mission	Launch/Landing	Orbiter	Primary Operation
STS-63	Feb. 3/Feb. 11	*Discovery*	**Spartan Satellite:** Deployment of a free-flying, reusable satellite equipped with ultraviolet-sensing telescopes to study interstellar gas and dust formations. Spartan was released for two days and then retrieved.
STS-67	Mar. 2/Mar. 18	*Endeavour*	**Astro-2:** Observatory containing three telescopes used to conduct ultraviolet studies of the Moon, quasars, and the gas and dust blown out from supernovas.
STS-71	June 27/July 7	*Atlantis*	***Mir* Docking:** Historic first docking with Russia's *Mir* space station. Two craft were joined in orbit for five days, forming the heaviest structure ever to orbit Earth.
STS-70	July 13/July 22	*Discovery*	**Tracking and Data Relay Satellite (TDRS):** Designed to track orbiting satellites and to serve as a communications relay between Earth and spacecraft. Deployed into an orbit 22,300 miles (35,900 kilometers) above Earth.
STS-69	Sept. 7/Sept. 18	*Endeavour*	**Wake Shield Facility:** Satellite designed to grow samples of high-grade semiconductor material while flying in the "ultravacuum" of the space shuttle's wake.
STS-73	Oct. 20/Nov. 5	*Columbia*	**U.S. Microgravity Laboratory (USML):** Numerous experiments to study the processing of metals, alloys, and semiconductor materials in microgravity conditions.
STS-74	Nov. 12/Nov. 20	*Atlantis*	***Mir* Docking:** Second joint docking with the Russian space station. During the three-day linkup, shuttle crew installed a 15-foot (4.6-meter) docking tunnel on *Mir*.

Remarks

- Russian cosmonaut Vladimir Titov was among the six-person crew.
- Astronaut Eileen Collins became the first woman to pilot a space shuttle.
- Shuttle flew within 37 feet (11 meters) of Russia's *Mir* station.

- A new shuttle endurance record was set on the 16-day, 15-hour, and 9-minute mission.
- Astronauts conducted experiments growing protein crystals in space for pharmaceutical applications.

- Medical lab on board permitted studies of long-term effects of microgravity on human physiology.
- Replacement crew of cosmonauts shuttled to *Mir*.

- Mission delayed five weeks due to extensive damage to shuttle's fuel tank caused by woodpeckers.
- The crew performed scores of medical, engineering, and materials-science experiments.

- Astronauts redeployed the Spartan satellite to study the characteristics of the solar corona.
- Redesigned suits, gloves, and other gear were extensively tested during long space walks.

- The mission's launch was delayed a record-tying six times.
- Spacelab facility used to conduct fluid-physics and materials-science experiments.

- Launch delayed one day due to inclement weather at emergency landing sites in Spain and Morocco.
- Solar-power arrays transferred to *Mir*.
- 73rd flight in the U.S. shuttle program.

During the February 1995 flight of the shuttle Discovery, *Air Force Lieutenant Colonel Eileen Collins became the first woman to pilot a U.S. spacecraft.*

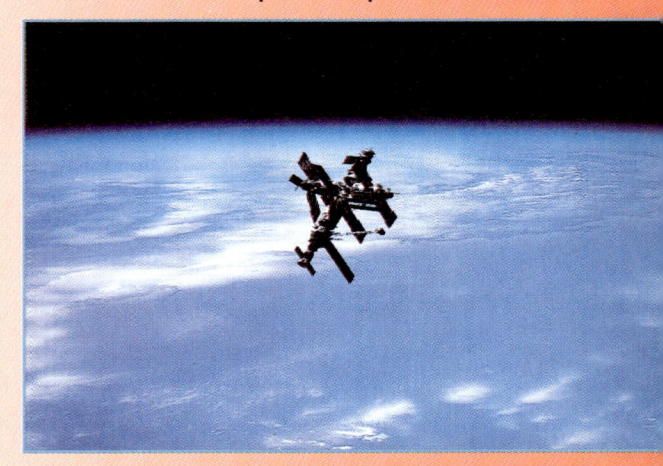

The shuttle Atlantis *made a historic docking with the Russian* Mir *space station (above, during approach) in June 1995. The two craft were joined for five days.*

Astronaut James Voss tested new thermal gear during the September 1995 mission of Endeavour.

SPACE SCIENCE 375

TRANSPORTATION

AIR TRAVEL

A new radar system that tracks planes as they taxi on runways is planned to be operational at the 37 busiest U.S. airports in 1997. By combining ground-radar technology and computer software to produce sharp images of taxiing planes (historically a difficult task for radar, because the ground contains visual "noise" that can lead to muddy and multiple images), the new system can provide audio and visual warnings of impending collisions to help controllers prevent such tragedies as the crash on a runway at the St. Louis, Missouri, airport in 1994 that killed two people.

A new communications network for air travel will soon span the globe, advancing the capabilities of pilots, ground controllers, and passengers. The Aeronautical Telecommunications Network is being developed by a consortium of 11 airlines and will resemble in some ways the Internet. The so-called "airnet" will connect pilots to pilots, pilots to traffic controllers, and passengers to computer networks on the ground.

Communication will be efficient, as the system software analyzes available linkages for each call and chooses the best. Transmissions might use radio or satellite routes. If successful, the system could conceivably lead to the eventual elimination of radar. In its place, planes themselves would pinpoint their locations (they already do this using satellites) and continually relay the information to ground controllers over the new communications network.

NAVIGATION

The Global Positioning System (GPS), the satellite navigation system used by planes, ships, and trucks for years to locate their positions to within 300 feet (about 100 meters), is becoming still more accurate. The GPS comprises a fleet of 24 satellites that receive and transmit signals from moving objects below. By measuring aspects of signals sent from a source (such as a plane) to a small number of GPS satellites, the system determines the source's location and sends notification. A new technology, the Wide Area Augmentation System, will improve GPS results to within 21 feet (about 7 meters).

The augmentation system will have 36 receiving stations on the ground in the United States. Each station will continuously monitor what the GPS satellites consider to be its location. Because each station knows its location precisely, it can calculate errors in satellite readings and inform a central station of the errors. Using an additional satellite, the central station will transmit adjusting factors to moving objects, which can use them to better establish their locations.

A consortium of contractors—including Wilcox Electric, Hughes Aircraft, and TRW—hopes to have the new system in operation by 1999. In the case of air travel, the improved system will allow ground controllers to direct planes to fly closer together without increasing risk. It also will allow planes to use more-direct, fuel- and time-saving routes. In the distant future, an expanded system of monitors at most airports could lead to tracking so precise that planes could land in zero visibility.

The Guidestar dashboard navigational system displays road maps, gives turning cues, records mileages, and otherwise assists motorists in rental cars in finding their way through unfamiliar areas.

At airports in a few large cities, the Avis car-rental company now offers vehicles that feature the Guidestar navigational system. In addition to using the global positioning satellite system to estimate a car's position,

Guidestar contains a computer program to process changing positional data and stored information. It displays maps, evolving route suggestions, turning cues, and mileages on a small screen attached to the dashboard. The cars are available at airports in Miami and in Los Angeles, San Francisco, and San Jose, California, with more on the way.

on the dashboard, an onboard system commands the car's cellular phone to dial the center. The center then uses global positioning satellites to establish the car's location, and sets in motion procedures for timely assistance. The onboard system also conveys to the control center the car's serial number and its last speed and direction of motion.

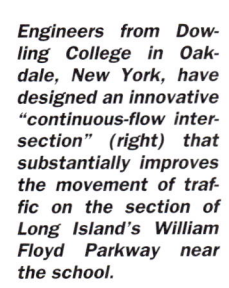

Engineers from Dowling College in Oakdale, New York, have designed an innovative "continuous-flow intersection" (right) that substantially improves the movement of traffic on the section of Long Island's William Floyd Parkway near the school.

Volvo's Dynaguide system uses satellite relays to bring data from a distant information center to the automobile's dashboard screen. Staff at the center collect data on traffic congestion, accidents, and other travel-related conditions from police, radio stations, and even roadside sensors, and transmit the data to cars. The information is displayed on small video screens along with local street maps. Volvo now offers the systems in Europe and plans to add them soon to American models.

Finally, Lincoln-Mercury offers a system that works in reverse. In some new Lincoln Continentals, drivers can communicate with a distant control center, sending distress signals and positional information in roadside emergencies. When the driver presses a button

ROAD STRATEGY

On a north-south highway in the middle of Long Island, New York, a traffic experiment is in progress. Road designers have built a novel intersection (with traffic light) through which traffic moves more quickly because left-turning vehicles do not hold up north- and southbound vehicles. Dowling College in Oakdale, New York, created the experimental intersection at a point of the William Floyd Parkway that passes near the college's center for careers in transportation.

At this intersection, north- and southbound vehicles cannot turn left at the light, which results in shorter waits for through traffic (which still must stop for vehicles entering the highway). Southbound vehicles wishing to turn left cross over the northbound lane well before the light—at the same time that north/south traffic is stopped for entering vehicles. The left-turning vehicles then make their way down an auxiliary road to meet up with the crossroad at a point east of the light.

The "continuous-flow intersection" is not really continuous, of course. Vehicles must stop, although for briefer periods. Depending on the amount of traffic, from 20 to 50 percent more vehicles can pass through the experimental intersection in a given time than can pass through a conventional intersection (with left turns at the light).

Railroads

Hope springs eternal for high-speed trains in the United States. Although plans for a high-speed connection between Dallas, Houston, and Fort Worth, Texas, fell through in 1994, plans for bullet trains that connect (1) Miami and Tampa, Florida, and (2) Washington, D.C., and Boston are proceeding.

Florida officials selected a consortium known as Florida Overland Express to proceed with its $4.8 billion project, which someday will whisk passengers from Miami to Tampa in about two and one-half hours. A Miami-Orlando link will be completed in 2004; an Orlando-Tampa link will be completed in 2006. The development group comprises Fluor Daniel Incorporated of California; Bombardier Incorporated of Montreal, Quebec; Odebrecht Contractors of Florida; and GEC Alsthom NV, which has built bullet trains that run in France. The consortium will own the tracks of the new system for 30 years, then cede ownership to the state of Florida.

The Northeast project belongs to Amtrak, which is awarding $700 million to a contractor who will produce a fleet of bullet trains that can use Amtrak's existing route. Amtrak also will upgrade its tracks and electrical system. As a result, trains will take passengers from New York City to Boston in under three hours, and from New York City to Washington, D.C., in about 2 hours and 15 minutes. The 150-mile (240-kilometer)-per-hour trains will use a tilting strategy when rounding corners at high speeds.

Donald W. Cunningham

Volcanology

More than a dozen significant eruptions occurred in 1995, mainly in subduction zones—restless seams in the globe where Earth's rigid crustal plates dive into the planet's hot interior to be recycled. Volcanoes showed signs of life in the Caribbean Sea, on the rim of the Pacific Ocean, in North America, and in Europe. Volcanologists rushed to the scene to monitor these events, hoping to advance their understanding of volcanoes while advising local officials on the potential hazards to life and property.

Pacific Rim Volcanoes

On New Zealand's North Island, Mount Ruapehu let loose the largest volcanic outburst in 400 years in that country. In September 1995, the 9,174-foot (2,797-meter) volcano cleared its throat through the caustic crater lake at its summit, triggering floods and mudflows. Later that month, an explosive eruption sent a column of gas and ash to a height of 12 miles (19 kilometers) above Ruapehu. The volcano erupted again on June 17, 1996—just six days after scientists declared that its activity had subsided.

Several underwater eruptions provided a dramatic demonstration of how volcanism continually alters Earth's surface. At Metis Shoals, a submerged volcano in the Tonga Islands of the Pacific, a new island rose from the sea in the space of a few weeks. The Tonga Islands, an arc of volcanoes about 1,250 miles (2,000 kilometers) east of Australia, mark the subduction zone where the Pacific plate dives down into a submarine trench and into the Earth's interior.

In some submerged chains, eruptions are insufficient to raise new land completely above sea level, as occurred in the eruption of the Ruby Seamount in the Mariana Islands. These islands in the West Pacific, like the Tonga Islands, are a volcanic arc. But the summit of the Ruby Seamount lies at an estimated depth of 750 feet (230 meters). The eruption was detected by fishermen, who could hear the malevolent rumbling of submarine explosions. Bubbles, dead fish, and a sulfurous smell also marked the site of the eruption.

The eruption of the Soufrière Hills Volcano on Montserrat disrupted the island's tourist industry and prompted several local evacuations.

Montserrat

Many of the eruptions of 1995 occurred in remote, relatively unpopulated areas, so hazards to people were minimal. Such was not the case on the Caribbean island of Montserrat. There the ongoing eruption of the Soufrière Hills Volcano directly threatened Plymouth, the island's capital.

Montserrat had an unlucky year. An unusually active hurricane season sent one tropical storm after another in the direction of the island, which lies in the volcanic chain of islands called the Lesser Antilles. Then, in July 1995, white ash began to spew 3,000 feet (915 meters) into the air from the volcano's crater—the first eruption in historical times. With a total area of only 39 square miles (102 square kilometers), Montserrat's 10,000 residents have few places to hide.

Over the months of intermittent eruptions of ash and steam from the volcano, a large fraction of the population was twice relocated to a safe area north of Plymouth. By early 1996, the volcano had gone into a "dome-building" phase as a plug of solidified magma built up in the active crater like a giant champagne cork. Should that cork grow large enough and collapse, avalanches of hot gas and ash called pyroclastic flows could pour down the slopes of the volcano into inhabited areas.

Soufrière Hills Volcano has infamous neighbors. Although largely quiescent now, the Caribbean is riddled with volcanoes. Indeed, the French Caribbean island of Martinique was the site of one of the worst volcanic disasters in recorded history: in 1902, a glowing avalanche of ash and debris swept down from the peak of Mont Pelée, killing some 28,000 people in the town of St. Pierre.

Mammoth Mountain

The U.S. Forest Service closed a recreational area on the flanks of Mammoth Mountain, a volcano in eastern California, after carbon dioxide gas built up in the soil to potentially dangerous levels. Officials were concerned that the odorless, colorless gas might collect in cabins or tents and possibly asphyxiate the people inside. Testing has shown the gas levels to present no risk of asphyxiation to campers, so the recreational areas will reopen.

Mammoth Mountain, 10,758 feet (3,281 meters) high, is on the southwestern edge of Long Valley caldera. This roughly oval depression, 19 miles (30 kilometers) across from east to west and 11 miles (18 kilometers) across from north to south, formed during an enormous eruption 760,000 years ago. In 1979, the caldera entered a period of unrest marked by a major earthquake and uplift of the crater floor.

The gas in Mammoth Mountain bubbles out of molten rock in the guts of the Long Valley system, from sources as deep as 12 miles (20 kilometers). Since Mammoth Mountain has no open, active crater, the gases seep up and collect in the soil. Last year, scientists say, a cap of winter snow and ice made carbon dioxide levels build up to unusually high concentrations.

The unrest under Mammoth Mountain Volcano continues to the present. The scientists monitoring it cannot say for certain when—or even if—Mammoth Mountain will erupt again. But nobody doubts that Long Valley remains a living, breathing volcanic hazard with the potential to do grievous harm to the communities in and around it.

Daniel Pendick

Zoology

Mystery of the Spectacled Eider

With white, black-ringed eye patches prominent on the male of the species, the spectacled eider is one of the most distinctive of Arctic waterfowl. Yet, though easily identified, many of the bird's habits are unknown. Specifically, biologists have been stumped by a fundamental question: Where does the spectacled eider go in winter?

That question was finally answered last year, thanks to a lone bird equipped with a radio transmitter. The mystery began unraveling in the spring of 1994, when biologists of the Fish and Wildlife Service and the National Biological Service captured 22 spectacled eiders in Alaska's Yukon-Kuskokwim Delta National Wildlife Refuge. They fitted the birds with tiny transmitters and tracked them until December 1994, when the batteries in the transmitters finally wore out. At that time, the eiders were spread out over a region of open water in the Bering Sea south of Saint Lawrence Island. Soon afterward, the birds disappeared—as they do every year—and with no more radio signals forthcoming, it was impossible to determine where they had gone.

Then, in February 1995, a transmitter that had been silent for months unexpectedly resumed sending signals. Surprised biologists tracked the signal to a location some 200 miles (320 kilometers) inside the Arctic ice pack, where they were certain that ice conditions made life impossible for waterfowl. Biologists Bill Larned and Greg Balogh decided to investigate anyway, and chartered a plane to fly them over the ice. To their amazement, they discovered tens of thousands of spectacled eiders swimming in small holes in the ice pack. The ducks were packed so tightly together that their body heat and the movement of their legs were apparently able to keep the surface water unfrozen, in spite of air temperatures that routinely fell to -20° F (-28° C). By early April, Larned and Balogh had documented that at least 140,000 spectacled eiders—about half the estimated world population—had spent the winter in the Bering Sea pack ice.

The spectacled eider, after passing the summer in relative comfort (above), winters in small holes in the Bering Sea pack ice.

Biologists hope that the discovery of the eider's wintering area will be a major step in understanding why the bird's population has plummeted more than 90 percent in the past three decades, causing it to be listed in 1993 as a threatened species. Fish and Wildlife Service biologist Russ Oates, leader of the Spectacled Eider Recovery Team, calls the discovery "[a] major step toward understanding how these birds live, what problems they may be facing, and other important questions we have about the Bering Sea ecosystem."

Life Blooms in the Ice

Ice surrounding Antarctica is home to life-forms very different from the spectacled eider: vast colonies of algae. Researchers who camped for five months on a 1-mile (0.6-kilometer)-long ice floe on the Weddell Sea found that algae have colonized an ecological niche in the ice, where they not only survive but thrive in a habitat so inhospitable that it had always been assumed to be virtually lifeless. The researchers discovered that the algae, which are trapped in the ice when it forms each autumn, are fed all winter by a steady broth of nutrient-rich seawater coming up through the bottom of the ice. As ice forms in seawater, salt is forced out of the ice-crystal lattices, forming dense brine that escapes through tiny channels into the sea. The brine sinks to the bottom, forcing nutrient-rich water on the bottom to well up and enter the channels in the ice, where it becomes a source of nutrition for algae. Meanwhile, sunlight filtering through the ice from above initiates photosynthesis, keeping the algae green and growing all winter.

The role of the algae in the Antarctic ecosystem is not yet fully understood, but it almost certainly has a place at the foundation of the food web. When the ice breaks up and melts in the summer, the algae are released into the sea and serve as food for krill and other organisms, or sink to the bottom and become part of the carbon-rich sediment.

TROUT TROUBLE

It was a tough year for trout in the Yellowstone National Park region. First came the news that rainbow trout in the Madison River, widely regarded as one of the great trout rivers in the world, were infected much more seriously with "whirling disease" than had first been reported. There is no known cure for the disease, which results when a parasite attacks the nervous system of young fish, causing them to swim in tight, whirling circles on the surface. Although not usually fatal, the disease makes the fish highly vulnerable to predators. It is being blamed for a 90 percent reduction in the rainbow population along 50 miles (80 kilometers) of the river's prime trout-fishing water.

The parasite, which lives during part of its life cycle in worms on the river bottom, is spread in many ways, including in downstream migration, in the droppings of birds that have ingested the parasite's spores, and by human interference. The parasite originated in Europe and probably arrived in North America in the 1950s in a batch of frozen fish.

Biologists at fisheries in Montana suspect that the parasite entered the Madison via an unauthorized planting of contaminated hatchery trout. The Madison rainbows are noted for being wild fish; hatchery rainbows have not been officially introduced in decades. In 1995, the extent of the infestation became apparent when biological surveys revealed that the Madison's population of rainbows had declined from an average 3,000 trout per mile of river to about 300. Perhaps most alarming is the potential for the disease to spread. Biologists say it is inevitable that the disease will spread downstream on the Madison into the Missouri system, and eventually into many other famous trout waters. Infected fish have already shown up in Montana's Beaverhead and Ruby River drainages and in isolated waters in Colorado and Utah. Biologists, frustrated in their efforts to contain the disease, say their only hope is that trout with a natural resistance to the disease will survive to reproduce and pass their resistant genes to their offspring.

Fishery biologists were also alarmed to discover last year that someone had introduced lake trout into Yellowstone Lake. At risk is the lake's famed population of Yellowstone cutthroats, a genetically isolated subspecies of cutthroat that has been evolving since the lake formed at the end of the last ice age, some 12,000 years ago. When introduced into other waters, lake trout have made quick work of smaller, less aggressive species like cutthroat trout, feeding on them and crowding them out of prime spawning areas. Biologists fear that the illegal release will doom that lake's native trout. The loss would have detrimental effects on grizzly bears, bald eagles, ospreys, otters, and other animals that depend on Yellowstone cutthroats for part or all of their food supply.

The Yellowstone cutthroat trout (above) is threatened by the larger and more aggressive lake trout that were introduced into the cutthroat's only habitat.

MORE TROUT TROUBLE

Rainbow trout, when they spend most of their life feeding and growing in oceans or large inland lakes, are known as steelhead. The best steelhead waters in the world are in the Pacific Northwest, where tumbling rivers serve as spawning grounds for the big,

Roundups of wild horses may become a thing of the past now that an injectable contraceptive vaccine can be used to control the horse population.

ocean-run trout. But rivers that hosted healthy runs only a decade ago are now mostly empty of steelhead, and environmentalists and anglers are looking for the cause.

Many fingers have been pointed at an unlikely villain—the California sea lion. As the sea lions have increased in number, they have become increasingly bold in their predations on steelhead, and have apparently been cutting a wide swath through them in places where the trout congregate before running up rivers to spawn.

One such place is the Ballard Locks, on the Lake Washington Ship Canal in upper Puget Sound. To reach spawning grounds in the Cedar River above Lake Washington, steelhead must climb a fish ladder constructed for that purpose at the locks. Since the early 1980s, sea lions have been waiting for the fish when they arrive at the bottom of the ladder. The number of steelhead returning to the Cedar River dropped from 3,000 in 1985 to 125 in 1995, and many people are blaming the sea lions for the decline.

Washington Department of Fish and Wildlife officials attempted to discourage the sea lions by shooting them with rubber-tipped bolts from crossbows, dropping firecrackers into the water, and installing a large-web net at the mouth of the fish ladder. When none of those efforts worked, a task force of representatives from universities, sport-fishing groups, Native American peoples, and environmental organizations convened to examine the problem. Some of the members of the task force recommended "lethal removal" of the sea lions. Other members dissented, arguing that factors besides the sea lions might be involved in the declining numbers of steelhead, including destruction of spawning grounds on the Cedar River and a general decline in productivity in the North Pacific caused by recent climate cycles. While the debate over what to do continues, steelhead numbers continue to fall, and the sea lions are being forced to go looking for more-productive fishing spots.

Birth Control for Wild Horses

Wild horses have roamed the American West for centuries, since they were brought there by early Spanish settlers, but in recent years their numbers have grown to stampede proportions. The problem is that the horses have adapted too well to life in the arid western states. They can get by on little food, are resistant to diseases, and reproduce so efficiently that, in some herds, half the mares give birth to foals each year. With natural predators on the decline, the population has grown steadily to a current total of about 42,000 animals.

That would not be a problem except that burgeoning development has created inevitable conflicts between humans and wildlife. The Bureau of Land Management (BLM) has traditionally kept the mustang population in check by capturing them with the aid of helicopters—a practice that has removed 140,000 horses during the past 17 years, but is under criticism for being expensive, dangerous, and inhumane to the horses. The captured horses are sold; some of them are given to people supporting the BLM's adopt-a-horse program, but many of them are auctioned off to commercial buyers who sell them for slaughter.

In November 1995, BLM officials in Nevada began final testing on a contraceptive vaccine called Porcine Zonae Pellucida (PZP), which can be injected into wild mares to block pregnancies. Initial testing of the vaccine proved its effectiveness, but it needed to be applied several times over a four-week period, and necessitated corralling the horses during the inoculation period. The new vaccine can be applied just once, is inexpensive, is effective for two years, has demonstrated no side effects, and does not affect the mating behavior of the horses. If the testing continues to be successful, and no long-term effects appear, BLM officials are confident they can cut the reproductive rate of the mustangs in half and eliminate the need to capture and destroy them.

THE MYTH OF THE MALE SEAL

It has long been assumed that seals are polygynists—that is, the males mate with numerous females during their lives—and that most of the mating is accomplished by a few dominant males, while subordinate males lurk forlornly on the fringes. But recent research on a population of gray seals that breeds every year on an island off the coast of Scotland suggests that there is probably a great deal more amorous activity taking place than had once been thought.

Between 1986 and 1989, the researchers, headed by geneticist Bill Amos of the University of Cambridge in England, took blood samples from 85 female and 88 male seals on the island. In the years since, the researchers discovered that about 70 percent of the females returned to the same island to breed, and that most of their offspring year after year were produced by the same fathers. By using molecular genetic techniques, Amos and his colleagues examined the parentage of the pups born to their original test group, and ruled out the possibility that most of them were fathered by a few dominant males. The scientists concluded, on the contrary, that a significant number of the pups were fathered by subordinate males that had formed long-term bonds with females.

Such behavior could have definite advantages for gray seals, according to Amos. He theorizes that less competition among males, especially less of the competitive thrashing and bashing indulged in between dominant males, could decrease the number of pups that are inadvertently crushed and killed. If so, it would be in the best interest of females and their offspring to mate with quieter, less aggressive males and to maintain fidelity with them year after year.

TREE-SAVING WOLVES

Another long-held assumption—that the abundance of animals in an ecosystem is determined by the abundance of plants—has been challenged by research on Isle Royale National Park in Lake Superior. The relationship between wolves and moose on the 210-square-mile (540-square-kilometer) island has been documented for decades and has become a textbook example of predator/prey dynamics. Ever since wolves crossed frozen Lake Superior to the island in 1949, they have served as an effective check on a moose population that until then was out of control. Now researchers have found that the population cycle of the moose and wolves has consequences reaching beyond the two mammals.

By studying growth rings on the island's balsam firs, biologists Rolf Peterson and Brian McLaren of Michigan Technological University in Houghton discovered that the trees, which had grown quickly in the 1960s and 1970s, went through periods of severely suppressed growth during the 1980s. During that same decade, a disease spread by domestic dogs, canine parvovirus, reached the island and decimated the wolf population, reducing it from 50 in 1980 to 12 in 1988. Without wolves to cull them, the moose increased in number and began browsing heavily on young balsam firs, causing their growth rate to slow. Recent evidence that the island wolves are again reproducing successfully could very likely mean healthier, faster-growing forests in the future—suggesting that at least some ecosystems are managed from the top of the food web down, not the other way around.

Jerry Dennis

In Memoriam – 1995

ALFVEN, HANNES (86), Swedish physicist who shared the 1970 Nobel Prize in Physics for founding the field of magnetohydrodynamics (MHD)—the study of electrically conducting fluids in a magnetic field. His research, often criticized and dismissed by other scientists, demonstrated the existence of magnetic-field lines in plasma, which came to be known as Alfven waves. d. Djursholm, Sweden, Jan. 1.

ANFINSEN, CHRISTIAN B. (79), U.S. biochemist whose work with protein structure and protein folding led to the 1972 Nobel Prize in Chemistry, which he shared with two other scientists. He headed the laboratory for the National Institutes of Health (NIH) in Bethesda, Maryland, from 1963 to 1982. d. Randallstown, Md., May 14.

ATANASOFF, JOHN V. (91), U.S. physicist and computer pioneer who, in the 1930s, together with graduate student Clifford Berry, invented a digital computing device—dubbed ABC—unlike the mechanical calculators of the era. At a cost of just $1,000, ABC solved algebraic problems and analyzed data, and is recognized by many computer historians as the first electronic digital computer of any kind. d. Frederick, Md., June 15.

BELZER, FOLKERT O. (64), Indonesian-born U.S. medical researcher who, with Dr. James Southard, developed "UW solution," a fluid that can keep donated organs viable for 18 to 30 hours. The extra time allows for better matching of organs with recipients, thus reducing rejection rates. d. Madison, Wis., Aug. 6.

BERRY, LEONIDAS H. (93), U.S. physician who specialized in diseases of the digestive tract and who brought medical care to health-care-needy black communities. Berry, who set up innovative drug-treatment clinics in Chicago, was one of the first black doctors admitted to the American Medical Association (AMA). d. Chicago, Dec. 4.

BIERMAN, EDWIN L. (64), U.S. physician who developed a diet low in fat and cholesterol and high in complex carbohydrates, an eating regimen that is widely considered helpful in preventing heart and vascular disease in diabetics. d. Seattle, Wash., July 5.

BRIDGE, HERBERT S. (76), U.S. scientist and expert on interplanetary plasma. He designed space-plasma experiments that flew aboard unmanned scientific missions to every planet except Pluto. d. Boston, Aug. 2.

BUTENANDT, ADOLF (91), German scientist whose pioneering work on sex hormones earned him the Nobel Prize in Chemistry in 1939. He isolated the male hormone testosterone and the female hormone progesterone. His research made possible the development of birth-control pills and such steroid drugs as cortisone. d. Munich, Germany, Jan. 18.

CAIN, STANLEY A. (92), U.S. conservationist who helped develop the science of ecology and, in doing so, fostered a national awareness of conservation. In 1965, President Lyndon B. Johnson named Cain assistant secretary of the Interior for Fish, Wildlife, and Parks, a position he held until 1968. d. Santa Cruz, Calif., April 1.

CHANDRASEKHAR, SUBRAHMANYAN (84), Indian-born U.S. astrophysicist whose discoveries about the evolution of stars pointed scientists toward the existence of "black holes" and "white dwarfs." For his pioneering work in astrophysics, he shared the 1983 Nobel Prize in Physics. d. Chicago, Aug. 21.

CORRIGAN, DOUGLAS (88), U.S. aviator who, in 1938, distracted Depression-weary Americans with his improbable journey: after filing a flight plan that should have ended in California, he took his rickety $310 airplane in the opposite direction on a nonstop solo flight from Brooklyn, New York, to Dublin, Ireland. (U.S. aviation authorities had repeatedly rejected his requests to make this ocean crossing.) His flight brought him instant celebrity—and the famous nickname "Wrong Way Corrigan." d. Orange, Calif., Dec. 9.

CRAIG, GEORGE B., JR. (65), U.S. entomologist and one of the world's foremost experts on mosquitoes. His research centered on effective ways to battle such mosquito-transmitted diseases as yellow fever, dengue fever, and encephalitis. d. Las Vegas, Nev., Dec. 21.

deSTEVENS, GEORGE (71), U.S. chemist who discovered two diuretics, hydrochlorothiazide and cyclopenthiazide. The former is the diuretic most widely prescribed to treat high blood pressure. d. Summit, N.J., Sept. 5.

DE VAUCOULEURS, GÉRARD (77), French-born astronomer who is widely acknowledged to have opened up the field of study that deals with galaxies beyond the Milky Way. d. Austin, Tex., Oct. 7.

DIETZ, ROBERT S. (80), U.S. geologist whose studies of the ocean floor confirmed the geological theory of continental drift. While working for the navy in the 1950s, Dietz arranged for the purchase of Aqua-Lungs, forerunners of scuba-diving equipment, developed in France by undersea explorer Jacques Cousteau. d. Tempe, Ariz., May 19.

ECKERT, J. PRESPER (76), U.S. electrical engineer who coinvented the ENIAC computer, believed to be the first digital electronic computer. Resembling an old-fashioned telephone switchboard, the mammoth computer could—in 30 seconds—complete calculations that took a clerk 20 hours. It was used extensively during World War II, particularly to confirm design calculations for the atomic bomb. d. Bryn Mawr, Pa., June 3.

EVANS, ROBLEY (88), U.S. nuclear physicist whose research set the standard for the maximum allowable human exposure to radiation. His work on radioactive forms of elements, called isotopes, led to the diagnosis and treatment of hyperactive thyroid. d. Paradise Valley, Ariz., Dec. 31.

FOWLER, WILLIAM A. (83), U.S. astrophysicist and corecipient of the 1983 Nobel Prize in Physics. His work led to a new field—nuclear astrophysics—that explained the formation of chemical elements in stars. d. Pasadena, Calif., March 14.

FUCHS, FRITZ F. (76), Danish-born U.S. obstetrician and gynecologist internationally known for developing an effective intravenous agent to prevent premature labor and for his pioneering work with amniocentesis, a procedure that detects fetal genetic disorders. d. Hamburg, Germany, Feb. 17.

GALILI, YISRAEL (72), Israeli weapons designer, known in the military as the "father of the rifle." He invented the Galil assault rifle and helped develop the Uzi submachine gun. d. Givatayim, Israel, March 9.

GALLAGHER, JAMES ROSWELL (92), U.S. physician who pioneered the field of adolescent medicine. Recognizing that patients from 12 to 21 years of age had unique medical problems, Gallagher organized the country's first comprehensive medical unit for teenagers. d. Lexington, Mass., Nov. 10.

GARDNER, BEATRIX T. (61), Austrian-born U.S. psychologist who, in the 1960s and 1970s, taught sign language to Washoe, a chimpanzee. Today, at age 31, Washoe still provides researchers with valuable clues to animal communication. d. Padua, Italy, June 5.

GIANTURCO, CESARE (90), Italian-born U.S. physician who pioneered the use of radiology to treat illness, rather than just as a tool for diagnosis. He also invented various medical devices, including coronary stents that are in worldwide use. d. Urbana, Ill., Aug. 25.

GLENNAN, T. KEITH (89), U.S. educator who headed the National Aeronautics and Space Administration (NASA) when it was established in 1958. He fought to keep NASA small and did not support the space race with the Soviet Union. d. Mitchellville, Md., April 12.

GOLDIAMOND, ISRAEL (76), Ukrainian-born U.S. psychologist whose research into human and animal behavior generated methods of altering such harmful habits as overeating, smoking, stammers, and a variety of phobias. d. Chicago, Nov. 19.

GOULD, LAURENCE M. (98), U.S. polar explorer and second in command of Admiral Richard E. Byrd's first Antarctic expedition (1928–30). Gould campaigned against any country laying territorial claims to the continent. He won the Congressional Gold Medal. d. Tucson, Ariz., June 20.

GUTHRIE, ROBERT (78), U.S. physician and scientist credited with developing a simple blood test to screen newborns for phenylketonuria (PKU), an inherited disease that can cause irreversible brain damage. d. Seattle, Wash., June 24.

HARGREAVES, ALISON (33), Scottish mountain climber and the first woman to reach the summit of Mount Everest alone and without supplementary oxygen. She perished during an ascent of K2, the world's second-highest peak. d. K2's Abruzzi Ridge, Pakistan, Aug. 13.

HESSBERG, RUFUS R. (74), U.S. physician and director of space medicine for NASA. His extensive "chimps in space" research into the effects of sudden acceleration, weightlessness, restraint, and isolation showed how well astronauts could live beyond the confines of Earth. d. Alexandria, Va., July 27.

HORSFALL, JAMES G. (90), U.S. plant pathologist and author of the landmark work *Fungicides and Their Actions* (1945). Horsfall's research changed the way the world's farmers combat blight and other fungal diseases. The fungicide arsenal—once consisting of sprays laced with copper, lead, and other heavy metals—now contains organic compounds, which are less expensive, more effective, and environmentally safer. d. Hamden, Conn., March 22.

JEN, CHIH-KUNG (89), Chinese-born U.S. scientist who pioneered the field of microwave spectroscopy, the measurement of an atom's or molecule's absorption or emission of radiation in the microwave portion of the electromagnetic spectrum. Jen's research in the 1920s provided early proof of the existence of the ionosphere, a layer of ionized air high above Earth's surface. d. Needham, Mass., Nov. 19.

KJELLSTROM, BJORN (84), Swedish inventor who developed the modern compass and is credited with introducing the sport of orienteering to North America. His 1955 book *Be Expert With Map and Compass* is still considered an accurate guide to orienteering. d. Stockholm, Sweden, Aug. 26.

KÖHLER, GEORGES (48), German biologist who shared the 1984 Nobel Prize in Physiology or Medicine for advances in the theory and technology of immunology. He developed new techniques for producing antibodies for use in such fields as cancer, hormone therapy, AIDS, and vaccine research. d. Freiburg, Germany, March 1.

KROGER, WILLIAM S. (89), U.S. psychiatrist and proponent of the medical use of hypnosis. In one demonstration, he removed a woman's thyroid gland using hypnosis as the sole anesthesia. In a well-publicized 1977 case, Kroger hypnotized a school-bus driver to help him remember the license plate of a van used to abduct 27 people. d. Los Angeles, Dec. 4.

KRUGMAN, SAUL (84), U.S. pediatrician who pioneered the field of childhood diseases. He established one of the country's first comprehensive children's health clinics and was a leader in the development and evaluation of vaccines against measles, rubella, and hepatitis. d. Fort Lauderdale, Fla., Oct. 26.

KUMON, TORU (81), Japanese inventor of a home- or after-school-teaching method for mathematics. Devised in 1954, the program is used by more than 2 million students in 29 countries. d. Osaka, Japan, July 25.

MAZAR, BENJAMIN (89), Israeli historian and archaeologist who directed large-scale excavations in the Old City of Jerusalem. d. Jerusalem, Sept. 8.

OLIVER, BERNARD M. (79), U.S. researcher and founder of Hewlett-Packard Laboratories. He is best known for designing the first programmable desktop calculator and various types of handheld calculators and computers.

Under the auspices of the National Aeronautics and Space Administration (NASA) and the SETI Institute, he also studied practical methods for detecting intelligent extraterrestrial life. d. Los Altos Hills, Calif., Nov. 23.

PATTERSON, CLAIR C. (73), U.S. geochemist who, in 1953, established Earth's age at 4.6 billion years old, far older than a previous estimate of 3 billion years. As a consequence of his geochemical research, Patterson also raised the alarm about the dangerous effects of lead in the environment, and was instrumental in the enactment of the Clean Air Act of 1970 and the phasing out of lead in gasoline. d. Sea Ranch, Calif., Dec. 5.

PILKINGTON, ALASTAIR (75), British inventor who developed a flotation process that produces glass with two smooth sides that do not require polishing. Such glass is used in automobiles and architectural applications. d. London, May 5.

RANDOLPH, THERON G. (89), U.S. allergist who founded the field of environmental medicine. For nearly half a century, Randolph contended that countless illnesses and allergies are the result of exposure to everyday chemical poisons. The scientific community, long skeptical of his theories and therapies, eventually embraced his ideology. d. Geneva, Ill., Sept. 29.

RAVIDAT, MARCEL (72), French garage mechanic who, in 1940, while walking with several companions in the woods, accidentally discovered the opening to the Lascaux cave, which contains vividly colored prehistoric paintings dating back 17,000 years. He remained a guide and an official guardian of the cave throughout his life. d. Montignac, France, March 29.

REDENBACHER, ORVILLE (88), U.S. agricultural scientist and advertising icon who experimented with hybrid popcorn. In 1965, he and his partner made the first innovation in popcorn in more than 5,000 years—a kernel that would expand to as much as 40 times its original size when popped. d. Coronado, Calif., Sept. 19.

REDINGTON, ROWLAND W. (70), U.S. physicist whose work contributed to the development of CT (computed-tomography)-scan and MRI (magnetic-resonance-imaging) equipment. Both diagnostic-imaging tools have indirectly improved the health of millions of people around the world. d. Troy, N.Y., June 22.

SALK, JONAS E. (80), U.S. microbiologist who, in 1955, developed the first effective vaccine against poliomyelitis, or infantile paralysis, a viral infection that can cause paralysis and death. During the 1950s, a polio epidemic raged in the United States, annually infecting some 25,000 people, most of them children and young adults. Many were permanently crippled; some spent the rest of their lives in tanklike "iron lungs." As a result of Salk's vaccine, the disease is on the verge of being eradicated worldwide. d. La Jolla, Calif., June 23.

SCHARRER, BERTA VOGEL (88), German-born U.S. biomedical scientist who pioneered the field of neuroendocrinology, the study of the interactions between the nervous and endocrine systems. She was also an authority on the South American cockroach. d. Bronx, N.Y., July 23.

SMITH, HUGH H. (93), U.S. virologist who helped develop the yellow-fever vaccine. For centuries, yellow fever, a disease transmitted by mosquitoes, was a leading cause of death in much of the world. Smith's research in Colombia and Brazil proved the effectiveness of the 17D vaccine, which has saved countless lives. d. Tucson, Ariz., Dec. 18.

STIBITZ, GEORGE R. (90), U.S. mathematician who, while working as a researcher at the Bell Telephone Laboratories, invented the first digital computer in 1940. d. Hanover, N.H., Feb. 2.

SUOMI, VERNER E. (79), U.S. meteorologist who developed imaging technologies that made modern weather satellites possible. His invention of the spin-scan camera in 1964, a device that transmits images back to Earth from 25,000 miles above the equator, revolutionized weather forecasting. d. Madison, Wis., July 30.

TIPPER, CONSTANCE (101), British metallurgist and mechanical engineer who discovered a crucial flaw in the material used on Liberty ships that sailed from the United States to Great Britain during World War II. The flaw produced a tendency for the 10,000-ton ships to break in two under certain conditions. Today engineers apply the "Tipper Test" to determine the brittleness of steel. d. Penrith, England, Dec. 14.

WALTON, ERNEST T.S. (91), Irish physicist and corecipient of the 1951 Nobel Prize in Physics for creating the first human-made atom-smashing machine, now called a particle accelerator. d. Belfast, Northern Ireland, June 25.

WEIL, GEORGE L. (87), U.S. physicist who assisted Enrico Fermi on the Manhattan Project, which produced the first atomic bomb. In later years, Weil cautioned against the risks of the nuclear era both as a consultant to the UN and in the private sector. d. Washington, D.C., July 1.

WHITAKER, CARL A. (83), U.S. psychotherapist who, in the 1960s, focused on treating the whole family rather than one patient in therapy. This novel approach, called family therapy, is standard procedure among psychotherapists today. d. Nashotah, Wis., April 21.

WILLIAMS, WALTER (76), U.S. aerospace engineer whose work spurred on the jet age and manned space exploration. He directed research programs for supersonic flight and oversaw the first American manned spaceflights, Project Mercury, in the early 1960s. d. Tarzana, Calif., Oct. 7.

WOOD, EVELYN (86), U.S. educator who developed speed-reading techniques that could improve the average American's reading rate from the usual 250 to 300 words per minute to as high as 6,000 words per minute. Together with her husband, she founded the Evelyn Wood Reading Dynamics Institute. d. Tucson, Ariz., Aug. 26.

ZIMMERMAN, HARRY M. (93), U.S. physician whose widely varied research included pioneering studies on the relationship between vitamin deficiency and nervous-system tumors; important work on Lou Gehrig's disease; and the development of an early vaccine for encephalitis. He was also a founder of the Albert Einstein College of Medicine. d. Bronx, N.Y., July 28.

INDEX

A

Absolute zero 361-62
Abu Simbel 177, 179 *illus.*
Accelerator, Particle 351, 363
Accidents *see* Trauma
Acetylcholine 310
Acid rain 96, 326
Acrobats 210, 214 *illus.*
Acts of God 204-5
Adaptation
    coats of animals 61
    pigeons 30-31, 34
    upright posture 286-87
    water needs of animals 54-59
Adenosine deaminase (ADA) 333
Adenovirus 333
Adolescents *see* Teenagers
Advanced Tissue Sciences (company) 151, 153
Advertising
    HIV 144
    smoking 366
    VCRs 317
    World Wide Web 260
Aerodynamics 236, 247, 249, 251, 267-68
Age of Exploration 197, 199, 201
Aging 157, 164
Agriculture 284-85
    asteroid impacts 98
    biological resistance 308
    book reviews 304
    brucellosis 17-18, 21
    endangered breeds 49-53
    explosive chemicals 310
    food and population 332
    mad-cow disease 340
    Mayan civilization 290
    nonindigenous pests 121
    silos 280
    Tasmanian devil 37
Agriculture, United States Department of 21, 51, 156, 355
AIDS 150, 338-40 *see also* HIV
Air bags 272, 274, 294, 318
Airflow *see* Aerodynamics
Air Force, United States 99
Airplanes *see* Aviation
Air pressure 118, 145-46
Air Transport Association (ATA) 250
Albatross 58 *illus.*
Alcoholic beverages 159, 355
Alexandria (Egypt) 177 *map*
Alfven, Hannes 384
Algae 123, 380-81
Allergies 302
Alvarez, Luis and Walter 95
Alves, Francisco J. S. 198, 200
Alvin (submersible) 277 *illus.*, 359
Alzheimer's disease 335, 339
American Geographic Society 192 *illus.*, 194-95
American Livestock Breeds Conservancy 49, 51
American Museum of Natural History 184-90
American River bridge 265 *illus.*
Americans with Disabilities Act (ADA) 138-39, 143
America Online 258
Amgen Corporation 334
Amino acids in diet 160-61
Amoco Corporation 278, 326
Amon, Temple of 178 *illus.*
Amos, Bill 383
Anatomy *see* Human body

Anemia 157
Anfinsen, Christian B. 384
Angioplasty 164 *chart*
Angra do Heroísmo 199
Animal research 51
Animals *see* Zoology
Animation, Computer 244
Antarctica 291, 330, 349, 380-81
Anthropology 286-88, 305
Antilock-braking system (ABS) 294, 318
Antlers 47
Ants 307
Apatosaurus 189, 189 *illus.*
Apes 286
Appian Way 206
Apple Computer (company) 316
Aqua regia 219
Archaea species 300-301
Archaeology 176-83, 197-201, 288-90
Archaeopteryx (bird) 361
Architecture *see* Art and architecture
Arctic 329, 380
Arden-Close, Sir Charles 194
Area codes (telephones) 313
Argentinosaurus 360-61 *illus.*
Arianespace (company) 372
Army, United States 240-41
Art and architecture
    bridges 262-69
    computer modeling 243-44
    creativity 297
    dinosaur restorations 184-85, 189
    Mayan civilization 290
    skyscrapers 311
    three-dimensional space 241-42
    Tomb of the Brothers 176-83
Asbestos 366
"Ask the scientist" (readers' questions)
    animals and plants 60-61
    astronomy 100-101
    Earth and the environment 134-35
    human sciences 172-73
    past, present, and future 206-7
    physical sciences 236-37
    technology 280-81
Asteroids 65, 94-99
Astrology 100
Astronauts 83-88, 100, 371-72
Astronomy 291-92 *see also* Space science
    asteroid impacts 94-99
    astrology 100
    book reviews 304
    liquid-mercury telescopes 71-74
    mass of the universe 363
    Rittenhouse, David 89-93
    starbirth 64-70
    Sun 75-81
    3-D computing 246
Aswan (Egypt) 177 *map*
Atacama plateau 193 *map*
Atanasoff, John V. 384
Atherosclerosis 341
Atlantic Ocean 194 *map*, 199, 328
Atlantis (space shuttle) 371
Atmosphere
    asteroid impact 96
    Jupiter 96, 373
    liquid-mercury telescope 72
    Moon 291
    ozone layer 330, 348-49
    thunderstorms 104, 107-8
Atoms 350-51, 361-62
Attention-deficit disorder (ADD) 341
Auger (submersible) 276 *illus.*
Auger tension-leg platform 278, 311
Aurora borealis 72

Australia
    geology 336-37
    wildlife 35-37, 54, 126
Australopithecines 288
Automation
    aviation 251-52
    highway-visibility monitor 311-12
    weather forecasting 118
Automobile crashes 270-74
Automotive technology 293-94
    early cars 280
    gasohol 135
    generator-engine combination 319
    natural gas 326
    navigation systems 376-77
    platinum 220
    security and tracking system 319
Aviation 295-96
    airplane crashes 247-53, 295-96
    computer models 246
    runway radar 376
    weather forecasting 118
Azidothymidine *see* AZT
Azores, Shipwrecks in 197-201
AZT (drug) 140-41, 143, 343

B

B-52 bomber 295 *illus.*
Babies *see* Infants
Baboons 45, 46 *illus.*, 340
Back injury 169, 172
Bacteria
    ancient microbes 361
    hyperbaric medicine 148
    optoelectronics 309
    selecting mild strains 170-71
    silver 219
    water purification 322
Balance, Sense of 100, 212-13
Ballard, Robert 289
Ballistics 236
Balogh, Greg 380
Banking 314-15
Barometers 118
Bats (animals) 366
Batteries, Electric 281, 293, 319
Bear attacks 60
Bed-wetting 335
Beef 20, 285
Bees 60-61, 244 *illus.*
Behavior, Animal *see* Zoology
Behavioral sciences 274, 297-99
Belted Galloway cows 50 *illus.*
Belzer, Folkert O. 384
Bends (Decompression sickness) 145, 147, 149
Berry, Leonidas H. 384
Beta-carotene 158
Bicarbonate 173
Bicycles 319
Bierman, Edwin L. 384
Big Bang 227, 232, 292
Bilham, Roger 128, 131, 133
Billings, Judith 139, 140 *illus.*
Bill Nye, the Science Guy (TV program) 368
Biodiversity 49-53, 120-26
Bioengineering *see* Biotechnology; Genetics
Biology 299-301, 352-54, 367
Biotechnology 151-55, 285, 302-3 *see also* Genetics
Bird, R. T. 190

Birds
  courtship displays 44-48
  dinosaurs 184
  fossil discoveries 361
  game birds 61
  Garamba National Park 40
  Madrean Archipelago 113
  peregrine falcon 323
  pigeons 30-34
  Project Feeder Watch 367
  seabird attacks 60
  spectacled eider 380
  toolmaking 301
  water needs of animals 55, 58
Birth defects 168, 221, 341, 353
Bison 16-21
Bjelkhagen, Hans I. 246
Black box (data recorder) 247-51, 274, 295
Bladder, Urinary 56, 154
Blood
  bone marrow cells 333, 340
  clotting 310
  Ehrlichiosis 365
  nutrition and diet 157
Blue jets (meteorology) *see* Red sprites and blue jets
Boeing Corporation
  airplane safety 248-50, 253, 295
  International Space Station 373
  new aircraft 296
  simulated-flight testing 246
Bombings 297 *illus.*, 310
Bone marrow 333, 340
Book reviews 303-6
Bopp, Thomas 291
Borra, Ermanno F. 72-74
Bose-Einstein condensate (BEC) 361-62
Botany 306-8 *see also* names of individual plants
  book reviews 303
  Isle Royal forests 383
  job opportunities 60
  Madrean Archipelago 111-13
  nonindigenous species 120-26
  science education 367-68
  sequoias 22-29
  water needs of animals 55
Bovine spongiform encephalopathy (BSE) *see* Mad-cow disease
Bowerbirds 48 *illus.*
Bradshaw, H.D. 306
Brain and nervous system
  Alzheimer's disease 339-40
  balance 212-13
  human genome 334
  mad-cow disease 340
  musical pitch 300
  neural-tube defects 341
  one-cell organisms 310
  schizophrenia 298
  trauma research 274
Brakes, Automobile 294, 318
Brass 221
Bray, Dennis 310
Breast cancer 155, 159, 168-69, 339
Breast-feeding 157, 160-61
Bridge, Herbert S. 384
Bridges 262-69, 306
Brontosaurus see Apatosaurus
Bronze 221
Brooklyn Bridge 265-66
Brown, George 29
Brown dwarfs (astronomy) 291
Brown tree snakes 121-22 *illus.*, 122
Browser programs 259, 314
Brucellosis 16-21
Bruises 145
Buffalo, American *see* Bison
Buffalo toads *see* Cane toads
Bullets 236
Bult, Carol 300
Bumblebee (computer model) 244 *illus.*

Burials, Ancient 176-83, 288, 290
Burns 150-51
Burton, James 179
Bush, Andreas 101
Butcher birds 126
Butenandt, Adolf 384
Bypass surgery 164 *chart*

C

Cable television 261, 313-14
Cain, Stanley A. 384
Cairo (Egypt) 177 *map*
Cajeput trees 120-21
Calcium 157, 160, 173
California 23-24, 28, 312, 379
Calories 161
Camcorders 318, 320
Camels 56 *illus.*, 57
Cameras 319-20
Canada
  crust of Earth 337
  International Space Station 373
  submersibles 276
Cancer 150, 158-59, 301, 339, 366
Cane toads 126 *illus.*
Canine distemper 325
Canine parvovirus 383
Caplen, Natasha 303
Caravels (ships) 198, 201
Carbohydrates 55, 160, 307-8
Carbon 292, 307
Carbon dioxide 225, 379
Carbon monoxide 148, 220, 317
Caribou 47
Car seats, Children's 271 *illus.*, 273, 294
Carter, Howard 179
Cartilage 154
Cartography *see* Maps and map making
Cassowaries 60
Catalytic converters 220
Cataracts 330
CAT scans 173
Cattle
  attacks on people 60
  brucellosis 16-18, 21
  livestock breeds 49-50
  mad-cow disease 285, 340
Cavendish, Henry 362
C-bus (electronics) 319
CD-4 cells 140, 338
Cells
  chemistry 322
  division 301, 339
  HIV infection 139-40, 142
  information transfer 310
  primitive organisms 300-301
  proliferation 158
  tissue engineering 151-55
Cellular phones 313, 319, 377
Censorship 261, 313, 316
CFCs *see* Chlorofluorocarbons
Challenger Deep (Pacific Ocean) 275, 279
Chandrasekhar, Subrahmanyan 384
Chapman, Sidney 348
Cheetahs 15 *illus.*
Chemistry 309-10
  cell processes 322
  earthquake precursors 370-71
  instant meals 318
  laundry 223
  Nobel prizes 348-49
  noble metals 216-21
  science education 367
  starbirth 68
  Sun 78
  T-ray analysis 321-22
  water purification 322
Cherokee language 25

Chessie the Manatee 324 *illus.*
Chevron Corporation 326
Chickens 52
Chicxulub (asteroid crater) 95-96
Children *see also* Infants; Teenagers
  automobile crashes 271, 273
  bed-wetting 335
  Ritalin 341
  television remote controls 320
  vegetarian diets 157-58, 160-61
Children's Television Act (1990) 368
China
  earthquakes 369
  flooding 347
  food and population 332
  measuring Mt. Everest 128, 130
  prehistoric humans 287
  space science 372-73
Chlorine 348
Chlorofluorocarbons (CFCs) 330, 348-49
Cholera 167-68
Cholesterol 158-59, 161, 356-57
Cholinesterase 310
Chromosomes 301, 339
Cigarettes *see* Smoking
Circulatory system 149-50
Circus performances 210-15
Cities
  falcons 323
  heat islands 204
  Mayan architecture 290
  pigeons 30-31
  transportation 207
  walled city 207
Civil engineering 262-69, 230, 311-12, 377-78
Civil rights
  HIV 138, 144
  Internet 261, 313, 316
Classification, Biological 299-300
Cleaning
  clothes processor 318
  dry cleaning 222-25
  gasoline spills 312
  self-cleaning coatings 310
Cleveland, Grover 235
Climate
  adaptation of animals 61
  asteroid impact 96
  forensic meteorology 203-5
  global warming 328-29
  Madrean Archipelago 111-12
  Mayan decline 290
  sequoias 24-25
Clocks 89, 230
Clouds
  ice crystals 101
  thunderstorms 107
  tornadoes 134
  ultraviolet radiation 358
  weather forecasting 117, 119
Coal 326
Cockpit voice recorders *see* Black box
Cocoons 57
Collagen 153, 215, 225
Collins, Eileen 375 *illus.*
Colon cancer 158-59, 339
Colonial Williamsburg 53
Colorado River 327-28
Columbia (space shuttle) 368
Comets 65, 94, 96, 98, 100, 291
Commercials *see* Advertising
Communications and Decency Act (1966) 261, 316
Communication technology 313-14
  Aeronautical Telecommunications Network 376
  automobile tracking 319
  Internet 254-61
  liquid-mercury telescope 73 *illus.*
  satellite phone 318
  sports data 320

388   INDEX

submersibles 278-79
Compact disks 315-16
Computer modeling 240-46
    automobile crashes 272 *illus.*, 274
    bridges 264, 268
    core of Earth 337
    space-shuttle emergencies 84
Computers 314-16
    book reviews 306
    communication technology 376
    display systems 317
    Internet 254-61
    laptops 318
    network transmissions 313
    optoelectronics 309
    patient-survival database 162-64
    power failures 281
    space shuttle 82-83, 88
    submersibles 278
    3-D computing 240-46
Condodonts 359-60
Confuciusornis (bird) 361
Conservation *see* Endangered species; Environment
Conservation of energy (physics) 350
Constellations 100
Constipation 159
Consumer technology 317-20
    computers 242, 261
    telescopes for amateurs 101
Continuous-flow intersection 377 *illus.*
Contortionists 215 *illus.*
Copper 157, 217
Corn 284-85
Cornell, Eric 361
Cornell Laboratory of Ornithology 367
Coronary disease *see* Heart disease
Corrigan, Douglas 384
Cosmic rays 81, 108
Cosmology 227, 232, 363
Cotswold sheep 51 *illus.*
Cotton-top tamarins 48
Cougar attacks 60
Craig, George B., Jr. 384
Crash-test dummies 86, 271, 273-74
Craters, Meteor 95, 101
Creativity 297
Crested auklets 46
Creutzfeldt-Jakob disease 340
Crutzen, Paul 348-49
Crystalline (plant) 121
Crystallography 337
Cuckoo wrasse fish 47
Cuesta Tunnel (California) 312
Cummins, Christopher C. 309-10
Cunningham, O.J. 146
Cutler, Horace 308
Cycliphora 299-300
Cygnus (constellation) 292
Cystic fibrosis 168, 303, 333

D

Dairy industry 49, 284
Darby, Abraham III 266
Darwin, Charles 45, 166 *illus.*, 306
Darwinian medicine 165-71
Da Silva, Rui Gomes 198
Data bus (electronics) 319
Datagrams (computers) 256
Data recorders *see* Black box
Dates
    International Date Line 192
    the year 2000 206
Day 338
DC-9 (airplane) 253
DDT (insecticide) 323
Death rates
    attacks by animals 60

Ebola-virus outbreak 364
patient-survival database 162-64
vegetarian diets 159
Decompression sickness *see* Bends
Deep Flight (submersibles) 275-76
Deep Rover (submersible) 276 *illus.*
Deer 47, 325
Deer ticks 365-66 *illus.*
Depression (psychology) 297, 299
Dermagraft 153
Deserts 54-56, 58, 111-12
DeSilva, Ashanthi 333
Des Plaines River 327
DeStevens, George 384
Detergents 223
DeVaucouleurs, Gérard 384
Developmental biology 352-54
Dew (meteorology) 117
Diabetes 51, 149, 153, 159
Diagnostic tests 173, 321
Diamonds 135
Diarrhea 167
Diet *see* Nutrition and diet
Dietemann, Chantal 324
Diet for a Small Planet (book, Lappe) 161
Dietz, Robert S. 384
Diffraction grating 92-93
Digestion 354-55
Digital cash 315
Digital Chart of the World 196
Digital Versatile Disc (DVD) 317
Dingus, Lowell 184, 190
Dinosaurs 95, 184-90, 359-61
Diphtheria 171
Discovery (space shuttle) 83 *illus.*
Dislocation of joints 215
Distance, Astronomical 91
Diverticular disease 159
Diving 147, 198-99
DNA 243 *illus.*, 334
Dogs in rescue work 320
Dooley, Calvin 29
Doppler radar 115
Dornan, Robert 143
Doxycycline (drug) 366
Drought
    global warming 328
    Mayan civilization 290
    plant survival 307-8
    year in weather 347
Drug abuse *see* Substance abuse
Drugs *see* names of individual drugs
Dry cleaning 222-25
Dryers, Clothes 225, 318
Duane, Tom 144
Duchenne muscular dystrophy 333
Ducks and geese 49 *illus.*, 52, 380
Dummies, Crash-test *see* Crash-test dummies
Dunham, Tom 115 *illus.*, 118-19
Dysentery 167, 171

E

$E=mc^2$ 66
Eagle nebula 292
Earle, Sylvia 276
Ears 212-13
Earth (Earth sciences) *see* Geology; Meteorology
Earthquakes *see* Seismology
Earth Summit (1992) 332
Eaton, Boyd 168-69
Ebola virus 165-66, 364-65
Eckert, J. Presper 384
Ecology *see* Environment
Edema 157
Education 144, 242 *illus.*, 367-68
Eggs in diet 161

Egypt, Ancient 176-83
Ehlers-Danlos syndrome 215
Ehrlichiosis 365-66
Eider ducks 380
Einstein, Albert 66, 228 *illus.*
El Amarna 177 *map*
Electoral college 233-35
Electric cars 293
Electricity
    energy resources 326
    generator 319
    space shuttle 85
    superconducting materials 309
    thunderstorms 107-8
Electrolyte balance 172-73
Electromagnetic earthquake warning 369, 371
Electronic bulletin boards 257
Electronics 321-22
    consumer technology 317-18
    optoelectronics 309
    semiconductors 236
    silver 219
Element 112 309
Elephants 39-40, 42 *illus.*, 43
Elevators 280-81
Elk 16-18, 19 *illus.*, 45, 47 *illus.*
E-mail 256
Emeralds 135
Endangered species 323-25
    breeding programs 303
    fish 124
    game birds 61
    Grand Canyon 327
    kangaroo rat 58
    livestock breeds 49-53
    northern savanna giraffe 39
    northern white rhinoceros 38-40
    spectacled eider 380
Endeavour (space shuttle) 371, 373
Energy 325-27
    mass conversion 66-67
    quantum theory 236
    Sun 77
Engines
    generator-engine combination 319
    space shuttle 82, 84, 87-88
Environment 327-30
    asteroid impacts 94-99
    book reviews 304
    cleaner-burning gasoline 325
    dry cleaning 224-25
    food and population 332
    food-chain inversion 383
    Garamba National Park 38-43
    gasoline-spill cleanup 312
    global warming 328-29
    greenhouse effect 108
    hog farms 284
    Madrean Archipelago 113
    manatee deaths 324-25
    Mayan civilization 290
    nonindigenous species 120-26
    nuclear waste 327
    ozone layer 348-49
    protein-based computers 309
    space satellites 372
    vegetarian diets 156
    water purification 322
Epidemiological studies
    automobile crashes 274
    Ebola-virus outbreak 364
    mathematics 342-43
Epiphytic plants 307
Erosion 111, 135, 290
Estrogen 169
Ethanol 135
Ethics, Medical 164
Eucalyptus trees 120
European Space Agency (ESA) 73, 372
Evans, Robley 384
Everest, Mount 127 *illus.*, 128-29, 133

Everglades (Florida) 121
Evolution
    animal-courtship displays 45
    asteroid impacts 95
    bird fossils 361
    book reviews 305
    Darwinian medicine 168-69
    dinosaurs 184, 188, 190
    genetic changes 306
    Madrean Archipelago 113
    primates 286
    primitive organisms 300-301
    tool use 286, 288
Ewald, Paul 165-67, 171
Exercise
    dietary guidelines 355
    HIV 139
    no-impact machines 320
Explosions 97, 310
Expo '98 201
Extinction (biology) 95
Exxon Corporation 278
Eyes and vision 219, 330

F

Falcons 323
FAO *see* Food and Agriculture Organization
Farming *see* Agriculture
Fat cells 155
Fats and oils
    dietary substitutes 354-56
    nutrition 156, 158-59, 161
    stain removal 223
FDA *see* Food and Drug Administration
Federal Aviation Administration (FAA) 250-51, 296
Feet
    automobile injuries 272-73
    diabetic ulcers 149, 153
    human evolution 286-87
Fertilizers 310
Fever 167
Fiat Cinquecento 293 *illus.*
Fiber, Dietary 158-59, 341
Fiber optics 322
Fibroblasts 152-53
File transfer protocol (FTP) 256
Fire
    asteroid impacts 96
    elevators 280-81
    hyperbaric medicine 150
    sequoias 26-29
    smoke inhalation 148
    space-shuttle launch 87-88
Fish 47-48, 58-59, 124, 341
Fleming, Patricia 143
Flexibility (physiology) 215
Flight-data recorders *see* Black box
Flight simulators 246, 253 *illus.*
Floods
    Colorado River management 327-28
    meteorology 116, 204
    wetlands restoration 327
    year in weather 346-47
Floriculturists 61
Florida 117, 121, 124, 378
Flour beetles 55
Flowers 307
Fluids, Accumulation of 157
Folate (Folic acid) 161, 341
Food and Agriculture Organization (FAO) (UN) 331-32
Food and Drug Administration (FDA) (U.S.)
    AIDS medications 338
    folate-fortified foods 341
    olestra 354-55
    tissue engineering 153, 155
    vegetarian diets 156, 158

Food and population 331-32
    book reviews 304
    endangered livestock breeds 51
    genetically engineered foods 302
    Mayan civilization 290
    vegetarian diets 156
Food chain
    asteroid impact 96
    Isle Royal National Park 383
    ultraviolet radiation 330
    zooplankton 329, 359
Food poisoning 366
Food supplements *see* Vitamins and minerals
Footprints, Fossil *see* Trackways
Forests
    deforestation 290
    Isle Royal National Park 383
    Madrean Archipelago 112
    sequoias 22-29
Fossils
    dinosaur exhibit 184-90
    lightning strikes 107 *illus.*
    paleontology 359-61
    prehistoric humans 286-87
    sequoias 24
    shark scales 359
Fowler, William A. 385
Fraunhofer, Joseph von 93
Free speech 261, 313, 316
Freezing rain 118
Friction 236
Frogs and toads 54-56, 113, 126
Fructans 307-8
Fruits in diet 159, 161
Fuchs, Fritz F. 385
Fuel, Rocket 82, 84-85, 88
Fulgurite 107 *illus.*
Funch, Peter 299
Fungi 308
Fusion, Nuclear 67, 77

G

G (gravitational constant) 362-63
Galaxies 292
Galileo (space probe) 373
Galili, Yisrael 385
Gallagher, James Roswell 385
Galloping Gertie *see* Tacoma Narrows Bridge
Gallstones 159
Game animals 61, 124
Gamma rays 81, 292
Garamba National Park (Zaïre) 38-43
Gardner, Beatrix T. 385
Gas (chemistry) 147, 309
Gas, Natural *see* Natural gas
Gasohol 135
Gasoline 220, 237, 293, 312, 325-26
Gemstones 135
General Sherman Tree 26 *illus.*
Generators, Electric 319
Genetics 333-35
    Alzheimer's disease 339
    cancers 339
    cardiac arrhythmia 341
    Darwinian medicine 168
    Ehlers-Danlos syndrome 215
    embryo development 352-54
    endangered livestock breeds 49, 51
    evolution 306-7
    gene transfers 302-3
    gray seals 383
    primitive organisms 300-301
    protein-information carriers 310
    wildlife studies 38
Genzyme Tissue Repair (company) 151, 153-55

Geography
    book reviews 305
    map of the world 191-96
    mountain elevation 127-33
Geoid 131-33
Geology 336-38
    asteroid impacts 94-95
    book reviews 305-6
    gemstones 135
    Madrean Archipelago 110-13
    measuring mountains 131-33
    meteor craters 101
    science education 367-68
George Washington Bridge 267
Geothermal energy 326
Geysers 103 *illus.*
Gianturco, Cesare 385
Gibraltar Strait bridge 262, 265 *illus.*, 269
Giganotosaurus carolinii 360 *illus.*
Gills (of fish) 59
Giraffes 39
Giza 177 *map*
Glen Canyon Dam 327
Glennan, T. Keith 385
Global cooling 95-96
Global Positioning System (GPS) 131, 285, 319, 376
Global warming 108, 328-29, 359
Glycerol 354-55
Glycolic acid 152
Goats 52, 121
Gold 216-19
Golden Gate Bridge 264 *illus.*, 265-66, 268
Goldiamond, Israel 385
Golf (sport) 320
Gophers (computer programs) 256-57
Gould, Laurence M. 385
Graham, Mount 113
Graham, Ron 214
Grains 159, 161, 284, 331-32
Grand Canyon 327-28
Grass anoles 48
Grasslands
    Garamba National Park 39-40
    Madrean Archipelago 111-12
Graupel 107
Gravity (physics)
    acrobatics 211, 213
    gravitational constant 362-63
    gravitational lens 292
    mass and distance 237
    measuring mountains 130-32
    ocean-platform design 269
    projectile motion 236
    starbirth 65
    time 226, 232
    weightlessness 100
Grayback beetles 126
Gray seals 383
Great Lakes 122-23
Great Trigonometric Survey 128, 129 *illus.*
Greenfield Village 53
Greenhouse effect 108
Groundwater 370-71
Guthrie, Robert 385
Gypsy moths 120 *illus.*

H

Hagfish 58
Hair 157, 335
Hale, Alan 291
Hallucinations 298
Hans Hedtoft (ship) 135
Hard drive (computer) 316
Hargreaves, Alison 385
Harps 281
Harrison, Benjamin 235
Hawaiian wildlife 121, 122, 126

Hawkes, Graham 275 *illus.*, 276
Hawking, Stephen 232 *illus.*
Hayes, Rutherford B. 235
HDL 357
Head injuries 272, 294
Health and disease 338-41 *see also* Public health
  animal research 51
  blood clotting 310
  book reviews 304-5
  computer-modeled anatomy 242-43
  Darwinian medicine 165-71
  HIV 138-44
  hyperbaric medicine 145-50
  mental illness and creativity 297
  nervous breakdown 172
  Nobel Prizes 352-54
  prosthetic hand 220 *illus.*
  tissue engineering 151-55
Health and Human Services, United States Department of 156, 355
Health-care costs
  HIV 143-44
  hyperbaric medicine 150
  patient-survival database 164
  technology 173
  tissue engineering 155
Heart disease
  arrhythmia 341
  exercise 173
  hyperbaric medicine 148-49
  nutrition and diet 158-59, 340-41
  patient-survival database 162, 164
  valve transplants 155
  vitamin E quinone 310
Heat islands (meteorology) 204
Heisenberg, Werner 230 *illus.*
Heisenberg uncertainty principle 229, 361
Helium
  Big Bang theory 292
  geology 338
  starbirth 65, 67-68
  Sun 77, 81
Heller, Adam 310
Hemorrhagic fever 364
Hepatitis-B vaccines 302
Herod the Great 289
Herpesvirus 338-39
Hessberg, Rufus R. 385
High-speed trains 378
Himalaya Mountains 128 *illus.*, 133, 323-24
Hippopotamuses 60
HIV 138-44, 170-71, 364 *see also* AIDS
Hogs 49, 52-53
Hokkaido (island, Japan) 192 *map*
Holography 245-46
Home pages (Internet) 260
Homo erectus 287
Homo sapiens 287-88
Honeybees 61
Honshu (island, Japan) 192 *map*
Hopkinson, Francis 92
Horoscopes 100
Horses 60, 288, 382-83
Horsfall, James G. 385
Horticulturists 60
Howland, Howard 169-70
http *see* Hypertext
Hubble Space Telescope (HST) 69, 291-92, 373
Human body
  Darwinian medicine 168-69
  electrolyte balance 172-73
  flexibility 215
  genetics 333-35
  genome 334
  tissue engineering 151-55
  upright posture 286-87
  water needs 55, 57
Human sacrifice 288
Humidity 118

Hummingbird (computer) 317
Hunger *see* Food and population; Nutrition and diet
Hurrians (ancient people) 289
Hurricanes 346
Huygens, Christian 93
Hydrochloric acid 219
Hydroelectric power 326
Hydrogen
  high-pressure chemistry 309
  inter-galactic clouds 292
  natural-laser star 292
  space shuttle 84, 88
  starbirth 65, 67-68, 70
  Sun 76-77
Hydrothermal vents 276, 300-301
Hygrometer 118
Hygrometers 118
Hyperbaric medicine 145-50
Hypertension 159
Hypertext 258, 314
Hypertrichosis 335

I

IBM Corporation 244, 309
Ice
  Antarctic algae 380-81
  aviation 251-52
  Bering Sea 380
  clouds 101, 117
  thunderstorms 107
Icebergs 134-35
i-Glasses (computer monitor) 317
Iguanodon (dinosaur) 187 *illus.*
Immunology 164, 302-3, 338
Incas 288
India 127-29, 131, 267, 336-37
Indinavir (drug) 141, 338
Infants
  automobile crashes 273-74
  heart defects 148
  human evolution 286
  silver nitrate eyedrops 219
  vegetarian diets 160-61
Infrared radiation 109, 292, 318, 372
Injury *see* Trauma
Insects
  attacks on people 60
  natural pesticides 308
  queens 61
  sequoias 35
  water needs 55-56
Instrumentation, Scientific
  liquid-mercury telescopes 71-74
  mercury 221
  mountain elevation 127-28, 130
  Rittenhouse, David 89
  weather forecasting 118
Intensive care 163-64
International Astronomical Union (IAU) 99
International Date Line 192
International Map of the World (IMW) 191-96
International Space Station 372-73
Internet 254-61
  free speech 261, 313, 316
  use 314
  weather information 345
Interstellar matter 65, 67-69, 81
Invirase (drug) 141, 143, 338
Io (moon of Jupiter) 291
Ireland, Snakes in 61
Iridium 95
Iron
  bridge design 266
  core of Earth 337
  in diet 157, 160-61
  meteors 98

Isle Royal National Park 383
Isopods 56
Ituri Forest 43

J

Jacob sheep 51 *illus.*
Jaguars 54 *illus.*
Japan
  bridges 264, 269
  Kobe earthquake 370-71
  nerve-gas attack 310 *illus.*
  space program 372-73
  submersibles 276, 279
Java (software language) 261, 314
Jen, Chih-Kung 385
Jensen, Elsa 358
Jewelry 218-20, 237
Johnson, Earvin (Magic) 138, 144
Joints 215, 320
Jolly, Jean-Baptiste 222
Jongsma, Maarten A. 308
Juan de Fuca Ridge 275, 359
Juggling 212 *illus.*
Jupiter (planet) 94, 96, 373

K

K2 (mountain) 127
Kaczynski, Theodore 297 *illus.*
Kaiko (submersible) 279
Kajiya, James 244
Kangaroo rats 58
Kaposi's sarcoma 338-39
Karnak, Temple at 176 *illus.*, 179
Kazakhstan oil fields 326
Keck Telescope 73
Keith, William 23 *illus.*
Kennedy Space Center (Florida) 83
Kenya 286, 325
Keratinocytes 153-54
Kidneys 56, 58-59, 159
KidSat 368
Kiwifruit 308
Kjellstrom, Bjorn 385
Knaus, William 163 *illus.*, 164
Knight, Charles R. 184, 189 *illus.*
Kobe earthquake 269, 370-71
Köhler, Georges 385
Kondakova, Elena 372
Koop, C. Everett 156
Korean space science 373
Kristensen, Reinhardt M. 299
Kroger, William S 385
Krugman, Saul 385
Kuala Lumpur (Malaysia) 311 *illus.*
Kuchar, Karel 230
Kudzu vine 122
Kuiper Belt (comets) 291
Kumon, Toru 385

L

Labeling regulations 158-59, 355
Lactic acid 152
Lake Farmpark 53
Lake Washington Ship Canal 382
Land use (U.S.) 28-29, 329
Langer, Robert 151, 152 *illus.*
Laplaza, Catalina E. 309-10
Lappe, Frances 161
Larned, Bill 380
Lasers 292, 309, 361-62

INDEX 391

Laundry machinery 223, 225, 318
Lava 336
LDL 158
Lead 309
Learning 32, 241-42, 274
    science education 367-68
Lefebvre, Louis 30-34
Legumes 161
Leopards 61
Leopold II (king of Belgium) 39
Leptin (protein) 334
Leptons (particle physics) 351
Lewis, Edward B. 352-54
Ligaments 215
Light 92-93, 309, 321
Lightning 104, 105 *illus.*, 107, 373
Limestone 135, 290
Lin, Tung-Yen 262, 265, 269
Lincoln, Abraham 203
Lions 45, 325
Liquid-mercury telescopes 71-74
Liquid-oxygen fuel 88
Lithium 362
Liuhua Floating Production System 278
Liver (anatomy) 155, 274
Lizards 48, 54, 61
Loch Ness monster 206-7
Locomotion, Biological 286-87
Logging 28-29, 329
Loma Prieta earthquake 265, 268
Longgupo Cave (China) 287
Low-density lipoprotein (LDL) 158
Lubin, Dan 358
Ludwig, Arnold M. 297
Lung cancer 159, 339
Lyme disease 365-66

M

Mad-cow disease 285, 340
Madrean Archipelago 110-13
Magnetic field
    Earth 336
    Sun 76, 80-81, 373
Magnetic resonance imaging *see* MRI
Malaria 168, 329
Mammoth Mountain 379
Manatees 324-25
Mandrills 48
Manic depression 297
Manned Space Flight Center 86-87
Maps and map making
    dashboard displays 377
    International Map of the World (IMW) 191-96
    mountain elevation 128, 132
    seafloor 357-58
Margarodes (insect) 56
Mariana Trench 275, 279
Marine biology
    algae in sea ice 380-81
    manatees 324-25
    ocean temperature 329
    water needs of animals 58
Marine toads *see* Cane toads
Marlboro Man (advertisement) 366
Mars Powell project 278
Marsupials 35, 37
Martian meteorites 291
Martinique 379
Marx, Robert F. 197, 198 *illus.*, 200
Mary Rose (ship) 199
Masked bobwhite quail 61
Mass (physics) 66-67, 237
Mathematics
    book reviews 305
    equations describing HIV 342-43
    juggling 214
    plane figure 208-9

quantum clocks 230
science education 367
square roots 237
weather forecasting 344-45
Mauna Kea Observatory 79 *illus.*
Mayan civilization 290
Mazar, Benjamin 385
McDonald's Corporation 330
McIlhenny, E.A. 125
McKenna, Joe 264
McKinley, Mount 128, 133
McLaren, Brian 383
McLaren, Wayne 366
McLean, David 366
Meat
    agriculture 284-85
    bison 19, 20 *illus.*
    disease transmission 340
    organic farming 52
    vegetarian diets 156-59
    water needs of animals 55
Medicine *see* Health and disease; Public health
Mediterranean archaeology 289
Melanin 334
Memory
    computers 316
    trauma 297-98
Menstruation 158, 167-69
Mental illness *see* Health and disease
Mercury (element) 71-74, 217, 220-21, 309
Meridians 192
Mesothelioma 366
Messier, Charles 67
Metabolism 55, 57-58, 339-40
Metals 216-21, 309
Meteor Crater (Arizona) 98 *illus.*
Meteorology 343-47
    forensic science 202-5
    red sprites and blue jets 104-9
    tornadoes 134
    weather forecasting 114-19
Meteors and meteorites 95-99, 101, 291
Methyl bromide 330
Methylphenidate *see* Ritalin
Mexico 95, 290
Mexico, Gulf of
    off-shore drilling 278, 326, 337
    sea level 328
Microclimate 203
Microscopes 321-22
Microsoft Corporation 244, 316
Midgley, Thomas 348
Military services (United States)
    HIV-positive personnel 143
    liquid-mercury telescope 72, 73 *illus.*
    meteor destruction 99 *illus.*
    war-games simulations 240-41
Milk
    of deer 47
    nutrition and diet 159, 161
    of seals 59
Mining 135, 219-21, 329
Minkowski, Hermann 227, 229 *illus.*
Mir (space station) 371-72, 374
Mir (submersibles) 276
Mirages 130
Mirrors, Telescope 71-74
Mississippi River 312, 327
Mobil Oil Corporation 326
Modem 255-56, 261, 314, 316
Molecular clouds (astronomy) 70
Molecular scaffold (biotechnology) 152
Molecules 223, 243 *illus.*
Molina, Mario 348-49
Molybdenum 310
Money 219, 315
Mongoose 126
Monkeyflowers 306
Monkeys 340, 364-65
Monorails 207

Monsoons 290, 347
Montserrat 379
Moon
    atmosphere 291
    forensic meteorology 203, 205
    gravity 237
    impact craters 94
    solar eclipse 101
    weather forecasting 101, 117 *illus.*, 119 *illus.*
Moose 383
Morning glory trees 112
Morning sickness 168
Mosaic (computer program) 259
Mosquitoes 46
Motion, Sense of 100, 212-14
Motion pictures 241 *illus.*, 245, 317
Motorcycle air bags 318
Mountains
    Madrean Archipelago 110-11, 113
    measuring elevation 127-33
Mount Graham red squirrels 111 *illus.*, 113
Mount Vernon 53
MRI 173
Mucus 56-57
Muir, John 23 *illus.*, 24, 26, 28
Mulefoot hogs 52-53
Multimedia 317
Mummies 286 *illus.*
Muscles
    flexibility 215
    gene therapy 333
    vegetarian diets 157
    weightlessness 100
Muscovy ducks 49 *illus.*
Muscular dystrophy 333
Museum of Natural History (New York City) *see* American Museum of Natural History
Music 281, 300, 315-16
Mustangs 382-83
Mutations 306, 335
Myoblast transfer 333-34
Myopia *see* Nearsightedness
Myotonic goats 51
Mythology of constellations 100

N

National Academy of Sciences 367
National Aeronautics and Space Administration (NASA)
    asteroid impacts 97, 99
    carbon-monoxide converter 317
    commercial shuttle operations 373
    launch-abort procedure 83
    liquid-mercury telescope 72
    missions of 1995 371
    natural-laser star 292
    rotation of Earth 338
    science education 368
National Basketball Association 320
National forests 28, 329
National Highway Traffic Safety Administration (NHTSA) 271, 274
National parks
    Garamba 38-43
    Isle Royal 383
    Sequoia 23, 25, 28
    Yellowstone 381
National Transportation Safety Board (NTSB) 248-51, 253, 295
National Weather Service 115-16, 118
Natural gas 278, 326, 337
Natural selection 45
Nausea 100, 168
Nautile (submersible) 276
Navajo-Churro sheep 51

Navigation systems 319, 376
Navy, United States 245, 357-58
Neanderthals 287-88
Near-Earth objects (NEOs) 99
Nearsightedness 169-70
Nebulae 65, 67-69, 292
Nerve gas 310
Nerves *see* Brain and nervous system
Nervous breakdown 172
Nesse, Randolph 166, 171
Netscape (computer program) 259
Neurological science *see* Brain and nervous system
Neutrinos 350-51, 363
Newcomb, Simon 93
Newsgroups (Internet) 257
Newton, Sir Isaac 71, 93, 227, 228 *illus.*, 362
NHTSA *see* National Highway Traffic Safety Administration
Niagara Falls bridge 266
Niagara Power Project 276
Nile Delta 195 *map*
Nitric acid 219
Nitrogen 72, 108, 307, 309-10
Nitrogen oxides 348
Nobel Prizes
 chemistry 348-49
 physics 350-51
 physiology or medicine 352-54
Noble metals 216-21
Nonindigenous species 120-26
North American anthropology 286 *illus.*
Northern lights *see* Aurora borealis
North Pole 336
NTSB *see* National Transportation Safety Board (NTSB)
Nuclear reaction
 neutrinos 350
 power plants 327
 starbirth 65, 67
 Sun 77
Nucleoside analogues 338
Nüsslein-Volhard, Christiane 352-54
NutraSweet 355
Nutrias 124-25 *illus.*
Nutrition and diet 354-57
 electrolyte balance 172-73
 folate 341
 genetically-engineered foods 302
 heart disease 340-41
 instant meals 318
 seaweed 366
 Tasmanian devil 35, 37
 vegetarian diets 156-61
 water needs of animals 55, 58
"Nutrition and Your Health: Dietary Guidelines for Americans" (booklet, U.S. government) 355

O

Obesity 159, 334, 355
Oceanography 357-59 *see also* Marine biology
 book reviews 304
 Expo '98 201
 global warming 328-29
 submersibles 275-79
 trenches 371
Oil (petroleum) *see* Petroleum
Oils *see* Fats and oils
Oklahoma City bombing 310
Olestra 354-56
Oliver, Bernard M. 385-86
Olympic Games 343-44
One-celled organisms 300, 310
On-line services 257-58
Oocyte fertilization 303

OPEC (Organization of Petroleum Exporting Countries) 325
Optoelectronics 309, 321
Orbital Debris Observatory 72 *illus.*
Organogenesis (company) 153
Origin of Species (book, Darwin) 45
Orion (astronomy) 66, 69 *illus.*
Orreries 89-90, 93
Oryx 57-58
Osborn, Henry Fairfield 38, 184
Osiris 182 *illus.*
Ossabaw Island hogs 51, 52 *illus.*
Osteoporosis 157, 159
Oster, Clint 248-49
Oviraptor (dinosaur) 359
Oxygen
 high-pressure chemistry 309
 hyperbaric medicine 145-46, 148
 ozone layer 348
 space shuttle 84, 88
 yawning 172
Ozone layer
 asteroid impact 96
 atmosphere 108
 liquid-mercury telescope 72
 Nobel Prizes 348-49
 Northern Hemisphere 330
 satellite mapping 358

P

Pacific Ocean
 submersibles 275-76, 279
 toxic seaweed 366
 volcanoes 378
 water temperatures 328-29, 359
Packaging 330, 348
Packet switching 256, 257 *illus.*
Pagers (electronics) 314, 320
Paint 310
Paleontology 184-90, 359-61
Palomar Observatory 73
Paper 281, 330 *illus.*
Paperbark trees 120-21
Paralysis 172
Parasites
 agriculture 50
 courtship displays 47
 rainbow trout 381
 toxoplasmosis 140
 water purification 322
Parvovirus *see* Canine parvovirus
Pastrana, Julia 335 *illus.*
Patterson, Clair C. 386
Pauli, Wolfgang 350
Paz, Yaron 310
Peacocks 44, 44 *illus.*
PEG-ADA (drug) 333
Pelée, Mont (Martinique) 379
Penck, Albrecht 191-93
Perchloroethylene 222-24
Peregrine falcons 323
Perfect pitch (music) 300
Periodic table 216-17
Perl, Martin L. 350-51
Perot, H. Ross 235
Personal communications services 313-14
Pesticides 308
Peterson, Rolf 383
Petrels 58
Petroleum
 Arctic National Wildlife Refuge 329
 deep-sea drilling 277-78
 energy resources 325-26
 platform design 246, 269, 311
 resources available 337
Petronas Towers 311 *illus.*
PET scans 298, 339-40

Phalaropes 46
Pheasants 61
Physical therapy 150
Physicians 140, 162, 164, 242 *illus.*, 243, 299
Physics 305, 361-63
 circus performances 210-15
 forces on bridges 264, 267-68
 Nobel Prizes 350-51
 science education 367
 time 226-32
Pianos 281
Pigeons 30-34, 323
Pigment in skin 334
Pilkington, Alastair 386
Pilon-Smits, Elizabeth 307
Pioneer (spacecraft) 373
Pitch (music) 281, 300
Pitching (baseball) 240 *illus.*
Planck length 230-31
Planck's constant 236
Plane figures (mathematics) 208-9 *illus.*
Planetariums 101
Planets 65, 291 *see also* names of individual planets
Plants *see* Botany
Plasma (physics) 80-81, 317
Plastic surgery 150
Plate tectonics 133, 336-37, 357, 371
Platinum 217, 220, 237
Platyfish 48
Plimoth Plantation 53
Polyakov, Valeriy 372
Ponds 134
Pont de Normandie, Le (bridge, France) 263 *illus.*
Population *see* Food and population
Porcine Zonae Pellucida (PZP) 383
Pork 49, 52, 284-85
Portugal 198, 200
 Azores shipwrecks 201
Positron-emission-tomography (PET) *see* PET scans
Potassium 173
Potatoes 285, 302
Poverty 331-32
POZ (magazine) 144
Pregnancy 157, 160-61, 168, 273, 303
Prehistoric humans 286-88
Presidency of the United States 233-35
Primates 48, 286
Prions 340
Procter & Gamble (company) 354-56
Profet, Margie 167-68
Projectile motion 236
Propellant gases 348
Proplyds 69
Prostate cancer 159
Prosthetic hand 220 *illus.*
Protease inhibitors 141, 143, 308
Proteins
 Alzheimer's disease 339
 cellular information 310
 obesity 334
 optoelectronics 309
 prions 340
 vegetarian diets 157, 160-61
Protoplanetary disks 65 *illus.*, 69
Psychology *see* Behavioral sciences
Public health 329, 364-66
Purple loosestrife 125-26 *illus.*

Q

Quail 61
Quantum theory 227, 229-32, 236
Quarks 351
Quasars 292
Quebec City (Canada) 207

INDEX 393

R

Rabies 366
Radar 115, 296, 376
Radio
    air-traffic control 296
    lightning 107-8
    liquid-mercury telescope 72, 73 *illus.*
    pre-earthquake signals 371
    surveying 131
    U.S. regulatory legislation 313
    wildlife tracking 41 *illus.*, 43
Radiology 150, 243
Radon 370
Railroads 266, 378
Rain 117, 328
Ramesses II 176-79
Ramsey Canyon leopard frogs 113
Randolph, Theron G. 386
Rare Breeds International 52
Rattlesnakes 111 *illus.*, 113
Ravidat, Marcel 386
Rectal cancer 158
Recycling 330 *illus.*
Red bird of paradise 45 *illus.*
Red dwarfs (stars) 67
Redenbacher, Orville 386
Red giants (stars) 78
Red hair and skin cancer 334
Redington, Rowland W. 386
Red River Waterway 312
Red sprites and blue jets (meteorology) 104-9
Redwoods 23-26
Reentry vehicles (spacecraft) 372
Refraction of light 130
Reines, Frederick 350-51
Relativity (physics) 227
Remote controls (television) 320
Reproduction
    animal-courtship displays 44-48
    biotechnology 303
    dinosaur nests 359
    human evolution 286
    pigeons 31, 33 *illus.*
    sequoias 24, 27
    social insects 61
    viviparous lizard 61
Reptiles 54, 61, 113
Rescue technology 320
Resistance (biology) 26, 50, 308
Respiration 58, 147
Retrouretal reflux 154
Revenge (ship) 199
Rhinoceros 40-43
Rhinoceroses 38-39
Riboflavin 160
Rice 285, 332 *illus.*
Rickets 157
Ride, Sally 368
Ritalin 341
Ritonavir 141, 338
Rittenhouse, David 89-93
Rivers 134
Roads and highways
    ancient Rome 206
    bridges 262-69
    continuous-flow intersection 377-78
    speed limit 325
    underground 312
    visibility-monitoring system 311-12
Robot submarines *see* Submersibles
Rockets, Space 82, 84, 88, 372-73
Roebling, John and Washington 266
Roman roads and bridges 206, 266 *illus.*
Rotation (physics) 74, 210
Rovelli, Carlo 231 *illus.*
Rowland, F. Sherwood 348-49
Royal Geographical Society 196
Ruapehu, Mount (New Zealand) 378
Rubidium 361
Rubies 135
Rule, Margaret 197-98, 200
Russia 73, 276, 371-73

S

Sacrifice, Human *see* Human sacrifice
Saddleback hog 52
Safety
    automobile crashes 270-74, 294
    automobile-locator systems 318, 377
    aviation 247-53, 296
    carbon monoxide 317
    highway-visibility monitor 312
    space-shuttle launch 82-88
    tick-borne diseases 365-66
Saint Elmo's fire 108-9
Sakkara 177 *map*
Salk, Jonas E. 386
Salmon 58 *illus.*, 59, 341
Salt 58-59, 152
Salvage (shipwrecks) 197-201
San Francisco (California)
    bridges 264-65, 268
    49ers 146 *illus.*
San Miguel Island 200
Saquinavir (drug) 338
Sarin 310
Satellites, Artificial
    Global Positioning System 376
    measuring mountains 131
    missions of 1995 372-73
    oil exploration 326
    ozone layer 349
    projectile motion 236
    seafloor map 357
    weather forecasting 116
Saturn (planet) 291
Schaller, George 325
Scharrer, Berta Vogel 386
Schizophrenia 298
Schlaug, Gottfried 300
Schrödinger, Erwin 229, 229 *illus.*
Schumann resonances 107-8
SCID *see* Severe combined immunodeficiency disease
Science education 367-68
Scrapie 340
Screen phones 314
Scythian horseman 288
Sea Cliff (submersible) 276
Seahawk (company) 200
Sea level 128, 130-32, 328-29
Sea lions 382
Seals 59, 383
Search programs (computers) 257, 260, 314
Seat belts, Automobile 272-74, 294
Sea turtles 58
Seawater 58-59, 380
Seaweed 366
Security systems 319, 321
Sedge warblers 46
Sedimentation 96, 111, 327-28
Seeds
    sequoias 24 *illus.*, 27
    vegetarian diets 161
    water needs of animals 58
Seim, Charles 263-69
Seismology 369-71
    book reviews 304
    bridges 265, 268-69
    core of Earth 337
    helioseismology 78 *illus.*
    Himalayas 133
    satellites, artificial 326
Self-cleaning coatings 310
Semiconductors 236
Sensors, Electronic
    automobiles 273-74, 294
    consumer technology 319
    weather forecasting 118
Sequoias 22-29
Sequoya 25 *illus.*
Severe combined immunodeficiency disease (SCID) 333
Sexual selection 45
Sharks 359
Sheep 50-53
Shell Oil Company 278, 311, 326
Shigella 167, 171
Shinkai 6500 (submersible) 276, 277 *illus.*
Ship design 241 *illus.*, 242
Shipwrecks 135, 197-201, 289
Shoemaker-Levy 9 (S-L 9) (comet) 94, 96
Shrews 113
Shrikes 126
Shulman, Mark D. 202-5 *illus.*
Sickle-cell anemia 168
Silicon 321
Silicon Graphics Incorporated 241
Silos (agriculture) 280
Silver 217, 219
Silver nitrate 219
Simpson, O.J. 202 *illus.*
Simulation, Computer *see* Computer modeling
Skeletons
    dinosaur 184-86, 359-61
    human 215, 240 *illus.*
Skiing 319
Skin
    cancer 330, 334, 349
    tissue engineering 151, 153-54
    water needs of animals 54, 59
    wildlife biopsies 41 *illus.*
Sky islands (geography) 110-13
Skyscrapers 311, 323
Skywalker (exercise machine) 320
Smart cards 315
Smith, Alvy Ray 244
Smith, Hugh H. 386
Smith, William Kennedy 205
Smoke inhalation 148
Smoking 366
Snakes 61, 111 *illus.*, 113, 121-22
Snowfall 346
Sodium 152, 173, 291, 361
Soil 27-28
Solar-powered camera 319
Solar system 89, 94, 97, 99 *see also* names of planets
Solar wind 76, 80
Solvents 348
Soufrière Hills Volcano (Montserrat) 379 *illus.*
South America 193 *map*, 195, 360-61
South Pole 336
Soybeans 161, 284-85, 356-57
Soyuz (spacecraft) 371
Space debris 72, 97, 99
Spaceguard Survey 99
Space science 371-75 *see also* Astronomy
    book reviews 304
    commercial exploration 207
    mass of the universe 363
    science education 367
    shuttle-launch emergencies 82-88
    space debris detection 72
    Sun 75-81
Space shuttle
    astronomy 292
    commercial operations 373
    launch-abort procedure 82-88
    missions of 1995 371, 374-75
    science education programs 368
    tourism 207
    weather forecasting 344-45
Space sickness 100
Space-time 227, 230

Spadefoot toads 56 *illus.*
Spain 287
Spectacled eiders 380 *illus.*
Sperm banks 140
Spiders (computers) 260
Spinal injuries 172
Spirit Cave man 286 *illus.*
Sports
    consumer technology 320
    hyperbaric medicine 145, 150
Sportstrax (sports data) 320
Spurling, Christian 207
Square roots (mathematics) 237
Squirrels 111 *illus.*, 113, 323-24
Sri Lanka 290
Stanford Linear Accelerator Center (SLAC) 351
Starlings 121 *illus.*
Stars 64-70, 100, 291-92
Steel 266, 268, 290
Steens Mountain (Oregon) 336
Steering, Automobile 294
Stem cells 333
Stereoscopy 245
Sterling silver 219
Stibitz, George R. 386
Stonehenge 174-75 *illus.*
Storms
    aviation 253
    forensic meteorology 204-5
    Moon 101
    thunderstorms 104-8
    weather forecasting 117, 119
    year in weather 346-47
Streams 134
Stress (engineering) 264, 267-68
Stroke 173, 310, 341
Subduction zones 371, 378
Submarine communication 73 *illus.*
Submersibles 199 *illus.*, 275-79, 289, 359
Substance abuse 341
Sucrose 354-55
Sudbury (Canada) 101, 363 *illus.*
Sugar 161, 307, 354-55
Sun 75-81
    eclipse 81 *illus.*, 101
    gravity 237
    starbirth 67-68
    weather forecasting 117
Sun, Wenn 302
Sunburn 334
Sunlight
    asteroid impact 96
    lunar atmosphere 291
    ozone layer 348
    self-cleaning coatings 310
    sequoias 27
    Vitamin D 160
Sunshine Skyway Bridge 268 *illus.*
Sunspots 75-76, 80
Suomi, Verner E. 386
Superconductivity 309
Surgery 149, 219, 243
Surveying 127, 130, 133
Suspended animation 57
Sweating 57
Sweeteners, Artificial 355
Swordtails (fish) 48
Symbion pandora (microscopic organism) 299-300 *illus.*
Symbiosis 307

T

Tacoma Narrows Bridge 264, 265 *illus.*, 267
Tadpoles 56
Tamarins 48
Tamarisk trees 121 *illus.*

Tasmanian devils 35-37
Tasmanian wolves *see* Thylacines
Tau lepton 350-51
Tau protein 339
Taurus-Auriga Complex 70
Taylor, Peter 268
T cells 343
Tears 58
Technology *see also* Automotive technology; Biotechnology; Communication technology; Computers; Consumer technology; Electronics; Space science
    book reviews 306
    coal burning 326
    crash-test dummies 273
    weather forecasting 344
Ted Williams Tunnel (Boston) 312
Teenagers 161, 341
Teeth 159, 219
Telecommunications Act (1966) 313
Telephones
    automotive tracking system 319, 377
    Internet 261
    market expansion 313
    satellite phone 318
Telescopes
    amateur astronomers 94, 101
    Earth-orbiting 291-92, 372-73
    gravity waves 232
    liquid-mercury telescopes 71-74
    Rittenhouse, David 92 *illus.*
    solar observatories 78
    starbirth 64, 67
Television
    child-friendly remote 320
    flat-screen plasma monitors 317
    HIV-positive portrayals 144
    Internet 261
    science education 368
    U.S. regulatory legislation 313
Telnet 256
Telomerase 301, 339
Temperature
    absolute zero 361
    global warming 328
    sea surface 359
    space-shuttle launch 85
    starbirth 65
    Sun 67, 75-77
    thunderstorms 107-8
Temperature, Body 167
Temples, Ancient
    Egypt 176 *illus.*, 178 *illus.*, 179
    Herod the Great 289
    Mayan civilization 290
Tendons 215
Terahertz waves *see* T-rays
Terceira (island, Azores) 199-201
Termites (insects) 61
Texaco Incorporated 326
Texas horned toads 54, 55 *illus.*
Textiles 224
Thagard, Norman 371
Thallium 309
Thebes 177 map
Theodolite 128-30
Thermonuclear reactions 65, 67, 77
3-D computing 240-46, 321
3TC (drug) 141
Thunderstorms 104-9, 116 *illus.*, 118, 253
Thylacines 37
Tibetan red deer 325
Ticks (parasites) 365-66
Tides 205, 237
Tigers 61
Time 226-32
    atomic clock 236
    rotation of Earth 338
    time zones 191-92
    the year 2000 206
Tipper, Constance 386

Tissue engineering 151-55
Titanic (ship) 134-35, 199-200, 276, 359
Titanium oxide 310
Tobacco crop 284
Tokyo (Japan)
    nerve-gas attack 310 *illus.*
Tomb of the Brothers 176-83
Tongass National Forest 329
Toolmaking
    birds 301
    human evolution 287-88
Tornadoes 115, 116 *illus.*, 134, 346
Torsion balance 362-63
Toulouse geese 49 *illus.*
Tourism 42 *illus.*, 43, 207
Toxicology
    dry cleaning 224
    mercury 74, 221
    seaweed 366
    vegetables 167 *illus.*
    water purification 322
Toxoplasmosis 140
Toy Story (film) 241 *illus.*
Trackways of dinosaurs 190
Traction 294, 318
Trade, Agricultural 284-85, 332
Trafalgar Square 31 *illus.*
Traffic accidents *see* Automobile crashes
Trains *see* Railroads
Transits (astronomy) 90-91, 93
Transmissions, Automobile 294
Transplants, Organ and tissue 151-55, 340
Transportation 376-78 *see also* Automotive technology
    ancient Rome 206
    aviation 295-96
    bridges 262-69
    civil engineering 311-12
    monorails 207
    motorcycles 318
Trauma
    automobile accidents 270-74
    hyperbaric medicine 145, 150
    memory loss 297-98
    nerve gas 310
    paralysis 172
    tissue engineering 153
T-rays 321-22
Tresder, Kathleen K. 307
Trichloroethylene 222
Trichlorofluoromethane 348
Trieste (submersible) 275
Triglycerides 354-55
Trogon 111 *illus.*, 113
Trout 327, 381-82
Tsunami 98, 370
T Tauri stars 70
Tuberculosis 146
Tumors 302
Tunguska explosion 97-98
Tutankhamen 176
Twin-spotted rattlesnakes 111 *illus.*
Twister (film) 134
Two thousand (year) 206
Tyrannosaurus rex 186, 188, 188 *illus.*, 190

U

Ulcers 149, 153
Ultraviolet radiation
    natural-laser star 292
    ozone layer 330, 348-49, 358
    space telescopes 292, 373
    starbirth 69
Ulysses (spacecraft) 79-81, 373
U.N. Conference on Environment and Development *see* Earth Summit

Unabomber 297 *illus.*
Underwater archaeology 197-201, 289
Undulant fever 16, 19
Uniform resource locator (URL) 258
United Airlines 250, 295-96
United Nations
 food and population 331-32
 Intergovernmental Panel on Climate Change 328
 International Map of the World (IMW) 195-96
United States
 agriculture 284
 communications legislation 314, 316
 energy resources 326
 food and population 156, 332
 International Map of the World (IMW) 195-96
 manned spaceflights 374-75
 poverty 332
 presidential elections 233-35
 space program 371, 373
 submersibles 276
UPS (uninterruptible power supply) 281
Uranus (planet) 93
Urinary system 56, 58-59, 154
URL (Internet) 258
USAir 247, 249-50, 253, 295
USDA *see* Agriculture, United States Department of
Utah, University of 246

V

Vacanti, Joseph 151, 152 *illus.*
Vaccines and vaccination
 birth control 382-83
 brucellosis 20 *illus.*
 edible vaccines 302
 rabies 366
Valley of the Kings 176-83
Vanishing point 242
VAN method of quake forecasting 369
Varatsos, Panayiotis 369
Vasculitis 150
V-chips (in televisions) 313
VCRs 317-18
Vegetables
 plant toxins 167 *illus.*
 vegetarian diets 156-61, 355
Venous ulcers 153
Venus (planet) 90-91, 93
Vertebrae 172
Vestibular system 212-14
Video cassette recorders *see* VCRs
Video games 242 *illus.*, 244, 246
Video technology 317-20
Virtual reality 243, 245
Viruses
 distemper 325
 Ebola-virus outbreak 364-65
 HIV 138-44
 Kaposi's sarcoma 338
 parvovirus 383
 vaccines 302
 water purification 322
 weaker strains 170-71
Vision *see* Eyes and vision
Vitamin A 158, 353
Vitamin B$_{12}$ 157, 160-61
Vitamin C 158, 160-61
Vitamin D 157, 160-61
Vitamin E 310, 341
Vitamin K 310
Vitamins and minerals
 blood chemistry 310
 olestra 354-55
 vegetarian diets 157-58, 160-61

Viviparous lizards 61
Voice-recognition software 317
Volcanology 378-79
 geomagnetic field 336
 Io 291
 mantle of Earth 337-38
 seafloor map 358

W

Walking catfish 124 *illus.*
Walking upright 169, 286, 288
Walled cities 207
Walton, Ernest T.S. 386
Wardenburg's syndrome 353
War games 240-41
Washburn, Bradford 128, 130, 132-33
Washing machines 223, 318
Wasps 60
Water *see also* Drought
 animal physiology 54-59
 California Aqueduct 312
 extrasolar planets 291
 Kobe earthquake 370-71
 purification 219, 322
 reservoirs 338
 sequoias 28-29
 wetlands 327
Water-holding frogs 56
Watts Bar nuclear power plant 327
Waves and wave motion
 absolute zero 361
 earthquakes 337
 Planck's constant 236
 Schrödinger equation 229
Weapons 99 *illus.*, 236, 295 *illus.*
Weather
 forecasting 114-19
 forensic meteorology 202-5
 global warming 328
 highway-visibility monitor 312
 measuring Mt. Everest 130
 Moon 101
Weddell Sea 380
Weeks, Kent 180 *illus.*, 181-83
Weightlessness 100
Weil, George L. 386
Wetlands 56, 121, 124-26, 327
Whales 59
Wheat 284, 331
Whitaker, Carl A. 386
White blood cells
 Ehrlichiosis 365
 gene therapy 333
 HIV 140
 hyperbaric medicine 148
White Cliffs of Dover (England) 135
White dwarfs (stars) 78
Widow birds 45 *illus.*, 46-47
Wieman, Carl 361
Wieschaus, Eric F. 352-54
Wilcox Electric (company) 376
Wild horses *see* Mustangs
Wildlife
 attacks on people 60
 bison 16-21
 book reviews 303
 endangered species 323-25
 Garamba National Park 38-43
 Madrean Archipelago 111, 113
 pigeons 30-34
 rabies vaccination 366
 spectacled eider 380
 Tasmanian devil 35-37
Wilkinson, John Gardner 180
William Floyd Parkway 377-78
Williams, George 166
Williams, Walter 386
Windows 95 316

Winds
 aviation 253
 bridges 266-67
 computer simulation 242
 electricity generation 326
 forensic meteorology 204-5
 Jupiter 373
 power for steel furnaces 290
 sailing the Atlantic 199
 solar wind 76, 80
 space shuttle 344-45
 thunderstorms 107
 weather forecasting 115, 117-18
Windshield wipers 280
Wireframes (3-D computing) 245
Wireless Operationally Linked Electronic and Video Exploration System *see* WOLVES
Wolves (animals) 383
WOLVES (rescue system) 320
Women
 breast-cancer incidence 169
 folate in diet 341
 HIV 140 *chart*
 oocyte preservation 303
 vegetarian diets 158
Wood, Evelyn 386
Wood, Robert W. 73
Wood lice 56
Woods Hole Oceanographic Institution 275, 337
Wool 52
Woolly-bear caterpillars 119
Woolly flying squirrels 323-24
World map 191-96
World Wide Web 242, 259-60, 314
Wounds *see* Trauma

X

X-rays
 cystallography 243
 galactic outburst 292
 space telescopes 373

Y

Yahoo (Internet) 260
Yang, Ning-Sun 302
Yawning 172
Yellow jackets (insects) 121
Yellowstone National Park 16-21, 381
Yosemite National Park 23, 28
Yucatán Peninsula 95

Z

Zahler, Peter 324
Zaïre
 Ebola-virus outbreak 165, 364
 Garamba National Park 38-43
 prehistoric tool use 288
Zebras 60
Zidovudine *see* AZT
Zimmerman, Harry M. 386
Zinc 158, 160, 309
Zodiac 100
Zoology 380-83 *see also* Wildlife
 courtship displays 44-48
 Ebola-virus transmission 364-65
 nonindigenous species 120-26
 water needs of animals 54-59
Zooplankton 329, 358-59

Acknowledgments

Sources of articles appear below, including those reprinted with the kind permission of publications and organizations.

WHERE THE BUFFALO ROAM, page 16: © Gary Turbak. Reprinted by permission of the author; article originally appeared in the December 1995 issue of *Wildlife Conservation* magazine, published by the Wildlife Conservation Society.

THE GOD OF THE WOODS, page 22: Originally published in the Spring 1995 issue of *Pacific Discovery*, a publication of the California Academy of Sciences.

IN PURSUIT OF PIGEONS, page 30: Copyright 1996 by the National Wildlife Federation. Reprinted with permission from *International Wildlife* magazine's January-February 1996 issue.

YOU TASMANIAN DEVIL, YOU!, page 35: Originally published in the Summer 1995 issue of *Pacific Discovery*, a publication of the California Academy of Sciences.

RHINO RELATIONS, page 38: Reprinted by permission of the author from *Wildlife Conservation* magazine, published by the Wildlife Conservation Society.

THE IMPORTANCE OF BEING FLASHY, page 44: Copyright 1995 by the National Wildlife Federation. Reprinted with permission from *International Wildlife* magazine's September-October 1995 issue.

THE QUEST FOR WATER, page 54: Copyright 1995 by the National Wildlife Federation. Reprinted with permission from *National Wildlife* magazine's June-July 1995 issue.

MERCURY MIRRORS: A NEW SPIN ON TELESCOPES, page 71: Copyright © 1995 by The New York Times Company. Reprinted by permission.

5..4..3..2..ABORT!, page 32: Reprinted by permission of the author; article originally appeared in the December 1995-January 1996 issue of *Air & Space Smithsonian*.

AMERICA'S FOREMOST EARLY ASTRONOMER, page 89: From *Sky & Telescope*. Copyright © 1995 Sky Publishing Corp. Reprinted by permission.

TARGET: EARTH!, page 94: Reprinted by permission. © *Astronomy* magazine, October 1995, Kalmbach Publishing Co.

STRANGE LIGHTS ABOVE THUNDERSTORMS, page 104: Originally published in the Summer 1995 issue of *Pacific Discovery*, a publication of the California Academy of Sciences.

MOUNTAIN ISLES, DESERT SEAS, page 110: Originally published in the Summer 1995 issue of *Pacific Discovery*, a publication of the California Academy of Sciences.

SKY READING, page 114: Reprinted by permission of the author; article originally appeared in the November-December 1995 issue of *Country Journal*, published by Cowles Magazines, Inc.

THE MEASURE OF A MOUNTAIN, page 127: © 1996, Kalmbach Publishing Co., *Earth* magazine. Reproduced with permission.

BEATING THE ODDS, page 138: Adapted with permission from the February 12, 1996, issue of *U.S News & World Report*.

TISSUE IN A TEST TUBE, page 151: Reprinted with permission from *Popular Science* magazine. © 1996, Times Mirror Magazines, Inc. Distributed by the Los Angeles Times Syndicate.

WHAT ARE MY CHANCES, DOC?, page 162: Reprinted by permission of *Forbes* Magazine © Forbes Inc., 1995.

DARWINIAN MEDICINE, page 165: Lori Oliwenstein/© 1995 The Walt Disney Co. Reprinted with permission of *Discover* Magazine.

BRINGING BACK THE DINOSAUR, page 184: Reprinted with permission from the May 1995 issue of *Natural History*. Copyright, the American Museum of Natural History.

AN ELUSIVE MAP OF THE WORLD, page 191: Reprinted by permission of the author; article originally appeared in the November/December 1995 issue of *Civilization* magazine.

WATERY GRAVE OF THE AZORES, page 197: Copyright © 1995 by The New York Times Company. Reprinted by permission.

THE WEATHER ON TRIAL, page 202: Copyright © 1995 by The New York Times Company. Reprinted by permission.

CIRCUS SCIENCE, page 210: Carl Zimmer/© 1996 The Walt Disney Co. Reprinted with permission of *Discover* Magazine.

CHANGING TIMES, page 226: Reprinted by permission of the author; article originally appeared in the November/December 1995 issue of *Technology Review* magazine.

3-D COMPUTING: A WHOLE NEW DIMENSION, page 240: Reprinted from the September 4, 1995, issue of *Business Week* by special permission. Copyright © 1995 by McGraw-Hill Inc.

WHY AIRPLANES CRASH, page 247: Reprinted with permission from *Popular Science* magazine. © 1996, Times Mirror Magazines, Inc. Distributed by the Los Angeles Times Syndicate.

STEEL MAGNOLIAS, page 262: Reprinted by permission of the author; article originally appeared in the November 1995 issue of *Destination Discovery*.

AFTER THE ACCIDENT, page 270: Vincent Lytle/© 1996 The Walt Disney Co. Reprinted with permission of *Discover* Magazine.

UNDERSEA EXPLORERS, page 275: Reprinted with permission from *Popular Science* magazine. © 1995, Times Mirror Magazines, Inc. Distributed by the Los Angeles Times Syndicate.

Manufacturing Acknowledgments

We wish to thank the following for their services:

Color Separations, Gamma One, Inc.;
Text Stock, printed on Champion's 60# Courtland Matte;
Cover Materials provided by Ecological Fibers, Inc.;
Printing and Binding, R.R. Donnelley & Sons Co.

ILLUSTRATION CREDITS

The following list acknowledges, according to page, the sources of illustrations used in this volume. The credits are listed illustration by illustration—top to bottom, left to right. Where necessary, the name of the photographer or artist has been listed with the source, the two separated by a slash. If two or more illustrations appear on the same page, their credits are separated by semicolons.

3 © Gilles Basignac/Gamma-Liaison; Courtesy, Genzyme Corporation; © Scott Andrews
8– American Museum of Natural History/© Scott Frances
9 Scott Frances
10 © Steve Barnett/Gamma-Liaison
11 NASA
12 © Chip Simons; © Frans Lanting/Minden Pictures; © Chamoux-Beynie/Gamma-Liaison
13 © Amos Nachoum Photography; Allsport
14– © Art Wolfe
15
16 © Charles Palek/Animals Animals
17 Both photos: © Henry H. Holdsworth
18 The Bettmann Archive
19 Both photos: © Henry H. Holdsworth
20 Both photos: © William Muñoz
21 © Henry H. Holdsworth
22 Larry Ulrich Photography
23 Dr. C. Hart Merriam, Special Collections, California Academy of Sciences
24 © Bill Evarts; © Lee Foster/Bruce Coleman
25 © Smithsonian Institution, Washington, D.C.
26 © E.R. Degginger/Earth Scenes/Animals Animals; © Bill Evarts
27 © N.H. Cheatham/Photo Researchers
28 © Bill Evarts
29 © Bill Evarts
30 © Robert Maier/Animals Animals; © Oxford Scientific Films/Animals Animals
31 © Mike Birkhead/Oxford Scientific Films/Animals Animals
32 © Doug Wechsler/VIREO/Academy of Natural Sciences, Philadelphia; Culver Pictures
33 Both photos: © Tim Gallagher
34 © Frances M. Roberts
35 © John Cancalosi; Photofest
36 Photo: © Dave Watts/Tom Stack & Associates
37 © Dave Watts/Tom Stack & Associates; © John Cancalosi
38 © Kes and Fraser Smith
39 © William B. Karesh
40– All photos: © Kes and Fraser Smith
43
44 © F.J. Hiersche/Okapia Pavo Cristatus/Photo Researchers
45 © Tom McHugh/Photo Researchers; © Nigel J. Dennis/Photo Researchers
46 © Toni Angermayer/Photo Researchers; © John Shaw/Tom Stack & Associates
47 © Phil A. Dotson/Photo Researchers
49 © Jeanne White/Photo Researchers; © Hans Reinhard/Okapia/Photo Researchers
50 © Matt Meadows/Peter Arnold; © Holt Studios International/Photo Researchers
51 © Mark Boulton/Photo Researchers; © William Muñoz
52 © Cameron Craig/Mount Vernon Ladies Association
53 © Gary Andrashko/Plimoth Plantation, Inc.
54 © Frans Lanting/Minden Pictures
55 © T.A. Wiewandt/DRK Photo
56 © T.A. Wiewandt/DRK Photo; © Thomas Dressler/DRK Photo
57 © Jim Brandenburg/Minden Pictures
58 © Frans Lanting/Minden Pictures; © Victoria McCormick/Animals Animals
59 © Frans Lanting/Minden Pictures
60 © Superstock
62– © Royal Observatory, Edinburgh/AATB/
63 Science Photo Library/Photo Researchers
64 © John R. Foster/Science Source/Photo Researchers
65 © Royal Observatory, Edinburgh/AATB/Science Photo Library/Photo Researchers; © STSCI
66 Illustration sequence: © Joseph M. Tucciarone/National Geographic Society Image Collection
67 NASA
68 © Roger Ressmeyer/Corbis
69 © Joseph M. Tucciarone/National Geographic Society Image Collection
71 © Chip Simons
72 © Chip Simons
73 New York Times Graphics/NYT Pictures
74 © Paul Hickson
75 © Tony Craddock/Science Photo Library/Photo Researchers
76 Top: The Granger Collection; photo: © Waros Scientific/Photo Researchers
78 Courtesy, NOAO
79 National Center for Atmospheric Research/University Corporation for Atmospheric Research/National Science Foundation
80 © Julian Baum/Science Photo Library/Photo Researchers
81 NASA/Science Source/Photo Researchers; © Fred Espenak/Science Photo Library/Photo Researchers
82– All photos: © Scott Andrews
83
84 Both photos: © Rich Mays
85 Rich Mays; © John F. Kennedy Space Center
86 John F. Kennedy Space Center
87 Both photos: © John F. Kennedy Space Center
89 Painting by Charles Willson Peale; photographed by David Rubincam, courtesy of Mr. and Mrs. Stanley P. Sax
90 Courtesy, Princeton University
91 Artwork from *Sky and Telescope*. Copyright © 1995 by Sky Publishing Corp. Reprinted by permission.
92 © American Philosophical Society
94 © Robert Eggleton
95 JPL
96 © Julian Baum/Photo Researchers
97 © Baker/Milon/Photo Researchers
98 © John Sanford/Photo Researchers
99 Lawrence Livermore National Lab
100 © Superstock
102– © Jeff Foott Productions
103
104– Crossover photo: © Keith Kent/Science
105 Photo Library/Photo Researchers; inset: © Dan Osborne/Geophysical Institute/UAF
106 Both photos: © Walter Lyons/ASTeR, Inc.
107 © Peter Menzel
109 © Keith Kent/Science Photo Library/Photo Researchers; inset: © Dan Osborne/Geophysical Institute/UAF
110 © Edward McCain
111 Top photos: © John Cancalosi; bottom: © J.H. Robinson/Animals Animals
112 Illustrations: Paula McKenzie Nelsen
113 © Edward McCain
114 © D. Cavagnaro/DRK Photo
115 © Paul Hazi Photography
116 © Nancy Sams/Photobank
117 © Robert P. Comport/Earth Scenes
118 © Barry L. Runk/Grant Heilman
119 © C. & S. Chattopadhyay/Photobank; © Charles Mann/Photo Researchers
120– Clockwise from bottom left: © Michael P.
121 Gadomski/Photo Researchers; © Richard R. Hansen/Photo Researchers; © S.R. Maglione/Photo Researchers; © Boyd Norton/Photo Researchers
122 © Michael Fogden/Animals Animals
123 © Mike Okoniewsk/Gamma-Liaison
124 © Tom Myers/Photo Researchers; © Eastcott/Momatiuk/Photo Researchers
125 © Andrew J. Martinez/Photo Researchers
126 © Suzanne & Joseph Collins/Photo Researchers
127 © Chamoux-Beynie/Gamma-Liaison
128– Crossover photo: © Galen Rowell/Mountain
129 Light; map: From the American Geographic Society Collection, University of Wisconsin-Milwaukee Library
130 NASA
131 © Roger Bilham/CIRES University of Colorado
132 © Galen Rowell/Mountain Light
134 © Superstock
136– © Hank Morgan/Photo Researchers
137
138 © Allsport
139 Table data: CDC; bottom illustration: © Christoph Blumrich/*Newsweek*
140 Table data: CDC; © Tim Crosby/Gamma-Liaison
141 © Tom McKitterick/Impact Visuals
142 © D. Vo Trung/Photo Researchers
143 Table data: CDC; © Nate Guidry/Impact Visuals
144 *POZ* Magazine
145 © Gregory G. Dimijian/Photo Researchers
146 © Andy Hayt/Focus on Sports; Courtesy, Hyox Systems
147 Historical Division, Cleveland Medical Library Association; © Steve Barnett/Gamma-Liaison
148 © SIU/Peter Arnold; Courtesy, University of Colorado at Boulder
149 Courtesy, Hyox Systems
151 Courtesy, Genzyme Corporation
152– All illustrations: Reprinted with permission
154 from *Popular Science* Magazine. © 1996, Times Mirror Magazines Inc. Distributed by Los Angeles Times Syndicate.
162 © Mark Richards/Photo Edit
163 Photo: © Alan Dorow; illustrations: courtesy, Apache Medical Systems
164 Courtesy, © Andy Christie/*Forbes*
165 Reuters/Corinne Dufka/Archive Photos
166 © Popperfoto/Archive Photos; © Eamonn McNulty/Photo Researchers
167 © Tony Freeman/Photo Edit
168 © J&M Studios/Gamma-Liaison
169 Courtesy, Vintage Books/Division of Random House, Inc.
170 © Yoav/Phototake/Peter Arnold
172 © Superstock
174– © Frank Rossotto
175
176 © Kurgan-Lisnet/Gamma-Liaison
177 Map by Lazlo Kubinyi
178 © Malcolm S. Kirk/Peter Arnold; © Erich Lessing/Art Resource
179 © Doris Licht/Peter Arnold
180 © Patrick Landmann/Gamma-Liaison; © Ovak Arslanian/Gamma-Liaison
181 *Newsweek*/Dixon Rohr. © 1995, Newsweek, Inc. All rights reserved. Reprinted by permission.
182 © Barry Iverson/Woodfin Camp & Associates
183 © Patrick Landmann/Gamma-Liaison
185 © James Joern
186 © James Joern
187 Bottom right photo: © American Museum of Natural History; all others: © James Joern
188 American Museum of Natural History/© Scott Frances
189 © James Joern
191 © The Granger Collection; background: From the American Geographical Society Collection, The University of Wisconsin-Milwaukee Library
192– All maps: From the American Geographical
195 Society Collection, The University of Wisconsin-Milwaukee Library
197 Photo: © Tony Arruza

ILLUSTRATION CREDITS 399

198	© Jenifer Marx; © Bob Marx	
199	© Bob Marx	
200	© Jenifer Marx	
201	© Bob Marx	
202-	Left: © Sam Mircovich, Pool/AP/Wide World	
203	Photos; crossover art by Vincent Caputo; right: © The Granger Collection	
204	Photo: © David Jennings/NYT Pictures; art by Vincent Caputo	
205	Photo: © Lannis Waters/*Palm Beach Post*, Pool; art by Vincent Caputo	
206	© Superstock	
208-	© Photonica	
209		
210-	© Tom Hanson/Gamma-Liaison	
211		
212-	Both photos: Bruce Curtis/© 1996 The Walt	
213	Disney Co. Reprinted with permission of *Discover* Magazine.	
214	© Tom Sobolik/Black Star	
215	Bruce Curtis/© 1996 The Walt Disney Co. Reprinted with permission of *Discover* Magazine.	
218	© Marc Deville/Gamma-Liaison	
219	© Tony Freeman/Photo Edit	
220	© SIU Biomedical Communications/Photo Researchers	
221	© Dr. Jeremy Burgess/Science Photo Library/Photo Researchers	
222	© Michael Newman/Photo Edit	
224	© Filmsmith/Garment Analysis Laboratory; © Michael Newman/Photo Edit	
226-	© Pedro Lobo/Photonica	
227		
228	Archive Photos; Yerkes Observatory	
229	Corbis-Bettmann; AIP/Emilio Segrè Visual Archives/Francis Simon Collection/American Institute of Physics	
230	Corbis-Bettmann	
231	Photograph by Frank Heny, courtesy Carlo Rovelli	
232	© J. Wilson/Gamma-Liaison	
233	© Sandra Baker/Gamma-Liaison	
236	© Superstock	
238-	© Andy Sachs/Tony Stone Images	
239		
240	Courtesy, Viewpoint Datalabs	
241	Courtesy, The Ralph M. Parsons Company; © Photofest	
242	Courtesy, Viewpoint Datalabs; courtesy, Paradigm Simulations	
243	© John Rosenberg/University of Pittsburgh	
244	© Philip Shaddock	
245	Courtesy, Viewpoint Datalabs	
247	© Charles Bennett/AP/Wide World Photos	
249	© John Grimwade; © Steve Leonard/Black Star	
250	© John Grimwade	
251	© Chris Sorensen	
252	© John Grimwade	
253	© Arnold Zahn/Black Star	
254-	Background photo: © John Wilkes/Photonica; insets, clockwise from upper right: © VeloNews; © The Really Useful Company, Inc.; © Hachette Filipacchi; © AltaVista; © AT&T; © CNN Interactive; © 1996 by YAHOO!, Inc. All rights reserved. YAHOO! and the YAHOO! logo are trademarks of YAHOO!, Inc.	
255		
256	Courtesy, Bolt Beranek and Newman; Space Telescope Science Institute	
257	Mila Grondahl/*The Philadelphia Inquirer*	
259	ESPNET SportsZone material provided courtesy of ESPN, Inc., and Starwave Corp. All rights reserved.	
260	Donna Cox, Bob Patterson/NCSA, University of Illinois at Champaign	
261	© B. Kraft/Sygma	
262-	Both photos: © Gilles Bassignac/Gamma-Liaison	
263		
264	© Sam Sargent/Gamma-Liaison	
265	Top: © AP/Wide World Photos; bottom photos: Courtesy, T.Y. Lin International	
266	© Alan Bonicatti/Gamma-Liaison	
267	© Tom Bell	
268	Courtesy, T.Y. Lin International	
269	© Torin Boyd/Gamma-Liaison	
270	Mirko Ilic/© 1996 The Walt Disney Co. Reprinted with permission of *Discover* Magazine.	
271	Courtesy, Ford Motor Company; © Jon Levy/Gamma-Liaison	
272	© Bill Swersey/Gamma-Liaison; Courtesy, Ford Motor Company	
273	© Rob Johns/Gamma-Liaison	
274	© Rob Johns/Gamma-Liaison	
275	© Amos Nachoum Photography	
276-	© David Teich	
277		
278	© Chip Clark; © F.S. Westmorland/Photo Researchers	
279	© David Teich	
280	© Superstock	
282	© Transport Research Laboratory; © Jeff Robbins/AP/Wide World Photos; © Lanmon Aerial Photography/AP/Wide World Photos	
283	Rich Alsopp and Gerry Shay/Southwestern Medical Center, Dallas; © Ted Mathias/AP/Wide World Photos	
285	© Jacob Sutton/Gamma-Liaison	
286	Sketch by Denise Sins, Nevada State Museum	
287	© Rachael Nador/Image Analysis Facility, University of Iowa	
289	Courtesy of Combined Caesarea Expedition	
291	J. Spencer/Lowell Observatory/NASA	
293	Courtesy, Fiat USA, Inc.	
295	© Chad Slattery	
297	Reuters/Rick Wilking/Archive Photos	
298	© Wellcome Dept. of Cognitive Neurology/Science Photo Library/Photo Researchers	
299	Copyright © 1996 Richard Gage, *U.S. News & World Report*	
300	© Russell D. Curtis/Photo Researchers	
301	Rich Alsopp and Gerry Shay/Southwestern Medical Center, Dallas	
302	Copyright © 1995 Richard Gage, *U.S. News & World Report*	
303	© 1996 Simon & Schuster; © Alfred A. Knopf, Publisher, New York	
304	© Plenum Publishing Corporation; © The University of Chicago Press; © G. P. Putnam's Sons; © Viking	
305	© 1995 Simon & Schuster; © William Morrow & Company; © Basic Books; © Walker and Company	
306	© Rutgers University Press	
307	© David Liittschwager and Susan Middleton/© 1996 The Walt Disney Co. Reprinted with permission of *Discover* Magazine.	
310	© Asahi Shimbun/Sipa	
311	© Robin Moyer/Gamma-Liaison	
314	Courtesy, SkyTel	
315	Courtesy of U.S. West Communications, Inc.	
317	Courtesy, Equator Corp.; © Transport Research Laboratory	
319	Courtesy, Canon	
320	Courtesy, Support Services, Inc.; courtesy, Proactive Sports	
321	Courtesy, AT&T	
323	© Peter Zahler	
324	© Ted Mathias/AP/Wide World Photos	
326	© Lanmon Aerial Photography/AP/Wide World Photos	
328	© Jeff Robbins/AP/Wide World Photos	
329	© Robert Huntzinger/The Stock Market	
330	© Stephanie Maze/Woodfin Camp & Associates	
332	© Betty Press/Woodfin Camp & Associates	
334	© Peter Vandermark/Stock Boston	
335	Illustration from *Anomalies and Curiosities of Medicine* (1896)/Reproduced from the Collections of the Library of Congress	
336	Joe Lemmonier/© 1996 The Walt Disney Co. Reprinted with permission of *Discover* Magazine.	
339	© Spencer Grant/Stock Boston	
340	© Susan Ragan/AP/Wide World Photos	
341	© David Young Wolff/PhotoEdit; © Joe Raymond/AP/Wide World Photos	
344	Courtesy, NOAA	
347	© Scott Olson/Reuters/Archive Photos; © Mark Wilson/AP/Wide World Photos; © Terry Ashe/Gamma-Liaison	
349	© Eric Roxfelt/AP/Wide World Photos; © Steven Senne/AP/Wide World Photos; © D. Groshong/Sygma	
350	© Erik Henriksson/AP/Wide World Photos; © P. Forden/Sygma	
353	© Mark Lenihan/AP/Wide World Photos; © Jeff Christensen/Gamma-Liaison; © DPA/Photoreporters	
355	Copyright © 1996 TIME Inc. Reprinted by permission.	
357	© Steven Needham/envision	
358	Courtesy, NOAA	
360	Both illustrations: Bill Parsons/© 1996 The Walt Disney Co. Reprinted with permission of *Discover* Magazine.	
362	© Mike Matthews, JILA Research Team/Univ. of Colorado at Boulder	
363	© Remi Benali/Gamma-Liaison	
364	© Reuters/Bettmann	
365	Courtesy, M. Fergione, Pfyzer Central Research	
368	© Tim Gallagher	
370	© AP/Wide World Photos	
372	NASA	
375	All photos: NASA	
376	Courtesy, Avis Inc.	
377	Courtesy, Dowling College	
379	© John McConnico/AP/Wide World Photos	
380	© Michael Francis/The Wildlife Collection	
381	© Michael Francis/The Wildlife Collection	
382	© Inga Spence/Tom Stack & Assoc.	

Photo credits for The New Book of Popular Science Edition

Front cover:
1. © Mike Okoniewski/Gamma-Liaison
2. NASA
3. © Henry H. Holdsworth
4. © Doris Licht/Peter Arnold
5. © Rob Johns/Gamma-Liaison
6. © Chip Simons
7. © Galen Rowell/Mountain Light

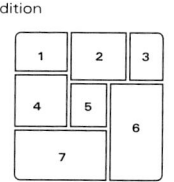

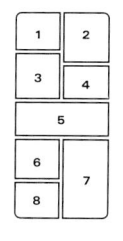

Back cover:
1. © Thomas Dressler/DRK Photo
2. NASA
3. © Mark Wilson/AP/Wide World Photos
4. © Tim Gallagher
5. © Lee Foster/Bruce Coleman
6. Courtesy, NOAO
7. © Robin Moyer/Gamma-Liaison
8. © Arnold Zahn/Black Star

Cover photo credit for Encyclopedia Science Supplement: © John Rosenberg/University of Pittsburgh